QUANTUM FIELDS ON THE COMPUTER

ADVANCED SERIES ON DIRECTIONS IN HIGH ENERGY PHYSICS

ISSN: 1793-1339

This is the best review series in high energy physics today. It comprehensively reviews the most important developments in each sector of high energy physics and is of lasting use to all researchers. All volumes are edited by eminent physicists — researchers who have themselves made substantial contributions to their respective fields of research.

Published

The complete list of titles in the series can be found at
http://www.worldscientific.com/series/asdhep

Advanced Series on
Directions in High Energy Physics — Vol. 11

QUANTUM FIELDS ON THE COMPUTER

Editor
Michael Creutz
Physics Department
Brookhaven National Laboratory
Upton, New York 11973
USA

World Scientific
Singapore • New Jersey • London • Hong Kong

Published by

World Scientific Publishing Co. Pte. Ltd.

5 Toh Tuck Link, Singapore 596224

USA office: 27 Warren Street, Suite 401-402, Hackensack, NJ 07601

UK office: 57 Shelton Street, Covent Garden, London WC2H 9HE

British Library Cataloguing-in-Publication Data

A catalogue record for this book is available from the British Library.

Advanced Series on Directions in High Energy Physics — Vol. 11
QUANTUM FIELDS ON THE COMPUTER

ISBN-13 978-981-02-0939-1
ISBN-10 981-02-0939-8
ISBN-13 978-981-02-0940-7 (pbk)
ISBN-10 981-02-0940-1 (pbk)

CONTENTS

INTRODUCTION

In 1948 Feynman showed that quantum field theory is equivalent to integration. In the more than 40 years since, particle theorists have occupied themselves with various ways to evaluate his sums over paths.

For most of this time the primary approach has been to consider only small perturbations about the free field theory limit, where the integrals become Gaussian. In the last decade, however, there has been an explosion of attempts to attack these integrals more directly, using the vast increases in raw computational power becoming available. These efforts are quite ambitious; indeed, they are attempts at first principle calculations for interacting relativistic quantum field theories.

Since the starting point is an infinite-dimensional integral, approximations are necessary to put it on the computer. The usual approach is to replace the space–time continuum by a discrete lattice in a finite volume. In this way the integrals at least become finite-dimensional, although still generally rather large.

When one is dealing with analytic calculations in the continuum theory, it is particularly difficult to be rigorous. This often allows meaningless calculations involving poorly defined quantities. On the lattice, if one tries to evaluate an undefined number, then the computer quickly complains. This forces an unusual honesty on the lattice gauge theorist.

Of course, one must still ask if the lattice model has anything to do with the continuum world of observation. If one can show that the continuum limit of a lattice theory exists, one might just as well use that as a definition of the continuum theory. If another definition of the continuum model makes sense but is different from the limit of the lattice form, then one must wonder if there is any uniqueness at all to the theory. Indeed, it appears a miracle that in four dimensions only a small class of theories appears to have a reasonable chance of surviving a continuum limit. In more dimensions none are known, while in less a much wider class of renormalizable theories is possible.

In perturbation theory, one quickly encounters the well-known divergences requiring renormalization. The various couplings and masses in the bare Lagrangian are not physical quantities, so predictions for cross sections etc. should be re-expressed in terms of observables, such as the long range electromagnetic force in quantum electrodynamics. These renormalization issues are not unique to perturbation theory, and appear as well in the lattice technique. The language here, however, sometimes appears quite different than in the continuum approach. In a perturbative discussion, it is usual to think of the bare coupling as a parameter which must be adjusted, or renormalized, as the cutoff controlling ultraviolet divergences is removed. Thus the bare coupling is a function of the cutoff. On the lattice, however, we often think of the lattice spacing, or cutoff, as a function of the bare coupling. To take a continuum limit one adjusts the bare coupling to make the lattice spacing small. Owing to asymptotic freedom, for the strong interactions this requires taking the bare coupling to zero.

The purpose of this book is to review the main results of numerical lattice gauge theory. From the initial studies relevant to the confinement problem, lattice calculations have grown into an industry involving hundreds of physicists and tackling a wide spectrum of problems. For a single author to do a thorough job reviewing this diversity would be near-impossible. Instead, we have divided the task and thus hope to have each of the subtopics treated in a reasonably complete manner.

The book consciously focuses on the computational aspects of the problem. While analytic approaches are also crucial to the subject, the results have so far been more limited. We have not attempted a thorough introduction to the basic foundations of lattice field theory. For this there are several reviews in the literature as well as my previous monograph, *Quarks, Gluons, and Lattices* (Cambridge University Press, 1983).

We assume that the reader is at least vaguely familiar with the general formulation of lattice gauge theory, wherein there is a system of group variables $U_{i,j}$ defined on the links $\{i, j\}$ of a four-dimensional hypercubic lattice. For the strong interactions, these elements are from the group SU(3), and we usually have them interact via the standard Wilson action,

$$S = -\sum_p \mathrm{Re\,Tr}\, U_p .\qquad(1.1)$$

Here U_p is the product of link variables around the elementary square, or plaquette, p, and the sum is over all such plaquettes. The Feynman path integral is then given by

$$Z = \int (dU) e^{-\beta S(U)} .\qquad(1.2)$$

Here the coupling constant dependence has been parametrized by the variable β. This is related to the more conventional perturbative coupling by

$$\beta = \frac{6}{g_0^2} ,\qquad(1.3)$$

where the subscript 0 on the coupling is to emphasize that this is the bare coupling, which must be renormalized in the continuum limit.

Note that these definitions emphasize an analogy with statistical mechanics. Solving the four-dimensional space–time field theory is equivalent to evaluating the thermodynamics of a four-dimensional classical system with Hamiltonian given by the action. As in condensed matter physics, it is not Z itself which is of primary interest, but the correlations between observables in the equilibrium ensemble. Correlation lengths are the inverse masses of physical particles, and coupling constants come from correlations between multiple fields.

Just as with the continuum formulation, the lattice action carries an invariance under gauge transformations. The Wilson approach has the remarkable property that this invariance remains an exact local symmetry. Indeed, if we introduce an arbitrary group element g_i associated with every site on our lattice, the action (1.1) is unchanged if we replace U_{ij} with $g_i U_{ij} g_j^{-1}$. Thus all issues related to gauge fixing can be discussed on the lattice as well as in the continuum. Indeed, this is necessary for a perturbative treatment. Nevertheless, Wilson's compact formulation makes the theory well defined even without a gauge choice.

From perturbative analysis we obtain some rather remarkable information on the continuum limit of the lattice theory. Indeed, the phenomenon of asymptotic freedom tells us that the coupling defined at increasingly shorter distances goes to zero logarithmically with the scale. As the bare coupling is an effective coupling at the scale of the lattice cutoff, it must be taken to zero as the lattice spacing is reduced. In this process we have the remarkable phenomenon of dimensional transmutation, wherein the dimensionless coupling constant is replaced by the scale of this logarithmic behavior. In the end, the pure gauge theory contains no dimensionless couplings, and all mass ratios should be determined. This is one of the main goals of lattice calculations: to determine the properties of the physical hadrons from fundamental principles and with no adjustable parameters beyond the quark masses.

We begin the book with two chapters on some of the primary physical results of these efforts. First T. DeGrand reviews the status of calculating hadronic masses. This in some sense represents the particle physicist's most fundamental desire: to predict the observed spectrum of nature. The second chapter turns to some experimental predictions that have not yet been verified. Here R. Gavai treats the behavior of hadronic matter at finite physical temperatures, and the predicted transition to a quark–gluon plasma. Indeed, it is the lattice approach which has given the best estimates of the temperature and properties of this transition.

The next two chapters push lattice methods beyond the gauge theory of the strong interactions. A particularly important topic for the near future is the Higgs boson, the study of which is a primary goal for the SSC. Lattice methods have been applied here, and given rather stringent nonperturbative bounds on the possible masses for this as-yet-undiscovered particle. This is the topic of the chapter by A. Hasenfratz. Extending these ideas to include the fermionic contributions to the standard model of weak interactions, R. Shrock then reviews chiral and Yukawa models. This represents an area where the lattice formulation is not yet fully understood; in particular there are fundamental difficulties in formulating a theory with fermions interacting in a chiral manner. Nevertheless, such work is essential for understanding weak interaction phenomenology on a nonperturbative level.

We then turn to two chapters on the basic numerical algorithms being used for lattice calculations. A. Sokal provides a detailed discussion on the basic Monte Carlo methods for lattice simulations, concentrating on bosonic fields. He points out the advantages and pitfalls in the various schemes, and covers recent develpments in overcoming computational limitations as the continuum limit is approached. I then present a discussion on the modifications of these methods developed to study quark fields. Here the anticommuting nature of the fundamental objects creates severe new difficulties, and probably the best approach is yet to be found.

For us to have confidence in any predictions for quantum field theory, the results must behave in a well-understood manner as one removes the cutoff. At the core of these analyses lies the renormalization group. Indeed, the verification of the appropriate scaling behaviors is an essential step towards confidence and future improvements in the basic lattice approach. These ideas form the basis of R. Gupta's chapter.

In the final chapter, C. Bernard and A. Soni review the use of lattice approaches to calculate the previously unknown hadronic corrections to weak interaction processes. This is a crucial step in our ability to comprehend the underlying fundamental weak forces. Indeed, these and related calculations represent the current dominant use of lattice gauge methods to describe real phenomena in particle physics.

We have enjoyed putting this book together, and hope it will be useful in clarifying what is known and pointing out directions for new research. Lattice field theory is a large subject, and necessarily some topics have been left out. We have tried to be reasonably up to date, but in any rapidly evolving field the true excitement lies in unanticipated new developments.

Michael Creutz

TECHNIQUES AND RESULTS FOR
LATTICE QCD SPECTROSCOPY

T. DeGrand

Department of Physics
University of Colorado
Boulder, Colorado 80309

ABSTRACT: This article describes how to compute the masses of hadrons in lattice simulations of QCD. I discuss the most important theoretical and calculational techniques for finding the masses of mesons, baryons, and glueballs. Part of this discussion is pedagogic and is intended for use by newcomers to this field. A second section describes some interesting open problems in spectroscopy: wave functions, resonances, orbitally excited states. I review recent developments in this subject (through Autumn 1991).

I. INTRODUCTION

This chapter of the book discusses how to compute the masses of the ordinary elementary particles in lattice simulation of Quantum Chromodynamics. (Most of the techniques which I discuss can be applied to other models in particle and condensed matter physics.) This is a problem of great current interest to the lattice gauge community as shown by the many large computer simulations devoted in whole or in part to the calculation of spectroscopy.

There are several reasons why this problem is important. First of all, we think QCD is the theory of the strong interactions. We cannot have confidence in our ability to compute in QCD if we cannot use it to calculate at least the low lying spectrum of the theory. We also wish to calculate hadronic matrix elements, either for their own sake or as ingredients to calculations which go beyond the strong interactions– perhaps to constrain parameters of the Standard Model. A necessary (but not sufficient) condition for trusting lattice calculations of matrix elements of complicated operators is the knowledge that matrix elements of simple operators (like the Hamiltonian) are under control. Thus we need to see good spectroscopy calculations first.

Actually, I believe that the goal of calculating the complete low-lying spectrum of QCD is presently unrealistic. The techniques at our disposal are too primitive. Lattice methods cannot compete with continuum models which are not QCD but are QCD-inspired which give an essentially perfect fit to light quark spectroscopy![1] However, calculations of matrix elements are much more model dependent. Since we believe that the lattice "model" is more fundamental than the other approaches, we want to use it to compute matrix elements. We want these calculations to be reasonably reliable. Poor qualitative agreement of QCD spectroscopy with experiment will indicate that these calculations are not trustworthy.

Lattice gauge theory simulations allow us to vary physical parameters (quark masses, number of flavors of sea quarks) in a way that experiment cannot. In principle, this gives us more information on confinement than we would get from real data. One even hopes that the behavior of QCD's with zero, 2 or 3 or 4 degenerate flavors of quarks are quite different from each other, and from the real world, in which all quark masses are different! (For future reference, simulations with zero flavors of dynamical fermions are done dropping the fermion determinant from the functional integral. This approximation is called the "quenched approximation.") However, present day simulations are presently limited by computer power and algorithms to unphysical values of the dynamic quark mass, and unphysical numbers of flavors or degeneracies. This means that if one's goal is to make a direct comparison of a simulation to the real world, one must make an extrapolation. It's important to remember that the extrapolation is not part of the lattice simulation. It certainly involves its own physical assumptions whose validity is independent of the validity of the simulation. One should try to keep extrapolation issues as separate as possible from simulation issues.

While the goal of this chapter is spectroscopy, a large component of the discussion will involve the study of wave functions of quarks and gluons inside hadrons. I believe that the major advances in spectroscopy have come through our ability to model and use realistic wave functions (mainly as interpolating fields), and that essentially all future progress in this field will center on the study of wave functions.

The outline of this review is as follows: In Sec. II I will present an overview of the basic techniques one uses in a lattice spectroscopy simulation, hopefully done at a level suitable for a beginner. Sec. III is devoted to a set of interesting open questions in lattice spectroscopy simulations. Secs. IV and V are reviews of recent progress in glueball and quark spectroscopy. Finally I make a few concluding remarks in Sec. VI. If you want to get an overview of the ingredients you "have to know" to do a lattice simulation, read Sec. II. If you want to see how interesting physics questions collide with lattice techniques, skip Sec. II and read Sec. III. If the year is later than 1994, ignore Secs. IV and V completely–they are (hopefully) obsolete.

II. BASICS OF LATTICE SPECTROSCOPY

The goal of this section is to provide as complete a description as possible of all the basic ingredients needed to carry out a spectroscopy calculation for hadrons or glueballs in QCD. (I will defer a description of interpolating fields for glueballs to Sec. IIID.)

A. Correlation Functions

Masses are computed in lattice simulations from the asymptotic behavior of Euclidean-time correlation functions. A typical (diagonal) correlator can be written as

$$C(t) = \langle 0|O(t)O(0)|0\rangle. \tag{2.1}$$

Making the replacement

$$O(t) = e^{Ht}Oe^{-Ht} \tag{2.2}$$

and inserting a complete set of energy eigenstates, Eq. (2.1) becomes

$$C(t) = \sum_n |\langle 0|O|n\rangle|^2 e^{-E_n t}. \tag{2.3}$$

At large separation the correlation function is approximately

$$C(t) \simeq |\langle 0|O|1\rangle|^2 e^{-E_1 t} \tag{2.4}$$

where E_1 is the energy of the lightest state which the operator O can create from the vacuum. In some cases that state is the vacuum itself, in which $E_1 = 0$. If the operator does not couple to the vacuum, then in the limit of large t one hopes to to find the mass E_1 by measuring the leading exponential falloff of the correlation function, and most lattice simulations begin with that measurement. If the operator O has poor overlap with the lightest state, a reliable value for the mass can be extracted only at a large time t. This makes the actual calculation of the energy E_1 more difficult. (Also, if the operator is not averaged over all sites or projected onto specific momentum eigenstates, it may have additional power law factors in its falloff.)

This is the basic way hadronic masses are found in lattice gauge theory. The many calculations differ in important specific details of choosing the operators $O(t)$. For the remainder of this subsection I will discuss some issues relating to the errors in measuring masses, and thus set the groundwork for later discussion.

In Monte Carlo we calculate the correlator $C(t)$ (or generically any observable Γ) by importance sampling: we do N measurements using the field variables $\phi^{(i)}$

appropriate to the sample.[2]

$$\langle \Gamma \rangle \simeq \bar{\Gamma} \equiv \frac{1}{N} \sum_{i=1}^{N} \Gamma[\phi^{(i)}] \qquad (2.5)$$

As the number of measurements N becomes large the quantity $\bar{\Gamma}$ will become a Gaussian distribution about a mean value. Its standard deviation is roughly

$$\sigma_{\bar{\Gamma}}^2 = \frac{1}{N}(\frac{1}{N} \sum_{i=1}^{N} |\Gamma[\phi^{(i)}]|^2 - \bar{\Gamma}^2). \qquad (2.6)$$

We can also calculate this quantity in field theory: it is just

$$\sigma_{\bar{\Gamma}}^2 = \frac{1}{N}(\langle |\Gamma|^2 \rangle - \langle \Gamma \rangle^2) \qquad (2.7)$$

This formula is very useful in analyzing the uncertainty in one's measurements. Let us now do that in a fairly general way.

There are (at least) three types of behavior a correlation function can display which are relevant to a spectroscopy calculation. The first case is one where the operator O has nonzero vacuum expectation value. Then the correlation function approaches a constant at large t:

$$\lim_{t\to\infty} C(t) \to |\langle 0|O|\vec{p}=0\rangle|^2 \exp(-mt) + |\langle 0|O|0\rangle|^2. \qquad (2.8)$$

The statistical fluctuations on $C(t)$ are given by Eq. (2.6) and we find after a short calculation that

$$\sigma \to \frac{C(0)}{\sqrt{N}}. \qquad (2.9)$$

Thus the signal to noise ratio collapses at large t like $\sqrt{N}\exp(-mt)$. This situation occurs in tha calculation of the mass of the 0^{++} glueball (or of any particle with vacuum quantum numbers).

The second two cases are similar in that we assume that the operator O has no vacuum expectation value. Then $C(t)$ shows a pure exponential behavior at large t. The fluctuations in $C(t)$ are given by Eqn. (2.7). The second term of Eqn. (2.7) falls off like $\exp(-2mt)$ and will give a constant signal-to-noise ratio at large separation. The first term of Eq. (2.7), however, is itself a correlation function of the operator $|O(t)|^2$ and will die away with a mass characteristic of the lightest particle $|O|^2$ can make from the vacuum.

Let us specialize temporarily to QCD for the rest of the discussion. If O is a meson operator (a $\bar{q}q$ pair) O^2 will be a $q^2\bar{q}^2$ state, which will most likely couple to a $\pi\pi$ pair. Its correlator will fall like $\exp(-2m_\pi t)$. In the baryon sector this

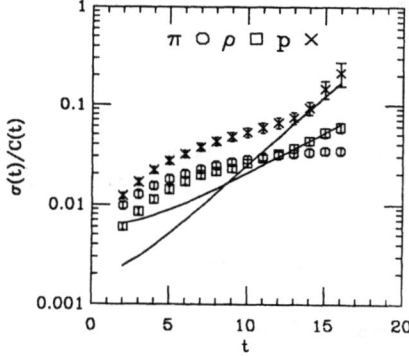

FIGURE 1

Ratio of the error on a propagator $\sigma(t)$ to the value of a correlation function $C(t)$ for the pion (crosses), rho (circles) and proton (diamonds), from a dynamical Wilson fermion simulation at $\beta = 5.3$, $\kappa = 0.1670$. The lines are the parameterization of Eqns. (2.10-2.11), with arbitrary overall normalization.

operator will make a $q^3 \bar{q}^3$ state, and the lightest such state is three pions. Thus we expect a signal to noise ratio to be at best a constant (for the pion) or a growing function of t:

$$\sigma/C_H(t) \simeq \exp\left(m_H - m_\pi\right)t \qquad (2.10)$$

for mesons, and

$$\sigma/C_H(t) \simeq \exp\left(m_H - 3/2m_\pi\right)t \qquad (2.11)$$

for baryons.

Fig. 1 shows an example of the behavior in the signal to noise ratio from a recent simulation[3] with two flavors of dynamical Wilson fermions with $\kappa = 0.1670$ and gauge coupling $\beta = 5.3$, with curves of Eqs. (2.10) and (2.11) drawn "by eye" through the large-t points.

Finally, observe that Eqn. (2.5) can also be used to find the correlation between fluctuations of $C(t)$ at different times:

$$N\sigma^2(t_1, t_2) = \langle (C(t_1) - \langle C(t_1) \rangle)(C(t_2) - \langle C(t_2) \rangle) \rangle \qquad (2.12)$$

$$= \langle O(t_1)O(0)O(t_2)O(0) \rangle - \langle O(t_1)O(0) \rangle \langle O(t_2)O(0) \rangle. \qquad (2.13)$$

In the limit when t_1 and t_2 become large, but the difference $t_1 - t_2$ remains small, the first term dominates the correlation and we expect

$$N\sigma^2(t_1, t_2) = \langle O(t_1)O(t_2)O(0)O(0) \rangle \qquad (2.14)$$

or

$$N\sigma^2(t_1, t_2) = C(t_1 - t_2)C(0) \qquad (2.15)$$

so

$$N\sigma^2(t_1, t_2) \simeq C(0)^2 \exp(-m|t_1 - t_2|). \qquad (2.16)$$

This large positive correlation means that if $C(t)$ on a lattice at one t value is high, all the $C(t)$'s within a t interval of roughly $1/m$ will also be high. These correlations are unavoidable and must be taken into account when the data is fit. A further discussion of this point is found in the description of error analysis later on in this section.

These results demonstrate that it will be very difficult to overwhelm a correlator by going to enormous separation. The only way to succeed in a spectroscopy calculation will be to find operators which couple very strongly to the desired state.

B. Lattice Fermions

While defining scalar fields on the lattice is relatively straightforward, defining fermions on the lattice presents a problem: doubling. The naive procedure of discretizing the continuum fermion action results in a lattice model with many more low energy modes than one originally anticipated. Let's illustrate this with free field theory.

The free Euclidean fermion action in the continuum (in four dimensions) is

$$S = \int d^4x [\bar{\psi}(x)\gamma_\mu \partial_\mu \psi(x) + m\bar{\psi}(x)\psi(x)]. \qquad (2.17)$$

One obtains the so-called naive lattice formulation by replacing the derivatives by symmetric differences: we explicitly introduce the lattice spacing a in the denominator and write

$$S_L^{naive} = \sum_{n,\mu} \bar{\psi}_n \frac{\gamma_\mu}{2a}(\psi_{n+\mu} - \psi_{n-\mu}) + m \sum_n \bar{\psi}_n \psi_n. \qquad (2.18)$$

Let's write down the propagator:

$$G(p) = (i\gamma_\mu \sin p_\mu a + ma)^{-1} = \frac{-i\gamma_\mu \sin p_\mu a + ma}{\sum_\mu \sin^2 p_\mu a + m^2 a^2} \qquad (2.19)$$

and identify poles in $G(p)$ with masses of particles. Let's suppose $m = 0$. Then, besides the pole at $p = (0,0,0,0)$ there are fifteen other poles, at $p = (\pi, 0, 0, 0)$, $(0, \pi, 0, 0,)$, $\ldots (\pi, \pi, \pi, \pi)$. This is a model for sixteen massless fermions, not one.

There are other ways to see what is happening. From the point of view of Minkowski space the dispersion relation is

$$\sinh^2 Ea = \sum_i \sin^2 p_i a + m^2 a^2 \tag{2.20}$$

and not

$$E^2 = p^2 + m^2. \tag{2.21}$$

We identify a light particle as one which has energy $O(1)$ as the lattice spacing a vanishes. Because the momentum appears in the dispersion relation as $\sin^2 p_i a$ rather than as p_i^2, there are extra solutions near $p_i \simeq \pi/a$. The end of the Brillouin zone has bent down to touch the $E = 0$ axis.

(a) Wilson Fermions

There are two ways to deal with the doublers. The first way is to alter the dispersion relation so that it has only one low energy solution. The other solutions are forced to $E \simeq 1/a$ and become very heavy as a is taken to zero. The simplest version of this solution (and almost the only one seen in the literature) is due to Wilson: add a second derivative-like term

$$S^W = -\frac{r}{2a} \sum_{n,\mu} \bar{\psi}_n (\psi_{n+\mu} - 2\psi_n + \psi_{n-\mu}) \tag{2.22}$$

to S^{naive}. The parameter r must lie between 0 and 1; $r = 1$ is almost always used and in common parlance "$r = 1$" is implied when one speaks of using "Wilson fermions." The propagator is

$$G(p) = \frac{-i\gamma_\mu \sin p_\mu a + ma - r \sum_\mu (\cos p_\mu a - 1)}{\sum_\mu \sin^2 p_\mu a + (ma - r \sum_\mu (\cos p_\mu a - 1))^2}.$$

It has one pole at $p = (im, 0, 0, 0)$, plus other poles at $p \simeq r/a$. In the continuum these states become infinitely massive and decouple (although this is not a trivial statement).

With Wilson fermions it is conventional not to use not the mass but the "hopping parameter" $\kappa = \frac{1}{2}(ma + 4r)^{-1}$, and to rescale the fields $\psi \to \sqrt{2\kappa}\psi$. The action for an interacting theory is conventionally written

$$S = \sum_n \bar{\psi}_n \psi_n - \kappa \sum_{n\mu} (\bar{\psi}_n (r - \gamma_\mu) U_\mu(n) \psi(n + \mu) + \bar{\psi}_n (r + \gamma_\mu) U_\mu^\dagger \psi(n - \mu). \tag{2.23}$$

The advantage of Wilson fermions is that they are closest to the continuum formulation there is a four component spinor on every lattice site for every color and/or flavor of quark. Constructing currents and states is just like in the continuum. (Operator mixing for weak matrix elements can get complicated; consult Soni's lectures.)

Unfortunately, the Wilson term explicitly breaks chiral symmetry. This has the consequence that the zero bare quark mass limit is not respected by interactions; the quark mass is additively renormalized. It was an open question for a long time what are the relics of chiral symmetry with Wilson fermions but now it is pretty well agreed that even at finite lattice spacing there is a critical value of the hopping parameter κ_c at which the pion mass vanishes, and near that point the violations of chiral symmetry are what one would expect from a nonzero quark mass. The value of κ_c is not known a priori before beginning a simulation; it must be computed. This is done in a simulation involving Wilson fermions by varying κ and watching the pion mass extrapolate quadratically to zero as $m_\pi^2 \simeq \kappa_c - \kappa$ ($\kappa_c - \kappa$ is proportional to the quark mass for small m_q).

There are a number of new variations on Wilson fermions which are designed to remove finite lattice spacing effects from matrix element calculations. See Ref. 4 for a discussion.

(b) Staggered Fermions

In this formulation one reduces the number of fermion flavors by using one component "staggered" fermion fields rather than four component Dirac spinors. The Dirac spinors are constructed by combining staggered fields on different lattice sites.

Let us call the staggered field on site n χ_n and define the matrix field

$$\psi_{\alpha k}(n) = \sum_b (\gamma^{n+b})_{\alpha k} \chi_{n+b} \qquad (2.24)$$

where

$$\gamma^{n+b} = \gamma_1^{n_1+b_1} \gamma_2^{n_2+b_2} \gamma_3^{n_3+b_3} \gamma_4^{n_4+b_4} \qquad (2.25)$$

and b runs over the sites of a hypercube $b = (0,0,0,0), (1,0,0,0), \ldots (1,1,1,1)$. $\bar{\psi}_{\alpha k}$ is defined analogously,

$$\bar{\psi}_{\alpha k}(n) = \sum_b (\gamma^{n+b})_{\alpha k}^\dagger \bar{\chi}_{n+b}. \qquad (2.26)$$

One interprets the α indices as spin (= 1 to 4) and k as flavor (= 1 to 4). The free action (with $m_q = 0$) is

$$S = \frac{1}{128} \sum \bar{\psi}_n \gamma_\mu (\psi_{n+\mu} - \psi_{n-\mu}) \qquad (2.27)$$

$$= \frac{1}{2} \sum \bar{\chi}_n \eta_{n,\mu} (\chi_{n-\mu} - \chi_{n+\mu}) \qquad (2.28)$$

where $\eta_{n,\mu} = (-1)^{n_1+n_2+n_3+n_4}$. Gauge fields can be added in the standard way, and explicit mass terms give an extra $m_q \sum \bar{\chi}\chi$ term.

Staggered fermions preserve an explicit chiral symmetry as $m_q \rightarrow 0$ even for finite lattice spacing, as long as all four flavors are degenerate. They are preferred over Wilson fermions in situations in which the chiral properties of the fermions dominate the dynamics–for example, in studying the chiral restoration/deconfinement transition at high temperature. They also present a computationally less intense situation from the point of view of numerics than Wilson fermions, for the trivial reason that there are less variables. However, flavor symmetry and translational symmetry are all mixed together. Construction of meson and baryon states (especially the Δ) is more complicated than for Wilson fermions[5]

(c) Nonrelativistic Fermions[6]

Finally, if one is interested in studying heavy quarks one should forget about the Dirac equation altogether and study the Schrodinger equation. Doing so explicitly excludes relativistic heavy particles from the theory. This is a sensible thing to do as long as the physics one is interested in is dominated by momenta $p << M$ the quark mass. Since we are dealing with a quantum field theory the relativistic states do affect low energy physics, but because their effects arise due to highly virtual intermediate states which cannot propagate long distances, they only introduce local interactions to the Lagrangian. Some of these interactions renormalize terms which are already present. The nonrelativistic theory has a cutoff (roughly $\Lambda \simeq M$) and so some of the new interactions are nonrenormalizable. To any order in (p/Λ) a finite number of nonrenoramlizable interactions are needed. Performing a Foldy -Wouthuysen transformation, one can write down a Lagrangian for nonrelativistic QCD

$$\mathcal{L}_{NRQCD} = -\frac{1}{2}\mathrm{Tr}F_{\mu\nu}F^{\mu\nu} + \psi^\dagger[iD_t + \frac{\vec{D}^2}{2M}]\psi + \psi^\dagger(c_1\frac{\vec{D}^4}{8M^3} + c_2\frac{g}{2M}\vec{\sigma}\cdot\vec{B})\psi + \ldots \quad (2.29)$$

The couplings $M, g, c_1, c_2, \ldots$ are specific to the cutoff and can be fixed by matching NRQCD and ordinary QCD calculations.

The lowest order Green's function obeys

$$(iD_t + \frac{\vec{D}^2}{2M})G(x, x') = \delta^4(x - x'). \quad (2.30)$$

On a Euclidean lattice there are several ways to implement this equation. One choice for the lowest order Green's function is to solve

$$G(\vec{x}, t+1) = U_0^\dagger(\vec{x}, t)(1 - \frac{1}{2M}\sum_j \nabla_j\nabla_{-j}G(\vec{x}, t) + \delta_{t,0}). \quad (2.31)$$

The quark propagator is much easier to calculate for NRQCD than for full QCD because it is the solution to an initial value problem, not a boundary value problem. It can be computed in one pass through a lattice.

All of the formalism of NRQCD for heavy quarks was introduced by Thacker and Lepage[6] and only their one paper has been published in this field so far. The subject of heavy quarks on the lattice is absolutely wide open at present.

C. Interpolating Fields for Mesons and Baryons

Let us see how to construct operators which couple to mesons and baryons by considering the simplest example of a current-current correlation function

$$C(x,t) = \langle J(x,t)J(0,0) \rangle \tag{2.32}$$

where

$$J(x,t) = \bar{\psi}(x,t)\Gamma\psi(x,t) \tag{2.33}$$

and Γ is a Dirac matrix. The intermediate states $|n\rangle$ which saturate $C(x,t)$ are the hadrons which the current J can create from the vacuum: the pion, for an axial current, the rho, for a vector current, and so on. Masses can be extracted exactly as in Sec. 2.1, from

$$C(t) = \sum_x C(x,t) = \sum_n |\langle 0|J|n\rangle|^2 \exp(-m_n t). \tag{2.34}$$

Now we write out the correlator in terms of fermion fields

$$C(t) = \sum_x \langle 0|\bar{\psi}_i(x,t)^\alpha \Gamma_{ij}\psi_j(x,t)^\alpha \bar{\psi}_k(0,0)^\beta \Gamma_{kl}\psi_l(0,0)^\beta|0\rangle \tag{2.35}$$

with a Roman index for spin and a Greek index for color. We contract creation and annihilation operators into quark propagators

$$\langle 0|T(\psi_j(x,t)^\alpha \bar{\psi}_k(0,0)^\beta)|0\rangle = G_{jk}^{\alpha\beta}(x,t;0,0) \tag{2.36}$$

so

$$C(t) = \sum_x G_{jk}^{\alpha\beta}(x,t,;0,0)\Gamma_{kl}G_{li}^{\beta\alpha}(0,0,x,t)\Gamma_{ij} \tag{2.37}$$

or

$$C(t) = \sum_x \mathrm{Tr}G(x,t;0,0)\Gamma G(0,0;x,t)\Gamma \tag{2.38}$$

where the trace runs over spin and color indices.

Baryons are constructed similarly. A local three quark operator which couples to a color singlet state can be written as

$$\epsilon_{\alpha\beta\gamma}\chi_{ijk}\psi_i^\alpha \psi_j^\beta \psi_k^\gamma \tag{2.39}$$

for some function χ_{ijk} which keeps track of spins. A correlation function which produces a baryon is

$$C_B(t) = \sum_x \epsilon_{\alpha\beta\gamma}\epsilon_{\alpha'\beta'\gamma'}\chi_{ijk}\chi^\dagger_{i'j'k'}G^{\alpha\alpha'}_{ii'}(x,t;0,0)G^{\beta\beta'}_{jj'}(x,t;0,0)G^{\gamma\gamma'}_{kk'}(x,t;0,0).$$

$$(2.40)$$

If two or more of the quarks have the same flavor, the signal may be enhanced by including both direct and exchange terms.

One final trick aids the evaluation of Eq. (2.39). One might think that one needs to separately construct quark and antiquark propagators. However, one can obtain the antiquark propagator from the quark propagator by the transformation

$$G(x,y) = \gamma_5 G(y,x)^\dagger \gamma_5. \qquad (2.41)$$

While this formalism looks like it is intended only for Wilson (or naive) fermions, it also carries over to staggered fermions. The entanglement of flavor, space and spin indices presents a few technical complications, but hadron propagators are computed similarly[7] One begins with an interpolating field for mesons

$$M_{AB}(y) = \bar{q}(y)(\Gamma_A \otimes \Gamma_B^*)q(y) \qquad (2.42)$$

where Γ_A and Γ_B are two of the 16 matrices $\Gamma_b = \gamma_1^{b_1}\ldots\gamma_4^{b_4}$ and b labels the location in the hypercube ($b_i = 0$ or 1). By convention, Γ_A acts on spin indices and Γ_B acts on flavor indices. If $\Gamma_A = \Gamma_B$ then the operator $M_A \equiv M_{AA}$ is local. Otherwise it involves combinations of field operators at different locations in the fundamental hypercube. One must either gauge fix before measurement or explicitly include link factors connecting the sites.

If the operator is local then

$$M_A(y) = \sum_b \epsilon_{Ab}\bar{\chi}_b(y)\chi_b(y) \qquad (2.43)$$

and ϵ is 1 or -1 depending on whether Γ_b and Γ_A commute or anticommute. In practice that means that a local channel tends to have two particles of opposite parity in it and one must fit the correlator to

$$C(t) = Z_1 e^{-\mu_1 t} + (-)^t Z_2 e^{-\mu_2 t} \qquad (2.44)$$

(plus boundary terms). There are four possibilities for local operators: they are (a) $\Gamma = \gamma_5$, $\epsilon_b = (-)^y$ (pseudoscalar) (b) $\Gamma = 1$ and $\gamma_0\gamma_5$, $\epsilon_b = 1$ (scalar and pseudoscalar) (c) $\Gamma = \gamma_3$ and $\gamma_1\gamma_2$, $\epsilon_b = (-)^{b_3}$ (vector and tensor) (d) $\Gamma = \gamma_0\gamma_3$ and $\gamma_5\gamma_3$, $\epsilon_b = (-)^{b_1+b_2}$ (vector and axial vector). The proton can be constructed by replacing ψ by χ in Eq. (2.39) and taking $\chi_{ijk} = 1$; the Δ cannot be made by a local operator[5] It has lately become state-of-the -art to look at nonlocal operators since the particles which would be Goldstone partners of the lightest pion in the continuum limit only couple to them. For examples of their use, see Ref. 8.

Finally, the analog of Eq. (2.41) for staggered fermions is

$$G(x,y) = (-)^x G(y,x)^\dagger (-)^y. \tag{2.45}$$

Local currents are not the optimal choice for interpolating fields for spectroscopy. In Section III.A we will discuss more modern techniques. They still involve products of Green's functions.

D. Constructing Propagators–Matrix Inversion Techniques

Most of the computational effort of a calculation involving fermions goes into constructing fermion propagators. That is, one wants to find the inverse $G(x,y)$ of one's latticized fermion differential operator $M(x,y)$, which obeys

$$M(x,y)G(y,z) = \delta(x-z). \tag{2.46}$$

Since M has order (volume) nonzero elements this is a problem of sparse matrix inversion. It is usually not possible to construct or store $G(x,x')$ for all x and y since this involves finding on the order of (volume)^2 numbers. Instead, one typically constructs $G(x,y)$ for all x and for some selected points y by solving

$$M(x,y)\tilde{G}(y,z) = S(z). \tag{2.47}$$

(that is, $\tilde{G}(x,y)$ is the vector $M^{-1}S$).

Matrix inversion in QCD is a hard problem for at least two reasons. A simple reason is that the matrices are very large. For example, state of the art calculations today use at least $16^3 \times 32$ lattices. Wilson fermions have four spins and three complex color degrees of freedom per site, so one is looking at $24 \times 16^3 \times 32 = 3.1$ million real variables in $\tilde{G}$. A more insidious problem is that the largest eigenvalue is on the order of $2\pi/a$ and the smallest eigenvalue is on the order of m_q. For small m_q this means that that M is ill conditioned: the ratio of its largest to smallest eigenvalue diverges as $m_q \to 0$. In practice, the more ill-conditioned the matrix, the harder it is to invert. This is why one generally does not work at the physical value of the quark mass. Instead, one is restricted to unphysically heavy values of the quark mass.

There have been a number of diagnostic studies of matrix inversion algorithms. A good one is Ref. [9]. The bottom line (at least for now) seems to be that the most robust algorithm, with the best behavior for small quark mass, is the conjugate gradient algorithm.

The conjugate gradient method[10] is designed to solve a set of N linear equations

$$Mx = y$$

where M is a symmetric positive-definite $N \times N$ matrix. If we had an initial solution vector x_0 we could try to approximate x by letting

$$x_m = x_0 + \sum_{i=1}^{m} \alpha_i M^{i-1}(Mx_0 - y) \qquad (2.48)$$

or

$$x_m - x_0 = (x_0 - x) + \sum_{i=1}^{m} \alpha_i M^i (x_0 - x), \qquad (2.49)$$

choosing the α's to minimize the "M-norm" $\|x_m - x\|_M \equiv \langle x_m - x | M | x_m - x \rangle$. The α's that do this can be found by orthonormalizing the set of vectors $M^i(x_0 - x)$ for $i = 1, 2 \ldots m$, to write

$$u_i = \sum_{j=1}^{i} c_{ij} M^j (x_0 - x) \qquad (2.50)$$

for $i = 1, 2 \ldots m$ with $\langle u_i | M | u_j \rangle = \delta_{ij}$. That is,

$$x_m = x_0 - \sum_{i=0}^{m} |u_i\rangle \langle u_i | Mx_0 - y \rangle. \qquad (2.51)$$

If it were necessary to orthoganalize each new vector $M^j(x_0 - x)$ with respect to all the previous ones, the method would be quite expensive. However, if we take as our jth vector $M(x_j - x_0)$ which is just a linear combination of $M^j(x - x_0)$ and all the previous ones, then the jth vector is orthogonal to the first $j - 2$ vectors, since

$$\langle M^i(x_0 - x) | M^2(x_{j-1} - x_0) \rangle = \langle M^{i+1}(x_0 - x) | M(x_{j-1} - x_0) \rangle; \quad i+1 \leq j-1 \quad (2.52)$$

but $x_{j-1} - x$ was already constructed to be M-orthogonal to $M^l(x_0 - x)$ for $l \leq j - 1$. Thus one only needs to orthogonalize the jth vector to the $(j + 1)$st as one goes along.

What one is doing is improving the solution (minimizing $\langle x_0 - x | M | x_0 - x \rangle$) by moving in a direction which is orthogonal to all the previously chosen directions. This is done by a kind of Gram-Schmidt orthogonalization procedure.

Here is a cookbook example. Given x_0, define the residual $r_0 = y - Mx_0$ and $p_0 = r_0$. Then iterate

$$\alpha_i = \langle r_i | r_i \rangle / \langle p_i | M p_i \rangle \qquad (2.53a)$$

$$x_{i+1} = x_i + \alpha_i p_i \qquad (2.53b)$$

$$r_{i+1} = r_i - \alpha_i M p_i \qquad\qquad (2.53c)$$

$$\beta_i = \langle r_{i+1} | r_{i+1} \rangle / \langle r_i | r_i \rangle \qquad\qquad (2.53d)$$

$$p_{i+1} = r_{i+1} + \beta p_i \qquad\qquad (2.53e)$$

The p's are just the un-normalized u's of Eq. 2.51.

One complication is that the conjugate gradient algorithm can only be used on symmetric positive-definite matrices, and our M is not. Instead of Eq. (2.47), one must solve

$$M^\dagger M G = M^\dagger S. \qquad\qquad (2.54)$$

Thsi is unfortunate because conjugate gradient, like all iterative matrix inversion schemes, solves $Mx = b$ by iterating trial solutions at the rate of one or two matrix multiplications per step. Calculating $M^\dagger M x$ usually takes twice as much work as calculating Mx. More seriously, the conditioning index of $M^\dagger M$ is the square of the conditioning index of M.

There are a class of algorithms, called Conjugate Residue or Minimum Residue, which solve $MG = S$ directly. These algorithms converge if the Hermetian part of M is positive-definite. Unfortunately, our M's involve the gauge field U and so we don't generally know whether our matrices are of this form. People have had success using these methods for larger values of the quark mass, and since they take half as much time per step as conjugate gradient (and seem to require about the same number of steps to converge), they may be the methods of choice there. However, I feel that one must be very careful with these algorithms at small quark mass.

So far I have discussed solving $MG = S$ directly. One would like to have a matrix inversion scheme which rewards a good guess. One way to do this is by preconditioning M, either by a matrix with a known inverse M_0 or by left and right multiplication by known invertable matrices L and R:

$$M_0^{-1} M G = S_1 \qquad\qquad (2.55)$$

(where $S_1 = M_0^{-1} S$) or

$$L^{-1} M R^{-1} G_2 = S_2 \qquad\qquad (2.56)$$

(where $S_2 = L^{-1} S$ and $G_2 = RG$). If $M \simeq M_0$ (2.55) or $M \simeq LR$ (2.56) the matrix which is being inverted is nearly diagonal and hopefully only a little work will need to be done to complete the construction of G. Of course, if preconditioning is to be successful the extra time spent on preconditioning must not be greater than the time saved because fewer iterations are needed.

Several preconditioning algorithms have been proposed. The ones most commonly used (not necessarily the best ones) are based on checkerboards.[9,11] Define a site on a lattice as even (odd) if its coordinate sum $(x + y + z + t)$ is even or odd. The Dirac equation can be written in a checkerboard basis as

$$M = \begin{pmatrix} m & \not{D}_R \\ \not{D}_L & m \end{pmatrix}. \tag{2.57}$$

Preconditioning by

$$M_0^{-1} = \begin{pmatrix} m & -\not{D}_R \\ -\not{D}_L & m \end{pmatrix}. \tag{2.58}$$

gives

$$M_0^{-1} M = \begin{pmatrix} m^2 - \not{D}_R \not{D}_L & 0 \\ 0 & m^2 - \not{D}_L \not{D}_R \end{pmatrix}. \tag{2.59}$$

This maps the problem onto one checkerboard. Notice that if one knows G on one checkerboard (G_1, say) one can reconstruct the other checkerboard G_2 from $\not{D}_1 G_1 + m G_2 = S_2$.

This trick is often applied to staggered fermions where $\not{D}_R = -\not{D}_L$. Then $M_0^{-1} = M^\dagger$ and $M_0^{-1} M$ is symmetric positive definite and can be directly inverted by conjugate gradient. This trick is also used more often than not to save space. One needs only to keep fermion fields defined on one checkerboard. For Wilson fermions this preconditioning saves about a factor of three in iterations for β near 6.0 and $m_\pi a$ from 0.3 to 1.0 or so. It has zero overhead compared to the unconditioned algorithm.

In the numerical analysis literature this trick is called working with the 'reduced system'.

A second (actually earlier) preconditioning using hyperplanes was invented by Oyanagi[12] It is claimed to give a factor of seven or so decrease in iteration number at a cost of a factor of three in time per iteration, compared to unconditioned conjugate gradient.

The Cornell group[13] proposed using Fourier acceleration: in one simple implementation one uses M_0^{-1} equal to the (fast) Fourier transform of $(\not{p} - m)^{-1}$. This conditioner is not gauge invariant and to use it, one must gauge fix to a smooth gauge. The motivation for this M_0 is that asymptotic freedom says that at short distances the quarks behave like free particles, so we should be guessing an approximate propagator which should be nearly right at short distances. For Wilson fermions at $\beta \simeq 6.0$ one can get a factor of three reduction in the number of iterations over unconditioned conjugate gradient. When applied on top of a checkerboard preconditioning (to save space) the number of iterations drops by only ten per cent. Fourier preconditioning would be exact if the Dirac propagator were diagonal in momentum space. That would be true deep in the perturbative regime but apparently it's not the case at $\beta = 6$.

Multigrid techniques have been touted as a great hope for matrix inversion. The idea behind multigrid is that one temporarily integrates out short distance physics, solves one's problem on a coarser grid, then interpolates this solution back to the fine grid. These techniques have been quite successful for solving boundary value problems for partial differential equations. Multigrid methods for lattice gauge theory are still in a primitive state and whether they will help with propagator construction is still unknown[14] They have not been used in any large scale QCD simulation as of this writing.

Finally, no one has successfully preconditioned staggered fermions on any lattice used to date. It seems[15] the reason is this: preconditioning works because it lowers the conditioning number of the matrix to a value nearer unity. On an equivalent size lattice the conditioning number for staggered M's is already much smaller than for Wilson M's so the potential gain for preconditioning is much less. Presumably that will not always be so.

We need better matrix inversion algorithms! Essentially all the computing budget in contemporary simulations is devoted to matrix inversion.

E. Error Analysis for Spectroscopy

There is a tendency for people working in the lattice field to belittle error analysis. I have heard many people tell me "My data are too noisy to warrant a detailed analysis" and then get into a tremendous fight over whether their results are consistent with someone else's. A very good physicist who shall remain nameless once tried to convince me that one of his results was meaningful even though his data set consisted of three lattices. Unfortunately, the place where most Monte Carlo simulations go awry is in the error analysis stage. If you are going to work in this field, my advice is to learn good habits early and be very careful with your error analysis[16]

The main difference between Monte Carlo simulation and the analysis of an ordinary experiment is that in Monte Carlo simulation all measurements are highly correlated.

There are two sources of correlations. The first arises from the updating algorithm. Most of the updating algorithms in use today make a change in a dynamical variable (a gauge field link variable, to be specific) which depends on the local environment of the variable. Correlation lengths at interesting values of coupling constants are typically large and correlation times tend to scale like the square of the correlation length so the whole simulation grinds to a halt and nothing ever changes. This is the problem of "critical slowing down" which is discussed by Sokal in another chapter of this book.

The second source of correlations is more insidious. As you recall from Sec. IIA (Eq. (2.16)), everything measured on the same lattice is highly correlated. This means that naive error analysis is at best misleading. It is an incorrect assumption to think that the "naive χ^2

$$\chi^2 = \sum_i \frac{(y_i - y(x_i))^2}{\sigma_i^2} \tag{2.60}$$

is a good indicator of goodness of fit. The right approach is to minimize a "correlated χ^2"

$$\chi^2 = \sum_{ij} \frac{(y_i - y(x_i))}{\sigma_i} \frac{(y_j - y(x_j))}{\sigma_j} H_{ij}^{-1} \tag{2.61}$$

where the correlation matrix is given by the data itself:

$$H_{ij} = \sum_l \frac{(y_i^l - \langle y_i \rangle)}{\sigma_i} \frac{(y_j^l - \langle y_j \rangle)}{\sigma_j} \tag{2.62}$$

(I have normalized $H_{ii} = 1$). In Fig. 2 I display a contour plot of H_{ij} for a typical hadron correlator $C(t) = \langle O(t)O(0) \rangle$. Notice that H_{ij} is above 50 per cent for $\Delta t = 4$. The value of χ^2 at the fit is a meaningful indicator of goodness of fit, in contrast to the value of the naive χ^2 ($H_{ij} = \delta_{ij}$).

This sort of error analysis is even more important for calculations of matrix elements. Typically one extracts values of matrix elements from ratios of correlation functions. For example, one correlator might behave like $Z \exp(-\mu t)$, a second correlator like $zZ \exp(-\mu t)$, and one wants to find z. Of course, it is wrong to fit the ratios of the correlators directly – the ratio of two Gaussianly-distributed random distributions is not Gaussianly-distributed. One correct thing to do is to make a correlated fit to the two propagators with three parameters (z, Z, and μ). Then one gets a meaningful χ^2 and can measure the correlation between the different fit parameters.

Finally, what can you do if you do not have enough data for a full correlated fit, or if you do not know how to disentangle the data to find a correlation matrix? One possibility is a "jacknife[17] fit." Here one drops J points from one's collection of N data points and computing the average value of the observable from the remaining $N - J$ pieces. The procedure is repeated N/J times and the error is determined from the N/J subensemble values, through the formula

$$\sigma_J^2 = ((N/J) - 1)\langle (x_J - \bar{x})^2 \rangle \tag{2.63}$$

where the x_J's are the results obtained from analyzing the subensemble with J elements removed. If the data set is uncorrelated, σ_J will not vary with J while if the data are highly correlated, doubling J will produce an increase in σ_J by $\sqrt{2}$.

If one is fitting n points to n parameters the χ^2 at the minimum will be zero. (In spectroscopy this is called determining an "effective mass." For example, one

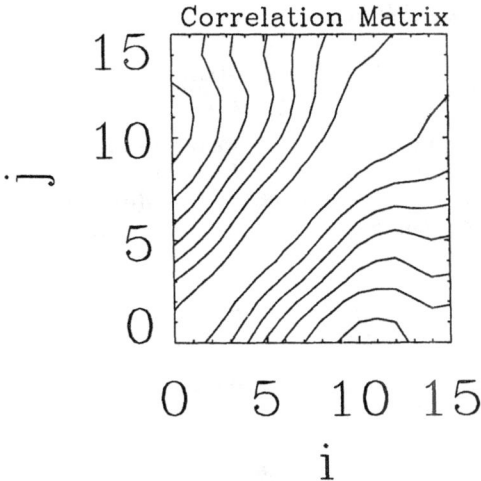

FIGURE 2

Contours of constant autocorrelation in a "typical" hadronic correlation function (an HEMCGC Wilson pion). Contours run from .95 at the center, decreasing by 0.1 per contour.

can determine the local mass from two successive values of a correlator, $\exp(-\mu) = C(t+1)/C(t)$ if $C(t)$ is a single exponential $C(t) = Z\exp(-\mu t)$.) All fitting methods should give the same parameters. When the number of degrees of freedom is greater than zero, however, I have seen correlated fits give different best-fit values than uncorrelated fits, and one should be very careful.

F. Taking The Continuum Limit (?)

Now I briefly describe how to take the lattice spacing a to zero. When we define a theory on a lattice we implicitly assume that a is an ultraviolet cutoff and $g(a)$ is the bare coupling defined with the cutoff. When we take a to zero we must also specify how $g(a)$ behaves. The proper continuum limit comes when we take a to zero holding physical quantities fixed, not when we take a to zero holding g fixed.

On the lattice, if all quark masses are set to zero, the only dimensionful parameter is the lattice spacing, so all excitations have masses

$$m = \frac{1}{a}f(g(a)) \tag{2.64}$$

where $f(g(a))$ is dimensionless. We must tune $g(a)$ so

$$\lim_{a \to 0} \frac{1}{a} f(g(a)) \to \text{constant}. \tag{2.65}$$

From the point of view of the lattice theory, we must tune g so that correlation lengths $1/ma$ diverge. This will occur only at the locations of second (or higher) order phase transitions in the lattice theory. How does $g(a)$ behave near a critical point? The renormalization group will tell us.

Recall that the β-function is defined by

$$\beta(g) = a \frac{dg}{da} g(a) = \frac{dg(a)}{d\ln(1/\Lambda a)}. \tag{2.66}$$

(Λ is a dimensional parameter introduced to make the argument of the logarithm dimensionless.) At a critical point $\beta(g_c) = 0$. Thus the continuum limit is the limit

$$\lim_{a \to 0} g(a) \to g_c. \tag{2.67}$$

In QCD $g_c = 0$ and in lowest order

$$\beta(g) = -b_1 g^3 \tag{2.68}$$

so integrating the renormalization group equation gives

$$g^2(a) = \frac{1}{b_1 \ln(1/\Lambda a)} \tag{2.69}$$

(the usual expression for the running charge) or

$$a\Lambda = \exp(-\frac{1}{b_1 g^2(a)}). \tag{2.70}$$

Actually, the two-loop β-function is prescription independent,

$$\beta(g) = -b_1 g^3 + b_2 g^5, \tag{2.71}$$

and so we want to observe

$$a\Lambda = (\frac{1}{g^2(a)})^{b_2/(2b_1^2)} \exp(-\frac{1}{b_1 g^2(a)}), \tag{2.72}$$

or equivalently ma showing the same dependence as a goes to zero, in order to be in an asymptotic scaling regime.

Thus to compute a physically interesting (i.e. continuum) result in lattice gauge theory one must do three things:

(1) Compute masses in lattice units at several g's (measure ma vs. $g(a)$).

(2) Observe ma scaling according to the β-function as g is varied.

(3) Use one dimensionful parameter to set the overall mass scale and predict all other dimensionful parameters in its units.

There is a less stringent form of scaling which is often discussed in the literature. The idea is that since mass ratios show less g dependence than the individual masses, one ought to require mass ratios which are g-independent even though individual masses do not scale correctly. That is,

$$m_1(a)/m_2(a) = m_1(0)/m_2(0)(1 + O(a^2/m^2)) \qquad (2.73)$$

or

$$m_1(a)/m_2(a) = m_1(0)/m_2(0)(1 + O(\exp(-1/(b_1 g(a)^2)))) \qquad (2.74)$$

This is called "scaling" as opposed to "asymptotic scaling" where the perturbative $f(g)$ is seen.

Having said these words, I must admit that this program is presently feasible only for glueballs in quenched QCD. (I will show examples in Section IV.) Simulations with fermions have another dimensionful parameter besides the lattice spacing, namely the quark mass. In principle, one must tune the bare quark mass along with the lattice spacing (or coupling) so that the physical quark mass is held fixed. I have not seen any analyses of hadron spectroscopy, either quenched or with dynamical fermions, which attempts an honest continuum limit. Instead, one commonly presents only figures involving mass ratios. Examples of such plots (which I will show in Chapter V) are the so-called Edinburgh plot (m_p/m_ρ vs. m_π/m_ρ) and the Rome plot (m_p/m_ρ vs. $(m_\pi/m_\rho)^2$). Here one displays mass ratios for a variety of quark masses, all heavier than the physical quark masses, and tries to tell whether or not an extrapolation to physical quark mass will yield correct spectroscopy. Of course, the use of such plots hides almost as much physics as it displays. One must have a model for the extrapolation. It may be hard to disentangle the variation in hadron mass with bare quark mass from variation due to the coupling constant.

Unfortunately, there doesn't appear to be a good solution to this problem.

III. MORE COMPLICATED ISSUES

Now we turn to a set of interesting state of the art problems whose solutions are still research topics.

A. Good Trial Wave Functions for Quark Bound States

The Euclidean time correlation function is

$$C(t) = \langle 0|O(t)O(0)|0\rangle = \sum_n |\langle 0|O|n\rangle|^2 e^{-E_n t} \tag{3.1}$$

and for large $(E_2 - E_1)t$ it becomes

$$C(t) \simeq |\langle 0|O|1\rangle|^2 e^{-E_1 t}(1 + R\ e^{-(E_2-E_1)t}) \tag{3.2}$$

where $R = |\langle 0|O|2\rangle|^2/|\langle 0|O|1\rangle|^2$. $C(t)$ is dominated by the exponential falloff of the leading term if $R\exp(-(E_2 - E_1)t) << 1$. If a poor interpolating field O is chosen, and R is not less than unity, a reliable value for the mass can be extracted only a large time t. In practice one must make the lattice larger in the time direction to compensate. Furthermore, since the signal to noise ratio can fall as the time separation of the measurement increases, higher statistics are needed. Both of these effects on lattice measurements of particle masses are expensive in terms of computer resources, so there is considerable incentive to find operators which couple very strongly to the ground state.

Let us now focus on the case of meson spectroscopy. Suppressing all color and spin indices, an operator which creates a meson is

$$O(t) = \sum_{x,y} c_q(x,t)^\dagger c_{\bar{q}}(y,t)^\dagger \Psi(x,y) \tag{3.3}$$

where $c_i(x,t)$ is an annihilation operator for a quark or an antiquark at space-time position (x,t) and Ψ is a c-number function, the wave function for the state. In Sec. II, we chose Ψ to be a delta function in the relative coordinate $x - y$ so that O is a current which can create the state from the vacuum. Then $O(t) = \sum_x J(x)$, where $J = \bar{\psi}\Gamma\psi$, Γ is a Dirac matrix, and we have converted temporarily back to implicit second-quantized language. However, especially for small quark masses or small lattice spacings, these pointlike wave functions perform very poorly as operators which generate hadronic states. This is not surprising, since the lightest physical hadrons are large objects (with radii on the order of 1 fermi) relative to the typical lattice spacings used in lattice QCD (≤ 0.1 fermi). One therefore expects a trial wave function which has physical extent over several lattice spacings to be evolved more readily into the true ground state by the projection operator e^{-Ht}.

There are at least two classes of trial wave functions. The first are wave functions which depend explicitly and separably on the relative (r) and center of mass (R) coordinates of the quarks, which are called "bound state" wave functions:

$$\Psi(x,y) = \Psi(R-r, R+r). \tag{3.4}$$

I call the second kind of wave function a "shell model" wave function. In these wave functions there is some externally imposed confinement volume; inside that volume the bound quarks and antiquarks move independently. For example, a "shell" meson wave function with equal mass quarks could be written as

$$\Psi(x,y) = \phi_R(x)\phi_R(y). \tag{3.5}$$

The first kind of wave function is obviously more physical. However, because of the way the correlation function is actually computed, the "shell" wave function may be preferred in actual calculations. This can be seen if we write out a meson correlation function as

$$C_{ij}(t) = \langle 0| \sum_{x_1,y_1} \sum_{x_2,y_2} \Psi_i(x_1,y_1)^\dagger G_q(x_1,t;x_2,0)G_{\bar{q}}(y_1,t;y_2,0)\Psi_j(x_2,y_2)|0\rangle. \tag{3.6}$$

The treatment of the two Ψ's in Eqn. (3.6) is quite different. One of them will be used as the source for propagators while the other will be used to weight the propagators at different locations on the lattice. With the bound state wave function as a source it turns out that one needs at least two matrix inversions (per color and spin index) to construct $C(t)$. (For example, to compute $\mathrm{TrG}(0;y)S(y;z)G(z;w)S(w;0)$ one can begin with $\Gamma_1 = G(z;w)S(w;0)$ from $M\Gamma_1 = S(w,0)$, then compute $\Gamma_2 = GSGS$ from $M\Gamma_2 = S(y;z)\Gamma_1$. See Ref. 18 for a discussion.)

However, when one uses a shell model wave function for the source the calculation of propagators factorizes since there is a separate source term for every propagator: one solves $M(G\phi) = \phi$ for the individual propagator and then just reconstructs the meson as if one had a source which was a local operator. One can use a shell model wave function at the sink, also. This gives a positive-definite correlation function and a variational bound on the ground state energy. A bound state wave function for the sink, perhaps even a point wave function (as is most commonly used) can be averaged over all spatial locations. This has the advantage that one is keeping only zero momentum states in the mode sum in Eqn. (3.1). The disadvantage of this scheme is that one's bound on the lightest energy is not variational; one must be sure one is measuring an asymptotic exponential before quoting a mass value.

If the points x and y are not coincident the wave function is not gauge invariant and care must be taken if the expectation value of the correlator is not to vanish due to Elitzur's theorem.[19] Two ways to insure that $C(t)$ is nonvanishing are either to gauge fix or to explicitly include the gauge links in the wave function, so that color information is parallel transported from point x to point y. Gauge fixing is a minor overhead compared to the time spent generating propagators and, as it allows one to use arbitrary Ψ's, is my method of choice (but not everyone's)[20]

Several kinds of wave functions have been used for spectroscopy. The earliest of these calculations was carried out by the APE group[21] They used shell model wave functions which are uniform cubes ($\phi = 1$ inside a cubic confinement volume and zero outside), with the gauge variables fixed to a smooth gauge.

Another class of shell model wave functions was introduced by the Wuppertal group[22] and more recently they have been used by Gupta, et. al[23] They take wave functions which are the solution to a three dimensional scalar field equation with a delta function source, i. e.,

$$(D^2 + m^2)\Psi(x - y) = \delta^3(x - y). \qquad (3.7)$$

The mass m can be used as a variational parameter. The use of the covariant derivative D makes the correlation function $C(t)$ gauge invariant.

Finally, several authors have used "wall" wave functions: the High Energy Monte Carlo Grand Challenge[24] (HEMCGC) and the Staggered Collaboration[8] have used shell model wave functions which are uniform over an entire timeslice of the lattice. They gauge fix to Coulomb gauge. These simple wave functions worked well for the parameters of these simulations and should do so as long as the lattice size is not too much bigger than a typical hadron diameter. (The HEMCGC actually concentrated on correlators of wall sources and pointlike sinks.) A comparison of wall and point trial wave functions is shown in Fig. 3.

Loft and I[25] have done diagnostic studies of Gaussian shell model wave functions. They are superior to point wave functions but on 16^3 spatial volumes at $\beta = 5.7$ to 6 they don't appear to be appreciably better than walls. Light hadrons are apparently so large that they nearly fill the confinement volume.

B. Direct reconstruction of wave functions

One can reconstruct quark wave functions (distribution amplitudes) directly from Monte Carlo simulations. The calculation is done in the following way for mesons: The wave function $\psi_G(r)$ of a hadron H in a gauge G is defined as

$$\psi_G(r) = \sum_{\vec{x}} \langle H | c_q^\dagger(\vec{x}) c_{\bar{q}}^\dagger(\vec{x} + \vec{r}) | 0 \rangle$$

where $c_q^\dagger(\vec{x})$ and $c_{\bar{q}}^\dagger(\vec{y})$ are quantum mechanical operators which create a quark and an antiquark at locations $\vec{x}$ and $\vec{y}$. One computes ψ from convolutions of quark and antiquark propagators $G(x, y)$ with the quark spins coupled appropriately

$$C(\vec{r}, t) = \sum_{\vec{x}} \langle 0 | \Psi(\vec{y}_1, \vec{y}_2) G_q(\vec{y}_1, 0; \vec{x}, t) G_{\bar{q}}(\vec{y}_2, 0; \vec{x} + \vec{r}, t) | 0 \rangle \qquad (3.8)$$

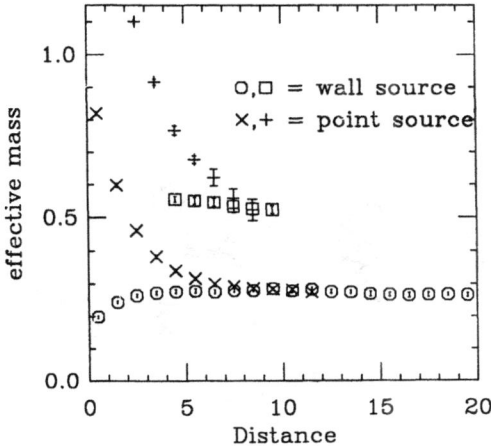

FIGURE 3

Pion and rho effective mass from point and wall sources (HEMCGC data).

where $\Psi(\vec{y}_1, \vec{y}_2)$ is some input trial wave function. At large t if the mass of the hadron is m_H, then

$$C(\vec{r}, t) \simeq \exp(-m_H t)\psi_G(\vec{r}) \qquad (3.9)$$

and so by plotting $C(\vec{r}, t)$ as a function of $\vec{r}$ we can reconstruct the wave function up to an overall constant. The convolutions can be done using (fast) Fourier transforms.

Until this year very little work on wave functions has been done. In 1985 Velikson and Weingarten[26] studied meson wave functions in SU(2) and Gottlieb[27] carried out the first study of wave functions with SU(3) Wilson fermions with what was then state of the art volume and statistics. Recently, Chu, Lissia and Negele have looked at gauge invariant wave functions (with a product of links connecting the quarks).[28] Some unpublished work with Kogut–Susskind fermions has been done by Kilcup, et. al. Wave functions have not been studied with dynamical fermions. At this year's lattice conference a number of groups[29] indicated that they are beginning studies of wave functions.

The goal of these studies is two-fold: First, visualizing wave functions is a powerful diagnostic for lattice studies. One can see whether the wave function of a hadron is squeezed by the simulation volume; if it is, then a calculation of spectroscopy may be compromised. A picture of the wave function provides a hint for a good trial wave function for spectroscopy.

Second, these wave functions are really the minimal Fock space states of zero

26

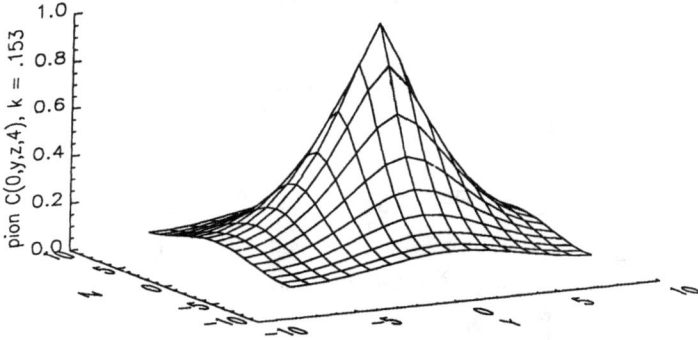

FIGURE 4

Coulomb gauge pion's wave function in the plane $z = 0$, from quenched $\beta = 6$ simulations with Wilson fermions at $\kappa = 0.153$.

momentum hadrons in a particular gauge. The value of the wave function at the origin, $\psi(0)$, is measured in vector meson electromagnetic decay or in leptonic decays of pseudoscalars, and derivatives of the wave function at the origin are related to form factors (through the formalism of Brodsky and Lepage). For heavy quark systems, overlaps of the wave functions themselves are related to electromagnetic transition rates.

Here are some examples. In Fig. 4 I show a snapshot of the Coulomb gauge pion's wave function and in Fig. 5 the wave function of the d quark in a proton, with the two u quarks offset by 0, 2, 4 and 6 lattice spacings, both in the plane $z = 0$, from quenched $\beta = 6$ simulations with Wilson fermions at $\kappa = 0.153$ ($m_\pi a = 0.42$). There is clearly a lot of qualitative science in these pictures.

C. Excited States

While it is conventionally claimed that one of the goals of lattice QCD is to compute the spectroscopy of the light hadrons, that claim cannot be taken seriously as long as one restricts ones attention to S-wave states. Most QCD simulations just consider the π, ρ, p, and Δ, and calculations of the Δ mass with staggered quarks have only appeared in the past year. Some P-wave states' masses have

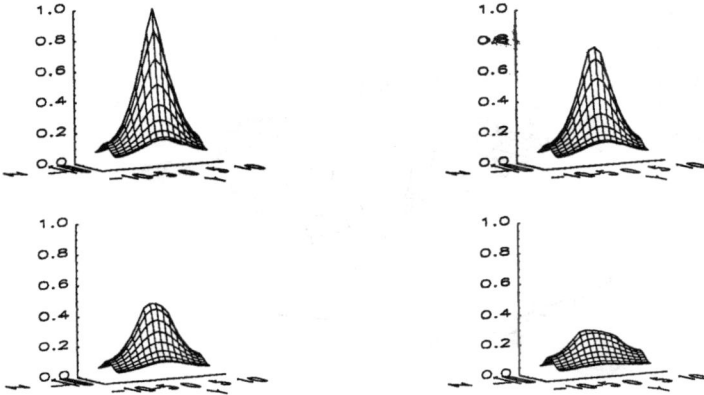

FIGURE 5

Coulomb gauge wave function of the d quark in a proton, with the two u quarks offset by 0, 2, 4 and 6 lattice spacings in the y direction, in the plane $z = 0$, from quenched $\beta = 6$ simulations with Wilson fermions at $\kappa = 0.153$.

been measured in staggered simulations because they are the odd parity partners of "ordinary" states: the a_1 and ρ are examples of such pairs. In NRQCD, Lepage and Thacker[6] have measured the masses of χ_C and χ_B states (without including spin effects).

These calculations are hard to do for several reasons: First, one needs a source with nonzero overlap onto the desired L sector and zero overlap on $L = 0$, otherwise the signal will be dominated at large t by the lighter $L = 0$ states. Second, the signal is going to be noisy. Recalling the discussion of Sec. II.1, the lightest state in the "squared source" channel is two (three) pions for mesons (baryons), so the fractional fluctuation in these channels scales like Eqs. (2.10) and (2.11). Finally, there are a lot of states: in the P-wave sector there are $^3P_2$, $^3P_1$, $^3P_0$, and $^1P_1$ mesons and a whole SU(6) $L = 1$ [70] of baryons[30] In the nonstrange $L = 1$ sector, there are five distinct nucleons, one of $j = 5/2$ and two each of $j = 3/2$ and $j = 1/2$, and $j = 3/2$ and $j = 1/2$ Δ's. I don't see how it will be possible to extract the mass of both nucleons of the same j since the lightest one will dominate the correlator in that channel.

M. Hecht and I[31] are making an attempt to see some of the $L = 1$ states, and our technique may be useful. We take as our source a shell model wave function with the quark in an S-wave and the antiquark in a P-wave. This state is a linear

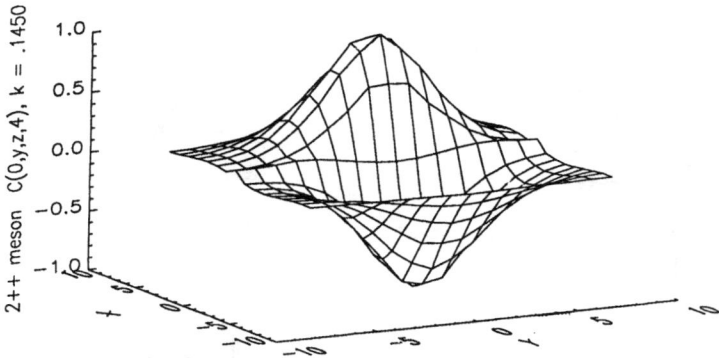

FIGURE 6

Real part of a Coulomb gauge wave function of a $^3P_2$ meson, in the plane $z = 0$ from quenched $\beta = 6$, $\kappa = 0.145$ simulations.

superposition of a $\vec{p} = 0$ $L = 1$ orbital excitation and a state whose center of mass momentum is nonzero. Convoluting quark propagators as in Sec. III.3 projects out the $\vec{p} \neq 0$ state and gives us the wave function of the $L = 1$ state, whose exponential falloff with t should give a mass. A picture of a $^3P_2$ meson is shown in Fig. 6. Can you think of a better way to measure P-wave (and higher) masses? How about radial excitations?

D. Trial Wave Functions for Glueballs

Trial wave functions for glueballs predate trial wave functions for quark states by several years. The physics is very similar. The simplest operator which can make a scalar glueball is the lattice analog of $\sum_{ij} \mathrm{Tr} F_{ij}^2$, a single plaquette averaged over orientations, but it has terrible contributions from ultraviolet modes.

We can do a fake calculation[6,32] in free scalar field theory which exposes this behavior by using the operator

$$\Gamma(t) = \frac{1}{L^{3/2}} \sum_X |\vec{\nabla}\phi(x,t)|^2. \tag{3.10}$$

Our fake glueball is a $\phi - \phi$ composite state and has a mass $M_G = 2m_\phi$, and the correlator is

$$G(t) = \langle \Gamma(t)\Gamma(0) \rangle \qquad (3.11)$$

which has a disconnected piece and a connected piece

$$\lim_{t \to \infty} G(t) \to \langle 0|\Gamma|0 \rangle^2 + G_c(t) \qquad (3.12)$$

where

$$\langle 0|\Gamma|0 \rangle = \frac{1}{L^{3/2}} \sum_{\vec{p}} \frac{\vec{p}^2}{2E_p} \simeq \frac{L^{3/2}}{a^4} \qquad (3.13)$$

and

$$G_c(t) = \frac{2}{L^3} \sum_{\vec{p}} \frac{\vec{p}^4}{2E_p} e^{-2E_p t} \qquad (3.14)$$

$$\simeq \frac{1}{t^5} \qquad (3.15)$$

for $t << L$. For $t \simeq L$ the glueball begins its asymptotic exponential falloff

$$G_c(t) \simeq (\frac{a}{L})^5 \exp(-m_G t). \qquad (3.16)$$

Thus the statistical noise overwhelms the signal due to the large factor of L^{-8}:

$$\frac{G_c(t)}{\sigma(t)} \simeq \sqrt{N} \frac{\exp(-m_G t)}{L^8}. \qquad (3.17)$$

The cure for this disease is once again to block or smear out the high momentum components of the interpolating field. If one can smear modes out to a size $a_b \simeq L$ one can convert the disconnected piece to $\langle \Gamma_b \rangle \simeq L^{3/2}/a_b^4$ or $\langle \Gamma_b \rangle \simeq L^{-5/2}$ and the signal to noise ratio becomes

$$\frac{G_c(t)}{\sigma(t)} \simeq \sqrt{N} e^{-m_G t}. \qquad (3.18)$$

Most of the successful methods for smearing do not involve gauge fixing. Rather, one defines new links which sample the local environment[33] For example, one can define a variable $V_\mu(x)$ on the sites of the lattice as a "fat link" which spans two lattice sites and is a linear combination of a product of links along the direct path $U_\mu(x)U_\mu(x+\mu)$ plus offset paths like $\sum_\nu U_\nu(x)U_\mu(x+\nu)U_\mu(x+\mu+\nu)U_\nu^\dagger(x+2\mu)$. One can also define "fat links" of unit length similarly: $V = U + UUU^\dagger + \dots$ One then uses various closed paths of the V's as interpolating fields for glueballs. One can project the V's back onto the gauge group or not (projecting keeps the size of the V's from growing) and one can repeat the procedure to define "super V's" which span four sites–or more. The idea is to make a glueball operator which is a big fuzzy ball of string whose physical size is 1/2 to 1 fermi across.

One produces the various spin-parity glueball states on the lattice with operators which are appropriate representations of the discrete rotation group.[34] a scalar glueball comes from $O_{xy} + O_{yz} + O_{zz}$ while a tensor glueball could be made by $O_{xy} + O_{yz} - 2O_{zz}$. Pseudoscalar glueballs are produced by some lattice version of $\epsilon_{\mu\nu\rho\sigma} F_{\mu\nu} F_{\rho\sigma}$. All excited state correlators are noisy because the "squared operator" is dominated by exchange of two scalar glueballs.

A few years ago it was fashionable to use explicit sources for glueballs (set a pattern of links equal to a constant throughout the simulation and extract a glueball mass from the system's response) but today diagonal correlators are preferred. They are a bit noisier, but give a variational bound on the lightest state in a channel.

I think that clever ideas for glueball spectroscopy remain to be found.

E. Systematic Effects due to Simulation Size or Lattice Spacing

a) Very Small Simulation Volume. If the simulation volume $V = L^3$ is less than or comparable in size to a hadronic bound state, then the wave function of the bound state will be squeezed and its energy will be pushed up. For sufficiently small L the energy will scale as $1/L$. This has been a problem in recent lattice simulations: groups want to do simulations with the smallest possible lattice spacing but are limited in computer power to the number of points in their lattices. The HEMCGC[24] encountered finite size effects in their baryon spectroscopy in staggered dynamical simulations at $\beta = 5.6$, $am_q = 0.01$ going from 12^3 to 16^3 spatial volumes. The smaller lattice was squeezing the baryon's wave function. The proton energy fell by about fifteen per cent. I show an example in Fig. 7. The pion mass is $am_\pi = 0.25$ so $m_\pi L \leq 3$ is too small an approximation to $m_\pi L = \infty$. Similar results have been seen recently in $\beta = 5.7$ staggered dynamical simulations comparing work of the Columbia[35] and KEK[36] groups.

If L is so small that $g^2(L)$ is small, all scales of physics in the box are accessible to perturbative calculations and one can compute the whole spectrum of excitations. Some people believe that it may be possible to extend these calculations from small L to larger L.[37]

(b) Larger Volumes, Stable Particles (π and p in full QCD, all $q\bar{q}$ and qqq states and the 0^{++} glueball in quenched approximation). If L is large but not infinite there will be a small shift in energies from vacuum polarization due to image effects. That is, vacuum polarization contributions to the particle propagator where an intermediate state propagates from location x to y will have an additional contribution where the intermediate state propagates "around the lattice" to $y + L$.

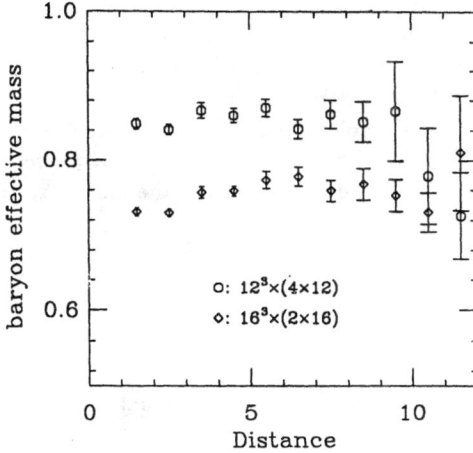

FIGURE 7

Baryon effective mass on 12^3 and 16^3 lattices from the HEMCGC (staggered fermions, $\beta = 5.6$, $am_q = 0.01$).

The formula for the energy shift has been given by Lüscher[38]

$$\frac{m(L) - m(\infty)}{m(\infty)} \simeq -\frac{\lambda^2}{m(\infty)L}e^{-\frac{\sqrt{3}}{2}m(\infty)L} \qquad (3.19)$$

Here λ is a low energy effective coupling which can be related to a scattering phase shift. The depression of the energy on a finite lattice is related to the lowering in energy of a particle in a potential when the potential is made periodic and the wave function delocalizes into a Bloch wave.

(c) Resonances and Finite Volumes. In all simulations of full QCD which have been performed to date particles which would be resonances in the real world are stable in the simulation. Often this is the case simply because the quark mass is unphysically heavy so that, for example, $m_\rho < 2m_\pi$. However, another potentially serious effect can occur if the lattice is too small. Angular momentum conservation prohibits the decay of a $\vec{p} = 0$ rho into two $\vec{p} = 0$ pions, and since on a finite lattice momentum is quantized, the rho will be stable unless $m_\rho < 2E_\pi$ where E_π is the energy of the lightest $\vec{p} \neq 0$ pion state. As the lattice size increases or the quark mass decreases the energy of this state will fall and the apparent mass of the rho will show irregular behavior as a result of avoided level crossings with the lightest $\pi\pi$ state. As the first $\pi\pi$ state falls onto the ρ, it will be pushed down. At bigger L the state which is mainly the ρ will be pushed up by the $\pi\pi$ state which is now beneath it, and then it will be pushed down again as the next $\pi\pi$ state falls onto it. It may be very hard to predict the mass of the ρ (or any other resonance) in an infinite system from simulations at finite L. It is certainly wrong to take the mass measured on a finite lattice as a lattice prediction for an infinite lattice!

As a way of clarifying this problem consider a toy model[39] in which the resonance has a bare mass m_R, the pions have an energy $\epsilon_p = \sqrt{m_\pi^2 + p^2}$, $\vec{p} = 2\pi\vec{n}/L$, and the only interaction is between the resonance and the pions, $\bar{\Delta}_n = (2\pi/L)^{3/2}\Delta_n$:

$$H = \begin{pmatrix} m_R & \bar{\Delta}_1 & \bar{\Delta}_2 & \bar{\Delta}_3 & \cdots \\ \bar{\Delta}_1 & \epsilon_1 & 0 & 0 & \cdots \\ \bar{\Delta}_2 & 0 & \epsilon_2 & 0 & \cdots \\ \bar{\Delta}_3 & 0 & 0 & \epsilon_3 & \cdots \\ \vdots & \vdots & \vdots & \vdots & \end{pmatrix} \qquad (3.20)$$

The eigenvalues of H, λ, are given by the secular equation

$$\lambda - m_R - \left(\frac{2\pi}{L}\right)^3 \sum_n \frac{\Delta_n^2}{\lambda - \epsilon_n} = 0. \qquad (3.21)$$

In the infinite volume limit this is the integral equation

$$\lambda - m_R - P \int d^3p \frac{\Delta^2(p)}{\lambda - \epsilon_p} = 0 \qquad (3.22)$$

which gives the shift in the resonance mass due to final-state interactions.

Far from any level crossing (but closer to one than the rest) one can write Eq. (3.21) in terms of the phase shift

$$\lambda - m_R = -\frac{4\pi}{m_R L^3} \tan \delta(p). \qquad (3.23)$$

However, near a level crossing this result breaks down.

It is now presently possible to do simulations where the ρ is degenerate with a $J = 1$, $T = 1$ $\pi\pi$ state. However, it is not known what is the best way to see the level crossing.[40] One might do separate simulations at several values of L, but noise in the ρ channel would make a shift hard to see. Another possibility is to somehow use the fact that in $\rho \to \pi\pi$ the pions must have a nonzero momentum component along the direction of the ρ's polarization. Thus a nonzero momentum ρ will have a slightly different energy depending on whether its momentum and polarization are parallel or not. Perhaps this energy difference can be measured.

There is a very real possibility that in the infinite volume continuum limit the shift in resonance masses from their values on small lattices could be small. The reason is phenomenological. The $\rho - \omega$ mass splitting is very small (14 MeV) and the ω is very narrow. Zweig's rule suggests that the mass shift of the ω from mixing with intermediate gluon states is small. In the quark model the ρ and ω are degenerate; perhaps neither particle receives much of a shift from final state interactions. (However, compare Ref. 41 for a phenomenological discussion of this point.)

In my opinion finding a practical way to deal with resonances is the most important problem in lattice spectroscopy today. One cannot claim that lattice calculations will approach the physical limit smoothly as computer power increases without understanding resonances.

F. Finite Lattice Spacing Effects

There is a lot of theory but very little in the way of concrete numerical results about the relation of nonzero lattice spacing to spectroscopy. When the lattice spacing is nonzero two things happen: rotational invariance is violated, and the whole lattice formulation of the model differs from the continuum formulation by terms of order the lattice spacing a, a^2,... The violation of rotational invariance means that states which are members of different representations of the cubic group which correspond to different m_J values in the continuum will not be degenerate. The order a terms in the action give a whole host of effects: asymptotic scaling is lost, universality is lost so that spectra depend on the kind of fermions (staggered or Wilson) and on the specific form of the gauge action. All of these effects can be seen in older lattice simulations, which were done at fairly strong coupling. However, more recent simulations are in a grey area: quenched glueball spectra are probably scaling at the ten per cent level (within errors) and quark spectroscopy is still too sparse to tell. When one looks at mass ratios (Edinburgh plots) the gross features of staggered and Wilson spectroscopy are quite similar. The immediate goal of simulations is of course to push to small lattice spacing and verify that one is doing a good job of simulating continuum physics.

Presently, most research on finite lattice spacing effects is concentrated on matrix element studies: for small lattice spacing ratios of expectation values of local operators $\langle a|J_\mu|b\rangle$ for different definitions of J_μ should not depend on the states $|a\rangle$ and $|b\rangle$. That is not presently the case. One tries to improve the calculation by adding extra terms to the action which cancel the leading a corrections. For more detail see Refs. 4,42 and Soni's chapter.

IV. RECENT RESULTS–GLUEBALLS

There was a lot of work on glueball spectroscopy in the mid-1980's but in the last few years there have only been three large studies of glueball spectra in quenched QCD and one large scale study in full QCD.

The quenched studies are those of APE,[43] Michael and Teper,[44] and Gupta, et al.[45] The first two studies use the Wilson action while Ref. 45 uses a more complicated action motivated by Monte Carlo renormalization group studies. All study blocked operators; Michael and Teper use the two-link blocking described in Sec. IIID, APE uses a one-link blocking, and Ref. 45 explores both methods. The end result seems to be that either method works well if carried far enough, though the factor of two blocking seems to require less numerical work to carry out.

The one full QCD simulation was performed by the HEMCGC[46] using techniques essentially identical to those of Michael and Teper. It suffers from low statistics.

In glueball studies it is conventional to present the glueball energy in units of the square root of the string tension ($\sqrt{\sigma}$), which is related to the coefficient of the linear term in the heavy $\bar{q}q$ potential. This allows us to perform a direct test of scaling. It is also conventional to present comparisons of the glueball energy versus lattice size L as a function of the dimensionless parameter $z = m(0^{++})L$ (compare Eq. 3.19).

The 0^{++} and 2^{++} data from the four groups I have chosen to highlight is shown with $m/\sqrt{\sigma}$ plotted vs. β in Fig. 8 and vs. z in Fig. 9. All the groups' quenched data for $\beta > 5.8$ look by eye to be a constant and if I treat them as such I find $m(0^{++})/\sqrt{\sigma} = 3.33(7)$ and $m(2^{++})/\sqrt{\sigma} = 5.31(15)$. I don't see any discernable z dependence in Fig. 9; apparently the lattices are large enough so that the glueball fits comfortably inside.

The three dynamical fermion simulation points are at $am_q = 0.025$, $L = 12$, and $am_q = 0.01$, $L = 12$ and 16. I have not differentiated among them, as they are all consistent within their large uncertainties. All one can gather from the figures is that (so far) dynamical fermions don't seem to affect glueball spectroscopy.

Masses of some states beyond the scalar and tensor have been reported by Michael and Teper. The signals are considerably noisier than in the scalar and tensor channels.

For the past year or so Teper has been studying the computationally easier case of SU(2).[47] That system can (probably) be overpowered by present day computers and the statistical fluctuations are small enough that one can see the approach of $m/\sqrt{\sigma}$ or $m(2^{++})/m(0^{++})$ to an asymptotic constant, although the string tension does not yet show asymptotic scaling.[48]

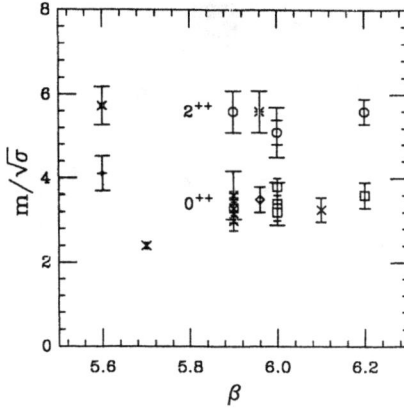

FIGURE 8

Scalar and tensor glueball masses in units of $\sqrt{\sigma}$, vs. β for quenched and full QCD. Data points are: APE [43] – cross, 0^{++}, fancy cross, 2^{++}; Michael and Teper [44] – square, 0^{++}, circle, 2^{++}; Gupta, et. al. [45] – diamond, 0^{++}, burst, 2^{++}; and the HEMCGC [46] – fancy diamond, 0^{++}, fancy square, 2^{++}.

V. RECENT RESULTS–MESONS AND BARYONS

Generic Features of Recent Calculations

The first lattice Monte Carlo studies of QCD spectroscopy were done in 1981[49] but only since 1988 have simulations been done with large enough data sets and good enough algorithms and trial wave functions that the statistical errors can be made small enough to begin compare different groups' work. A standard methodology has since evolved. All groups use the conjugate gradient algorithm for matrix inversion. All use some extended source for spectroscopy. Quenched simulations tend to use the overrelaxed algorithm[50] and Hybrid Molecular Dynamics (HMD)[51] or Hybrid Monte Carlo (HMC)[52] are presently the algorithms of choice for dynamic fermion simulations. They have the advantage of shorter autocorrelation times, compared to the Langevin algorithm, and HMD has the advantage over the pseudofermion algorithm that there is a simple parameter Δt which regulates the

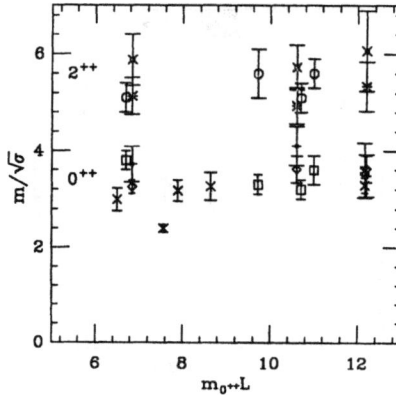

FIGURE 9

Scalar and tensor glueball energies in units of $\sqrt{\sigma}$, vs. $z = m(0^{++})L$ for quenched and full QCD. Data points are labeled as in Fig. 8.

exactness of the simulation[53] HMC has no Δt biases but is restricted to unphysical numbers of degenerate flavors of dynamical fermions (0, 2, 4). The real world has three light nondegenerate flavors. Unfortunately all these algorithms are small timestep algorithms and so one needs large data sets. At present quark masses, autocorrelation times are tens to hundreds of simulation time units. Propagators show longer autocorrelation times than simple operators like $\bar{\psi}\psi$ or the plaquette. We need large data sets because we are looking for subtle physics effects, like the influence of sea quarks on spectroscopy, or subtle simulation effects. The HEMCGC[24] ran a 1500 time unit test of dynamic fermion conjugate gradient accuracy and saw a 1 σ change in hadron masses. They think this is a fluctuation but more running would be needed to show this convincingly.

Most simulations these days present their results similarly. Most results can be summarized in one plot, an Edinburgh plot of the ratio of particle energies (m_p/m_ρ vs. m_π/m_ρ). In the continuum limit and on an infinite lattice these mass ratios will be a smooth function of the quark mass. However, on a finite lattice that need not be the case because of the behavior of the rho meson, as discussed earlier. One should be very cautious in attempting to extrapolate lattice results from heavy to light quark mass.

For future reference, most collaborations schedule presentation of their results for the Lattice Conference, which is usually held in late fall. You should consult

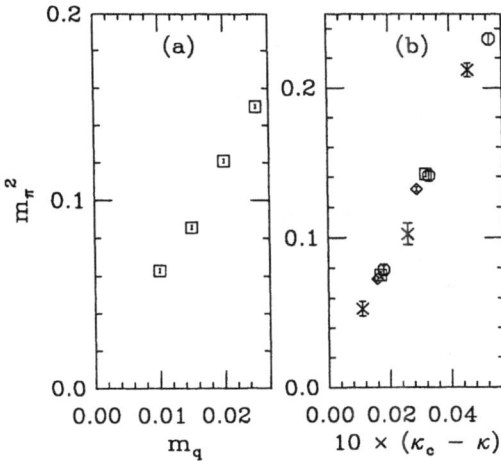

FIGURE 10

Variation in pion mass with quark mass for dynamical staggered fermions
(a) and with the hopping parameter κ for Wilson fermions (b), where the
labels are: crosses and octagons for dynamical staggered fermions at $\beta = 5.6$,
$am_q = 0.01$ and 0.025, and squares and diamonds for quenched simulations
at $\beta = 5.85$ and 5.95. Staggered data are from the Columbia Collaboration[35]
(two flavors, $\beta = 5.7$) and Wilson data from the HEMCGC (Ref. 54).

the latest Proceedings of that series for the most up to date results. The newest
data I present here are through Autumn 1991. This review is not is not intended
to be historically complete; I have exercised some editorial license, but I believe am
showing a good image of the state of the art. I am concentrating on spectroscopy
of S-wave hadrons since that is where most of the effort of the groups has gone.

Pions

All QCD simulations, quenched or full, with staggered or Wilson fermions, have
shown the expected $m_\pi^2 \simeq m_q$ behavior at small quark mass. (Of course, for Wilson
fermions the relation between bare hopping parameter and renormalized mass is
not known, but one can see the square of the pion mass interpolating linearly to
zero at some critical value of the hopping parameter.) An example of this behavior
from recent data is shown in Fig. 10.

Present day staggered fermion simulations are run at parameter values where

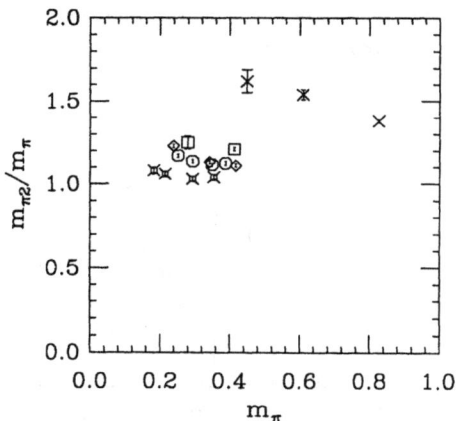

FIGURE 11

Ratio of non-Goldstone pion to Goldstone pion mass for $n_f = 2$ dynamic simulations (Columbia (circle), HEMCGC (squares) and the old Santa Barbara results[55] (crosses)) and quenched results from the Staggered collaboration at $\beta = 6.0$ (diamonds) and $\beta = 6.2$ (decorated squares).

flavor symmetry is nearly restored. In Fig. 11 I show a plot of the ratio of the mass of one of the non-Goldstone pions to the mass of the would-be Goldstone pion, for quenched and dynamic data. The quenched $\beta = 6.2$ data from the Staggered collaboration shows flavor symmetry breaking reduced to a few per cent.

Simulations with Dynamical Fermions

(a) Dynamic Staggered Fermions, $n_f = 2$: There have been four large scale simulations. The pioneer was the work of the (mainly) Santa Barbara group[55] on 6^3, 8^3 and 10^3 by 24 lattices, quark masses in the range $am_q = 0.1$ to 0.025, and β in the range 5.28 to 5.47. Since then the HEMCGC[24] has run at $\beta = 5.6$ on 12^4, 16^4 and $16^3 \times 32$ lattices with $am_q = .01$ and .025, and the Columbia group[35] has run at $\beta = 5.7$ on $16^3 \times 32$ lattices with dynamic mass $am_q = .01$, .015, .02, and .025. The KEK group[36] has also presented spectroscopy at $\beta = 5.7$ on up to 20^4 lattices with $am_q = 0.01$ and 0.02. All simulations used HMD, and gauge fixed to a smooth gauge and used wall sources and pointlike sinks. All calculations feature high statistics and careful error analysis. The latter three collaborations continue

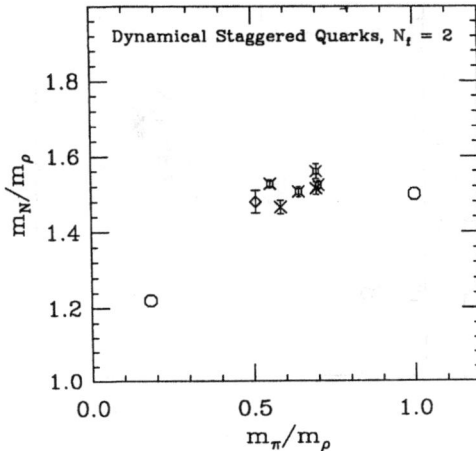

FIGURE 12

Edinburgh plot for $n_f = 2$ dynamic staggered fermions. In all Edinburgh plots the circles are the physical ratio and infinite quark mass limit. Here data are from HEMCGC (on a 16^3 lattice) (diamond), Columbia data (squares), Tsukuba data (crosses).

to take data. The HEMCGC has up to 5000 time units for some parameters and up to 500 lattices stored for analysis. The Columbia group[35] has up to 3000 time units per mass value. These simulations are of unprecedented size and accuracy in lattice QCD. In Fig. 12 I show the Edinburgh plot for these simulations. The errors are tiny even compared to quenched simulations.

The 20^3 masses of the KEK group[36] are consistently smaller than the 16^3 masses of the Columbia collaboration. The nucleon mass decreases by twelve per cent, for example. If these results hold up, wave function squeezing might continue beyond lattice size $m_\pi L > 5$.

(b) Dynamic Staggered Fermions, $n_f = 4$: The MT_c collaboration[56] is performing simulations on $16^3 \times 24$ lattices at $\beta = 5.2$ and 5.35 with $am_q = 0.01$ (using HMC) and at $\beta = 5.15$ and 5.35 using pseudofermions[57] Their data has shown considerable evolution from the 1990 to the 1991 Lattice Conferences and so I feel it would be inappropriate to show any of it here.

(c) Dynamic Wilson Fermions, $n_f = 2$: Two groups are currently active in this area, both using HMC and both with heavier quarks than in the staggered simulations. The first is a Los Alamos-based collaboration[23] They mainly run on 16^4 lattices,

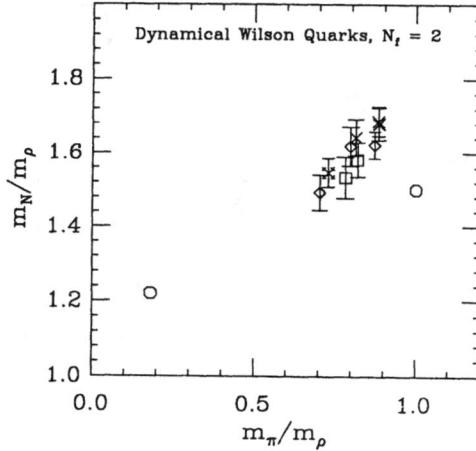

FIGURE 13

Edinburgh plot for $n_f = 2$ dynamic Wilson fermions. Data are from the Los Alamos collaboration: $\beta = 5.4$ (crosses), $\beta = 5.5$ (diamonds), $\beta = 5.6$ (squares), and the HEMCGC, $\beta = 5.3$ (fancy cross).

$\beta = 5.4 - 5.6$, and low statistics, T about 600 units, about 30 lattices analyzed. They use a gauge invariant extended trial wave function. The second group is the HEMCGC[3] which has done a simulation at $\beta = 5.3$, $\kappa = .1670$ (so $m_\pi = .45$) on a $16^3 \times 32$ lattice. Their results are still preliminary. The Edinburgh plot is shown in Fig. 13.

Quenched Simulations

(a) Quenched Staggered Fermions: Three groups have published large simulations: the APE collaboration[58,59] at $\beta = 6.0$ on $24^3 \times 32$ lattices, the Staggered collaboration,[8] mainly at $\beta = 6.0$ and 6.2, on $16^3 \times 40$ and $24^3 \times 40$ lattices, and the HEMCGC[54] at $\beta = 5.85$ and 5.95 on $16^3 \times 32$ lattices (they are trying to match their dynamic simulations to quenched data). The data are all in general agreement with each other. All use extended sources (APE uses cubes, the others, walls). The Edinburgh plot is shown in Fig. 14.

(b) Quenched Wilson Fermions: I show the results of two large scale simulations: APE[60,61] ($\beta = 5.7$, $12^3 \times 24$ lattices and $\beta = 6.0$, $24^3 \times 32$ lattices) and HEMCGC[54] ($16^3 \times 32$ lattices at $\beta = 5.85$ and 5.95). APE uses cube sources and HEMCGC

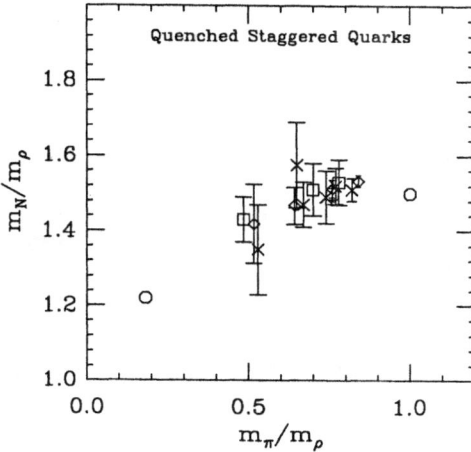

FIGURE 14

Edinburgh plot for quenched staggered fermion simulations. Data are APE (diamonds), Staggered collaboration (crosses) and HEMCGC (squares)

used wall sources. Fig. 15 is an Edinburgh plot.

Theory of the Edinburgh Plot

One can apply quark models to lattice spectroscopy as well as to real experimental spectroscopy. For heavy quarks, the nonrelativistic quark model should be a good guide. There, the mass of a hadron is given by a sum of constituent quark masses plus a color hyperfine term,

$$M_H = \sum_i m_i + \xi_H \sum_{ij} \frac{\sigma_i \cdot \sigma_j}{m_i m_j} \tag{5.1}$$

where ξ is twice as great for mesons as for baryons because of color. At small quark mass chiral models require that the pion mass vanishes as the square root of the quark mass; all other particles extrapolate linearly in the quark mass. One feature which is usually missing is the presence of cusps in the rho mass as it crosses $\pi\pi$ thresholds. A "cartoon" which shows all these effects is shown in Fig. 16. It is clear that one could tune parameters in these models and reproduce the results of lattice QCD simulations. (Try it yourself!)

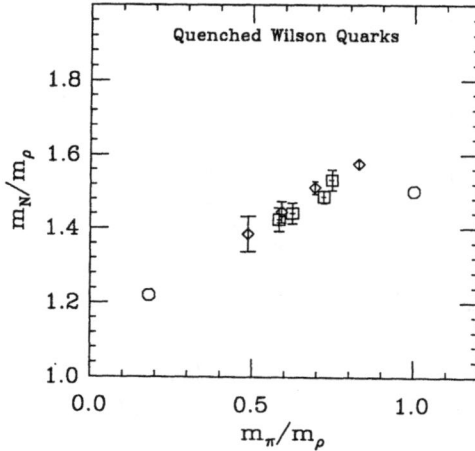

FIGURE 15

Edinburgh plot for quenched Wilson fermion simulations. Data are APE (diamonds), HEMCGC (squares).

The main question one is trying to address with dynamic fermion spectroscopy simulations is, "Do the sea quarks make a difference in spectroscopy?" Unfortunately, the answer to that question is still No. Other effects (lattice volume or just statistics) are still large enough to mask any obvious differences between quenched and dynamic quark spectroscopy at dynamic fermion mass values studied to date. As a fairly dramatic example, Fig. 17 shows spectroscopy from two Wilson simulations I am involved in, a quenched simulation at $\beta = 6.0$, and Wilson valence spectroscopy with dynamical staggered fermions at $\beta = 5.6$, from the HEMCGC.[24] I have plotted the rho, proton, and delta mass in lattice units as a function of the lattice pion mass. Finding observables which change when one passes from the quenched approximation to full QCD remains an open problem.

Other Particles

There is other spectroscopy information beyond the Edinburgh plot: One can display a plot of baryon hyperfine splitting vs. meson hyperfine splitting (a "Boulder" plot), Fig. 18. Here I compare the two dimensionless quantities

$$R_M = \frac{m_\rho - m_\pi}{3m_\rho + m_\pi} \tag{5.2}$$

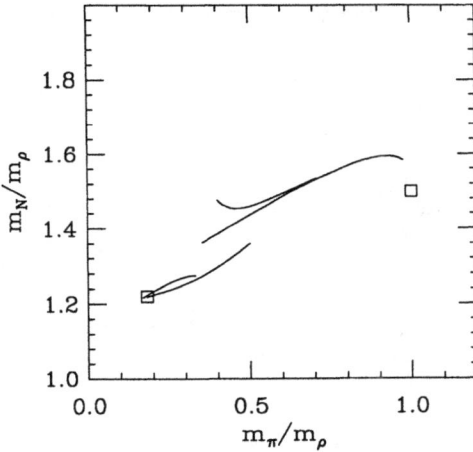

FIGURE 16

A "theoretical" Edinburgh plot. The two lines are a simple quark model which includes an avoided $\rho\pi\pi$ level crossing, and a chiral model.

and

$$R_B = \frac{m_\Delta - m_p}{m_\Delta + m_p}. \tag{5.3}$$

Each of these quantities is the ratio of hyperfine splitting in a multiplet divided by the center of mass of the multiplet. Error bars are still too large for much phenomenology.

Very few calculations have attempted to look at the full flavor SU(3) spectrum of the lattice. In my opinion the rather scanty data appears to reproduce the gross features of the Gell-Mann Okubo formula for octets and the equal-spacing rule for decuplets, but the splittings (often as extrapolated to the zero quark mass limit) are smaller than experiment. I have never seen a lattice simulation with a statistically significant $\Sigma - \Lambda$ mass splitting. Simulations studying orbital or radial excited states have scarcely begun.

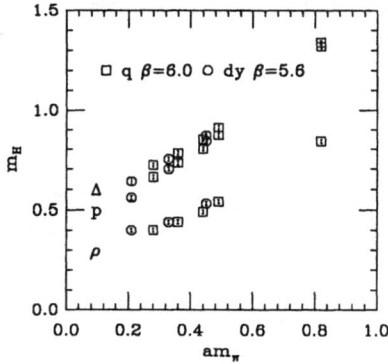

FIGURE 17

Comparison of quenched Wilson spectroscopy at $\beta = 6.0$ (squares) with Wilson valence spectroscopy from a dynamical staggered fermion simulation at at $\beta = 5.6$ (octagons).

VI. PROSPECTS

The goal of lattice QCD spectroscopy studies is to answer the question, "What are the QCD predictions for the masses and other properties of the lightest hadrons?" Progress in this field will come from two directions.

The first will come as large-scale simulations become larger and larger, both in terms of simulation volume and length of simulation. Larger volume is needed both to decrease the lattice spacing (to approach continuum physics) and to increase the physical volume (to make certain that the states are not being squeezed). Present day simulations have barely enough statistics to allow one to begin to worry about the inherent long autocorrelation times of the algorithms. Dynamical fermion simulations need to be run at lighter, more physical, values of the quark mass. All these changes are computationally very costly.

Thus we see physicists moving onto new generations of supercomputers as (or before) they become available, and in Europe, Japan, and America constructing or proposing to construct new special purpose machines for lattice QCD.

Unfortunately, new supercomputers have drawbacks. Software is usually primitive, making it hard to do more than recode pre-existing algorithms. These ma-

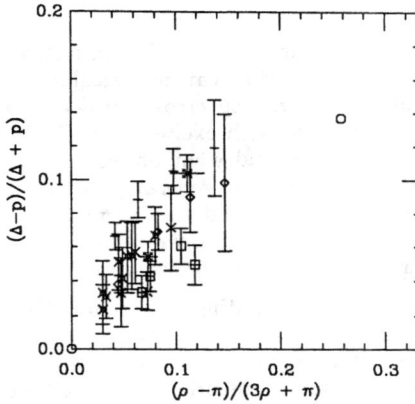

FIGURE 18

Baryon hyperfine splitting vs. meson hyperfine splitting, normalized to the center of mass of the multiplet. Data are HEMCGC Wilson valence quarks with staggered sea quarks (fancy cross) Wilson sea quarks (burst) and quenched (squares), APE (diamonds), APE quenched Wilson quarks (diamonds) and staggered quarks (plus signs), Gupta, et. al. (Ref. 23) Wilson sea quarks (fancy square) and the staggered fermion result of the Staggered collaboration (crosses).

chines are quite expensive, so physicists have to form large groups in order to get the money to use them. Large groups have their own form of "critical slowing down." Finally, because large projects are so expensive, they are extremely visible. This means that they cannot be allowed to fail – which further discourages innovation.

However, the open questions in the field are not restricted to problems of adapting one's program to a new computer architecture. Indeed, if the future of computational QCD is governed entirely by more and more powerful computers redoing today's calculations, then the field deserves to die. Fortunately, that is not the case: Nature continues to provide us with a set of interesting "thinking" questions.

Algorithmic issues are mainly of interest to the lattice community, to statistical mechanicians also doing simulations, and to applied mathematicians concerned with solving systems of partial differential equations. These questions are almost all focussed on ameliorating critical slowing down, either for matrix inversion or for fermionic updating algorithms in general: How can we calculate propagators at small quark mass? Is there a large step size algorithm for systems containing dynamic fermions?

The second set of questions is directed at bringing lattice QCD back into the mainstream of high energy physics: How can we calculate the properties of resonances in a lattice calculation? Since spectroscopy does not just mean S-wave states, what is the best way to deal with excited states? What are the best interpolating fields for hadronic states and what do we learn about the real world from them? Finally, there are a set of more general questions: What are the effects of sea quarks on spectroscopy and on bound states, and the related problem, why does the quark model work so well? How can we combine lattice simulations with continuum models to address increasingly more complex questions?

My feeling is that present day computing power and algorithms have brought us to the brink of interesting results in lattice QCD spectroscopy. There are a large number of questions which will have interesting answers and can be answered–even with relatively small amounts of supercomputer time. The best ideas have still not been found.

ACKNOWLEDGMENTS

I would like to thank my friends and colleagues whose discussions and insights I have included in this review: Claude Bernard, Steve Gottlieb, Rajan Gupta, Anna Hasenfratz, Matthew Hecht, Peter Lepage, Rich Loft, Bob Sugar, Michael Teper, and Doug Toussaint. I would also like to thank Mike Creutz and Reba Katz for a careful reading of the manuscript. This work was supported by the U.S. Department of Energy.

REFERENCES

1. Cf. N. Isgur and G. Karl *Phys. Rev. D* **18**, 4187 (1978); and *Phys. Rev. D* **20**, 1191 (1979).

2. This discussion is stolen from G. P. Lepage, in "From Actions to Answers–Proceedings of the 1989 Theoretical Advanced Summer Institute in Particle Physics," T. DeGrand and D. Toussaint, eds., (World, 1990).

3. K. M. Bitar, et al., in preparation.

4. G. Heattlie, G. Martinelli, C. Pittori, G. C. Rossi, and C. Sachrajda, *Nucl. Phys.* **B352**, 266 (1991).

5. M. Golterman and J. Smit, *Nucl. Phys.* **B255**, 328 (1985).

6. B. Thacker and G. P. Lepage, *Phys. Rev.* **D43**, 196 (1991). This work is based on ideas of Caswell and Lepage, *Phys. Lett.* **167B**, 437 (1986).

7. Compare H. Kluberg-Stern, A. Morel, O. Napoly, and B. Petersson, *Nucl. Phys.* **B220**, 447 (1983).

8. R. Gupta, G. Guralnik, G. Kilcup, S. Sharpe, *Phys. Rev.* **D43**, 2003 (1991).

9. P. Rossi, C. T. H. Davies, and G. P. Lepage *Nucl. Phys.* **B297**, 287 (1988).

10. M. R. Hestenes and E. Stiefel, *Nat. Bur. Standards J. Res.* **49**, 409 (1952).

11. T. DeGrand, *Comput. Phys. Commun.* **52**, 161 (1988); R. Gupta, A. Patel, C. Baillie, G. Guralnik, G. Kilcup, and S. Sharpe *Phys. Rev.* **D40**, 2072 (1989).

12. Y. Oyanagi *Comput. Phys. Commun.* **42**, 333 (1986).

13. G. Katz, G. Batrouni, C. Davies, A. Kronfeld, P. Lepage, P. Rossi, B. Svetitsky, K. Wilson *Phys. Rev.* **D37**, 1589 (1988); C.T.H. Davies , G. Batrouni, G. Katz, A. Kronfeld, P. Lepage, K. Wilson, P. Rossi, B. Svetitsky *Phys. Rev.* **D37**, 1581 (1988); G. Batrouni, G. Katz, A.S. Kronfeld, P. Lepage, B. Svetitsky, K. Wilson *Phys. Rev.* **D32**, 2736 (1985).

14. See, for example, recent work by R. Brower, R. Edwards, C. Rebbi, and E. Vicari, preprint FSU-SCRI-91-54 (1991), and A. Hulsebos, J. Smit, and J. Vink, in the Proceedings of Lattice '90, U. Heller, A. Kennedy and S. Sanielevici, eds. *Nucl. Phys.* **B (Proc. Suppl)** **20**, (1991) 94. A good introductory review from the applied mathematics point of view is W. Briggs, "A Multigrid Tutorial," SIAM, 1987.

15. P. Rossi, C.T.H. Davies, *Phys. Lett.* **202B**, 547 (1988).

16. For a good introduction to error analysis see D. Toussaint, in "From Actions to Answers–Proceedings of the 1989 Theoretical Advanced Summer Institute in Particle Physics," T. DeGrand and D. Toussaint, eds. (World, 1990).

17. For a review, see B. Efrom, *SIAM Review* **21**, 460 (1979). For an example of its use, see S. Gottlieb, P. Mackenzie, H. Thacker, and D. Weingarten, *Nucl. Phys.* **B263**, 704 (1986).

18. C. Bernard, T. Draper, G. Hockney, and A. Soni, in B. Bunk, K. H. Mütter and K. Schilling, eds., "Lattice Gauge Theory, a Challenge for Large Scale Computing," Plenum, New York, 1986; L. Maiani and G. Martinelli *Phys. Lett.* **B178**, 265 (1986).

19. S. Elitzur, *Phys. Rev.* **D12**, 3978 (1975).

20. A good gauge fixing algorithm is overrelaxation and is described in J. E. Mandula and M. C. Ogilvie, *Phys. Lett.* **B248**, 156 (1990).

21. E. Marinari, in the Proceedings of the 1988 Symposium on Lattice Field Theory, A. Kronfeld and P. Mackenzie, eds., *Nucl. Phys.* **B (Proc. Suppl) 9**, (1989) 209, P. Bacilieri, et al. *Phys. Lett.* **B214**, 115 (1988)

22. S. Gusken, et. al. *Phys. Lett.* **B227**, 266 (1989) , S. Guskin, *Nucl. Phys.* **B17 (Proc. Suppl)** , 361, (1989).

23. R. Gupta, C. Baillie, R. Brickner, G. Kilcup, A. Patel and S.Sharpe, *Phys. Rev.* **D44**, 3272 (1991).

24. K. M. Bitar, et al. *Phys. Rev.* **D42**, 3794 (1990).

25. T. DeGrand and R. Loft, *Comp. Phys. Comm.* **65**, 84 (1991).

26. B. Velikson and D. Weingarten, *Nucl. Phys.* **B249**, 433 (1985)

27. S. Gottlieb, in "Advances in Lattice Gauge Theory," D. Duke and J. Owens, eds. (World Scientific, 1985).

28. M.-C. Chu, M. Lassia and J. W. Negele, *Nucl. Phys.* **B360**, 31 (1991).

29. Talks on related work were presented at Lattice '90 by (among others) E. Eichten and K. Yee.

30. J. Kokkedee, "The Quark Model " (W. A. Benjamin, New York, 1969).

31. M. Hecht and T. DeGrand, Colorado preprint COLO-HEP 265 (1991), Phys. Lett. B, in press.

32. A. Kronfeld, in the Proceedings of the 1988 Symposium on Lattice Field Theory, A. Kronfeld and P. Mackenzie, eds. (Nucl. Phys. B (Proc. Suppl.) 9 (1989)).

33. Compare M. Falcioni, M. Paciello, G. Parisi, and B. Taglienti, *Nucl. Phys.* **B251**, 624 (1985); M. Teper, *Phys. Lett.* **183B**, 345 (1987); *Phys. Lett.* **185**, 121 (1987); T. DeGrand, *Phys. Rev. D* **36**, 176 (1987).

34. J. Mandula, G. Zweig, J. Govaerts, *Nucl.Phys* **B228**, 91 (1983); *Nucl.Phys.* **B228**, 109 (1983).

35. F. Brown, et al., *Phys. Rev. Lett.* **67**, 1062 (1991).

36. M. Fukugita, H. Mino, M. Okawa, A. Ukawa, preprint KEK-TH-271 (1990).

37. See the talk of P. van Baal, in the Proceedings of Lattice '90, U. Heller, A. Kennedy and S. Sanielevici, eds. *Nucl. Phys.* **B (Proc. Suppl)** **20**, (1991) 3.

38. M. Lüscher, Cargèse lectures (1983), in "Progress in Gauge Field Theory," eds. G. 't Hooft, et al. (Plenum, New York, 1984); *Commun. Math. Phys.* **104**, 177 (1986); *Commun. Math. Phys.* **105**, 153 (1986); *Nucl. Phys.* **B354**, 531 (1991); Compare also N. Fukuda and R. Newton, *Phys. Rev.* **103**, 1558 (1956).

39. This argument was shown to me by G. Bertsch.

40. U-J. Wiese, in the Proceedings of the 1988 Symposium on Lattice Field Theory, A. Kronfeld and P. Mackenzie, eds., (Nucl. Phys. B (Proc. Suppl.) 9 (1989)); P. Geiger and N. Isgur, *Phys. Rev. D* **41**, 1595 (1990); C. Michael, *Nucl. Phys.* **B327**, 515 (1990); T. DeGrand, *Phys. Rev. D* , 43, 2296 (1991). Resonance masses have been measured most cleanly in spin models: see C. R. Gattringer and C. B. Lang, Graz preprint UNIGRAZ-UTP-08-10-91 (1991).

41. P. Geiger and N. Isgur, *Phys. Rev. Lett.* **67**, 1066 (1991).

42. These techniques go back to K. Symanzik, *Nucl. Phys.* **B226**, 187 (1983). Recently these techniques have been applied to QCD simulations of matrix elements by G. Martinelli, C. Sachrajda, and A. Vladikas, in the Proceedings of Lattice '90, U. Heller, A. Kennedy and S. Sanielevici, eds. *Nucl. Phys.* **B (Proc. Suppl)** **20**, (1991) 448.

43. P. Bacilieri, et al., *Phys. Lett.* **205B**, 535 (1988).

44. C. Michael and M.Teper, *Nucl. Phys.* **B314**, 347 (1989).

45. R. Gupta, et al., *Phys. Rev.* **D43**, 2301 (1991).

46. K. Bitar, et al., *Phys. Rev.* **D44**, 2090 (1991).

47. See the talk of Teper in the Proceedings of Lattice '90, U. Heller, A. Kennedy and S. Sanielevici, eds. *Nucl. Phys.* **B (Proc. Suppl)** **20**, (1991) 159.

48. The UKQCD collaboration, S. P. Booth, et al., Liverpool preprint LTH 271 (1991).

49. D. Weingarten and D. Petcher, *Phys.Lett.* **99B**, 333 (1981); H. Hamber and G. Parisi, *Phys. Rev. Lett.* **47**, 1792 (1981).

50. F. Brown and T. Woch, *Phys. Rev. Lett.* **58**, 2394 (1987) ; M. Creutz, *Phys. Rev.* **D36**, 55 (1987) . For a review, see S. Adler in the Proceedings of the 1988 Symposium on Lattice Field Theory, A. Kronfeld and P. Mackenzie, eds. (Nucl. Phys. B (Proc. Suppl.) 9 (1989)).

51. H. C. Andersen, *J. Chem. Phys.* **72**, 2384 (1980); S. Duane, *Nucl. Phys.* **B257**, 652 (1985); S. Duane and J. Kogut, *Phys. Rev. Lett.* **55**, 2774 (1985); S. Gottlieb, W. Liu, D. Toussaint, R. Renken and R. Sugar, *Phys. Rev. D* **35**, 2531 (1987).

52. S. Duane, A. Kennedy, B. Pendleton, and D. Roweth, *Phys. Lett.* **194B**, 271 (1987).

53. There is also a stepsize with pseudofermions that can be tuned and extrapolated to exactness.

54. K. Bitar et al. in the Proceedings of Lattice '90, U. Heller, A. Kennedy and S. Sanielevici, eds. *Nucl. Phys.* **B (Proc. Suppl)** **20**, (1991) 362.

55. S. Gottlieb, W. Liu, D. Toussaint, R. Renken and R. Sugar, *Phys. Rev.* **D38**, 2245 (1988).

56. E. Laermann, in the Proceedings of Lattice '90, U. Heller, A. Kennedy and S. Sanielevici, eds. *Nucl. Phys.* **B (Proc. Suppl)** **20**, (1991) 380.

57. R. Altmeyer, et al., Minnesota preprint UMN-TH-838-90, June 1990.

58. P. Bacilieri, et al. *Nucl. Phys.* **B343**, 228 (1990).

59. S. Cabasino, et al., *Phys. Lett.* **B258**, 202 (1991).

60. P. Bacilieri, et al. *Nucl. Phys.* **B317**, 509 (1989).

61. S. Cabasino, et al., *Phys. Lett.* **B258**, 195 (1991).

Lattice Field Theories at High Temperatures and Densities

R. V. GAVAI

Theoretical Physics Group
Tata Institute of Fundamental Research
Homi Bhabha Road
Bombay 400005, India

ABSTRACT

After a brief introduction to the formalism of Euclidean lattice field theories at finite temperature and density, with special emphasis on Quantum Chromo Dynamics (QCD), the physics of the deconfinement and chiral symmetry restoration order parameters and other useful thermodynamic observables is discussed. Recent numerical results on the phase diagram of QCD at finite temperature are surveyed. Attempts to learn more about the nature of the high temperature phase through several interesting observables are summarized. Finally, comments are offered on QCD at finite density and the $SU(2)$ gauge-fundamental Higgs system at finite temperature.

1. Introduction

It is now widely believed that the Standard Model based on the gauge group $SU(3) \times SU(2) \times U(1)$ provides an excellent description of the strong, the weak and the electromagnetic interactions observed in nature. Once we have a theory of all these interactions a natural question to ask is about its behaviour under unusual conditions such as high temperatures and/or high densities. Such considerations can have immense astrophysical and cosmological consequences. Thus, e.g., one may encounter very dense matter in the core of neutron stars or very hot matter immediately after the Big Bang. The physics of nucleosynthesis or the inflationary Universe may critically depend on these considerations. Further the relativistic heavy ion collision programs at Brookhaven and CERN, both the ongoing ones at AGS and SPS respectively and the proposed ones at RHIC and LHC, offer exciting possibilities of creating strongly interacting matter at high temperatures and densities under laboratory conditions. A variety of the predictions of hot quantum chromodynamics (QCD) may be checked against the data from these collisions.

Investigating the Standard Model in unusual environments can, in addition, lead to further theoretical understanding. Indeed, most of the experimental evidences that led to the establishment of the Standard Model comprise of verifications of its perturbative predictions against experimental data. However, there are several situations where the coupling is not small. Confinement of quarks and gluons is one such instance. Spontaneous breaking of the chiral symmetry is another, perhaps, related example. Studying these phenomena at high temperatures, one may learn more about them at zero temperature. In the electroweak sector, various aspects of the Higgs sector are intriguing and information gleaned from finite temperature studies may help in understanding some of them. Entirely new physics may become accessible at high temperatures. It has been argued that nonperturbative effects may cause substantial baryon number violation at high temperatures. If true, it will have to be taken into account in any picture which explains the observed baryon asymmetry in the universe.

From a purely theoretical point of view, there are several interesting questions that arise in considering the theory of high temperatures/densities. These relate to the existence of new phases, such as the quark-gluon plasma, possible phase transitions to these phases and nature of these phases and the phase transitions. There are, furthermore, quantitative issues such as the critical temperature/density, critical indices, the equation of state, the mass spectra or screening lengths and so on. Indeed, such questions have been raised, and attempts have been made to answer them, for quite some time now. One of the first attempts in treating the thermodynamics of hadrons was made by Hagedorn[1]. Using a statistical bootstrap hypothesis, he argued for the existence of a limiting temperature above which hadronic matter cannot exist. Later, Cabbibo and Parisi[2] used the quark model of hadrons with confinement of quarks to suggest the existence of a quark-liberating phase transition. With the advent of the theory of strong interaction, namely quantum chromodynamics (QCD), and the subsequent discovery of its property of asymptotic freedom, Collins and Perry[3] showed that at asymptotically large densities the equation of state of strongly interacting matter can be computed perturbatively, with the leading term corresponding to an ideal relativistic gas of quarks and gluons. A lot of work has since been done using perturbative techniques[4]. Unfortunately, however, the perturbation theory breaks down due to the severe infra-red problems of QCD[5]. Moreover, in the interesting regions around the phase transitions, be it the deconfining transition in QCD or the symmetry restoring phase transition in the electroweak theory[6], typically the coupling is large and non-perturbative methods are called for.

The lattice regularization of field theories[7] provides such a non-perturbative tool. Due to its strong similarity with models of statistical physics, a variety of known calculational techniques can be borrowed from that field[8]. Indeed, using the strong coupling expansion, one can show the presence of confinement in a non-abelian gauge theory at zero temperature which, however, gets lost at high temperature[9]. The existence of a deconfinement transition has now been proved rigorously on the lattice[10]. Numerical simulations of lattice field theories take us even further: one can approach the continuum limit and extract answers to several of the questions posed above, which is the major part of this chapter. A significant advantage of this approach is that it involves no arbitrary assumptions or parameters as an input. All one needs is the Lagrangian and the bare parameters in it, such as quark masses. Thus starting from basic principles, one can explore the existence of a phase transition in any given theory, obtain the critical parameters and investigate the nature of the new phases. On the other hand, the technique of numerical solutions has its characteristic problems. In order to obtain reliable continuum results, the grid of the lattice needs to be fine. Thermodynamically meaningful results necessitate large volumes, i.e., large lattices. In addition, there is the problem of fermion doubling on the lattice[8] which, coupled with the nonlocality fermions introduce in the effective action, forces additional approximations in the simulations. Despite all this, however, a host of interesting results of both qualitative and quantitative nature have been obtained, as we will review below.

In the next section we briefly review the formalism of field theory at finite temperature and density and comment on the relevant symmetries. The observables which are commonly used are defined here. Next we introduce the lattice regularization with special emphasis on the additional features due to finite temperature. Transcription of the definitions of the physical observable to the lattice language is collected here. Section 3 is also of general theoretical nature. A brief review of the calculational techniques is presented there. Again the emphasis is more on the finite temperature aspects. Sections 4 and 5 are devoted to the review of the recent results obtained for $SU(N)$ gauge theories with and without fermions. A review of the earlier results can be found in Ref. 11. Furthermore, the focus is mainly on quantum chromodynamics and other results are included with an aim to contrast them with those for QCD. Section 6 deals with the challenging yet somewhat stagnated subject of finite chemical potential while the results obtained for the Higgs models are summarized in section 7. Finally, concluding remarks are presented in the last section.

2. Formalism

Thermodynamic functions for a many particle system can be obtained from its partition function Z:

$$Z = \text{Tr}\,[\exp(-(\hat{H} - \mu\hat{N})/T)]. \tag{1}$$

Here $\hat{H}$ is the Hamiltonian of the system and $\hat{N}$ is the number operator for a conserved charge, say, the baryon number. μ is the corresponding chemical potential and T is the temperature of the system. Tr denotes trace over a complete set of physical states. Throughout, we will use $\hbar = c = k = 1$ units. The thermodynamic potential Ω is related to the partition function by

$$\Omega = -T \ln Z, \tag{2}$$

in terms of which entropy density, number density and pressure are respectively given by

$$S = -V^{-1}(\partial\Omega/\partial T)_{V,\mu}, n = -V^{-1}(\partial\Omega/\partial\mu)_{T,V} \text{ and } P = -(\partial\Omega/\partial V)_{T,\mu} \tag{3}$$

In general, the thermal expectation value of a physical observable $\mathcal{O}$ is defined by

$$\langle\mathcal{O}\rangle = \text{Tr}\,[\mathcal{O}\exp(-(\hat{H} - \mu\hat{N}))/T]/Z. \tag{4}$$

Thus, e.g., the substitution of $\hat{H}/V$ for $\mathcal{O}$ in Eq. (4) results in the energy density whose derivative with respect to temperature yields the specific heat. If one knew how to evaluate the trace in Eq. (4) for a given system then physical quantities such as the energy density or the specific heat can be obtained as a function of T and μ. The presence of a phase transition will show up as a discontinuity or a singularity in these quantities. However, except for very simple systems, we do not know how to evaluate the trace exactly. Lattice field theories defined at finite temperatures (and densities) offer a framework to evaluate the thermal expectation values of physical observables by using Eq. (4).

2.1. Field Theory at $T \neq 0$ and $\mu \neq 0$

The first step towards the finite temperature studies of lattice field theories consists of rewriting the partition function in Eq. (1) as a path integral over all the fields of the exponential of the (Euclidean) Lagrangian density of the theory. In order to do that one chooses a complete set of orthogonal states to evaluate the trace in Eq. (1). Writing $T^{-1} = \epsilon M, \epsilon \to 0, \exp(-(\hat{H} - \mu\hat{N})/T)$ can be written as $[\exp(-\epsilon\hat{H} - \epsilon\mu\hat{N})]^M$. Sandwiching completeness relations between successive terms of this product, the partition function Z becomes a product of matrix elements of the operator $\exp(-\epsilon\hat{H} - \epsilon\mu\hat{N})$. Thus, for a simple case of a harmonic oscillator, with $\hat{H} = \hat{p}^2/2m + m\omega^2\hat{x}^2/2$ and $\mu = 0$, Z can be expressed as

$$Z = \int_{-\infty}^{\infty} \prod_{i=1}^{M} dx_i \left\langle x_{i-1} | \exp(-\epsilon\hat{H}) | x_i \right\rangle, \tag{5}$$

where we have used the eigenstates of the position operator $\hat{x}$ to evaluate the trace. The M integrations in Eq. (5) result from $M - 1$ uses of the completeness relation and the trace itself. A notation has been introduced: $x_o = x_M = x$ is the integration variable for the trace. It can be easily shown that in the limit $\epsilon \to 0$

$$\left\langle x_{i-1} | \exp(-\epsilon\hat{H}) | x_i \right\rangle \longrightarrow$$
$$\left\langle x_{i-1} | \exp\left(-\frac{\epsilon}{4} m\omega^2 \hat{x}^2\right) \exp\left(-\frac{\hat{p}^2}{2m}\right) \exp\left(-\frac{\epsilon}{4} m\omega^2 \hat{x}^2\right) | x_i \right\rangle$$
$$= \exp\left(-\frac{\epsilon}{2} m\omega^2 \frac{x_i^2 + x_{i-1}^2}{2}\right) \int_{-\infty}^{\infty} dp \exp\left[ip(x_{i-1} - x_i) - \frac{\epsilon p^2}{2m}\right] \tag{6}$$
$$= \exp\left[-\frac{m}{2\epsilon}(x_{i-1} - x_i)^2 - \frac{\epsilon}{2} m\omega^2 \frac{x_i^2 + x_{i-1}^2}{2}\right] \cdot [2\pi MmT]^{\frac{1}{2}}$$

Substituting in Eq. (5), we have

$$Z = (2\pi M m T)^{M/2} \left\{ \int_{-\infty}^{\infty} \prod_{i=1}^{M} dx_i \right\}$$

$$\exp \left\{ \sum_{i=1}^{M} \left[-\frac{m}{2\epsilon}(x_{i-1} - x_i)^2 - \frac{\epsilon}{2} m\omega^2 \frac{x_i^2 + x_{i-1}^2}{2} \right] \right\} \quad . \tag{7}$$

As $\epsilon \to 0$, we obtain the desired result[†]:

$$Z = \int_{x(0)=x(1/T)} [dx(t)] e^{-S} \quad \text{with} \quad S = \int_0^{1/T} dt \left[\frac{m}{2} \left(\frac{dx}{dt} \right)^2 + \frac{m\omega^2}{2} x(t)^2 \right] \tag{8}$$

Generalizing the above derivation to a free scalar field theory is straightforward. Technical complications arise, however, in case of fermionic theories and gauge theories. Due to the anticommuting nature of the fermionic operators, introduction of Grassmann variables and their calculus[13] becomes necessary. The natural choice of the space of states for gauge theories $\{A_\nu\}$, where A_ν are the gauge potentials, is not gauge invariant. One further needs to impose[4] Gauss's law to obtain a space of physical states. Once due care is taken of these differences, essentially the same steps as above lead to the path integral formulation of the partition function of a gauge theory with interacting fermion or scalar (Higgs) fields. Thus for an $SU(N)$ gauge theory with N_f flavours of quarks the partition function is given by

$$Z = \int_{bc} \mathcal{D}A_\nu \mathcal{D}\psi \mathcal{D}\bar{\psi} \, e^{-S(A_\nu, \psi, \bar{\psi}; \, g^2, m_f, \mu_f, T)} \tag{9}$$

with boundary conditions, bc, given by

$$A_\nu(\mathbf{x}, 0) = A_\nu(\mathbf{x}, 1/T); \quad \psi(\mathbf{x}, 0) = -\psi(\mathbf{x}, 1/T) \quad \text{and}$$
$$\bar{\psi}(\mathbf{x}, 0) = -\bar{\psi}(\mathbf{x}, 1/T), \quad \forall \mathbf{x}, \nu. \tag{10}$$

Note that these boundary conditions are simply due to the trace in Eq. (1) and are thus analogous to those in Eq. (8). The action S in Eq. (9) is given by the

---

[†] We have dropped a temperature dependent infinite constant. It cancels in thermal expectation values or can be shown to be irrelevant. For more discussion on it, including a demonstration that it also cancels in a careful evaluation of Z for a free bose gas, see Ref. 12.

usual standard form[14]. For the Euclidean metric, it is

$$S(A_\nu, \psi, \bar{\psi}; g^2, \mu_f, m_f, T) = \int_0^{1/T} dt \int_{-\infty}^{\infty} d^3x \, \frac{1}{4} \, \text{tr} \, F_{\mu\nu} F_{\mu\nu} +$$
$$\int_0^{1/T} dt \int d^3x \sum_{f=1}^{N_f} \{ \bar{\psi}_f (\slashed{D} + m_f)\psi_f - \mu_f \bar{\psi}_f \gamma_0 \psi_f \} \qquad , \qquad (11)$$

where $F_{\mu\nu}$ is the field tensor, tr denotes sum over the colour index and $\slashed{D}$ is the covariant derivative. The space-time indices μ and ν run over 1 to 4. μ_f and m_f are the chemical potential and the quark mass of the flavour f. The thermal expectation value of a physical observable $\mathcal{O}$, given by Eq. (4), can be re-expressed in the path integral form as

$$\langle \mathcal{O} \rangle = Z^{-1} \int_{bc} \mathcal{D}A_\nu \mathcal{D}\psi \mathcal{D}\bar{\psi} \, \mathcal{O} \, e^{-S(A_\nu, \psi, \bar{\psi}; \, g^2, m_f, \mu_f, T)} \qquad (12)$$

Comparing Eqs. (9)–(12) with their $T = 0$, $\mu_f = 0$ analogues one identifies the effects of nonzero temperature in the finite integration limit on the variable t in Eq. (11). Moreover, the boundary conditions (10) change the R^4 space-time manifold to $R^3 \times S^1$. Nonzero chemical potential, on the other hand, introduces an extra interaction which can be thought of as being due to a constant imaginary electrostatic potential (in a fictitious $U(1)$ gauge theory) which couples with strength μ_f. Adding scalar fields, coupled to both the gauge fields and the fermion fields in the usual way, to the action S in Eq. (11), one can study the physics of spontaneously broken gauge theories at nonzero temperature which is relevant for the Higgs sector of the standard model. We will postpone the discussion on nonzero chemical potential and on the fate of the Higgs sector at finite temperatures to sections 6 and 7 and concentrate below on QCD at finite temperature since it is by far the most studied subject.

The similarity of the Euclidean partition functions at finite temperature and at zero temperature suggests that various known approximation schemes to evaluate expectation values for the latter may be relevant for the former as well. In particular, one may employ perturbation theory to evaluate the thermodynamic potential Ω of Eq. (3). At present, this has been done[15] up to $\mathcal{O}(g^4 \ln g^2)$. For an $SU(N)$ gauge theory coupled to massless fermions, $m_f = 0$, $\forall f$, one has

$$\Omega = -VT^4 \left[\frac{\pi^2}{45} (N^2 - 1 + \tfrac{7}{4}N_f N) - \frac{g^2}{144} (N^2 - 1)(N + \tfrac{5}{4}N_f) \right.$$
$$\left. + \frac{g^3}{12\pi} (N^2 - 1)(\tfrac{1}{3}N + \tfrac{1}{6}N_f)^{3/2} + \frac{g^4 \ln g^2}{32\pi^2} N(N^2 - 1)(\tfrac{1}{3}N + \tfrac{1}{6}N_f) \right] \qquad . \qquad (13)$$

If the number of massless flavours is sufficiently small ($N_f < 11N/2$) then the property of asymptotic freedom tells us that $g^2(T) \to 0$ as $T \to \infty$ according to

$$g^2(T) = \frac{1}{2b_0 \ln \frac{T}{\Lambda}} - \frac{b_1}{2b_0^3} \frac{\ln \ln T/\Lambda}{(\ln T/\Lambda)^2} + O\left(\frac{1}{\ln T/\Lambda}\right)^2, \tag{14}$$

where $b_0 = (11N - 2N_f)/48\pi^2$ and

$$b_1 = [17N^2 - 5NN_f - 3(N^2 - 1)N_f/2N]/384\pi^4 \tag{15}$$

are the two universal coefficients of the β-function. Thus, for sufficiently high temperatures ($T \gg \Lambda$) Eq. (13) will be applicable with g^2 replaced by $g^2(T)$ of Eq. (14). Using Eq. (3), one has the following results for thermodynamic quantities in the high temperature limit:

$$s = -\frac{4\Omega}{VT} - \frac{1}{V}\frac{\partial\Omega}{\partial g} \cdot \frac{\partial g}{\partial T}$$

$$P \equiv -\Omega/V \quad \text{and}$$

$$\epsilon = \frac{T^2}{V}\frac{\partial \ln Z}{\partial T}\bigg|_{V,\mu} \tag{16}$$

$$= -\frac{3\Omega}{V} - \frac{T}{V}\frac{\partial\Omega}{\partial g} \cdot \frac{\partial g}{\partial T} = 3P - \frac{T}{V}\frac{\partial\Omega}{\partial g} \cdot \frac{\partial g}{\partial T}$$

From Eqs. (13)–(16), one finds that at very high temperatures the strongly interacting matter behaves as an ideal gas of quarks and gluons. As the temperature is lowered, it becomes a weakly interacting gas. However, with decreasing temperature or increasing $g^2(T)$, higher orders become important. Unfortunately, the severe infra-red problems of these theories render already terms of $O(g^6)$ in Ω incalculable[5]; even their finiteness is really not proven and depends on the existence of a nonvanishing magnetic mass for the gluons. It has even been[16] suggested that the perturbative vacuum is really a false vacuum and that in the true vacuum the fourth component of the gauge fields acquires a nonzero vacuum expectation value. If true, the expression in Eq. (13) needs to be recalculated. However, conventional perturbative techniques are unlikely to be of any use in that case.

It thus appears that even at very high temperatures where perturbation theory ought to work in QCD-like theories one may not have a reliable expansion in $g^2(T)$. It is definitely then a much too meager a tool to investigate what are usually nonperturbative effects such as the presence of a phase transition. Nevertheless, it does serve a useful purpose, as we shall see later, in checking results obtained with other techniques.

2.2. Symmetries and Order Parameters

Finite temperature QCD has extra global symmetries in the limits of vanishing quark mass and infinite quark mass. These enable us to obtain some ideas about the nature of possible phases. Furthermore, one can write down effective theories using these symmetries which can be analysed following standard Landau-Ginzburg techniques in the theory of phase transitions. The order parameters for these symmetries are extremely useful in numerical simulations as well. Let us therefore consider these limits.

It is clear from the Lagrangian in Eq. (11) that if $m_f = 0$, $\forall f$, the theory has a $SU(N_f) \times SU(N_f)$ chiral symmetry. If all m_f are equal to some nonzero m then the theory has a usual $SU(N_f)$ flavour symmetry which may be broken, if some masses are unequal. An order parameter which can test whether the vacuum respects the chiral symmetry or not is $\langle \bar{\psi}_f \psi_f \rangle$; a nonvanishing $\langle \bar{\psi}_f \psi_f \rangle$ indicates that the vacuum breaks the corresponding chiral symmetry spontaneously. It is widely believed, and the lattice studies support such a belief, that the chiral symmetry is spontaneously broken at low temperatures, giving rise to (almost) massless pions but heavy nucleons. On the other hand, the perturbative high temperature phase should be chirally symmetric, leading one to suspect the presence of a chiral symmetry restoring phase transition. Pisarski and Wilczek[17] considered such a phase transition in a linear σ model with $SU(N_f) \times SU(N_f)$ symmetry. Arguing that its critical behaviour is governed by an effective three dimensional chiral model, they suggested a first order chiral phase transition on the basis of renormalization group arguments (and numerical simulations of chiral models), if $N_f \geq 3$. For smaller values of N_f, the amount of breaking of the axial $U_A(1)$ symmetry becomes crucial. As is well known, instantons break the $U_A(1)$ (unless the number of colours is infinite). For $N_f = 1$, they cause an explicit breaking of the chiral symmetry just as an external magnetic field breaks the symmetry in an Ising model. Since at least a few instantons are present at all temperatures, there is no chiral phase transition for $N_f = 1$ and a finite number of colours N. No prediction for $N_f = 2$ results out of these considerations; a first (second) order chiral phase transition may be an indication that instantons are (not) depleted around the chiral phase transition.

In the limit of infinite quark mass, the quark terms act as an irrelevant multiplicative constant to the partition function Z and the thermodynamics is governed by the gluons alone. From Eq. (9) one can then deduce the presence of an additional symmetry. Due to the boundary conditions on the gauge

60

fields, the set of allowed gauge transformations is also restricted to be periodic in $1/T : V(\mathbf{x},0) = V(\mathbf{x},1/T)$, where $V \in SU(N)$. For a purely gluonic theory, this set is, however, enlarged; $V(\mathbf{x},0) = z\, V(\mathbf{x},1/T)$ is also allowed, where $z \in Z(N)$, the center of the gauge group. It is straightforward to check that the action is invariant under this global $Z(N)$ symmetry. The order parameter $\langle L(\mathbf{x}) \rangle$, where $L(\mathbf{x}) = N^{-1}\mathrm{tr}\, P \exp(i \int_0^{1/T} A_4(\mathbf{x},t)dt)$ and P denotes path ordering, tells us whether vacuum respects this symmetry, since $L(\mathbf{x}) \to zL(\mathbf{x})$ under the global $Z(N)$. L is called the Polyakov loop or the thermal Wilson line since it is a Wilson loop closed due to the boundary conditions in the temperature direction. $\langle L \rangle$ can be shown[18] to be related to the free energy of a static quark immersed in a gluonic bath at temperature T: $\langle L(x) \rangle = \exp(-F_q(x)/T)$. Thus, spontaneous breaking of the global $Z(N)$-symmetry implies a finite free energy for the static quark or deconfinement of colour, in other words. At low temperatures, one expects $F_q = \infty$ and hence $\langle L \rangle = 0$, i.e. an unbroken $Z(N)$. In the heavy quark limit, $\langle L \rangle$ is therefore an order parameter for confinement. Svetitsky and Yaffe[19] have related the deconfinement transition in finite temperature $SU(N)$ gauge theories in d dimensions to the order-disorder transition in $Z(N)$ spin systems in d dimensions by contemplating an effective action for $L(\mathbf{x})$ obtained by integrating out other degrees of freedom from the path integral for Z. This allowed them to predict various critical exponents for the deconfinement phase transition provided it is a continuous transition and only a single fixed point exists in the space of effective couplings for the corresponding global center symmetry. A lack of knowledge of any fixed point for three dimensional systems with $Z(3)$ symmetry led to the prediction of first order phase transition in the $SU(3)$ gauge theory.

It is perhaps clear from the discussion above that in a realistic world of two light quarks and a heavier quark neither of the two symmetries are exact. The antiperiodic boundary conditions for dynamical quarks force $z = 1$, breaking thus the global center symmetry, $Z(N)$, for any finite quark mass. Similarly, the mass term $m\bar{\psi}\psi$ explicitly breaks the chiral symmetry $SU(N_f) \times SU(N_f)$ to $SU(N_f)$ for any nonzero mass m. In an effective action picture, the symmetry breaking term in either case plays the role of an external magnetic field for spin systems which weakens first order phase transitions and wipes out second order ones. This correspondence can be made exact for small symmetry breaking terms. One can then treat them perturbatively to find out how a line of first order transition enters the m-T phase diagram from its extremes ($m = 0$ or $m = \infty$) and predict its endpoint.

2.3. Lattice Regularization

Just as a simple integral can be evaluated as a limit of a sum, one can envisage doing the formidable path integral in Z, defined by Eq. (9), by discretizing it. In order to preserve the internal symmetries of the theory, such as gauge invariance, it turns out convenient to introduce a hypercubic space-time lattice. Thus for an $N_\sigma^3 \times N_\beta$ lattice with lattice spacings a_σ and a_β in the spacelike and timelike (or inverse temperature-like) the temperature is $T = (N_\beta a_\beta)^{-1}$ and the volume is $V = N_\sigma^3 a_\sigma^3$. Defining the gauge fields and the quark fields on the lattice, one reduces the path integral to a product of ordinary integrals. Further, the lattice spacings act as ultra-violet cut-offs and provide a regularization scheme necessary for any quantum field theory. Indeed, one can perform the usual perturbation theoretical computations with the lattice regularization scheme although its main advantage lies in dealing with non-perturbative aspects, as we will discuss below.

A gauge invariant theory can be defined on the lattice by introducing link variables U_x^μ and site variables $\psi(x), \bar\psi(x)$. $U_x^\mu \in SU(N)$ and is associated with a directed link leaving site $x = (\mathbf{x}, x_4)$ in the direction $\hat\mu$, $\mu = 1, 2, 3$ and 4. The site variables ψ and $\bar\psi$ are associated with each site x of the lattice and they carry the colour, flavour and the spin indices. They are Grassmann variables, i.e.

$$\{\bar\psi(x), \psi(y)\} \equiv \bar\psi(x)\psi(y) + \psi(y)\bar\psi(x) = 0,$$
$$\{\psi(x), \psi(y)\} = 0 \text{ etc. } \forall x, y \quad . \tag{17}$$

The local gauge transformations $V(x) \in SU(N)$ change these fields in the colour space as below:

$$\psi(x) \to \psi'(x) = V(x)\psi(x); \; U_x^\mu \to U_x^{\mu'} = V(x)U_x^\mu V^\dagger(x + \hat\mu) \quad . \tag{18}$$

The partition function Z is then given by

$$Z = \int_{bc} \prod_{x,\hat\mu} dU_x^\mu \prod_x d\psi(x) \, d\bar\psi(x) \exp\left[-S\left(U_x^\mu, \psi, \bar\psi; \; g^2, m_i, a_\sigma, a_\beta\right)\right], \tag{19}$$

with the boundary conditions

$$U_{\mathbf{x},0}^\mu = U_{\mathbf{x},1/T}^\mu \quad \text{and}$$
$$\psi(\mathbf{x}, 0) = -\psi(\mathbf{x}, 1/T), \quad \bar\psi(\mathbf{x}, 0) = -\bar\psi(\mathbf{x}, 1/T), \quad \forall x, \hat\mu. \tag{20}$$

The action S in Eq. (19) has to satisfy the following two requirements: i) it should be invariant under the gauge transformations defined by Eq. (18) and ii) in the continuum limit of $a_\sigma \to 0, a_\beta \to 0$, it should reduce to the familiar QCD action of Eq. (11). These requirements are not very stringent, leaving open infinitely many possible choices. In fact, this is a usual feature of any discretization. Indeed, we could even have chosen a different lattice structure in place of the hypercubic one. In stead of causing any trouble, this arbitrariness comes handy in checking the reliability of the results, as we shall see later. In what follows, however, we will primarily restrict ourselves to the most popular choices only.

2.3.1 Lattice Action

The Wilson form[7] of the action for gauge variables at finite temperature[20] is given by

$$S_G = \frac{2N}{g_\sigma^2} \frac{a_\beta}{a_\sigma} \sum_{\substack{x \\ \mu < \nu < 4}} P_x^{\mu\nu} + \frac{2N}{g_\beta^2} \frac{a_\sigma}{a_\beta} \sum_{\substack{x \\ \mu < 4}} P_x^{\mu 4} \qquad (21)$$

where

$$P_x^{\mu\nu} = 1 - \text{Re} \, \frac{1}{N} \text{tr} \, U_x^\mu U_{x+\hat{\mu}}^\nu U_{x+\hat{\nu}}^{\mu\dagger} U_x^{\nu\dagger} \qquad (22)$$

is a square Wilson loop, called plaquette, made from four link variables in the $\mu\nu$ plane and it has the site x as one of its corners. The first term in Eq. (23) contains only space-like plaquettes, i.e., plaquettes without any link variables in the time direction, whereas the second contains only time-like plaquettes. The gauge invariance of this action is obvious from Eqs. (18) and (22). Further, substituting $a_\sigma = a_\beta$ and $g_\sigma = g_\beta$, one recovers the usual zero temperature gauge action, as expected; finite temperature effects in that case stem purely from $N_\sigma > N_\beta$. Relating the continuum gauge potential $A_\mu^a(x)$ to the link variable U_x^μ as

$$U_x^\mu = \exp\left(i \, g \, a \sum_{j=1}^{N^2-1} A_\mu^j(x) t^j \right) \qquad (24)$$

one can obtain[8] the continuum gauge action in Eq. (11) in the limit $a = a_\sigma = a_\beta \to 0$.

Choosing a fermionic lattice action is more difficult due to the well-known conceptual problem[8], called the doubling problem. A "No-Go" theorem[21] tells us that either one has to sacrifice continuous chiral symmetries or one ends up having too many (2^d) flavours in the continuum limit unless one is willing to use non-local actions. We will confine ourselves to two oft-used choices, referring the reader to excellent review articles[22,23] for more details. Both the staggered fermions[24] and the Wilson fermions[25] have local nearest neighbour actions. The former has a continuous $U(1) \times U(1)$ symmetry on the lattice while the latter breaks all continuous chiral symmetries. On the other hand, flavour is not a well-defined concept for the staggered fermions on the lattice.

The staggered fermion action at finite temperature can be written down as below,

$$S_F^{(S)} \equiv \sum_{f=1}^{n} \sum_{x,x'} \bar{\chi}_f(x) Q_S^f(x,x') \chi_f(x')$$

$$= \sum_{f=1}^{n} \sum_{x,x'} \bar{\chi}_f(x) \left[\sum_{\mu=1}^{3} D^{(\mu)}(x,x') + \frac{a_\sigma}{a_\beta} \gamma_F \, D^{(4)}(x,x') + m_f a_\sigma \delta_{x,x'} \right] \chi_f(x')$$

(25)

where

$$D^\mu(x,x') = \tfrac{1}{2}(-1)^{x_1+x_2+\dots+x_{\mu-1}} \left[U_x^\mu \delta_{x,x'-\hat{\mu}} - U_{x'}^{\mu\dagger} \delta_{x,x'+\hat{\mu}} \right]$$

(26)

for $\mu = 1, 2, 3$ and 4. Here $\chi_f, \bar{\chi}_f$ are single component spinors and γ_F is an extra coupling[26] which is unity for $a_\sigma = a_\beta = a$. Gauge invariance is assured by the presence of U's in Eq. (26). It is clear that χ_f cannot be identified with the quark fields in the continuum limit. It turns out[27], however, that $16n$ χ-fields on the corners of an elementary hypercube can be combined to define usual continuum quark fields of 4-component spinors which for $m_f = 0$, have $U(4n) \times U(4n)$ symmetry in the continuum limit of $a_\sigma = a_\beta = a \to 0$. For finite a, however, the chiral symmetry is broken down to $U(n) \times U(n)$. This facilitates a straightforward investigation of the physics of the spontaneous breaking of the chiral symmetry at zero temperature and its restoration at finite temperature. The presence of discrete chiral symmetries, in addition, forbids mass counter terms for these fermions, thus rendering the bare quark mass a straightforward input in the calculations with staggered fermions. The disadvantage is that the full flavour symmetry is restored only for very small a, leaving the number of light flavours uncertain for coarser lattices.

The Wilson fermions have a clear definition of flavours on the lattice. Their action is given by

$$S_F^{(W)} \equiv \sum_{f=1}^{N_f} \sum_{x,x'} \bar{\psi}_f(x) Q_W^f(x,x') \psi_f(x)$$

$$= \sum_{f=1}^{N_f} \sum_{x,x'} \bar{\psi}_f(x) \left[\delta_{x,x'} - \kappa_\beta^f M^{(4)}(x,x') - \kappa_\sigma^f \sum_{\mu=1}^{3} M^{(\mu)}(x,x') \right] \psi_f(x'),$$

(27)

where

$$M^{\mu}(x, x') = (1 - \gamma_{\mu})U_x^{\mu}\delta_{x,x'-\hat{\mu}} + (1 + \gamma_{\mu})U_{x'}^{\mu\dagger},\delta_{x,x'+\hat{\mu}}, \text{ for } \mu = 1, 2, 3 \text{ and } 4. \quad (28)$$

It is easily recognized that the $\psi, \bar{\psi}$ are the usual 4-component spinors. The matrices $M^{(\mu)}$ thus carry color, spin and space-time indices although the compact notation of Eq. (28) hides the former two. Notice the similarity in the structure of Eqs. (26) and (28). Indeed, the gauge invariance in both the cases is due to the presence of link variables in the lattice covariant derivatives. However, the continuum limit is more straightforward in this case. For $U_x^{\mu} = 1$ and $a_{\sigma} = a_{\beta}, \kappa_{\sigma}^f = \kappa_{\beta}^f = \kappa^f$, one obtains

$$\frac{1}{\kappa^f} - \frac{1}{\kappa_c} = 2m_f a \text{ with } \kappa_c = \frac{1}{8}. \quad (29)$$

Thus even for $m_f = 0$ or $\kappa^f = \kappa_c, S_F^{(W)}$ has an explicit $\bar{\psi}\psi$ term, breaking all chiral symmetries. Furthermore, the quark mass becomes difficult to define once interactions are switched on. These difficulties have made Wilson fermions a disfavoured choice for finite temperature studies. However, eventually we need to obtain the same results irrespective of the choice of fermions in order to be sure of their relevance to continuum physics. A possible definition of the quark mass, even for an interacting theory, is Eq. (29) itself but with a replacement of κ_c by $\kappa_c(g^2)$ which is defined as the value of κ where $m_{\pi}^2 = 0$. Alternatively, it has been proposed[28,29] to define $m_f a$ through a PCAC relation. Numerical simulations[30] have shown that such a definition yields a quark mass which i) obeys Eq. (29) and the κ_c so defined agrees with $\kappa_c(m_{\pi}^2 = 0.)$ and ii) yields a unique definition of the quark mass in both the low and the high temperature phases.

In both the cases, a convenient way to deal with the anticommuting nature of the basic fermionic variables is to integrate them out in Eq. (19) to obtain the following expression for Z:

$$Z = \int \prod_{x,\hat{\mu}} dU_x^{\mu} \exp(-S_G) \det{}^{\ell}Q, \quad (30)$$

where we have used $S = S_G + S_F$. In the above expression $\ell = N_f$ for Wilson fermions and $\ell = n$ for staggered fermions. Anticipating the restoration of the flavour symmetry in the continuum limit, the effective ℓ for the staggered fermions can be thought of as $4n$. Although the allowed number of flavours for these fermions thus has to be a multiple of four, it has been proposed[31] to interpolate the above expression to intermediate values as well by allowing fractional values of n. Thus $n = \frac{1}{2}$ would be interpreted as a theory with two flavours. While it can be argued why this interpretation may be acceptable for thermodynamics, we wish to emphasize here its shaky theoretical basis despite its widespread acceptance. A little algebra shows that for both types of fermions $\gamma_5 Q^\dagger \gamma_5 = Q$. Since $\gamma_5^2 = 1$, this implies det Q is real. One can, therefore, rewrite Eq. (30) as

$$Z_W = \int_{bc} \prod_{x,\hat{\mu}} dU_x^\mu \exp\left(-S_G\left(U_x^\mu\right) + \frac{N_f}{2} \operatorname{Tr} \ln Q_W^2\right) \tag{31}$$

for the Wilson fermions and

$$Z_S = \int_{bc} \prod_{x,\hat{\mu}} dU_x^\mu \exp\left(-S_G\left(U_x^\mu\right) + \frac{N_f}{8} \operatorname{Tr} \ln Q_S Q_S^\dagger\right) \tag{32}$$

for the staggered fermions, where Tr denotes trace over all indices.

2.3.2. Physical Observables

Having defined the partition function for an $SU(N)$ lattice gauge theory with dynamical fermions above, we will collect here the lattice definitions for the order parameters, introduced in sec. 2.2, and various thermodynamical quantities, such as the energy density. The thermal expectation value of a physical observable $\mathcal{O}$, defined by Eq. (12) takes the following obvious form on the lattice

$$\langle \mathcal{O} \rangle = Z^{-1} \int_{bc} \prod_{x,\hat{\mu}} dU_x^\mu \mathcal{O} \, \exp\left(-S_G\left(U_x^\mu\right) + \frac{N_f}{8} \operatorname{Tr} \ln Q Q^\dagger\right) \tag{33}$$

From now on we will use the staggered fermions, unless specified otherwise, and have dropped therefore the subscript S. As discussed already in sec. 2.3.1, there exists a remnant $U(1) \times U(1)$ chiral symmetry for all values of the lattice spacing for Z of Eq. (32). Following sec. 2.2, the corresponding order parameter is $\langle \bar{\psi}\psi \rangle$ given by

$$\langle \bar{\psi}\psi \rangle = \left\langle \sum_{i,x} \bar{\chi}^i(x)\chi^i(x) \right\rangle \bigg/ N_\sigma^3 N_\beta N$$

$$= \left\langle \frac{N_f}{4} \operatorname{Re} \operatorname{Tr} Q^{-1} \right\rangle \bigg/ N_\sigma^3 N_\beta N \tag{34}$$

The boundary conditions on the link variables imply in the infinite mass limit the $Z(N)$ global center symmetry discussed in sec. 2.2. The corresponding order parameter $\langle L \rangle$ can be defined on the lattice as

$$\langle L \rangle = \left\langle \sum_{\mathbf{x}} L(\mathbf{x}) \right\rangle \Big/ N_\sigma^3 \qquad (35)$$

with

$$L(\mathbf{x}) \equiv \text{tr } W(\mathbf{x}) = \text{tr} \prod_{t=1}^{N_\beta} U_{\mathbf{x},t}^{(4)} \Big/ N. \qquad (36)$$

The thermodynamic quantities on the lattice are obtained by substituting $T = (N_\beta a_\beta)^{-1}$ and $V = N_\sigma^3 a_\sigma^3$ in the derivatives in the standard expressions, given by Eq. (3). Introducing $\xi = a_\sigma/a_\beta$, one has for the energy density ϵ the following expression:

$$\begin{aligned}
\epsilon &= -V^{-1} \frac{\partial}{\partial(1/T)} \ln Z \\
&= \frac{\xi^2}{N_\sigma^3 N_\beta a_\sigma^4} \frac{\partial}{\partial \xi} \ln Z
\end{aligned} \qquad (37)$$

Using Eq. (32) along with Eqs. (21) and (25) the expression for ϵ at $\xi = 1$ can be rearranged as follows:

$$\epsilon/T^4 \equiv (\epsilon_G + \epsilon_F)/T^4, \qquad (38)$$

where

$$\frac{\epsilon_G}{T^4} = 3\beta N_\beta^4 \left[\langle \overline{P}_\sigma \rangle - \langle \overline{P}_\beta \rangle - g^2 \left(\frac{\partial g_\sigma^{-2}}{\partial \xi} \Big|_{\xi=1} \langle \overline{P}_\sigma \rangle + \frac{\partial g_\beta^{-2}}{\partial \xi} \Big|_{\xi=1} \langle \overline{P}_\beta \rangle \right) \right] \qquad (39)$$

and

$$\begin{aligned}
\frac{\epsilon_F}{T^4} &= \frac{N_f}{4} \left(\frac{N_\beta}{N_\sigma} \right)^3 \left(1 + \frac{\partial \gamma_F}{\partial \xi} \Big|_{\xi=1} \right) \langle \text{Re Tr } D^{(4)} Q^{-1} \rangle \\
&\quad + N N_\beta^4 \frac{\partial m_f a_\sigma}{\partial \xi} \Big|_{\xi=1} \langle \bar{\psi}\psi \rangle .
\end{aligned} \qquad (40)$$

Here we have introduced some additional notation: $\overline{P}_{\sigma(\beta)}$ stands for average space-like (timelike) plaquette defined by

$$\begin{aligned}
\overline{P}_\sigma &= \sum_{\substack{x \\ \mu < \nu < 4}} P_x^{\mu\nu} \Big/ 3N_\sigma^3 N_\beta, \\
\overline{P}_\beta &= \sum_{\substack{x \\ \mu < 4}} P_x^{\mu 4} \Big/ 3N_\sigma^3 N_\beta,
\end{aligned} \qquad (41)$$

where the sum runs over all independent spacelike (timelike) plaquettes. $\beta = 2N/g^2$, where g^2 is the coupling on symmetric lattice. The energy density so obtained contains a contribution from the vacuum[12], which can be eliminated by subtracting $\epsilon(T = 0)$. For sufficiently large N_σ, ϵ evaluated on a symmetric N_σ^4 lattice is a good approximation to $\epsilon(T = 0) \equiv \epsilon^{vac}$. From Eqs. (39) and (40), one finds

$$
\epsilon_G^{vac}/T^4 = -6NN_\sigma^4 \left(\frac{\partial g_\sigma^{-2}}{\partial \xi}\bigg|_{\xi=1} + \frac{\partial g_\beta^{-2}}{\partial \xi}\bigg|_{\xi=1} \right) \langle \bar{P} \rangle
$$

$$
\epsilon_F^{vac}/T^4 = \tfrac{1}{4}NN_\sigma^4 \left[\tfrac{1}{4}N_f - m\langle\bar{\psi}\psi\rangle_{T=0}\right] \left(1 + \frac{\partial\gamma_F}{\partial\xi}\bigg|_{\xi=1} \right) \tag{42}
$$

$$
+ NN_\sigma^4 \frac{\partial m_f a_\sigma}{\partial\xi}\bigg|_{\xi=1} \langle\bar{\psi}\psi\rangle_{T=0} \quad,
$$

with $\bar{P} = \bar{P}_\sigma = \bar{P}_\beta$ being the average plaquette on the symmetric N_σ^4 lattice. Subtracting off these vacuum contributions from Eqs. (39) and (40) and substituting the resulting expressions in Eq. (37), one obtains the true physical energy density.

The pressure can be obtained in essentially the same way from Eq. (3):

$$
P = \frac{1}{3N_\sigma^3 N_\beta a_\sigma^2 a_\beta} \frac{\partial}{\partial a_\sigma} \ln Z \quad, \tag{43}
$$

which at $\xi = 1$ can be shown to be

$$
\frac{P}{T^4} = \frac{\epsilon}{3T^4} - 2NN_\beta^4 \left[\frac{\partial g_\sigma^{-2}}{\partial \ln a}\bigg|_{\xi=1} \langle \bar{P}_\sigma \rangle + \frac{\partial g_\beta^{-2}}{\partial \ln a}\bigg|_{\xi=1} \langle \bar{P}_\beta \rangle - \frac{1}{6}\frac{\partial m_f a_\sigma}{\partial \ln a}\bigg|_{\xi=1} \langle\bar{\psi}\psi\rangle \right]
$$

$$
+ \frac{N_f}{12} \left(\frac{N_\beta}{N_\sigma}\right)^3 \frac{\partial\gamma_F}{\partial \ln a}\bigg|_{\xi=1} \left\langle \mathrm{Re}\,\mathrm{Tr}\, D^{(4)}Q^{-1} \right\rangle \quad.
$$

$$\tag{44}$$

Again a subtraction of $P(T = 0)$ is necessary to obtain the physical pressure from Eq. (44). Having obtained P and ϵ, one can construct other quantities such as the entropy density, $s = (\epsilon + P)/T$, the so-called interaction measure $\Delta = \epsilon - 3P$, or the velocity of sound $v_s^2 = dP/d\epsilon$.

In order to evaluate the energy or the pressure, one needs to know the various derivatives occurring in Eqs. (38)–(44). In general, these derivatives can be obtained by demanding rotational invariance of physical quantities on an asymmetric lattice at zero temperature. If g is small enough, one can evaluate them using perturbation theory. For a pure $SU(N)$ gauge theory, such a perturbative estimate was obtained by Karsch[32]. Subsequently, fermions were included[33] and recently, the need for the coupling γ_F was recognized by Karsch and Stamatescu[26] who computed its derivative with respect to ξ for small g^2. Collecting all these results, one has the following predictions from perturbation theory:

$$\left.\frac{\partial g_\sigma^{-2}}{\partial \xi}\right|_{\xi=1} = 0.1466\frac{N^2-1}{2N} + 0.002N - 0.0003N_f$$

$$\left.\frac{\partial g_\beta^{-2}}{\partial \xi}\right|_{\xi=1} = -0.1466\frac{N^2-1}{2N} + 0.0212N - 0.003 N_f \quad . \tag{45}$$

$$\left.\frac{\partial \gamma_F}{\partial \xi}\right|_{\xi=1} = -0.1599g^2\frac{N^2-1}{2N}$$

In addition, the partial derivatives of $g_{\sigma,\beta}^{-2}$ can be replaced by the usual perturbative β-function for small enough g:

$$2N\left.\frac{\partial g_\sigma^{-2}}{\partial \ell n\, a}\right|_{\xi=1} = 2N\left.\frac{\partial g_\beta^{-2}}{\partial \ell n\, a}\right|_{\xi=1} = \frac{d\beta}{d\ell n\, a} \quad . \tag{46}$$

The remaining two partial derivatives occurring in Eq. (44) have not yet been computed perturbatively, although the γ_F-derivative is formally $O(g^4)$ and hence negligible. As one may anticipate, substitution of these expressions in Eqs. (39) and (40) or Eq. (44) shows that they add a correction proportional to g^2. Unfortunately, the values of g^2 at which numerical simulations are performed are, however, rather large, as we shall see later. Therefore, the reliability of Eqs. (45) and (46) in obtaining the energy density or the pressure from numerical simulations appears to be questionable. In fact, one even obtains inconsistencies, making it necessary to determine these entities non-perturbatively. Nevertheless, Eqs. (45) and (46) do provide a reasonable first estimate of the significance of their contribution.

If the deconfinement phase transition or the chiral phase transition is of first order then the corresponding phases can coexist in a thermal equilibrium at the critical point with a stable interface in between. The surface tension of such an interface is an important observable that governs nucleation rates in a cooling plasma. These rates are a crucial input to an alternative scenario for the light element nucleosynthesis[34]. Lattice techniques offer us a way to obtain the surface tension, α, in a non-perturbative manner. Two different methods have been proposed to obtain α for QCD[35,36]. One of them involves[35] first creating an interface and then removing it by locally varying some controlling parameters such as the temperature or the coupling. The difference of the free energy with and without the interface, obtained by integrating suitable derivatives of the free energy along the paths, yields the surface tension after dividing by the area of the interface. The other method[36] for obtaining the surface tension is based on the methodolgy described above to obtain the energy density. In the presence of an interface of area A, the thermodynamic potential Ω depends on the area A in addition to T and V:

$$d\Omega = -sV\,dT - p\,dV + \alpha\,dA \ . \tag{47}$$

Thus the surface tension can be obtained as a partial derivative of Ω: $\alpha = (\partial\Omega/\partial A)_{V,T}$. Transcribing this to the lattice language necessitates introduction of further asymmetry in lattice spacings. Introducing a_T as the lattice spacing in the two directions, say 1 and 2, of a planar interface and following essentially the same steps of taking the partial derivatives first and equating all lattice spacings later, α can be easily evaluated. It again involves partial derivatives of the couplings with respect to the asymmetry as well. For the $SU(3)$ gauge theory, one has[36]

$$\alpha\frac{A}{T} = \beta\sum_x\left[1 + \frac{g^2}{2}\left(\frac{\partial g_\beta^{-2}}{\partial\xi}\Big|_{\xi=1} - \frac{\partial g_\sigma^{-2}}{\partial\xi}\Big|_{\xi=1}\right)\right] \ . \tag{48}$$
$$\langle 2P_x^{34} - P_x^{14} - P_x^{24} - 2P_x^{12} + P_x^{13} + P_x^{23}\rangle$$

Note that both the first integral method and the second differential method require a careful extrapolation to extract the surface tension. This is because the interface is not stable in any finite volume at equilibrium unless some controlling parameter, such as the coupling, is tuned differently on the two sides of the interface. Typically, equilibrating one side of the lattice at $\beta_c - \Delta\beta$ and the other at $\beta_c + \Delta\beta$ a stable interface can be achieved for large enough $\Delta\beta$. The surface tension has to be then extracted by taking the limit $N_\sigma^3 \to \infty$ first and the limit $\Delta\beta \to 0$ later.

Finally, a lot of information can be obtained by studying correlation functions of various operators at nonzero temperatures. Indeed, the physical correlation length can be an important tool to distinguish the order of the phase transition or to learn about the nature of the medium being probed. In QCD, one can construct correlations of operators with definite hadronic quantum numbers and extract the corresponding screening lengths. In general, a connected correlation function for a translationally invariant theory is usually given by

$$C(A(x), B(y)) = \langle A(x)B(y)\rangle - \langle A(0)\rangle\langle B(0)\rangle \ . \tag{49}$$

Using operators summed over all spatial coordinates but one, the corresponding screening length can be conveniently extracted by fitting the correlation function with an *ansatz*, $\exp(-z/\xi)$.

3. Calculational Techniques

The partition function for finite temperature QCD, defined in various forms in Eqs. (30)–(32), bears a strong resemblance to the partition functions one comes across in various statistical systems, such as the Ising-like spin models. In fact, this is true for all field theories on the lattice. It is, therefore, perhaps natural that most of the well-known calculational techniques from statistical mechanics have been found useful in the analysis of lattice field theories at both zero and finite temperature. Analogous to the high and low temperature expansions there, one has strong and weak coupling expansions, the latter being simply the usual perturbation theory but with a lattice regularization. In addition, techniques such as the $1/d$-expansion (the mean field approximation) or the $1/N$-expansion, where d is the space-time dimensionality and N is the number of colours, have also been employed. All these techniques have the advantage of being analytical which means that limitations of the approximations and corresponding remedies are usually known; typically the error due to neglected higher orders can be estimated. However, their main disadvantage is that they are too inadequate to carry through the continuum limit, $a \to 0$. The only method which provides a possibility of achieving the continuum limit is that of numerical simulations, since it permits computations of expectation values at all couplings. The renormalization group tells us how to take the continuum limit. One demands that the physical observables remain unchanged, as one reduces the lattice spacing. By simultaneously changing the bare coupling g^2, this can be achieved. For an asymptotically free theory such as QCD, or an $SU(N)$ theory with moderate numbers of light flavours of quarks ($N_f < 11N/2$) in general, the perturbatively computed β-function implies that in the limit $a \to 0$ the bare coupling g^2 must be adjusted to zero according to the following relation:

$$a\Lambda_L(N, N_f) = \left(b_0 g^2\right)^{-b_1/2b_0^2} \exp\left(-1/2b_0 g^2\right)\left[1 + \mathrm{O}(g^2)\right], \qquad (50)$$

where the coefficients b_0 and b_1 are defined in Eq. (15) and $\Lambda(N, N_f)$ is the lattice Λ-parameter for N colours and N_f (massless) flavours.

Using Monte Carlo simulations Creutz[37] found that for $N = 2, 3$ and $N_f = 0$ the above equation holds already on small lattices and rather large couplings.

Operating, therefore, in the range of g^2 where Eq. (50) holds, one can obtain the lattice spacing $a(g^2)$ in units of Λ_L^{-1}. If any experimentally known physical quantity such as the mass of the proton is also known at a given g^2 in lattice units of a^{-1} then Λ_L, and hence $a(g^2)$, can be determined in appropriate physical units such as MeV. Once the lattice spacing is thus known then one can obtain all the physical quantities at finite temperature, as discussed in sect. 2.3.2. In particular, the critical temperature and other critical parameters can be thus determined. In view of this attractive feature of the numerical simulations, we focus in later sections on them.

We will not discuss the intricacies of either the analytical or the numerical methods here. An excellent introduction to these techniques in the context of lattice field theories can, for example, be found in Refs. 8 and 23. In addition, more general expositions can be obtained from textbooks on statistical field theories[38]. In the remaining part of this section, we intend to discuss some of them in brief for both completeness and to emphasize the finite temperature aspects. The next subsection is devoted to a lightning review of the analytical results on the deconfinement phase transition in both pure gauge theory and the full QCD. The aim is to familiarize the reader with the theoretical expectations arising out of these considerations. More details, including discussions of other gauge theories can be found in Ref. 39. In the second subsection, we wish to comment on various algorithms, used in numerical simulations, and the methods employed to look for the order of a phase transition. More about fermion algorithms, a still developing field, can be learnt from the article by Creutz in this book.

3.1. Analytical methods

The deconfinement phase transition in pure $SU(N)$ gauge theories and the chiral phase transition in massless QCD can be studied with the aid of the respective order parameters, $\langle L \rangle$ and $\langle \bar{\psi}\psi \rangle$, as we saw in sect. 2.2. The analytical methods are mostly aimed at obtaining a Landau-Ginzburg action for these order parameters, starting from the corresponding lattice action. Indeed, the approach is very similar to the usual effective field theory approach that utilises the underlying symmetry structure. The main difference is that the effective action is derived by a systematic approximation from the basic lattice action. Since the relation of the chiral symmetry on the lattice with that in the continuum is itself not very clear, there is a scarcity of realistic lattice results on the chiral phase transition. The deconfinement transition, on the other hand, can be studied using various techniques such as mean field theory and/or strong coupling.

In the strong coupling limit, $\beta \to 0$, the integrand in Eq. (32) can be expanded as a sum of products of various plaquettes. Ignoring the fermion term in the so-called quenched approximation (or equivalently in the limit $N_f \to 0$ or $m_f \to \infty$) and using the fact that $\int U \, dU = 0$, one sees that the lowest order term to contribute to the effective action for L will consist of a tower of plaquettes in the temperature direction. It yields a term $\text{tr } W(\mathbf{x}) \cdot \text{tr } W^\dagger(\mathbf{x} + \hat{\mu})$, where W is defined in Eq. (36). The partition function at the leading order is then given by[40]

$$Z = \int \prod_{\mathbf{x}} dW(\mathbf{x}) \, \exp\left(-\lambda \sum_{\mathbf{x},\mu=1}^{3} \left[\text{tr } W(\mathbf{x}) \cdot \text{tr } W^\dagger(\mathbf{x} + \hat{\mu})\right]\right) \quad , \tag{51}$$

with the coupling $\lambda \sim g^{-2N_\beta}$. A mean-field analysis of this effective theory leads[40] to the prediction of a second order phase transition for $N = 2$ and a first order phase transition for $N \geq 3$. One also obtains the critical coupling in this manner as well. However, the strong coupling approximation turns out to be valid for small N_β only, restricting its quantitative utility. Recently, a mean field analysis of the $SU(N)$ gauge theories has been performed at finite temperature without any strong coupling expansion[41]. The values of the critical coupling obtained there compare rather favourably with the numerical results even beyond the strong coupling regime, especially for the $SU(2)$ gauge theory.

A natural way to extend the above considerations to full QCD with dynamical quarks is to regard the above effective action as a leading term in an expansion of the fermionic trace term in Eq. (32) in powers of m_f^{-1}. The next non-trivial term can then be seen to be analogous to an external magnetic field in a corresponding spin model, with a form

$$S_f = -h \sum_{\mathbf{x}} \text{Re tr } W(\mathbf{x}) \quad , \tag{52}$$

with $h \sim m_f^{-N_\beta}$. A mean field analysis of the theory with this additional term predicts[42] a deconfinement phase transition to be washed out for any value of the quark mass, if it is of second order for $m_f = \infty$. On the other hand, a first order deconfinement phase transition weakens as m_f is decreased from ∞, perhaps turning into second order at some critical mass m_D. While these analyses further predict a wash-out for quark masses below m_D, it is possible that higher order terms of the expansion are needed to arrive at any reliable conclusion. Fig. 1 depicts schematically the expected line of deconfinement phase transitions in an

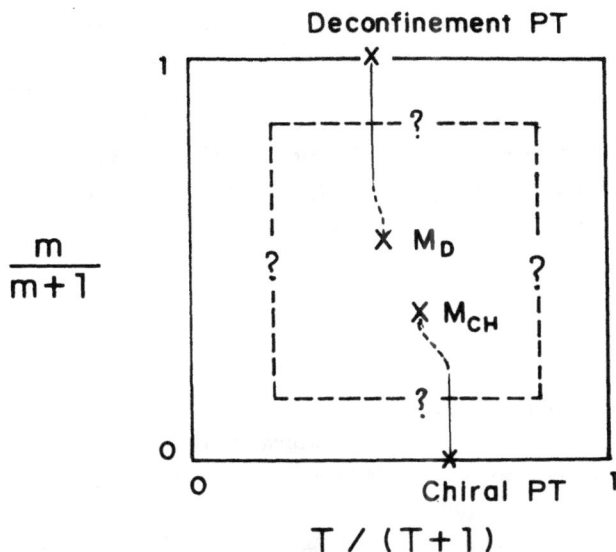

Fig. 1: A schematic phase diagram for QCD with 3 or more flavours of quarks. The upper line is for the deconfinement phase transition while the lower one is for the chiral phase transition. The analytical predictions are unreliable in the interior of the box, as suggested by the inner box of dashed lines.

m_f–T phase diagram for QCD where a similarly expected line of first order chiral phase transitions, with its possible end point m_{CH}, is also shown at the low mass end. Note, however, that the latter is based on a simple model with the appropriate symmetries and is thus not obtained from Eq. (32) in any approximation. Furthermore, one should also remember that Fig. 1 depends on the number of flavours, although, as drawn, it is generically valid for three or more flavours. There are several things one would like to know more precisely in this phase diagram. Apart from making sure that the full QCD does indeed have such a phase diagram even in the continuum limit, one would like to know the locations of the end points of both the phase transition lines. One may hope to shed some light on the relation between the chiral symmetry and the confinement if the two lines meet. Alternatively, if the real world quark masses lie in a gap between the two endpoints, one may not have a phase transition at all.

3.2. Computer Simulations

The capability of carrying out the desired continuum limit of the lattice field theories in the neighbourhood of a critical point has selected out computer simulation techniques as the primary tool to obtain non-perturbative results for various continuum field theories. Basically, they consist of creating sequences of (independent) configurations of all fields such that each configuration occurs in the space of configurations with a probability proportional to the Boltzmann factor, $\exp(-S)$, where S is the lattice action for the theory. The expectation value of an observable $\mathcal{O}$ is then simply obtained by averaging the value of $\mathcal{O}$ over N such configurations:

$$\langle \mathcal{O} \rangle = \sum_{i=1}^{N} \mathcal{O}(i)/N \ . \tag{53}$$

It has the usual statistical error, $\sqrt{(\langle \mathcal{O}^2 \rangle - \langle \mathcal{O} \rangle^2)/(N-1)}$ so that one needs large N for better precision on $\langle \mathcal{O} \rangle$.

There are many known algorithms which enable one to obtain configurations with the desired probability distribution. Thus, at least in principle, the task of looking for a range of the coupling where Eq. (50) is valid is rather straighforward. However, in practice, taking the continuum limit still turns out to be a nontrivial problem. There are a number of reasons for that. As the lattice spacing becomes smaller, the continuum physics can be kept invariant only if the lattice correlation length increases simultaneously, i. e., if the bare coupling moves closer to its critical value. In order to minimize the finite size effects, the lattice size also has to grow corrspondingly. This puts a progressively stringent requirement on the available computer resources. Further, the computer algorithms generate sequences of configurations which are typically correlated, especially more so near the critical coupling. Assuming an exponential fall-off for these autocorrelations, one can hope to obtain an independent configuration only after skipping several autocorrelation lengths in the sequence of configurations, thus requiring more computer time. The non-locality of the fermion determinant, which appears in the effective action in Eq. (32), complicates the problem even more. Its straighforward evaluation would necessitate $O(M^3)$ operations and $M = 3N_\sigma^3 \times N_\beta$ is 98304 on a moderate size($16^3 \times 8$) lattice. In sect. 3.2.1 we describe briefly some of the popular approximation schemes to include the determinant.

In addition, there are the usual problems inherent to numerical techniques, e. g., how large N in Eq. (53) needs to be or what is an optimal lattice size or how close one needs to be of $g_c^2 = 0$ for asymptotically free theories in order to simulate continuum physics adequately is not fully clear. Consequently, the field has witnessed a gradual refinement of the results with a somewhat slow approach towards the continuum limit, as more and more powerful computers became available over the years. Recently, the power of finite size scaling methods has been recognised, especially in the studies of the order of the phase transitions. In sect. 3.2.2, we summarize the various ways one can look for the order of a phase transition in numerical simulations.

3.2.1 Algorithms

The simplest algorithm which generates configurations according to the Boltzmann weight, $\exp(-S)$, is that of Metropolis et al.[43] Given a configuration of the link variables U_x^μ, $\forall\, x,\mu$, one generates a tentative replacement $U_{x_0}^{\prime\mu_0}$ for the link $U_{x_0}^{\mu_0}$ by a stochastic and reversible rule: Prob $\left(U_{x_0}^{\mu_0} \to U_{x_0}^{\prime\mu_0}\right) = $ Prob $\left(U_{x_0}^{\prime\mu_0} \to U_{x_0}^{\mu_0}\right)$. Computing the change of action δS due to this proposed change of the link $\delta S = S(U') - S(U)$, one accepts the change with a probability $\min(1, \exp(-\delta S))$. Repeating this procedure for each link of the lattice, one completes a Monte Carlo iteration (or sweep). Typically, at least thousands of such sweeps are necessary to obtain reasonable statistical accuracy.

One can now appreciate in more detail the enormity of the problem caused by the nonlocality of the fermion determinant in Eq. (32), since it needs to be evaluated for each proposed update of a link. Thus even on a tera-flop computer (i. e., a computer which can perform 10^{12} floating point operations per second), one sweep with straightforward evaluations of the determinant would take about 10 years on a $16^3 \times 8$ lattice. Evidently, some better method is called for. There have been a lot of proposals in the literature, which deal with the problem of inclusion of fermions in numerical simulations. Referring the interested reader to the article by Creutz in this book for more details, we concentrate below only on those algorithms which have been used in the simulation of finite temperature field theories. Even for them our aim is more to provide a broad picture so that the reader can appreciate better the results to be discussed in section 4.

A common intermediate step to many of the algorithms is to re-write the fermionic term as a path integral over pseudo-fermion fields[44] ϕ:

$$\det QQ^\dagger = \int \prod_{x,i} d\phi_i(x)d\phi_i^\dagger(x) \exp\left(-\phi^\dagger Q^{-1}\left(Q^\dagger\right)^{-1}\phi\right) \qquad (54)$$

Applying now the Metropolis algorithm to the extended field space $\{U_x^\mu,\ \phi(x)\}$, the task of evaluating the determinant can be seen to have reduced to that of obtaining the inverse of Q. Further, if one proposes to change all links simultaneously then even the inverse needs to be computed only once. However, in general

δS will be large, being typically $O(4N_\sigma^3 N_\beta)$ and hence the acceptance of proposed changes will be exponentially small. The Hybrid Monte Carlo (HMC) algorithm[45] gets over this problem in an elegant manner.

Introducing momenta $\pi_{x,\mu}^\alpha$, $\alpha = 1,\ldots,8$ ($N^2 - 1$ for $SU(N)$) conjugate to the link variables U_x^μ, the partition function in Eq. (32) can be written as

$$Z = \int_{bc} \prod_{x,\mu} dU_x^\mu \prod_{\substack{even \\ x}} d\phi_e(x)d\phi_e^\dagger(x) \prod_{x,\mu,\alpha} d\pi_{x,\mu}^\alpha \exp(-H) \qquad (55)$$

where

$$H = \frac{1}{2}\sum_{x,\mu,\alpha}(\pi_{x,\mu}^\alpha)^2 + S_G(U_x^\mu) + \frac{1}{2}\phi_e^\dagger(Q^\dagger Q)^{-1}\phi_e \qquad (56)$$

It is clear from Eqs. (32) and (54) that one needs 8 flavours of staggered fermions in order to proceed as above (the corresponding number of flavours for Wilson fermions is 2). It turns out, however, that $Q^\dagger Q$ connects even (odd) sites with only even (odd) sites, where an even (odd) site is defined by $x+y+z+t =$ even (odd). Setting $\phi = 0$ on all odd sites, one can therefore reduce the number of flavours to four. The subscript e in Eqs. (55) and (56) is a reminder for this. For a given $\{\phi_e\}$, obtained by drawing gaussian complex random numbers η and by defining $\phi_e = (Q^\dagger\eta)_e$ to satisfy Eq. (54), one can evolve the $\{U_x^\mu\}$ by using classical hamiltonian equations for U_x^μ and $\pi_{\mu,x}^\alpha$. $\pi_{x,\mu}^\alpha(\tau = 0)$ can again be initially chosen from a gaussian distribution. Taking $\{U_x^\mu, \pi_{x,\mu}^\alpha\}$ at the end of the classical trajectory to be a new proposed set for the Metropolis test, one sees that the new configuration is always accepted, since $\delta H = 0$. In practice, the evolution on the computer necessitates a discretization of classical equations of motion. It can be shown[45] that a suitable discretization scheme leads to $\delta H = O(\Delta\tau^2)$, where $\Delta\tau$ is the time step and further it satisfies the conditions required of Metropolis test.

The HMC algorithm has the advantage of being free of systematic errors [†]. Its premier disadvantage, however, is that it can be used for only multiples of four flavours of staggered fermions or for even number of flavours of Wilson fermions. For other cases, one has to employ approximate algorithms. One such approximate algorithm is the hybrid molecular dynamics (HMD) algorithm[46] which differs from the HMC algorithm only in not having the final accept/reject step. Depending on the size of $\Delta\tau$, the HMD algorithm will clearly have systematic errors due to the

---

[†] A possible source of systematic errors is the deviation of the numerically obtained $Q^{-1}\phi_e$ from its true value. Its impact on the numerical simulations is not really known.

inclusion of even those configurations which will be rejected in HMC. Since for small $\Delta\tau$ the acceptance will be close to 100% in HMC, the influence of the systematic errors in the HMD will be minimal for such $\Delta\tau$. On the other hand, one may now introduce only $\pi_{x\mu}^{\alpha}$ in Eq. (32) and obtain the molecular dynamics equations for U_x^{μ}, $\pi_{x,\mu}^{\alpha}$. The fermion determinant causes these equations to have terms of the type $\frac{N_f}{8} Tr \left[(Q^{\dagger}Q)^{-1} \partial (Q^{\dagger}Q)/\partial U \right]$. Since N_f thus enters in the molecular dynamics equations as a mere coefficient of these terms, the HMD algorithm can be used to simulate a theory with arbitrary number of flavours. Complex gaussian random η can be used to estimate the trace terms.

There are many other approximate algorithms which too enable one to treat arbitrary N_f. The Langevin[47] and the pseudo-fermion[31], are two amongst these which have been used to study finite temperature QCD. The former is similar to the HMD algorithm described above except that the evolution of the U-fields is governed by a diffusion equation. The pseudo-fermion method consists of expanding the ratio of the determinants, $\det Q(U')/\det Q(U)$, in powers of $\delta U = U' - U$ and evaluating the coefficient of the leading term using the pseudo-fermion fields similar to those of Eq. (54).

To end this subsection, let us recall that the error in the expectation value of an observable in a numerical simulation can come from systematic biases or statistical limitations. The approximate algorithms for full QCD can be faster but contain (frequently unknown) systematic errors. Since all of them become exact in some limit (e. g., $\Delta\tau \to 0$ or $\delta U \to 0$) one can hope to eliminate them by tuning them appropriately and then extrapolating to the right limit. The HMC algorithm does not have such problem, in principle. However, there too one needs to tune parameters to have a decent acceptance but no extrapolation is necessary. It is, on the other hand, incapable of dealing with arbitrary number of flavours. The issue of statistical errors is dominated by the considerations of auto-correlation length. Naively, one expects that the "small-step" approximate algorithms would lead to more correlations from sweep-to-sweep than the HMC algorithm. However, no thorough investigations of this crucial aspect have been done to support the naive expectation. In other words, the quoted errors in the fermionic simulations need to be treated with caution.

3.2.2 Looking for a phase transition

Locating a phase transition, determining its order and obtaining the critical parameters precisely can be greatly facilitated by its characteristic discontinuities or divergences. However, one necessarily deals with a finite lattice in a numeri-

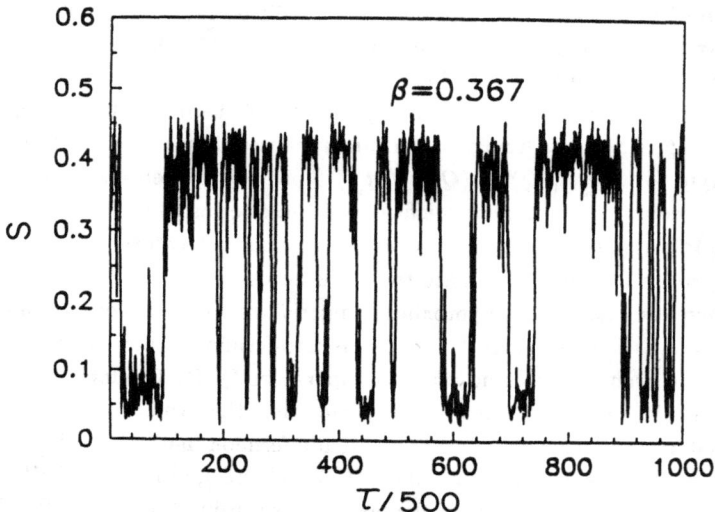

Fig. 2: Monte Carlo time evolution of the order parameter for the three-dimensional three-states Potts model on a 36^3 lattice in its critical region ($\beta = 0.367$). From Ref. 48.

cal simulation. Consequently discontinuities are smoothed out and there are no divergences, making the task of looking for a phase transition harder. This is more so in distinguishing a weak first order phase transition (with a small latent heat) from a second order phase transition. Intuitively, one expects a coexistence of two phases at the transition point of a first order phase transition. In a numerical simulation this shows up as tunnellings from one phase to another, where the tunnelling rate is expected to go down as the volume becomes larger. Fig. 2 shows the evolution of the order parameter[48] for the three-dimensional three-state Potts model on a 36^3 lattice in its critical region, where the tunnellings are clearly visible. The corresponding probability distribution, compared with those on 30^3 and 48^3 lattice in Fig. 3, shows a clear two peak structure. Note that the increase in volume by factors of about 1.7 and 4 makes the relative valley in between the peaks deeper. Further, the distance between two peaks shows no decrease with the increase in volume, tying up well with an intuitive expectation for a first order phase transition. Evidently, the above qualitative method of identifying the order of the phase transition and obtaining T_c becomes unreliable if the gap in the order parameter is too small.

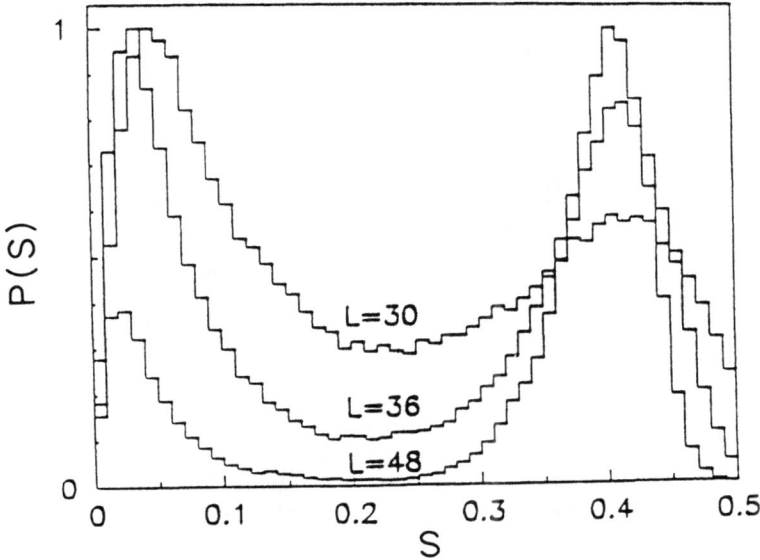

Fig. 3: The probability distribution of the order parameter in the critical region of the three-dimensional three-state Potts model on 30^3, 36^3 and 48^3 lattices ($\beta = 0.367$). From Ref. 48.

Finite size scaling[49] provides a quantitative tool for investigating phase transitions in such cases. Originally proposed[50] for second order phase transitions, it has now been extended[51] to first order phase transitions as well, albeit the latter is more phenomenological in nature. Finite size scaling predicts the shift in the transition temperature T_{c,N_σ}, and the width of the transition region $\Delta T_{c,N_\sigma}$ to be

$$T_{c,N_\sigma}^{-1} - T_{c,\infty}^{-1} \propto N_\sigma^{-y_T}$$
$$\text{and} \quad \Delta T_{c,N_\sigma} \propto N_\sigma^{-y_T}, \tag{57}$$

while the peaks of the susceptibility, $\chi_{\max}$ ($\chi = \langle L^2 \rangle - \langle L \rangle^2$ where $\langle L \rangle$ is the order parameter), and the specific heat, $C_{v,\max}(C_v \sim \partial\epsilon/\partial T)$ to be

$$\chi_{\max} \sim N_\sigma^{\rho_s}$$
$$C_{v,\max} \sim N_\sigma^{\rho_{sp}}. \tag{58}$$

The exponents y_T, ρ_s and ρ_{sp} characterize the order of the phase transition. For a first order transition, they are trivial, being governed by the naive dimensionality,

$$y_T = \rho_s = \rho_{sp} = 3 \quad (\text{or } d, \text{ in general}), \tag{59}$$

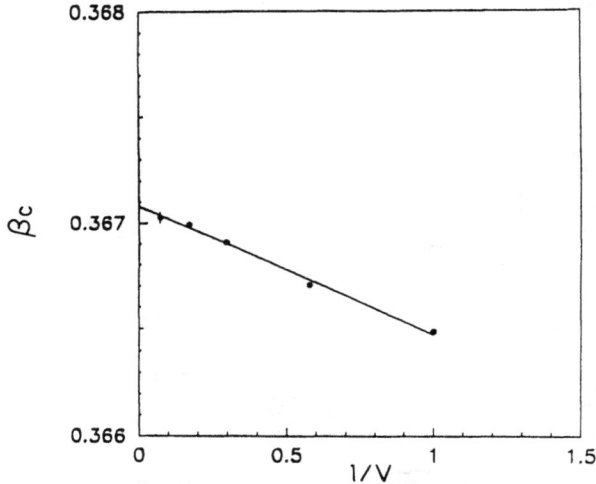

Fig. 4: Finite size scaling of T_{c,N_σ}^{-1} for the three-dimensional three-state Potts model. The volume is measured in units of 20^3. From Ref. 48.

while for a second order phase transitions,

$$y_T = 1/\nu, \; \rho_s = \gamma/\nu \text{ and } \rho_{sp} = \alpha/\nu \tag{60}$$

where γ, α and ν are the standard critical exponents[49] for susceptibility, specific heat and the correlation length. In general, they have values which are smaller than those in Eq. (59) and are arbitrary. Fig. 4 shows the data[48] for the 3-d, 3-state Potts model for T_{c,N_σ}^{-1} as a function of $N_\sigma^3 = V$. A similar linear scaling with volume is also seen for $\Delta T_{c,N_\sigma}$. They together clearly support the first order nature seen in Figs. 2 and 3 since the corresponding exponents, with great precision, are seen to coincide with Eq. (59). The data in Fig. 4 further provides a precise determination of $T_{c,\infty}$.

One can exploit the double-peak structure of the probability distributions of physical observables to derive further tests for a first order phase transitions. Thus, e. g., the gap ΔL between the two peaks in Fig. 3, yields[52] the proportionality factor in Eq. (58) : $\chi_{max}/V = (\Delta L/2)^2$. Testing this relation, one can also check how reliably the discontinuity estimated from the double-peak structure approximates the infinite volume limit. Higher order cumulants also share the above property. One particularly useful cumulant[51] is given by

$$V_E = \frac{\langle H^4 \rangle}{\langle H^2 \rangle^2} - 1, \tag{61}$$

where H is the hamiltonian. It can be easily checked that $\lim_{V \to \infty} V_E(T_c) = \left[\Delta E^2 / (E_+^2 + E_-^2) \right]^2$, where $\Delta E = E_+ - E_-$ is the latent heat and E_+, E_- are the expectation values of H in the upper and lower phase respectively. Thus, V_E exhibits a peak which persists as volume grows if the phase transition is of first order; absence of a peak in the limit of infinite volume signals a second order phase transition.

4. Phase Diagram of Quantum Chromodynamics

Due to the conceptual and technical difficulties with fermions on the lattice, discussed above in sects. 2.3.1 and 3.2.1, the simulations of QCD at finite temperature started out neglecting the quarks entirely. The self-interacting nature of gluons gives a hope that this quenched approximation still yields a reasonable picture of the true thermodynamics. Formally, the limit $N_f \to 0$ or $m_f a \to \infty$ in Eqs. (32) and (25) reduces the partition function of the full theory to that of the pure gluons. Thus, a logical next step to the quenched approximation is to obtain the phase diagram in the (N_f, m_f) space. In principle, we need to know the nature of the phase transition and the corresponding T_c at only one point of this multi-parameter space. However, we do not know the parameters m_f and N_f so precisely. Moreover, one would like to compare the simulation results with the theoretical expectations, discussed in sect. 2.2, to enhance our understanding.

In the next three subsections we will discuss the results for the phase diagram of the quenched QCD or the $SU(3)$ gauge theory, the full QCD with staggered fermions and the Wilson fermions respectively. More emphasis will be given on the recent results. Review of earlier work, including that for other gauge groups such as $SU(2)$ and $SU(4)$, can be found in Refs. 11 and 39.

4.1. Pure Gauge

The ease of simulating pure $SU(3)$ gauge theories on the computer has resulted in the use of both the biggest lattices and high statistics in the investigations of pure gauge thermodynamics. Consequently, these results are the best in terms of control of finite size effects and finite lattice spacing effects. Further, several qualitative issues about the determination of the order of the phase transition have been addressed in these simulations. Early results[11], obtained on small lattices and with moderate statistics, indicated a first order phase transition for $N_\beta = 2 - 4$. With the advent of QCD-machines, substantially larger spatial lattices were employed for numerical simulations. Although, for a while this lead to a controversy about the order of the phase transition, it seems that at least for $N_\beta = 4$ the transition is of first order.

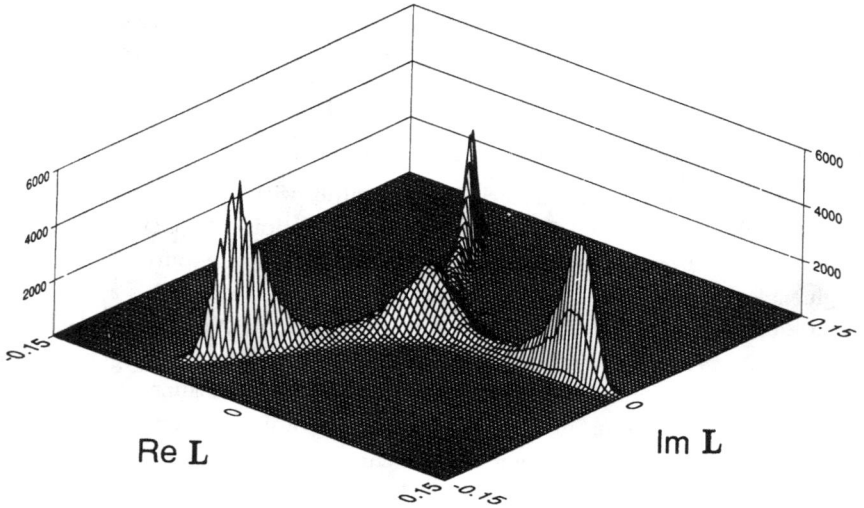

Fig. 5: Probability distribution of the order parameter L in the complex L plane on a $24^2 \times 36 \times 4$ lattice at $\beta = 5.6925$. From Ref. 53.

Fig. 5 shows the probability distribution[53] of the order parameter L at $\beta = 5.6925$ on a $24^2 \times 36 \times 4$ lattice in the complex L plane. Recall that the ordered low temperature phase is $Z(3)$-symmetric with $\langle L \rangle = 0$ whereas the disordered high temperature phase breaks $Z(3)$ and has $\langle L \rangle \neq 0$. One sees in Fig. 5 one peak near $L = 0$ and three $Z(3)$-symmetric peaks equidistant from the origin. This four peak structure is characteristic of a first order phase transition in a finite volume where the system tunnels from one $Z(3)$-vacuum to another and spends roughly the same time in all. In fact, this causes $\langle L \rangle = 0$ even in the deconfined phase. One can lift the degeneracy of the broken $Z(3)$ vacua by applying an external symmetry breaking field, as in the case of the Ising model. Alternately, one obtains $\langle L \rangle$ by projecting it on the nearest cuberoot of unity before averaging

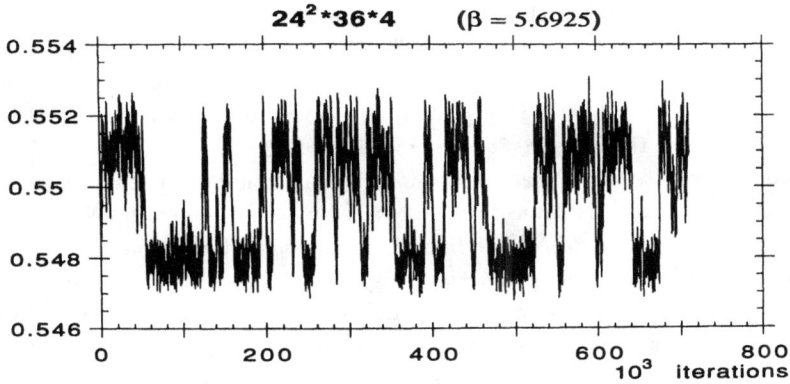

Fig. 6: Monte Carlo evolution of the average plaquette on a $24^2 \times 36 \times 4$ lattice at $\beta = 5.6925$. From Ref. 53.

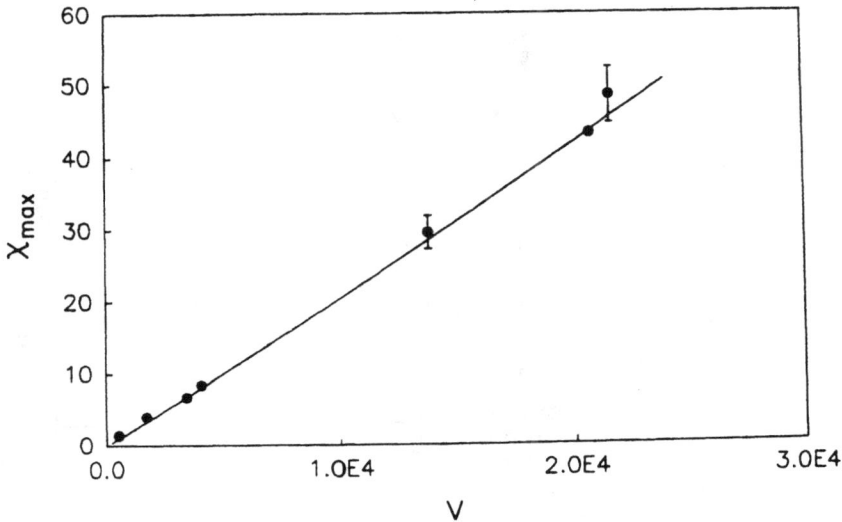

Fig. 7: The peak height of the susceptibility $\chi(L_{proj})$ as a function of volume. Data from Refs. 53 and 54.

over the MC sweeps. Fig. 6 shows the evolution of the plaquette in the same run where a large number of tunnellings, as seen in Fig. 2, are clearly evident. Similar two state signals have also been seen by other groups[54,55], albeit with a somewhat lower statistics. One can obtain the susceptibility, $\chi(L_{\mathrm{proj}})$, from these runs and check for the exponents of its peak. Fig. 7 shows $\chi_{\max}$ as a function of V/N_β^3, where the data from Refs. 53 and 54 is displayed. The data of Ref. 56, which is in agreement with those in Fig. 7, is not shown for clarity. A linear trend is evident; a fit to the form in Eq. (58) yields $\rho_s = 3.114 \pm 0.045$ for $V/N_\beta^3 > 50$, which is very close to the space dimensionality. A further consistency check on the first order nature of the transition is provided by Fig. 8 where $\beta_c(N_\beta = 4, N_\sigma)$ is shown as a function of N_β^3/V. Here β_c is defined as the coupling at which the L_{proj}-susceptibility or the plaquette-susceptibility has a peak. The data is from Refs. 53, 54 and 57; comparing Fig. 8 with the corresponding figure for the 3-state Potts model, Fig. 4, one sees, however, that the current precision in determining the transition coupling β_c is still not good enough to go beyond a mere consistency check on the order of the phase transition.

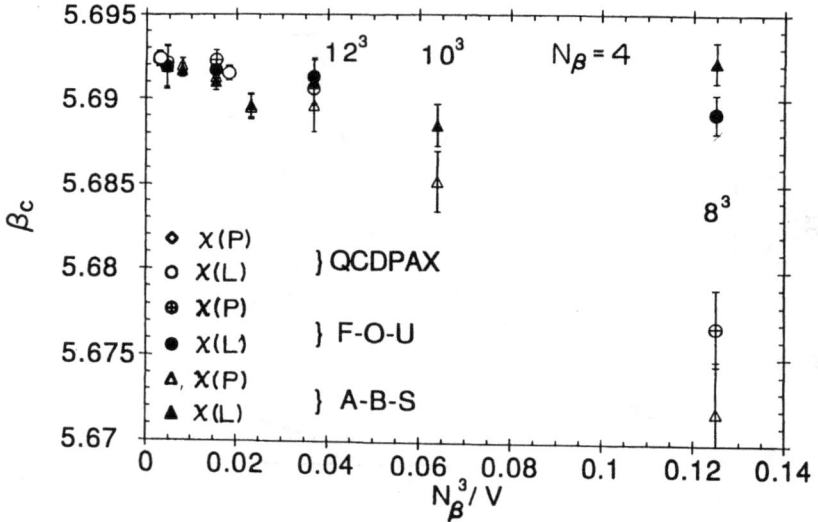

Fig. 8: Finite size scaling of β_{c,N_σ}^{-1} for $N_\beta = 4$. Data from Refs. 53, 54 and 57.

It may be remarked here that on a finite lattice the transition coupling (or temperature) can be defined in various different ways. In general, these different definitions may yield different values for β_c. In the thermodynamic limit of infinite volume, however, all of them must lead to a unique $\beta_c(N_\sigma = \infty)$. Such a check has been carried out for the $SU(3)$ spin model[52] by extrapolating sets of β_c's obtained by different definitions. In view of the precision of the data in Fig. 8, a similar check for the $SU(3)$ gauge theory cannot be very stringent. Nevertheless, in addition to the peak positions of other susceptibilities[54,57], shown in Fig. 8, the structure of probability distributions of the order parameter[55,58] (e. g., β at which there is equal population in the two peaks, or a certain fraction r in the deconfined peak) has also been employed to obtain alternative estimates of β_c. Assuming that $\beta_c(N_\sigma)$ scales with the exponents given by Eq. (59), a consistent value for $\beta_c(N_\sigma = \infty, N_\beta = 4) = 5.69243(20)$ is obtained[53] from extrapolations of all these different estimates[54,57,58], provided data from lattices with $N_\sigma \geq 14\text{-}16$ are used for the extrapolation. Finally, several groups[54,57,59] have attempted to determine the order of the $SU(3)$ phase transition by studying the cumulant V_E, defined in Eq. (61), by choosing $H = P \equiv (\bar{P}_\sigma + \bar{P}_\beta)/2$ where $\bar{P}_\sigma$ and $\bar{P}_\beta$ are defined in Eq. (41). All of them find a peak in V_E in the infinite volume limit, signalling a first order phase transition for $N_\beta = 4$. Further, the height of the peak agrees with the estimate obtained by using ΔP and P_+, P_- in the corresponding formula in sec. 3.2.2.

A familiar characteristic of a second order phase transition is a diverging correlation length: $\xi \sim (T - T_c)^{-\nu}$. On the other hand, ξ is expected to stay finite at a first order transition. On finite lattices, ξ is limited by the size of the lattice. The APE group[60] measured the correlations, $\langle L(0)L^\dagger(r)\rangle$, and obtained the correlation length by fitting them to an exponential, $\exp(-r/\xi)$, on $N_\sigma^2 \times 4N_\sigma \times 4$ lattices for $N_\sigma = 8, 12$ and 16. They found $\xi \propto N_\sigma$ suggesting a second order phase transition. This spurred a lot of activity in the field. In fact, most of the results described above for bulk thermodynamic quantities, except those of Ref. 55, were obtained after the APE results and they all seem to indicate a first order transition, consistent with all the early results[11].

The resolution of the discrepancy between the APE result and the early results came first through investigations of correlation lengths near first order phase transitions in simpler systems: the $Z(3)$ spin model[48] and the $SU(3)$ spin model[52]. For both these models, the finite size scaling analysis of the bulk thermodynamic quantities revealed a first order phase transition. However, the analysis of correlation lengths turned out to be subtle due to the presence of a tunnelling correlation length[61]. For $\beta > \beta_c$, the tunnelling correlation dominates the above correlation functions. Close to β_c, the correlation function has contributions from both the physical correlation length, to which the above remarks about finite size scaling apply, and the tunnelling correlation length. One needs to exercise caution in drawing conclusions since the two have presumably different finite size scaling behaviour. Indeed, one found[48] $\xi \propto N_\sigma$ in the $Z(3)$-spin model at a certain β for $N_\sigma \leq 36$ but not if $N_\sigma = 48$. One can envisage a variety of prescriptions[48,62,63] to extract the physical correlation length from the correlation function data. All of them indicate a finite physical correlation length with a growing lattice size. However, it remains unclear whether with all the uncertainties, both theoretical and computational, such an analysis can yield a clear answer about the order of the phase transition. Subsequently, such an analysis was carried out for the $SU(3)$ gauge theory as well[54] and remarkably similar results were obtained. In particular, $\xi \propto N_\sigma$ was found to be true at $\beta = 5.69$, in agreement with the results of the APE collaboration, provided $N_\sigma \leq 16$. ξ decreased, however, for $N_\sigma = 24$ and 28, a result now confirmed by the APE group[64].

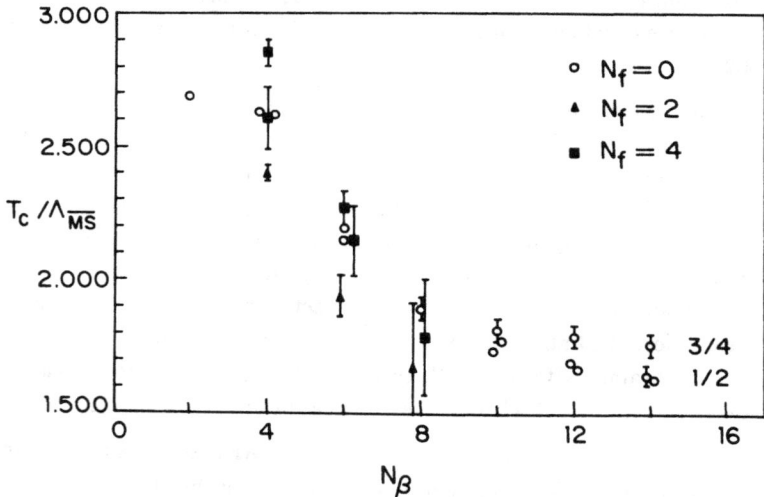

Fig. 9: $T_c/\Lambda_{\overline{MS}}$ as a function of the temporal lattice size N_β for the quenched QCD (circles) and the full QCD with 2 (triangles) and 4(squares) of dynamical staggered quarks. Data for quenched QCD from Refs. 58 and 65, for 2-flavour QCD from Refs. 66 and 67 and for 4-flavour QCD from Refs. 68 and 69.

Having obtained $\beta_c(N_\sigma = \infty, N_\beta = 4)$ and having determined that the phase transition for $N_\beta = 4$ is of first order, one would like to know how relevant these results are to the continuum physics. As discussed earlier, one needs to ensure $a \to 0$ and $N_\beta \to \infty$ to check whether $T_c = \left[N_\beta a \left(\beta_{c,N_\beta}\right)\right]^{-1}$ has a well-defined value in that limit. This means that the entire exercise above needs to be repeated for various large N_β. Noting the range of N_σ needed for the $N_\beta = 4$ analysis, $N_\sigma \geq 4N_\beta$, one sees how rapidly the computational task becomes a formidable one. Not surprisingly, a satisfactory check of the scaling limit has not been done even for the pure gauge theories so far. There have been two attempts[58,65] to simulate lattices with N_β up to 14 but both were confined to N_σ/N_β in the range 1.125 – 2, with lower values for larger N_β. In addition to the large finite volume effects, the β_c determination in these simulations had to face the fact that $\langle L \rangle$ becomes smaller as N_β increases (due to an ultraviolet divergence). Refs. 58 and 65 defined β_c by demanding that a fraction $r(= 1/2$ or $3/4)$ of the configurations in the Monte Carlo run at β_c should be in the deconfined phase, corresponding to a significant value for the Polyakov loop L. Given a set of N_β and the corresponding β_{c,N_β}, there are various ways one can check for scaling. Assuming Eq. (50) to be valid, Fig. 9 shows $T_c/\Lambda_{\overline{MS}}$ as a function of N_β. Here we have used one loop perturbation theory[70] to relate Λ_L to $\Lambda_{\overline{MS}}$:

$$\frac{\Lambda_{\overline{MS}}}{\Lambda_L} = \begin{cases} 28.81 & N_f = 0 \\ 43.88 & N_f = 2 \\ 56.75 & N_f = 3 \\ 76.45 & N_f = 4 \end{cases} \tag{62}$$

The data in Fig. 9 seems to attain a plateau for $N_\beta = 10 - 14$, although even for those values, the $r = 1/2$ definition still yields a significant slope to $T_c/\Lambda_{\overline{MS}}$. The plateau can be interpreted as an evidence that the asymptotic scaling relation in Eq. (50) is valid for $N_\beta \geq 10$ for $\beta \geq 6.2$.

Another way to check scaling is to construct dimensionless ratios of physical quantities at a given coupling and see if they are independent of the lattice artifacts. While this must be true for small enough lattice spacing and large enough lattices, if the continuum limit exists, it has the potential of probing for continuum physics beyond the regime of validity of Eq. (50). The string tension, σ, has been studied on a similar range of β and found to exhibit the same behaviour as $T_c/\Lambda_{\overline{MS}}$ (see Fig. 3.7 of Ref. 11 and Fig. 8 of Ref. 71). Indeed, $T_c/\sqrt{\sigma}$ is constant over a much larger range of β indicating a wider scaling region. From the asymptotic values of T_c/Λ_L and $\sqrt{\sigma}/\Lambda_L$ from Refs. 58 and 72, one obtains

$$T_c = (0.56 \pm 0.10)\sqrt{\sigma} = 235 \pm 42 \ \text{MeV}, \tag{63}$$

which is in agreement with the value in Ref. 71 and is true for $N_\beta \geq 4$ already. Alternatively, one can use hadron masses to construct such ratios. There are,

88

however, additional problems in that case since[73] i) the results from simulations with staggered fermions agree with those for Wilson fermions only for $\beta \geq 6.2$ and ii) M_N/M_ρ is $\sim 10-20\%$ higher than the experimental value in the current simulations. Thus T_c in MeV will change depending upon which of the two is used for normalization. Nevertheless, using the data from Ref. 73 and the experimental values of the ρ and the nucleon mass, one obtains

$$T_c^\rho = (0.325 \pm 0.039)\, M_\rho = 250 \pm 30 \ \text{MeV}$$
$$T_c^N = (0.240 \pm 0.048)\, M_N = 226 \pm 45 \ \text{MeV},$$

(64)

both of which are consistent with Eq. (63) above.

4.2. Staggered Fermions

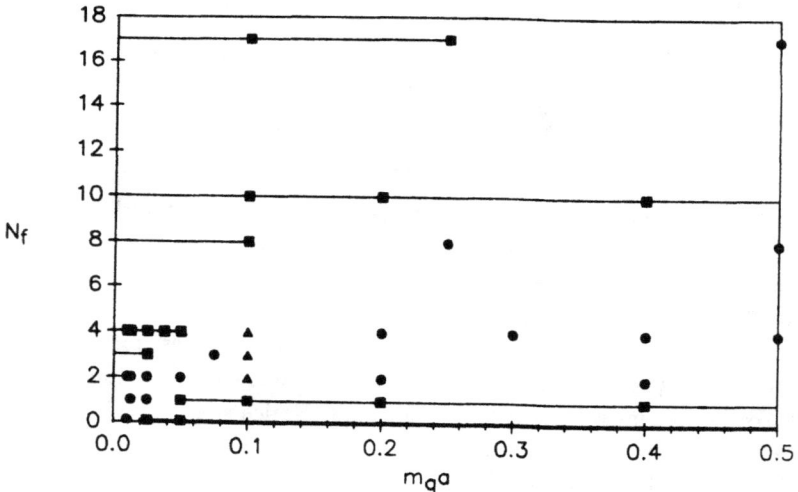

Fig. 10: Phase diagram of full QCD on $N_\beta = 4$ lattices in the critical surface spanned by the quark mass ma and the number of flavours N_f. Updated version of the figure in Ref. 74. The explanation of the symbols and the references for them are given in the text.

Due to the expectation that dynamical fermions will affect the thermodynamics of quantum chromodynamics in a significant way, as sketched in Fig. 1, a lot of simulations of full QCD at finite temperature have been carried out. Motivated by the search for chiral phase transition and its detailed investigation, most of the simulations have used staggered quarks. Due to the difficulties of simulating fermions, only moderate size lattices, typically $8^3 \times 4$, have been used, although recently spatial lattices up to 16 sites in each direction and temporal lattices with 8 sites have also been used in order to employ finite size scaling techniques and to move deeper into the scaling region. The main attempt so far has been to obtain a glimpse of the true phase diagram and to compare it with the expected phase diagram in Fig. 1. Fig. 10 summarizes our present knowledge of the QCD phase diagram on an $N_\beta = 4$ lattice. The squares denote first order transitions seen in numerical simulations while the circles indicate a lack of a first order transition at that $(N_f, m_q a)$. The triangles correspond to disputed first order phase transitions, which have been claimed as first order phase transitions by only some authors, as explained below. Also shown in Fig. 10 are the anticipated first order chiral or deconfinement phase transition lines, where the low(high) mass first order phase transitions have been suggestively joined to $ma = 0(\infty)$. Before we discuss the phase diagram in detail, and the subsequent issue of the continuum limit for some of the points in it, we recall the discussion in sec. 2.3.1 regarding the flavour interpretation for the staggered fermions and remark that the effective number of light flavours is expected to change in the continuum limit and $N_\beta = 4$ may be too coarse a lattice.

4.2.1. $N_f = 4$

The primary reason that this physically not-so-interesting case has been extensively investigated is because $N_f = 4$ is more natural for the staggered fermions (see sect. 2.3.1) and is therefore amenable to the exact HMC algorithm described in sec. 3.2.1. So far, the phase diagram has been explored more completely for small N_β; only small quark masses have been employed in simulations on lattices with larger N_β.

There is now substantial evidence on $N_\beta = 4$ that a first order chiral phase transition exists for small enough $m_q a$ which weakens as $m_q a$ is increased. Further, no first order transition exists for some larger $m_q a$, signalling the presence of a gap in the two lines in Fig. 1, although results on smaller lattices[78] obtained by using the HMC algorithm, suggested no such gap. Fig. 11 shows the order parameters, $\langle \bar\psi\psi \rangle$ and $\langle L \rangle$, on an $8^3 \times 4$ lattice for $ma = 0.1$, obtained by using the pseudofermion[75], the Langevin[76] and the microcanonical[77] method to include the fermionic effects. There are also results obtained by using the HMD[66] and the HMC[78,79] algorithms which are consistent with those in Fig. 11 but are not shown for the sake of clarity. The general qualitative features, which are common

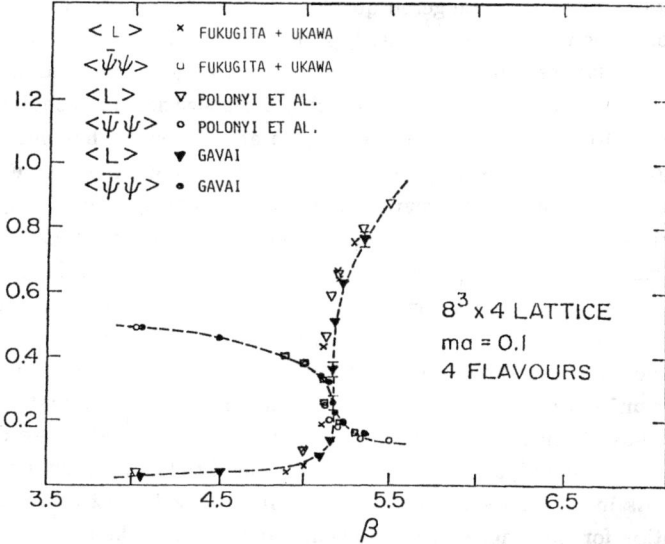

Fig. 11: The chiral condensate $\langle\bar\psi\psi\rangle$ and the average Polyakov loop $\langle L\rangle$ as a function of $\beta = 6/g^2$ on an $8^3 \times 4$ lattice for full QCD with four staggered flavours of mass $ma = 0.1$ Data is from Refs. 75, 76 and 77.

also for $N_f = 2$, 3 and $ma \leq 0.1$, are i) $\langle L\rangle$ rises from ~ 0 to nonzero values at the same β_c where $\langle\bar\psi\psi\rangle$ drops significantly, ii) an overall agreement between the different algorithms despite their differences in systematic biases and iii) a relative shift of ~ 0.04 in β_c for the pseudofermionic simulations. The behaviour of $\langle\bar\psi\psi\rangle$ is suggestive of a chiral transition although the change in $\langle L\rangle$ is suggestive of deconfinement as well. The order of the phase transition at $ma = 0.1$ is controversial. Simulations on 4^3, 6^3 and 8^3 lattices have yielded[78,79] a first order phase transition using the HMC algorithm, although its weakening with volume is perhaps also indicated by the data. Other simulations[66,75,77], using approximate algorithms to incorporate fermions found no metastabilites, except Ref. 76. As we describe below, this may be due to the proximity of $m_{CH}a$ near 0.1, which necessitates bigger lattices and more careful finite size scaling analysis.

For ma smaller than 0.1, all simulations seem to suggest a clear first order chiral phase transition. In particular, clear metastabilities have been observed in both L and $\bar\psi\psi$ for $N_\beta = 4$ at $ma = 0.01^{[68]}$, $0.0125^{[80]}$, $0.025^{[66,68,80,81,82,83]}$, $0.0375^{[68,80]}$ and $0.05^{[68,81]}$. In a finite size scaling analysis Ref. 83 used high statistics data for

L for $ma = 0.025$ and $N_\sigma = 4, 6, 8, 10$ and 12 and found an increasingly strong separation of the two phases with growing N_σ. Analysing further the susceptibility $\chi(L)$ and the cumulant V_L(obtained by using $\langle L \rangle$ in Eq. (61)), a quantitative evidence indicative of a first order phase transition was obtained: for $4 \leq N_\sigma \leq 8$, $\chi_{\max} \sim N_\sigma^3$ and the decrease of $V_L^{\max}$ was consistent with a nontrivial thermodynamic limit. The simulations of Ref. 68 were done for the same quark mass and N_β but even bigger spatial size: $N_\sigma = 16$. Their results are in agreement with the trend seen by Ref. 83. Ref. 68 also finds a decreasing $\Delta\langle\bar{\psi}\psi\rangle$ with increasing ma from their simulations on a $16^3 \times 4$ for $ma = 0.01, 0.025, 0.0375$ and 0.05. A linear extrapolation of their data leads[84] to an estimate of $m_{\mathrm{CH}}a \sim 0.07 - 0.08$ for $N_\beta = 4$, where $m_{\mathrm{CH}}a$ is the endpoint of the line of first order chiral phase transitions shown in Fig. 1.

These simulations for low ma have been extended to $N_\beta = 6$ and 8, although the corresponding N_σ ranged between 12 and 16. In both the cases one found no metastabilities at $ma = 0.025$[68,69] and a two-state signal at $ma = 0.01$[68,69] although again for smaller N_σ the $ma = 0.025$[80,85] data were interpreted as corresponding to a first order transition. Note that the decrease in $m_{\mathrm{CH}}a$ with N_β is natural if m_{CH}/T_c is to scale. In fact, from the world data on $\langle\bar{\psi}\psi\rangle$, one can infer[84] that m_{CH}/T_c scales for $N_\beta = 6-8$ and is about 0.12. A natural question is whether the $T_c/\Lambda_{\overline{MS}}$ shows asymptotic scaling or whether a ratio like T_c/m_ρ stays constant in this range of N_β. Ideally, one would like to check scaling by holding the quark mass constant in physical units. However, due to the present paucity of the data one prefers to extrapolate the existing (β_c, ma) data to $ma = 0$ for a given N_β and then to check for scaling for massless quarks. Extrapolating the $\beta_c(ma)$ values for the largest N_σ for each N_β to $\beta_c(0)$, and using Eqs. (50) and (62) one obtains $T_c/\Lambda_{\overline{MS}}$ as a function of N_β, shown in Fig. 9 by squares. It is seen to be remarkably close to the quenched QCD data, suggesting an onset of asymptotic scaling at similar N_β here as well. Using the hadron masses calculated at the relevant β, Ref. 69 obtained

$$T_c^\rho = (0.12 \pm 0.02) \, m_\rho = 93 \pm 15 \ \ \mathrm{MeV}$$
$$T_c^N = (0.09 \pm 0.01) \, m_N = 85 \pm 11 \ \ \mathrm{MeV}$$

(65)

for $ma = 0.01$ and $N_\beta = 8$. Unfortunately, a lack of the knowledge of the hadronic spectrum at the β_c for $N_\beta = 4$ and 6 prevent us from checking whether these ratios scale. Note, however, that the estimates in Eq. (65) are down by a factor of about two compared to the quenched QCD estimates in Eq. (64) and it also implies an $m_{\mathrm{CH}} = 10\text{-}12$ MeV for four flavours.

For $ma > 0.1$, the earlier claims[76,78] of a first order transition at $ma = 0.2, 0.5$ and 1.0, on typically $8^3 \times 4$ or smaller lattices, could not be substantiated in more recent[86,87] works on lattices up to $16^3 \times 4$. Instead, large fluctuations and very long correlations were seen[87] at $ma = 1.0$ on a $16^3 \times 4$ lattice. Thus, it appears that the first order line of deconfinement transition ends somewhere above $ma = 1$ but no precise estimate of m_D is available. Nevertheless, a clear gap between the two endpoints, m_{CH} and m_D, is indicated for $N_\beta = 4$. It would be interesting to determine if this persists on lattices with larger N_β and if they scale.

4.2.2 $N_f = 2$

This physically interesting case has a further appeal: the effective σ-model considerations do not lead to any prediction for the corresponding chiral phase transition. Again, not much work has been done to locate $m_D a$ in this case: a pseudo-fermion simulation[88] found metastabilities for $ma = 2.5$ and 1.67 on an $8^3 \times 2$ lattice, indicating $m_D a < 1.67$, while a Langevin simulation[89] on an $8^3 \times 4$ lattice found $0.4 < m_D a < 1.0$. Allowing for the change in N_β, these two estimates are consistent with each other. However, in view of the expected decrease in $\langle \Delta L \rangle$, as m_q decreases from infinity, a proper finite size scaling analysis is necessary to determine at which ma the phase transition is still first order. On the low mass end, $0.01 < ma < 0.1$ there has been some controversy about the order of the phase transition on $8^3 \times 4$ lattices at various quark masses. A Langevin[89] and an HMD[90] simulation found a first order for $ma = 0.1$ while, using the same algorithms, Refs. 66 and 74 found no evidence for this. This is an indication of the subjective elements in such analyses, and call for a thorough finite size scaling analysis. Again Refs. 74, 91 and 92 claimed a first order chiral transition for $ma = 0.025, 0.02$ and 0.0125 using respectively a Langevin, an exact and a HMD algorithm to include dynamical fermions while Refs. 66 and 93 found none for $ma = 0.0125$ and 0.025. Recent simulations[83,94] on larger lattices have resolved the controversies for $ma = 0.0125$ and 0.025: no metastabilities are seen on $12^3 \times 4$ and $16^3 \times 4$ lattices for $ma = 0.0125, 0.025$ and 0.01, 0.025 respectively, although the data[83] on 6^3 and 8^3 lattices could be interpreted to have a double-peaked histogram structure for $ma = 0.025$. Further, finite size scaling tests were applied to the $\chi_{max}(L)$ and V_L. While the lattice sizes employed for these tests are still not large enough to rule out $\chi_{max} \propto N_\sigma^3$ conclusively, there is now a considerable evidence (see Fig. 12 of Ref. 95) which suggests a lack of a discontinuous transition for the mass range $0.01 \leq ma \leq 0.025$. As seen in Fig. 10, the point at $ma = 0.1$ now stands out in the diagram, as the simulations on its both sides indicate a lack of first order but no attempt has been made to resolve the controversy there. Confirmation of a first order phase transition at $ma = 0.1$ would be an unexpected feature compared to Fig. 1, assuming that m_D lies in the range mentioned above.

In order to check scaling, and obtain T_c in MeV, simulations[66,67] were made on a $10^3 \times 6$ lattice for quark masses $ma = 0.025$, 0.05 and 0.1 and on a $12^3 \times 8$ lattice for $ma = 0.025$ and 0.0125. Once again, one can obtain $\beta_c(ma = 0)$ from the cross over points determined in these simulations for $N_\beta = 4$, 6 and 8. Using Eqs. (50) and (62) one converts them to $T_c/\Lambda_{\overline{MS}}$. These data are displayed in Fig. 9 by triangles and they too exhibit the same trend as the $N_f = 0$ and 4 data discussed earlier. It thus appears that the change in T_c due to the shift in β_c in the case of the dynamical fermions is compensated by the corresponding changes in Eqs. (50) and (62), resulting in a rather mild N_f dependence in $T_c/\Lambda_{\overline{MS}}$. However, the resulting change in T_c in MeV appears to be substantial, as we saw in sect. 4.2.1 and will see below. In order to obtain T_c in MeV, hadron masses were obtained[96] at the critical couplings for $N_\beta = 4$ and 6 on $6^3 \times 24$ and $8^3 \times 24$ lattices respectively at the corresponding quark masses. Extrapolating the data for the ρ and the nucleon mass to the physical quark mass, Ref. 96 obtained

$$T_c^\rho = \begin{cases} 142 \pm 6 & \text{MeV} \\ 164 \pm 25 & \text{MeV} \end{cases} \quad \text{and} \quad T_c^N = \begin{cases} 109 \pm 6 & \text{MeV} \\ 117 \pm 14 & \text{MeV} \end{cases} \quad (66)$$

for $N_\beta = 4$ and 6 respectively. Recent improved estimates[97] for T_c^ρ for $N_\beta = 6$ and 8 are also in good agreement with Eq. (66). Comparing with Eqs. (64) and (65), one notices that the critical temperature for $N_f = 2$ is roughly in between the two. Note further that the estimate in Eq. (66) translates $ma = 0.01$, where a lack of first order transition has been observed[94], to a physical quark mass $\sim$ 4-6 MeV which is tantalisingly close to the real value. However, taking the behaviour of the data in Fig. 9 as a guide, one anticipates the asymptotic scaling region to commence for much larger N_β and therefore the estimates in Eq. (66) may suffer from finite lattice spacing effects.

4.2.3. $N_f = 3$ and $2 + 1$

A better approximation to the real world may be to assume $m = m_u = m_d \ll m_s$, where u, d and s denote up, down and strange quark respectively. This $2 + 1$ flavour theory can be simulated by having 2 fermion terms in the action, corresponding to ma and $m_s a$. Clearly, $r = ma/m_s a \to 0$, for $ma \neq 0$, will correspond to $N_f = 2$ discussed above, while $r = 1$ corresponds to $N_f = 3$. Simulations for $(r, ma) = (1, 0.025)$ have lead to the conclusion of a first order chiral phase transition on smaller[74,92,98] lattices. This has now been confirmed[94] with much better statistics on a $16^3 \times 4$ lattice. For larger quark masses simulations have been made only on $8^3 \times 4$ lattices and no metastabilities were seen[99] for $(r, ma) = (1, 0.075)$ and $(1,0.1)$. This suggests that the end point of the line of chiral phase transitions for 3 flavours lies in the range, $0.025 < m_{CH}a < 0.075$, although a study[100] on a smaller lattice, $6^3 \times 4$, did find metastabilities in this range.

For intermediate values of r, an early evidence[98] for a first order phase transition for $(0.25, 0.025)$ on an $8^3 \times 4$ lattice seems to be absent in simulations on a $16^3 \times 4$ lattice with much higher statistics[94]. Similarly, the metastabilities observed for $(0.05, 0.0125)$ on an $8^3 \times 4$ lattice[101] are absent[102] on a $12^3 \times 4$ lattice. It thus appears that there is no definitive evidence, which survives in the large volume limit, for a first order phase transition for any (r, ma) with $r < 1$ even for $N_\beta = 4$. No results for larger N_β are available.

4.2.4 $N_f = 0.1$ and $N_f = 1$

These simulations are of purely theoretical interest: $N_f = 0.1$ to study the quenched approximation limit of $N_f \to 0$ and $N_f = 1$ to check the prediction of the σ-model[17]. From the presence of two-state signals in the Polyakov loop L, a first order transition was inferred[74] for $N_f = 0.1$, $ma = 0.05$ and 0.025 on an $8^3 \times 4$ lattice. No metastabilities were seen at $ma = 0.01$, leading to the interpretation that the deconfinement phase transition extends up to $ma = 0.025$ and that m_D — the end point of the line — lies between 0.025 and 0.01. Two-state signals and double-peaked structure of the histograms at β_c were employed[89,90,74] to infer a first order phase transition also for $N_f = 1$ and $ma = 0.05$, 0.1, 0.2 and 0.4 on an $8^3 \times 4$ lattice but no evidence of metastabilities was seen[74] for $ma = 0.025$ and 0.0125, again leading to an interpretation that $0.025 < m_D a < 0.05$. Furthermore, these results are also consistent with the absence of a chiral phase transition, as predicted by the σ-model analysis.

4.2.5. $N_f > 4$

The interest in these simulations comes from an interesting speculation[103]. From Eq. (15), one finds $b_0 < 0$ if $N_f > 11N/2$. However, $b_1 < 0$, already for $N_f > 34N^3/(13N^2 - 3)$. Thus, at least perturbatively, the β-function can have a zero at $g^2 > 0$ for $11N/2 > N_f > 34N^3/(13N^2 - 3)$, or for $16 \geq N_f > 8$ if $N = 3$. If true, one can expect the existence of new phases already at zero temperature. On the other hand, the trend for lower N_f suggests an even stronger chiral phase transition which, in turn, will have a larger m_{CH}. This raises the possibility of having a first order phase transition for all ma for some N_f or having two phase transitions in a range of quark masses. Note that a finite temperature transition is expected to shift in β according to Eq. (50), as N_β is changed, while a zero temperature transition should have only a finite volume shift. All large N_f simulations[75,89,103,104] done so far have found strong first order transitions, as shown in Fig. 10. While a shift in β consistent with Eq. (50) has been found[89] for $N_f = 10$, unlike the $N_f \leq 4$ cases, the β_c obtained from $\langle \bar{\psi}\psi \rangle$ was found[104] to be different than that from $\langle L \rangle$ for $N_f = 8$. The former even showed very little shift in going from $N_\beta = 6$ to 8. Similar results, showing no shift in β_c, have also been reported for larger N_β for the staggered fermions[105] of 8 flavours, and interestingly enough, for Wilson fermions[106] of $N_f > 7$ in extreme strong coupling region. While they do suggest that the β-function may indeed have a zero for such large N_f-values, further investigations clarifying the physical number of flavours in the staggered fermion simulations and the nature of the chiral symmetry in the Wilson fermion simulations are clearly needed to be sure of the significance of these results.

4.3. Wilson Fermions

As a test of the reliability of the lattice results, one needs to obtain them with various different forms of action and check that they all yield the same physics. In view of the complementary nature of the Wilson fermions to the staggered fermions discussed above, and the fermion problem, mentioned in sec. 2.3.1, it is even more necessary to perform finite temperature simulations with Wilson fermions and compare them with those in sec. 4.2. However, formidable difficulties associated with the Wilson fermions have made such a check elusive and inconclusive so far.

The qualitative feature of weakening of the deconfining transition as κ is increased from zero (see Eq. (29)) has been long seen[107,108] and is in agreement with the large mass behaviour of the staggered fermions. On an $N_\beta = 2$ lattice, even quantitative agreement with the results of Ref. 107 was found[88] in regions where the corresponding methods to incorporate fermions were expected to be similar. Perhaps not so unexpectedly, the attempts[108,109,110] to look for a chiral transition yielded different results. On a $5^3 \times 4$ lattice[108] with $N_f = 4$ and later on a $8^3 \times 4$ lattice[111,112] with $N_f = 2$, it was found that in the (κ, β) plane the line of phase transitions did not cross the $\kappa_c(\beta)$ line, defined by $m_\pi^2 = 0$ on a symmetric lattice; further no sign of a first order phase transition was evident at

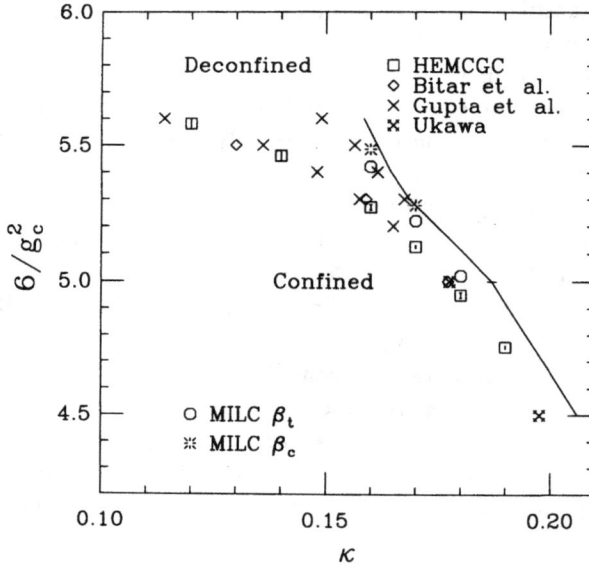

Fig. 12: A summary of the current results on the phase diagram for QCD with Wilson fermions. The critical coupling $6/g_c^2$ is shown as a function of κ for $N_\beta = 4$ (lower curve of data) and $N_\beta = 6$ (upper set of data). The line shows κ_c, corresponding to massless quarks. Data from Refs. 111-115.

lower κ. Subsequently, the $8^3 \times 4$, $N_f = 2$ calculation has been re-done at more κ values[113,114], confirming these results. Fig. 12 summarizes all the $N_\beta = 4$ results in the β-κ plane where the $\kappa_c(\beta)$ line for massless quarks is also shown. The HEMCGC collaboration[114] also calculated the hadronic spectrum on an $8^3 \times 16$ lattice and looked at T_c/m_ρ as a function of m_π/m_ρ at the same κ value. Very little overlap with the corresponding range of m_π/m_ρ for the staggered fermions was found. In contrast to the simulations of staggered fermions with light quarks, the range of κ studied so far thus corresponds to rather heavy quarks. Consequently, the T_c estimates[110] are also close to the pure gauge theory. Since increasing κ shifts the transition to lower β, it is possible that one enters the strong coupling region on an $N_\beta = 4$ lattice as $\kappa \to \kappa_c$. Recall that the chiral symmetry of the Wilson fermion is restored only in the continuum limit and one may therefore need larger N_β to see a chiral transition. Results[112,115] on $N_\beta = 6$ lattices are in agreement with such an expectation, as can be seen from Fig. 12.

5. Thermodynamics of Quantum Chromo Dynamics

In the previous section, we learnt about the phase diagram of QCD with the aid of the order parameters, defined by Eqs. (34) and (35). Their behaviour at high temperatures suggests the existence of a novel phase at high temperature. In order to investigate the nature of this phase, and to compare its properties with those expected from high temperature perturbation theory, given by Eqs. (13)–(16), various physical observables have been measured in finite temperature lattice QCD simulations across the corresponding phase transition (or cross over). The main features to emerge are i) for $T \gtrsim T_c$ several observables behave *as if* the high temperature phase were a weakly interacting gas of quarks and gluons and ii) no qualitative difference is evident in the simulations of the quenched QCD and the full QCD. There are, however, strong hints of non-perturbative effects for $T \gtrsim T_c$ as well.

5.1. Energy, Pressure, Entropy and Velocity of Sound

The main features of the energy density ϵ, defined by Eqs. (38)–(42), have been known in both the quenched approximation and the full theory since the early simulations[11]: for $T < T_c$, ϵ/T^4 is small and at T_c, it discontinuously, or at least rapidly, jumps over to a value corresponding to that of an appropriate mixture of the ideal quark and gluon gases. Moreover, the approach to the ideal gas limit is from below (above) for the quenched (full) QCD. While confirming these qualitative features, recent[55,116] high statistics simulations on bigger spatial lattices have brought forth important quantitative changes. The earlier estimates of latent heat in the quenched QCD have come down significantly and a precise determination of the pressure has underscored the fact that the coupling g^2 for current simulations is too large to justify the use of the perturbation theory estimates in Eqs. (45) and (46).

Fig. 13 displays the $SU(3)$ results of Ref. 116 on a $16^3 \times 4$ lattice for the energy density ϵ/T^4, the pressure P/T^4 and the quantity $(\epsilon - 3P)/T^4$, obtained by using Eqs. (39) and (44) along with Eqs. (42) and (45) and the one loop perturbative value for $d\beta/d\ell n\ a$. Note that i) the pressure is negative around T_c and apparently discontinuous and ii) the energy density has a rather smooth rounding, unseen in previous simulations, resulting in a lot lower estimate of the latent heat. These findings have been confirmed by other groups[53,54], who determined the discontinuities in $(\epsilon - 3P)/T_c^4$ and the entropy density $(\epsilon + P)/T_c^4$ by exploiting the two-phase structure in their data at the critical point. Their results along with

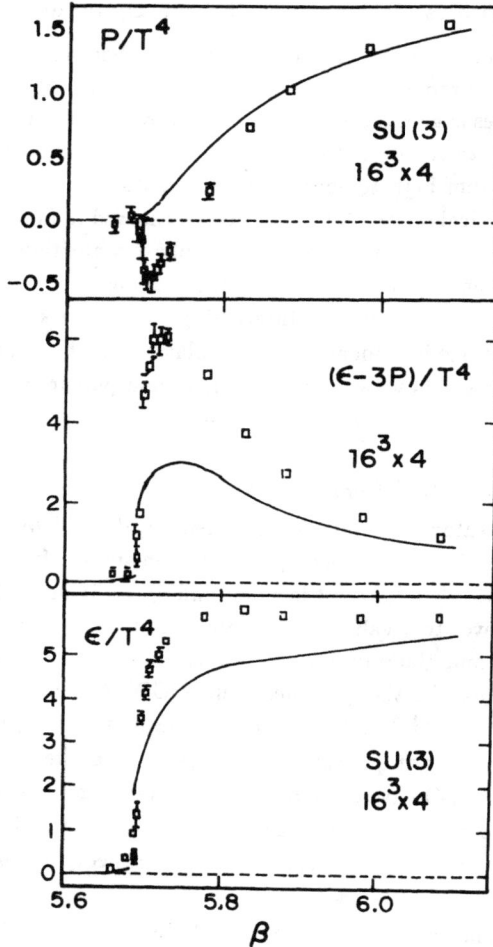

Fig. 13: The energy density ϵ/T^4, the pressure P/T^4 and $(\epsilon - 3P)/T^4$ as a function of the inverse coupling β (or equivalently temperature T) on a $16^3 \times 4$ lattice. Data from Ref. 116 and the curves from Ref. 117, as explained in the text.

those of Refs. 55, 57 and 116 are shown in Table 1. Ref. 57 used finite size scaling to obtain their infinite volume estimates of $\Delta(\epsilon - 3P)/T_c^4$. Comparing them with the rest, one finds a good agreement with the direct estimates for $N_\beta = 4$ while still larger spatial volumes appear to be necessary to achieve a similar agreement

TABLE 1

Estimates of discontinuities on large lattices at the first order phase transition in $SU(3)$ gauge theory at finite temperature. Asterisks indicate non-availability of the data.

Lattice	β	$\Delta(\epsilon + P)/T_c^4$	$\Delta(\epsilon - 3P)/T_c^4$	Reference
$24^3 \times 4$	5.6925	2.54(12)	3.78(20)	Ref. 55
$24^3 \times 4$	5.6925	2.927(97)	4.200(95)	Ref. 54
$28^3 \times 4$	5.6920	2.826(37)	4.11(12)	Ref. 54
$24^2 \times 36 \times 4$	5.6925	2.773(55)	4.062(85)	Ref. 53
$36^3 \times 4$	5.6925	2.743(66)	4.016(96)	Ref. 54
$14^3 - 28^3 \times 4$	**	**	4.005(11)	Ref. 57
$24^3 \times 6$	**	2.48(24)	**	Ref. 116
$36^2 \times 48 \times 6$	5.8936	1.835(51)	2.395(63)	Ref. 53
$20^3 - 36^2 \cdot 48 \times 6$	**	**	2.185(55)	Ref. 57

for the larger N_β. Two aspects of the data in Table 1 are striking. Both the dimensionless ratios in columns 3 and 4 go down significantly as N_β changes from 4 to 6. Further, at each N_β, they suggest $\Delta P(T_c) < 0$; if pressure were to be continuous at T_c, then the two sets of discontinuities should have been identical. Using a non-perturbative estimate for $d\beta/d\ell n \, a$ but still assuming Eq. (46) to hold, Ref. 53 finds that $\Delta(\epsilon - 3P)/T_c^4$ goes down by $\sim 40\%$ and $\sim 25\%$ for $N_\beta = 4$ and 6 respectively. However, it still shows scaling violations in going from $N_\beta = 4$ to 6, suggesting that even Eq. (46) may not be valid at the relevant couplings. Unfortunately, in the absence of non-perturbative estimates for the derivatives in Eqs. (45) and (46), one cannot yet check for the validity of scaling and a reliable estimate of the latent heat still eludes us.

The negative and discontinuous pressure has been attributed[116,117] to the failure of the perturbation theory since $g^2 \sim 1$. It has been recently proposed[117] to obtain pressure by using the relation $P = -(\Omega - \Omega_{\text{vac}})$. Ω, in turn, is obtained by numerically integrating the equation

$$\frac{\partial \Omega}{\partial \beta} = 3\left(\overline{P}_\sigma + \overline{P}_\beta\right) \ , \tag{67}$$

where $\overline{P}_\sigma, \overline{P}_\beta$ are defined in Eq. (41). Ω_{vac} is approximated by using Eq. (67) on a symmetric, N_σ^4, lattice. Combining the pressure P thus obtained with the

$(\epsilon - 3P)/T^4$, obtained by assuming Eq. (46) to be valid and by substituting the non-perturbative[118] $d\beta/d\ell n\ a$ in Eq. (44), Ref. 117 obtained the results for ϵ/T^4, P/T^4 and $(\epsilon - 3P)/T^4$ shown in Fig. 13 by solid lines. As mentioned earlier, the new estimate of $(\epsilon - 3P)/T^4$ drops by a factor of $\sim$2 near T_c but converges to the perturbative estimate at larger β, as expected. Further, the pressure, by construction, is positive everywhere and continuous at T_c. The resulting difference in ϵ with that obtained by using Eqs. (39) and (42) with Eq. (45) is remarkable. The new estimate of ϵ/T^4 appears to be far below the ideal gas value ($8\pi^2/15 = 5.26$ which becomes 7.84 after including the finite N_β correction) just above T_c and, unlike the old one, exhibits a significant variation in temperature. Since the $O(g^2)$ correction to the ideal gas value of ϵ/T^4 brings it down[119] by $\sim$10% at the relevant g^2-values, one sees that even at the highest temperature in Fig. 13, neither estimate is in agreement with even the $O(g^2)$ perturbation theory. This suggests the presence of strong non-perturbative effects and/or sizeable higher order contributions. Of course, the huge difference between the two estimates of ϵ must go down on larger N_β-lattices, where a better agreement with finite temperature perturbation theory may also be expected. The latent heat, $\Delta\epsilon/T^4$, suggested in Fig. 13, is naturally the same as obtained by assuming $\Delta(\epsilon - 3P)/T^4 = \Delta\epsilon/T^4$. Using non-perturbative $d\beta/d\ell n\ a$ to modify the value of $\Delta(\epsilon - 3P)/T^4$ in Table 1, the latent heat is

$$\Delta\epsilon/T_c^4 = 1.828 \pm 0.038 \quad \text{for} \quad N_\beta = 4. \tag{68}$$

Fig. 14 exhibits the results of Ref. 66 for ϵ/T^4 and P/T^4 for the full theory with $N_f = 2$ dynamical flavours of mass $ma = 0.025$ on an $8^3 \times 4$ lattice. The formulae given in sec. 2.3.2 have been used, except that the derivative terms in Eq. (40) have been left out. The solid line indicates the ideal gas limit on the same size lattice. A considerable overshooting of the Monte Carlo results is evident. A proper inclusion of the γ_F-derivative term in ϵ_F is expected to reduce the overshooting significantly but then inclusion of the $O(g^2)$ term in the perturbation theory will also reduce the ideal gas value by a similar amount. The pressure appears to be positive everywhere and a non-trivial $(\epsilon - 3P)/T^4$ is evident over the entire β-range, signalling non-perturbative effects again. Similar results have been obtained for a $2 + 1$ flavour theory with $(r, ma) = (0.05, 0.0125)$ on a $12^3 \times 4$ lattice[102], where, however, the pressure is found to be negative near T_c, pointing again to the inadequacy of Eq. (45). Further, it is also found that the heavy fermions contribute about half as much as each of the light flavours near T_c;

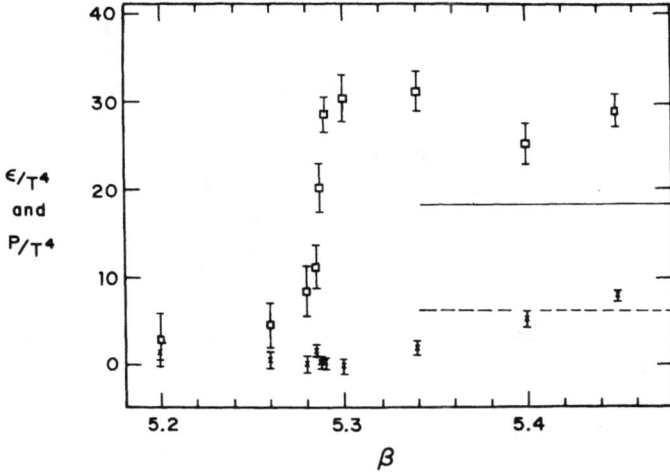

Fig. 14: The energy density ϵ/T^4 and the pressure P/T^4 for the full theory with 2 staggered flavours of mass $ma = 0.025$ on an $8^3 \times 4$ lattice. From Ref. 66.

for the corresponding ideal gases, one expects a factor ~ 0.9 instead. As in the case of the quenched QCD, significantly different estimates for both the energy density and the pressure may be anticipated, if one employs the non-perturbative procedure outlined above to obtain pressure P instead of the perturbative formulae of sec.2.3.2. This has not yet been done.

Given ϵ and P, the entropy density $S = (\epsilon + P)/T$ can be obtained straight-forwardly. In both the pure gauge[116] and the full theory with two[26,66] and four[26,69] flavours, the entropy density behaves[120] similar to the energy density. There have also been attempts[121] to obtain the velocity of sound, v_s^2, from the ϵ and P data, although one needs to assume scaling to estimate T and obtain $P(\epsilon)$. The velocity of sound is then given by $dP/d\epsilon$. Assuming a nonrelativistic ideal gas of mass m_G at low temperatures and an ideal relativistic gas of massless particles at high temperatures, one obtains $v_s^2 \propto T/m_G$ as $T \to 0$ and $v_s^2 = 1/3$ as $T \to \infty$. A phase transition at T_c can cause $v_s^2 \to 0$ or it can result in a discontinuity in v_s^2 at T_c. Exploratory simulations[121] of the $SU(3)$ gauge theory and the full QCD with moderately heavy Wilson fermions have confirmed these expectations although a reliable quantitative determination of $v_s^2(T)$ is still awaited.

5.2. Screening Lengths

One intuitive way of characterizing the two phases separated by the phase transitions discussed above is the behaviour of the heavy quark potential or various screening lengths in general. For the $SU(3)$ gauge theory, one expects a linearly rising potential in the confined phase and a Debye-screened Coulomb potential in the deconfined phase. It has been argued[18] that the correlation function of the Polyakov loops measures the free energy of a heavy quark-antiquark $(Q\overline{Q})$ pair,

$$\langle L(r)L^\dagger(0)\rangle = \exp\left(-F_{Q\overline{Q}}/T\right) \tag{69}$$

and thus yields a color-averaged $Q\overline{Q}$ potential. At high temperatures it can be computed in leading order perturbation theory and one finds

$$\langle L(r)L^\dagger(0)\rangle - |\langle L\rangle|^2 = \frac{g^4(T)}{16\pi^2 T^2}\frac{e^{-2r/r_D}}{r^2} \tag{70}$$

where the Debye screening length r_D is given by $r_D^{-2} = g^2(T)T^2(\frac{1}{6}N_f + 1)$ and $g^2(T)$ is given by Eq. (14). Unfortunately, the infra-red problems[16] make it difficult to push these calculations beyond leading order, making a non-perturbative calculation of the above correlation function more necessary. The Polyakov loop correlation function has been investigated in both pure $SU(3)$ theory[54,60,122,123,124] and the full theory[93,125]. In both cases[122,125], it is found that for $T/T_c \geq 4$ the correlation function data is well fitted by the perturbative form in Eq. (70), however, for $1 \leq T/T_c < 4$, a phenomenological form, $A\exp(-r/r_D)/r$, fits the data better. Further, $r_D(T)T$ decreases as T increases in both cases and the Debye screening length for the full theory is $\sim 30 - 40\%$ smaller at similar T/T_c. The latter observation has been interpreted[125] as additional screening caused by the presence of dynamical fermions. The precise value of r_D in the full theory has a great significance to a proposed signature for the discovery of the new phase in heavy ion collisions. Taking the current estimates seriously, they imply[126] that the screening will prohibit all charmonium states from forming for $T/T_c \gtrsim 1.3$. On the other side of the phase transition, the correlation functions for the pure $SU(3)$ theory yield a string tension which decreases[54,60,122] as $T \to T_c$. In fact, it was its finite size scaling behaviour which caused a controversy about the order of the phase transition, as discussed already in sect. 4.1.

Another interesting set of screening lengths is obtained by considering spatial correlation functions of hadronic operators with quantum numbers of mesons like π and σ and baryons of both parity. These have been studied rather extensively at low temperatures in both the full theory and the quenched theory, as can be seen from the article by DeGrand in this book, where precise definitions of the corresponding lattice operators can also be found. The broken chiral symmetry of the low temperature phase is reflected in the thus obtained spectrum. It is interesting to study how the restored chiral symmetry manifests itself in the high temperature phase and the hadronic screening lengths offer us a tool to do precisely that. Using arguments based on analytic continuation, it has been suggested[127] further that these hadronic screening lengths correspond to real time excitations of the plasma.

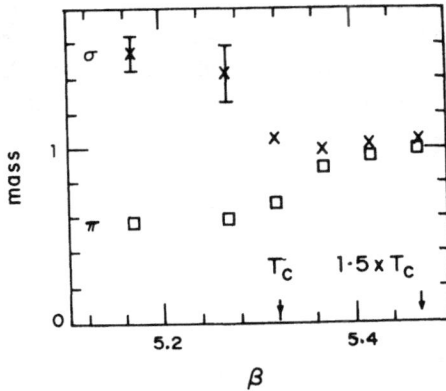

Fig. 15: Inverse screening lengths for π (squares) and σ (crosses) mesons in lattice units as a function of β for the full theory with 2 staggered flavours of mass $m/T = 0.2$. From Ref. 128.

Fig. 15 shows inverse screening lengths[128] for π and σ mesons in lattice units as a function of β. They were obtained by simulating the full theory with two dynamical staggered quarks of mass $ma = 0.05$ on a $8^2 \times 24 \times 4$ lattice. Assuming Eq. (50) to be valid for the β-range concerned, the points $T/T_c = 1$ and 1.5 are also marked on the β-scale. For $T < T_c$, a light π meson, denoted by squares, is indicated by the data while the σ-meson, denoted by crosses, is about 2.5 times heavier. For $T > T_c$, the two inverse screening lengths come together, indicating the restoration of chiral symmetry in the high temperature phase. Note that the quarks are rather heavy in these simulations: $m/T = 0.2$ or $m \sim 20 - 30$ MeV, assuming Eq. (66) to be valid. Similar degeneracy of parity partners in the high temperature phase is also seen for other mesons and baryons. Furthermore, such degeneracies have been observed for both the $N_f = 0$ theory[129] and the $N_f = 4$ theory[127,130] as well. Interestingly enough, these results have also been confirmed using Wilson fermions[30,111] for $\kappa \approx \kappa_c$, although, as discussed in sect. 4.3, one has no convincing evidence of a chiral phase transition in that case. This may also be true for $N_f \to 0$ in the case of the staggered fermions (see sec. 4.2.4). In this context, the observation of Ref. 130 that the screening lengths in the ρ and N channel at $T \gtrsim 1.2T_c$ are close to those in a free quark gas for both $N_f = 0$ and $N_f = 4$ theories may be significant. However, its simple minded interpretation is in conflict with the results on the spatial structure[131] of the screening propagators, as shown in Fig. 16 below, where the ρ-wavefunction at $T = 0$ is compared with the screening ρ-correlation function at $T \simeq 1.5T_c$. The two are remarkably close to each other. If the high temperature phase were to be an ideal gas, the latter would be a constant. More work is clearly needed to understand the close resemblance of the low and high temperature results and to clarify the nature of binding in the high temperature phase.

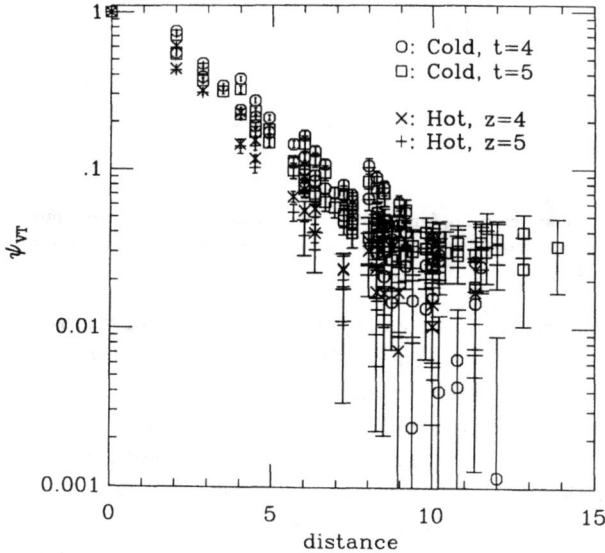

Fig. 16: The ρ-wavefunction ψ_{VT} at $T = 0$ on an $16^3 \times 24$ lattice (cold) and the corresponding correlation function at $T \simeq 1.5T_c$ on an $16^2 \times 24 \times 4$ lattice (hot) at the same coupling and quark mass. Taken from Ref. 131.

5.3. Surface Tension

Numerical determinations of the surface tension, α, have so far been attempted for only quenched QCD. Using a lattice $8^2 \times 40 \times 2$, with an interface created midway of the longer side by equilibrating its two sides at couplings $\beta_c - \Delta\beta$ and $\beta_c + \Delta\beta$, Ref. 36 measured αa^3 as a function of $\Delta\beta$ for $0.05 \leq \Delta\beta \leq 0.3$ using the differential method of Eq. (48). Extrapolating to $\Delta\beta = 0$, they obtained

$$\alpha/T_c^3 = 0.24 \pm 0.06. \tag{71}$$

Ref. 35 employed instead the integral method on $N_\sigma^2 \times 2N_\sigma \times 2$ lattices with $N_\sigma = 6, 8$ and 10 and obtained the surface tension for $0.005 \leq \Delta\beta \leq 0.3$. They found that the $N_\sigma = 8$ and 10 results were consistent with each other and yielded

$$\alpha/T_c^3 = 0.12 \pm 0.02 \tag{72}$$

after a linear extrapolation to $\Delta\beta = 0$. The discrepancy between the two results may again be due to the use of Eq. (45) for too large g^2 in getting the estimate in Eq. (71). On the other hand, simulations[132] on a $16^2 \times 32 \times 4$ lattice, using an external field h to create the interface by favouring the ordered phase on its one side and the integral method to evaluate α, resulted in an even lower value:

$$\alpha/T_c^3 = 0.024 \pm 0.004. \tag{73}$$

While this confirms the observation that $N_\beta = 2$ is too small to exhibit scaling behaviour, the decrease in the dimensionless α/T_c^3 in going from $N_\beta = 2$ to 4 is astonishingly large when compared with that in $T_c/\Lambda_{\overline{MS}}$ (see Fig. 9) or the latent heat[133] $\Delta\epsilon/T_c^4$. The latter even showed no signs of decrease at all. Nevertheless, a possible inference from the current data can be the following. Nonvanishing estimates for the surface tension are indicative of a first order phase transition in the $SU(3)$ gauge theory while the considerable fall in its value in going from $N_\beta = 2$ to 4 suggests a weakening of the transition. Using Eq. (63) along with Eq. (73), one obtains $\alpha \approx 7$-8 MeV/fm^2, which is rather small. However, in view of the remarks above, a reliable quantitative estimate of the surface tension of the pure gauge theory has to await future simulations with larger N_β.

Due to the spontaneously broken $Z(3)$ symmetry of the high temperature phase of the $SU(3)$ gauge theory, one can also define an order-order surface tension, α_{oo}. On fairly general grounds[132], one can show that $\alpha_{oo} \geq 2\alpha$, where α is the order-disorder surface tension discussed above. The theory is said to have "perfect wetting" if $\alpha_{oo} = 2\alpha$. Perfect wetting has been argued[134] to give rise to a different nucleation scenario in cosmology. Both the differential and the integral method above have been employed to determine α_{oo} by numerical simulations with the following results: the integral method yielded

$$\alpha_{oo}/T_c^3 = \begin{cases} 0.17 \pm 0.02 & \text{for}^{135} N_\beta = 2 \\ 0.043 \pm 0.010 & \text{for}^{132} N_\beta = 4, \end{cases} \tag{74}$$

while the differential method[135] resulted in $\alpha_{oo}/T_c^3 = 0.61 \pm 0.14$ for $N_\beta = 2$. Within the statistical errors, the above results are consistent with perfect wetting, provided the comparison is made for the same method and lattice size. However, the strong variations of α_{oo} with the temporal lattice size or the method again suggest that the above conclusion should be regarded as tentative.

6. Nonzero Baryon Density

Several physical and astrophysical applications need to consider strongly interacting matter at high densities and low temperatures. The baryon rich fragmentation region in heavy ion collisions or the speculative strange stars, containing 33% strange quarks, are two prime examples of such applications. In order to treat such situations, one needs to introduce chemical potential, μ, as in Eq. (11). Indeed, one can again use perturbation theory[136] to investigate the theory at large μ and zero temperature since then the coupling $g^2(\mu)$ is given by Eq. (14) with μ replacing T. However, for the same reasons as for the finite temperature case, the perturbation theory is inadequate for dealing with phase transitions, calling for lattice techniques. Noting that the additional term in Eq. (11) for $\mu_f \neq 0$ is simply the number operator corresponding to a conserved flavour charge f, it is straightforward to write down the modified partition function on the lattice: it is still given by Eq. (32) for staggered fermions but with Q_S modified from Eq. (25) as below.

$$Q_S^f(x, x', \mu) = Q_S^f(x, x') + \frac{\mu_f a_\sigma}{2} \left[U_x^4 \delta_{x, x'-\hat{4}} + U_{x'}^{4\dagger} \delta_{x, x'+\hat{4}} \right] \tag{75}$$

Defining two functions f and g by

$$f(\mu_f a_\beta) = 1 + \mu_f a_\beta \quad \text{and} \quad g(\mu_f a_\beta) = 1 - \mu_f a_\beta, \tag{76}$$

Eq. (75) can be written in the same form as Eq. (25) but with the functions f and g multiplying respectively the first and second term in the definition of $D^4(x, x')$

in Eq. (26). Comparing Eqs. (11) and (75), one notes the point split form of the number density operator on the lattice, and the consequent appearance of the gauge fields in it to ensure gauge invariance.

Unfortunately, the above modifications to introduce μ lead to technical problems, as we discuss below. Recall that the finite temperature simulations employed a symmetric lattice, $a_\sigma = a_\beta$, and correspondingly needed only one equation to look for asymptotic scaling, namely Eq. (50). On the other hand, even for the non-interacting case of $U_x^\mu = 1$, $\forall x, \hat{\mu}$, Eq. (75) yields a finite energy density only if the limit $a_\beta \to 0$ is taken first and the limit $a_\sigma \to 0$ is taken later; in the symmetric limit, $a_\sigma = a_\beta = a \to 0$, the energy density has a quadratically divergent μ^2/a^2 term[137,138]. This divergence is also present in the continuum formulation and is eliminated there by a contour integration prescription. In the lattice formulation, it can be gotten rid of by demanding[139] $f(\mu a_\beta) \cdot g(\mu a_\beta) = 1$. Physically, this condition can be interpreted[140] as following. If the chemical potential enhances the quark propagation by a factor f, then it must dampen the antiquark propagation by f^{-1} so that meson-like $nq\bar{q}$ states will be unaffected by it but baryons are enhanced over antibaryons. Two proposals have been made in the literature which satisfy the restriction above and generalize Eq. (76):

$$f(\mu_f a_\beta) = \begin{cases} \exp(\mu_f a_\beta) & \text{Refs. 137 and 140} \\ (1 + \mu_f a_\beta)/\sqrt{1 - \mu_f^2 a_\beta^2} & \text{Ref. 138.} \end{cases} \tag{77}$$

As in the case of the fermionic actions, Eqs. (25) and (27), universality can be, and indeed must be, checked by using both the forms above since in the continuum limit they both are expected to yield the same physics.

A further technical problem, which all the proposals have in common, is the loss of the antihermiticity of the covariant derivative D for nonzero μ. As a result $\det Q(\mu)$ in Eq. (30) is complex in general, although the partition function is still real. Since all the analytical as well as the numerical methods exploit the positive definiteness of the integrand in Eq. (30), one needs newer methods, capable of dealing with complex actions, for nonzero μ. Following the finite temperature case, one may be tempted to try the quenched approximation which effectively sets the determinant to a constant. Since the partition function Z is then completely independent of the chemical potential μ, one may worry whether it is too drastic an approximation in this case. Indeed, toy model calculations[141] suggest that the quenched approximation works only for small μ when the phase acquired by the fermion determinant is still small. As μ gets larger, the phase fluctuates rapidly and the results of the full theory disagree strongly with the quenched approximation. This is presumably the reason that simulations of QCD performed in the quenched approximation[142] gave rise to the absurd result that chiral symmetry is restored in the massless quark limit for any $\mu > 0$. Also for the same reason simulations done with partial quenching[143] or with a complex extension of the Langevin algorithm[142] or with a hybrid algorithm which treats[144] the phase as a part of the observable seem to work for only small μa, $\mu a \lesssim 0.2$, as can be seen from Fig. 5 of Ref. 144, for instance.

At present the phase diagram of QCD with nonzero chemical potential, μ, has been investigated for the full range of μ either in the extreme strong coupling limit[145,146,147] or on rather small lattices[148] which still may be in the strong coupling region. Using the first form for the function f in Eq. (77), all of them find a strong first order phase transition in $\langle \bar{\psi}\psi \rangle$ at $\mu_c a \approx 0.6$ for $\beta = 0$ in the massless quark limit, in agreement with the prediction[145] of the mean field theory. The chiral condensate, however, appears to be a much smoother[147] function of n, the baryon number density, defined in Eq. (3). The methods of Refs. 146 and 148 for simulating QCD with $\mu \neq 0$ are not restricted to infinite coupling, $\beta = 0$, since the former employs the spectral density method to keep track of the phase and the latter expands the grand canonical partition function in terms of positive canonical partition functions for a fixed particle content on the lattice. However, both are rather time-consuming and therefore limited to small lattices still. Nevertheless, Ref. 148 performed simulations on a 4^4 lattice and found a transition for $\beta = 4.9$, $ma = 0.025$ at $\mu_c a \approx 0.35$ in the number density. One still needs to study this transition on bigger lattices, check for the β-dependence of the $\mu_c a$ and establish its order. Furthermore, these results are presumably in the strong coupling region where the two forms in Eq. (77) may lead to different results. We are thus long way off from any definitive knowledge of the phase diagram of QCD at finite baryon density.

6.1. Quark Number Susceptibility

The quark number susceptibility is analogous to the familiar magnetic susceptibility or $\chi(L)$ of sec. 3.2.2. In the general case of N_f flavours of quarks of mass ma, corresponding to an $SU(N_f)$ flavour symmetry, one can define N_f conserved densities by obviously generalized relations analogous to Eq. (3). The baryon number density, discussed in the previous subsection, is then the $SU(N_f)$-singlet combination of these. By taking derivatives of these densities with respect to the N_f

chemical potentials, $N_f(N_f + 1)/2$ susceptibilities can be defined: $\chi_{ij} = \partial n_i/\partial \mu_j$, $i, j = 1, N_f$. Again the singlet susceptibility can be formed out of them by taking a linear combination and it will correspond to the quark number or the baryon number susceptibility. Each of these susceptibilities will clearly contain first and second derivatives of the functions f and g, introduced above. They will therefore depend on the prescription to introduce the chemical potential. However, it can be easily shown that the constraints[139] on these functions imply

$$f'(0) = -g'(0) = 1 \quad \text{and} \quad f''(0) = g''(0) = 1. \tag{78}$$

Thus all the susceptibilities are independent of the prescription for f and g if one evaluates them either for $\mu = 0$ or in the continuum limit $\mu a \to 0$ for finite μ.

The susceptibilities at $\mu = 0$ have been proposed[149,150] as a probe to study the dominant modes of the hot QCD phase. In order to see the basic argument, consider the baryon number susceptibility:

$$\chi_0(T) = \frac{1}{N_f} \sum_{i=1}^{N_f} \frac{\partial n_i}{\partial \mu_i}(\mu_i = 0). \tag{79}$$

On simple dimensional grounds $\chi_0 = F(T) \cdot T^2$, where the dimensionless function $F(T)$ contains the information about the dynamics of the system. For a non-interacting baryon (or quark) gas of mass m, one expects $F(T) \sim \exp(-m/T)$ if $m \gg T$ and $F(T) \approx 1$ if $m \ll T$. Thus by studying various χ's, one can check whether the dominant mode of a given quantum number, corresponding to the susceptibility under study, is light or heavy compared to the temperature. Of course, the interpretation becomes difficult if a non-trivial F, other than the cases discussed, is obtained. Moreover, even for those cases the simple minded interpretation is clearly far from being unique.

Fig. 17 shows the baryon number susceptibility, $\chi/N_f T^2$ as a function of T/T_c for the full QCD[149](circles) with 2 flavours of staggered quarks and the quenched QCD[151](daggers). In both cases the simulations were done on an $8^3 \times 4$ lattice and the susceptibility measurements were made with quarks of $m/T = 0.2$. The HMD algorithm was used to simulate the full theory while the standard Metropolis algorithm was used for the quenched case. Although the results are shown as a function of T/T_c, instead of β as in Figs. 13 and 14, in neither case the β-range covered is perhaps in a region where Eq. (50) is valid. Thus there are unknown systematic errors in T/T_c in both the cases. However, they are unlikely to affect

110

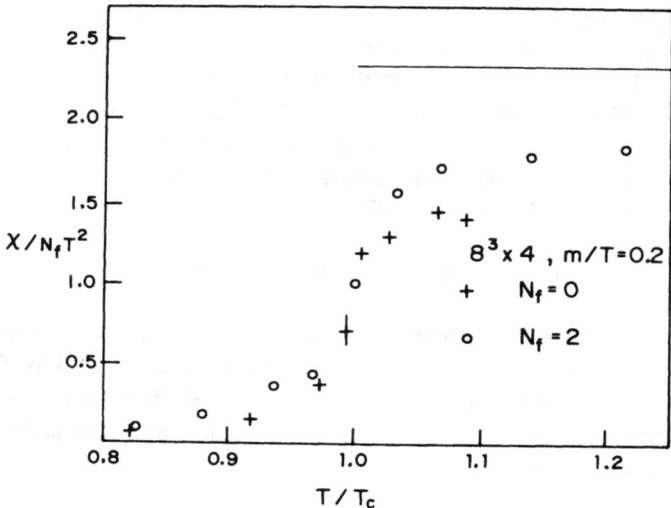

Fig. 17: Susceptibility in units of $N_f T^2$ as a function of T/T_c for quark mass $m/T = 0.2$ on an $8^3 \times 4$ lattice. Data from Refs. 149 and 151.

the qualitative conclusions we wish to discuss. From Fig. 17 a sudden rise is seen in the baryon number susceptibility near $T = T_c$ in both the cases. The qualitative behaviour is remarkably similar to the energy density, seen in Figs. 13 and 14. χ is rather small below T_c and after a sudden rise at T_c it approaches slowly the ideal gas value on an $8^3 \times 4$ lattice, shown by the horizontal line in the figure. In line with the simple arguments above, Ref. 149 interpreted these results as due to heavy baryons below T_c and due to light quarks and antiquarks in the high temperature phase. On the other hand, the quenched results are remarkably close to those of the full theory and the expected dynamics of the high temperature phase in that case is that of gluons only. Indeed, as we discussed in sec. 4, the phase transition in the quenched case has more to do with deconfinement while the full theory transition has its roots in an apparently unrelated chiral symmetry restoration. Furthermore, one needs to understand the deviation from the ideal gas limit in, say, perturbation theory in order to be able to ignore the difference between the horizontal line and the data. This is especially so since other thermodynamic quantities in that range are indicative of non-perturbative deviations from an ideal gas. Of course, an attractive explanation could be that the two theories, and consequently the deconfinement and the chiral phase transition, are the same, except for a change of a scale. More work is, therefore, needed to support the interpretation in the full theory or to shed sufficient light on the relation between the respective transitions in the two theories. It could come in form of a comparison of the nonsinglet susceptibilities which for lower quark masses, m/T, should yield differences between the full theory and the quenched theory.

7. Higgs and Related Models at Finite Temperature

Finite temperature effects in the electroweak theory or the grand unified theories (GUTs) can play an important role in the physics of the early Universe. The inflationary Universe is a prime example of this and the baryon asymmetry in the Universe may have an important influence of the electroweak phase transition. The common element in these phenomena, which distinguishes them from the QCD-like theories reviewed above, is the scalar-gauge interaction and the Higgs mechanism responsible for the spontaneous breakdown of the gauge symmetry. Using perturbative methods, it has been argued[6] that the spontaneously broken gauge symmetry is restored at high temperatures in a phase transition: the nonzero vacuum expectation value of the scalar field at zero temperature becomes zero at the phase transition much like the loss of the spontaneous magnetization of an Ising ferromagnet at its critical point T_c. It is clearly desirable to investigate such symmetry restoring phase transitions in the gauge-Higgs systems by non-perturbative lattice techniques. Indeed, this motivation has attracted the attention of both analytical[152] and numerical[153,154,155] efforts towards the general problem of gauge-Higgs systems at finite temperature. All investigations seem to sugggest a softening of the symmetry restoring phase transition by non-perturbative effects into a crossover, at best, if the Higgs multiplet is in the fundamental representation of the gauge group. Recall that the $SU(2) \times U(1)$ gauge group of the standard electoweak theory is broken by Higgs fields in the fundamental representation while adjoint Higgs are usually employed for summetry breaking in GUTs. In view of the fact that the non-perturbative investigations of the adjoint (or higher representation) Higgs models[152,153] so far have been limited to models of academic interest only, we will only review the results obtained for the former in more detail below.

Extending the finite temperature formalism of Eqs. (9)–(11) to the electroweak theory is straightforward. As may be anticipated, the functional integral has to be extended over the complex scalar doublets Φ and $\Phi^\dagger$, both of which satisfy periodic boundary conditions similar to the A-fields in Eq. (10). There are two gauge fields belonging to $SU(2)$ and $U(1)$ groups respectively and the action of Eq. (11) has to be modified accordingly. Further, Yukawa-type fermion-Higgs coupling terms and a gauge invariant Higgs action have to be added to Eq. (11), just as in the zero temperature case. Following the basic principles outlined in sec. 2.3, one can write down the lattice version of the standard electroweak model without any further technical problems. Since for the sake of computational simplicity the first set of simulations[154,155] , and the only ones so far, have made the approximation of neglecting i) the fermion sector and ii) the $U(1)$ part, we will begin by writing down the lattice version of the $SU(2)$ gauge theory with fundamental Higgs at finite temperature. As we will see below, the basic physics is still retained under these approximations, although quantitative details may be affected by them.

The corresponding partition function Z is still given by Eq. (19) but with ψ and $\bar{\psi}$ fields replaced by Φ and $\Phi^\dagger$ which satisfy periodic boundary conditions, similar to those on the U-fields in Eq. (20). Parametrising the complex Higgs doublet as $\Phi_x = \rho_x \sigma_x$, the measure can be seen to be

$$\prod_x d\Phi_x d\Phi_x^\dagger = \prod_x d\rho_x d\sigma_x, \quad \text{with} \quad 0 < \rho_x < \infty \text{ and } \sigma_x \in SU(2) \ . \tag{80}$$

The action S for the gauge-Higgs theory is given by $S = S_G + S_H$, where S_G is given by Eq. (21) with $N = 2$. The Higgs action S_H is given by

$$S_H = -\kappa\,\xi \sum_x \rho_x \rho_{x+\hat{4}} \ \text{Re Tr}(\sigma_x^\dagger U_x^4 \sigma_{x+\hat{4}}) - \frac{\kappa}{\xi} \sum_{x,\hat{\mu}=1}^{3} \rho_x \rho_{x+\hat{\mu}} \ \text{Re Tr}(\sigma_x^\dagger U_x^\mu \sigma_{x+\hat{\mu}})$$
$$+ \frac{\lambda}{\xi} \sum_x (\rho_x^2 - 1)^2 + \frac{1}{\xi} \sum_x \rho_x^2 \ ,$$

$$\tag{81}$$

where $\xi = a_\sigma/a_\beta$, as in sec. 2. Thus, in addition to the gauge coupling $\beta = 4/g^2$, the model has two more couplings, namely, the hopping parameter κ and the quartic self-coupling λ. It can be easily shown that they are related to their continuum counterparts, the bare mass m_0 and λ_0, by

$$\kappa = \frac{1 - 2\lambda}{2\xi^2 + 6 + m_0^2 a_\sigma^2} \quad \text{and} \quad \lambda_0 = \frac{6\lambda}{\kappa^2} \ . \tag{82}$$

Note that $\kappa = 0$ and ∞ correspond to large positive and negative m_0^2 respectively.

The model has been extensively studied at zero temperature, $i.\,e.$, on symmetric lattices and is found[156,157,158] to have only one phase divided into two physically different regions: the confining region and the Higgs region. For small β and large λ these two regions are analytically connected[156] to each other[†]. For larger β, they are separated by a phase transition, called the Higgs phase transition, which is either a weak first order or higher order phase transition for $\lambda \gtrsim 0.5$. The only known second order phase transition in the model occurs at $\beta = \infty$ where the theory is reduced to an $O(4)$ spin model. However, the theory is expected to be trivial in this limit. One can derive bounds on the Higgs mass, utilising the triviality, as can be seen from the article by Hasenfratz in this book. We will assume here that by working in the appropriate scaling region around the Higgs phase transition an effective low energy theory can be defined and its high temperature behaviour can be studied.

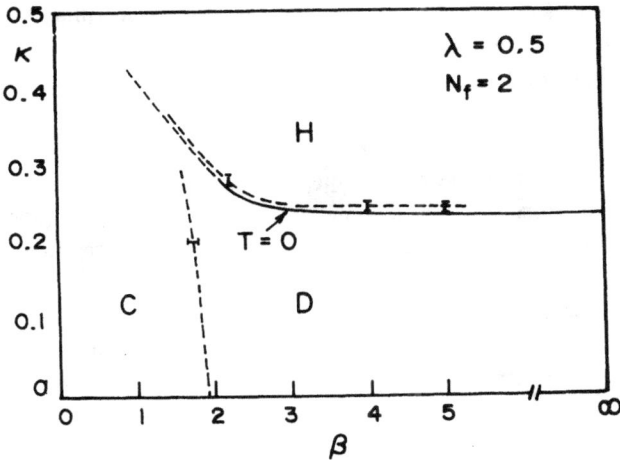

Fig. 18: The phase diagram of the fundamental $SU(2)$ higgs theory at finite tempera-ture for $\lambda = 0.5$ and $N_\beta = 2$ from Ref. 154. The bold dashed lines are crossover lines on which the data points are also marked. The thin solid and dashed line indicates the Higgs phase transition at zero temperature.

Fig. 18 displays the phase diagram obtained by Ref. 154 for $\lambda = 0.5$ on $8^3 \times 2$ and $16^3 \times 2$ lattices along with the zero temperature Higgs phase transition line. They find only one phase at high temperatures which is separated by crossovers into three physically different regions, marked by the letters C(confinement), D

[†]   Bringing back the fermions may change the zero temperature phase diagram, eliminating the analytical connection between the two regions[157,159].

(deconfinement) and H(Higgs) in Fig. 18. The vertical dashed line, separating the confinement and deconfinement regions is an extension of the $SU(2)$ deconfinement phase transition from $\kappa = 0$. This phase transition has also been investigated[11] in great detail like the $SU(3)$ case. It is found to be of second order, with various critical exponents[160] matching with those of the three-dimensional Ising model, as predicted by universality arguments[19]. Introducing scalar fields by turning on κ has the same effect as bringing in fermions. Indeed, as in sec. 3.1, here too a nonzero κ corresponds to an explicit symmetry breaking magnetic field since $\kappa \sim m^{-1}$ for small κ. It is therefore expected that the phase transition will be washed out for any nonzero κ. Numerical simulations[154,155] are in agreement with this expectation, although a shadow of the phase transition seems to be still visible for very small κ. From the observed absence of any finite size effects, for $\lambda = 0.5$ and $\beta = 2.25$ and 4.0, one sees that the Higgs phase transition also turns into a crossover at finite temperature. The finite temperature effects cause the crossover to shift by about 0.01 in κ at $\lambda = 0.5$. No such shift was seen[155] for $\lambda = 0.1$ presumably because it was too tiny there and needed much better simulations. Note that due to this shift at finite tempearture there exists a small range of κ near κ_c in the Higgs phase at zero temperature where heating the system causes it to go through a crossover to the symmetry restored deconfined region, as expected on the basis of perturbation theory. The smoother crossover takes place of the perturbatively predicted phase transition.

Following essentially the same steps as in sect. 2.3.2 one can construct the energy density $\epsilon = \epsilon_G + \epsilon_H$, with ϵ_G given by the expressions in sect. 2.3.2 with $N = 2$, and ϵ_H given[154,155] by

$$\epsilon_H/T^4 = N_\beta^4 \left[2\kappa \left(\langle \Phi^\dagger U^4 \Phi \rangle - \sum_{\mu=1}^{3} \langle \Phi^\dagger U^\mu \Phi \rangle \right) + \lambda \langle \rho^4 \rangle + (1 - 2\lambda - 4\kappa) \langle \rho^2 \rangle \right] .$$
(83)

Subtracting off the zero temperature contribution by evaluating the same expression on a symmetric N_σ^4 lattice, one can study the variation of ϵ across both the crossovers. For constant κ, it shows[154,155] features similar to that of the pure $SU(2)$ gauge theory: the energy density rises across the deconfinement crossover rapidly to a value close to that of the ideal gas. However, it shows[154] a spurious peak when studied as a function of κ for constant β, i. e., across the Higgs crossover. The peak has been attributed[154] to the prescription of using ϵ on a symmetric lattice as $\epsilon(T = 0)$ since it occurs at the κ_c corresponding to the zero temperature Higgs phase transition. In order to investigate further its origin, or the approach to the ideal gas at high temperature or to obtain the crossover temperature in physical units, one needs to know the lines of constant physics in the model at zero temperature, i. e., the renormalization group flow lines. Only then can one be sure that the temperature of the *same* physical system is being varied. Furthermore, the crossover must be in the scaling region in order for the results to be relevant to the (effective) continuum theory. This means that the Higgs boson mass in lattice units should be small, forcing one much more close to the critical point at $\beta = \infty$. Indeed, as in the case of the bound on the Higgs mass, one may choose to investigate[161,162] the $\beta = \infty$ limit, i. e., the $O(4)$ model itself at finite temperature. The relative computational simplicity in that case allows one to use bigger lattices and get deeper into the scaling region and one finds that the symmetry restoration temperature T_{SR} is bounded[162] below by $(0.54 \pm 0.04)\ m_H$.

8. Concluding Remarks

The standard model of particle physics, based on the gauge group $SU(3) \times SU(2) \times U(1)$, can behave very differently at finite temperatures and/or densities compared to the low temperature and density regime around us. Perturbation theory provides a glimpse of the new phenomena, such as deconfinement or restoration of symmetries broken either spontaneously or dynamically at low temperatures. It is, however, woefully inadequate to gain any quantitative understanding since its very foundation appears to be shaky due to infra-red problems and the large relevant couplings. The lattice formulation, along with the computer simulation techniques which enable us to approach (eventually) the continuum limit of vanishing lattice spacing, is the only non-perturbative tool we have to investigate these interesting phenomena quantitatively from first principles. Unfortunately, conceptual and technical problems make our task formidable. Notable amongst them are the problems of fermions with the right chiral symmetry and flavour content, and their non-hermitean covariant derivative for nonzero chemical potential. Any simulations incorporating dynamical fermions have to face a tremendous algorithmic slowdown, especially for low physical quark masses, and have to cope up with unknown systematic errors. The Hybrid Monte Carlo algorithm seems to be the best so far but it cannot be employed for the physically interesting situation of two light flavours and one relatively heavy flavour. Remarkably, the field has still witnessed a lot of progress in the recent past, thanks partly to the home-grown computers.

The quenched QCD, or the $SU(3)$ gauge theory, has been firmly established to have a first order deconfinement phase transition, at least for $N_\beta = 4$. Finite size scaling analysis of high statistics data on large lattices has convincingly demonstrated that the peak of the susceptibility scales like the volume in agreement with the observed qualitative characteristics of a first order phase transition, such as the coexistence of two states. Such strong evidence is at present lacking for $N_\beta > 6$. Asymptotic scaling has been tested by studying lattices up to $N_\beta = 16$ and there are indications that it is valid for $N_\beta \geq 10$. Simulations of the full theory with dynamical staggered quarks have just begun to emulate the quenched ones. For four flavours of mass $m/T \leq 0.2$ a first order chiral phase transition is seen for $N_\beta = 4$, although investigations with $N_\beta = 6$ and 8 suggest $m_{CH}/T \approx 0.12$, where m_{CH} is the critical quark mass beyond which no first order chiral phase transition exists. Going down in flavours, a first order chiral phase transition is also seen for three flavours of $m/T = 0.1$ but for the realistic case of 2 light flavours and one heavy flavour no signal for a first order phase transition was visible on large spatial lattices. These investigations need to be extended to smaller quark masses and larger lattices, especially in the temporal direction. Estimates of the critical temperature depend unfortunately on the scale used, indicating that the continuum limit is not reached. Nonetheless, for QCD with zero and two dynamical flavours, the best evidence indicates $T_c \approx 235$ MeV and 150 MeV respectively.

In order to learn more about the high temperature phase various thermodynamical quantities and interesting physical observables have been studied for both the quenched theory and the full theory. These include the energy density, pressure, screening lengths and the baryon number susceptibility. Since most investigations employed lattices with $N_\beta \leq 6$, the numerical results may still be significantly affected by lattice artifacts. Remarkably, viewed as a function of T/T_c both the quenched and the full theory behave very similarly for all of them; dynamical fermions do appear to make quantitative differences but they need to be properly examined on larger lattices. The high temperature phase appears to be chirally symmetric and several observables behave *as if* it were a mixture of ideal gases of quarks and gluons. However, strong non-perturbative effects are visible in many observables just above T_c. In particular, the Debye screening length for heavy quarks is consistent with leading order perturbation theory only for $T/T_c \geq 4$.

Exploratory investigations of QCD at finite density and the fundamental $SU(2)$ Higgs-gauge system point towards the naive expectations. Chiral symmetry appears to get restored for a critical μa for the former while the $SU(2)$

gauge symmetry seems to be restored for the latter in a crossover. Larger lattices need to be employed in the respective scaling regions in order to check the validity of these findings. It would be then interesting to see whether the crossover in the Higgs case ever turns into a phase transition, as suggested by perturbation theory. Dynamical fermions may play a crucial role also since they do seem to alter the zero temperature phase diagram. In both cases, it may be more useful to investigate anisotropic lattices with $a_\beta \neq a_\sigma$.

REFERENCES

1. R. Hagedorn, *Nuovo Cim. Suppl.* **3**, 147 (1965).
2. N. Cabbibo and G. Parisi, *Phys. Lett.* **B59**, 67 (1975).
3. J. Collins and M. Perry, *Phys. Rev. Lett.* **34**, 135 (1975).
4. D. Gross, R.D. Pisarski and L.G. Yaffe, *Rev. Mod. Phys.* **53**, 43 (1981).
5. A. Linde, *Rep. Prog. Phys.* **42**, 389 (1979).
6. D. Kirzhnits and A. Linde, *Phys. Lett.* **B42**, 471 (1972); S. Weinberg, *Phys. Rev.* **D9**, 3557 (1974).
7. K.G. Wilson, *Phys. Rev.* **D10**, 2445 (1974).
8. M. Creutz, *Quarks, gluons and lattices*, Cambridge University Press, Cambridge 1985.
9. A. Polyakov, *Phys. Lett.* **B72**, 477 (1978); L. Susskind, *Phys. Rev.* **D20**, 2610 (1979).
10. C. Borgs and E. Seiler, *Nucl. Phys.* **B215[FS7]**, 125 (1983); *Comm. Math. Phys.* **91**, 329 (1983).
11. J. Cleymans, R. V. Gavai and E. Suhonen, *Phys. Rep.* **130**, 217 (1986).
12. C. Bernard, *Phys. Rev.* **D9**, 3312 (1974).
13. F. A. Berezin, *The Method of Second Quantization, Pure and Applied Physics, Vol. 24*(Academic Press, N.Y., 1966).
14. W. Marciano and H. Pagels, *Phys. Rep.* **36**, 137 (1978).
15. T. Toimela, *Phys. Lett.* **124B**, 407 (1983); *Int. J. Theor. Phys.* **24**, 901 (1985) and erratum **26**, 1021 (1987).
16. S. Nadkarni, *Phys. Rev. Lett.* **60**, 491 (1988); *Proceedings of the International Conference on Physics and Astrophysics of Quark-Gluon Plasma*, Ed. B. Sinha and S. Raha, World Scientific 1988, p.608.
17. R. Pisarski and F. Wilczek, *Phys. Rev.* **D29**, 338 (1984).
18. L. McLerran and B. Svetitsky, *Phys. Rev.* **D24**, 450 (1981); *Phys. Lett.* **98B**, 195 (1981); J. Kuti, J. Polónyi and K. Szlachányi, *Phys. Lett.* **98B**, 199 (1981).

19. B. Svetitsky and L. Yaffe, *Nucl. Phys.* **B210[FS6]**, 423 (1982); *Phys. Rev.* **D26** , 963 (1982).

20. J. Engels, F. Karsch, I. Montvay and H. Satz, *Nucl. Phys.* **B205[FS5]**, 545 (1982).

21. H.B. Nielsen and M. Ninomiya, *Nucl. Phys.* **B185**, 20 (1981); *Phys. Lett.* **105B**, 219 (1981).

22. J. Kogut, *Rev. Mod. Phys.* **55**, 775 (1983).

23. M. Creutz, L. Jacobs and C. Rebbi, *Phys. Rep.* **93**, 201 (1983).

24. J. Kogut and L. Susskind, *Phys. Rev.* **D11**, 395 (1975); L. Susskind, *Phys. Rev.* **D16**, 3031 (1977).

25. K. Wilson, in *New Phenomena in Subnuclear Physics*, Ed. A. Zichichi(Plenum Press, N. Y.)

26. F. Karsch and I. O. Stamatescu, *Phys. Lett.* **B227**, 153 (1989).

27. H. Kluberg-Stern, A. Morel, O. Napoly and B. Petersson, *Nucl. Phys.* **B220** **[FS8]**, 447 (1983); J. Kogut *et al.* , *Nucl. Phys.* **B225[FS9]**, 326 (1983); A. Duncan, R. Roskies and H. Vaidya, *Phys. Lett.* **118B**, 439 (1982).

28. M. Bochicchio *et al.* , *Nucl. Phys.* **B262**, 331 (1985).

29. S. Itoh, Y. Iwasaki, Y. Oyangi and T. Yoshié, *Nucl. Phys.* **B274**, 33 (1986).

30. Y. Iwasaki, Y. Tsuboi and T. Yoshié, *Phys. Lett.* **B220**, 602 (1989); Y. Iwasaki, K. Kanaya, S. Sakai and T. Yoshié, *Phys. Rev. Lett.* **67**, 1494 (1991).

31. H. W. Hamber, E. Marinari, G. Parisi and C. Rebbi, *Phys. Lett.* **121B**, 99 (1983).

32. F. Karsch, *Nucl. Phys.* **B205[FS5]**, 285 (1982).

33. R. C. Trinchero, *Nucl. Phys.* **B227**, 61 (1983).

34. G. M. Fuller, G. J. Mathews and C. Alcock, *Phys. Rev.* **D37**, 1380 (1988); J. H. Applegate, C. Hogan and R. J. Scherrer, *Phys. Rev.* **D35**, 1151 (1987).

35. S. Huang J. Potvin , C. Rebbi and S. Sanielevici, *Phys. Rev.* **D42**, 2864 (1990); J. Potvin and C. Rebbi, *Phys. Rev. Lett.* **62**, 3062 (1989).

36. K. Kajantie, L. Kärkkäinen and K. Rummukainen, *Nucl. Phys.* **B333**, 100 (1990); K. Kajantie and L. Kärkkäinen, *Phys. Lett.* **B214**, 595 (1988).

37. M. Creutz, *Phys. Rev. Lett.* **43**, 553 (1979); *Phys. Rev.* **D21**, 2308 (1980).

38. G. Parisi, *Statistical Field Theory*, Addison-Wesley, Redwood, California, USA, 1988.

39. B. Svetitsky, *Phys. Rep.* **132**, 1 (1986).

40. J. Polónyi and K. Szlachányi, *Phys. Lett.* **110B**, 395 (1982); F. Green and F. Karsch, *Nucl. Phys.* **B238**, 297 (1984).

41. S. Naik, *Nucl. Phys.* **B334**, 611 (1990).

42. J. Bartholomew, D. Hochberg, P. Damgaard and M. Gross, *Phys. Lett.* **133B**, 218 (1983); T. Banks and A. Ukawa, *Nucl. Phys.* **B225[FS9]**, 145 (1983); F. Green and F. Karsch, *Phys. Rev.* **D29**, 2986 (1984); M. Ogilvie, *Phys. Rev. Lett.* **52**, 1369 (1982).

43. N. A. Metropolis, M. N. Rosenbluth, E. Teller and J. Teller, *J. Chem. Phys.* **21** 1087 (1953).

44. D. H. Weingarten and D. N. Petcher, *Phys. Lett.* **99B**, 333 (1981).

45. S. Duane, A. D. Kennedy, B. J. Pendleton and D. Roweth, *Phys. Lett.* **195B**, 216 (1987).

46. S. Duane, *Nucl. Phys.* **B257[FS14]**, 652 (1985); S. Duane and J. B. Kogut, *Nucl. Phys.* **B275[FS17]**, 398 (1986); S. Gottlieb, W. Liu, D. Toussaint, R. L. Renken and R. L. Sugar *Phys. Rev.* **D35**, 2531 (1987).

47. G. G. Batrouni *et al.* , *Phys. Rev.* **D32**, 2736 (1985); A. Ukawa and M. Fukugita, *Phys. Rev. Lett.* **55**, 1854 (1985).

48. R. V. Gavai, F. Karsch and B. Petersson, *Nucl. Phys.* **B322**, 738 (1989).

49. M. N. Barber, in *Phase Transitions and Critical Phenomena*, Vol. 8, Eds. C. Domb and J. Lebowitz, (Academic Press, N. Y., 1983), p. 145.

50. M. E. Fisher and M. N. Barber, *Phys. Rev. Lett.* **28**, 1516 (1972).

51. M. S. S. Challa, D. P. Landau and K. Binder, *Phys. Rev.* **B34**, 1841 (1986).

52. S. Gupta, A. Irbäck, B. Petersson, R. V. Gavai and F. Karsch, *Nucl. Phys.* **B329**, 263 (1990).

53. K. Kanaya *et al.* , in *Proceedings of "Lattice 90"*, *Tallahassee, U.S.A*, Eds. U. M. Heller, A. D. Kennedy and S. Sanielevici, *Nucl. Phys.* **B(PS)20**, 300 (1991); The QCDPAX collaboration, Y. Iwasaki *et al.* , University of Tsukuba Preprint UTHEP-218, 1991.

54. M. Fukugita, M. Okawa and A. Ukawa, *Nucl. Phys.* **B337**, 181 (1990).

55. F. R. Brown *et al.* , *Phys. Rev. Lett.* **61**, 2058 (1988).

56. S. Cabasino *et al.* , in *Proceedings of "Lattice '89"*, *Capri, Italy*, Eds. N. Cabibbo *et al.* , *Nucl. Phys.* **B(PS)17**, 218 (1990).

57. N. A. Alves, B. A. Berg and S. Sanielevici, *Phys. Rev. Lett.* , **64**, 3107 (1990) and SCRI preprint FSU-SCRI-91-93.

58. A. D. Kennedy, J. Kuti, S. Meyer and B. J. Pendleton *Phys. Rev. Lett.* **54**, 87 (1985).

59. S. Gottlieb *et al.* , in *Proceedings of "Lattice '89"*, *Capri, Italy*, Eds. N. Cabibbo *et al.* , *Nucl. Phys.* **B(PS)17**, 173 (1990).

120

60. P. Bacilieri et al. , APE Collaboration, *Phys. Rev. Lett.* **61**, 1545 (1988); *Nucl. Phys.* **B318**, 553 (1989).

61. K. Jansen et al. , *Phys. Lett.* **B213**, 203 (1988).

62. M. Fukugita and M. Okawa, *Phys. Rev. Lett.* **63**, 13 (1989).

63. R. V. Gavai, in *Proceedings of "Lattice '89", Capri, Italy*, Eds. N. Cabibbo et al. , *Nucl. Phys.* **B(PS)17**, 335 (1990).

64. A. Ukawa, in *Proceedings of "Lattice '89", Capri, Italy*, Eds. N. Cabibbo et al. , *Nucl. Phys.* **B(PS)17**, 118 (1990).

65. N. H. Christ and A. E. Terrano, *Phys. Rev. Lett.* **56**, 111 (1986).

66. S. Gottlieb et al. , *Phys. Rev.* **D35**, 3972 (1987); *Phys. Rev. Lett.* **59**, 1513 (1987).

67. S. Gottlieb et al. , *Phys. Rev.* **D41**, 622 (1990).

68. F. R. Brown et al. , *Phys. Lett.* **B251**, 181 (1990).

69. The MT_c Collaboration, R. V. Gavai et al. , *Phys. Lett.* **B241**, 567 (1990); *Phys. Lett.* **B232** 491 (1989).

70. A. Hasenfratz and P. Hasenfratz, *Nucl. Phys.* **B193**, 210 (1981);
    H. S. Sharatchandra, H. J. Thun and P. Weisz, *Nucl. Phys.* **B192**, 205 (1981).

71. N. H. Christ, in *Proceedings of "Lattice '89", Capri, Italy*, Eds. N. Cabibbo et al. , *Nucl. Phys.* **B(PS)17**, 267 (1990).

72. D. Barkai, K. J. M. Moriarty and C. Rebbi, *Phys. Rev.* **D30**, 1293 (1984).

73. R. Gupta, in *Proceedings of "Lattice '89", Capri, Italy*, Eds. N. Cabibbo et al. , *Nucl. Phys.* **B(PS)17**, 70 (1990) and *Phys. Rev.* **D36**, 2813 (1987).

74. R.V. Gavai, J. Potvin and S. Sanielevici, *Phys. Rev.* **D38**, 3266 (1988); *Phys. Rev. Lett.* **58**, 2519 (1987); *Phys. Lett.* **200B**, 137 (1988).

75. R. V. Gavai, *Nucl. Phys.* **B269**, 530 (1986).

76. M. Fukugita and A. Ukawa, *Phys. Rev. Lett.* **57**, 503 (1986).

77. J. Polonyi et al. , *Phys. Rev. Lett.* **53**, 644 (1984).

78. R. Gupta, G. W. Kilcup and S. R. Sharpe, *Phys. Rev.* **D38**, 2188 (1988).

79. K. Bitar et al. , *Nucl. Phys.* **B313**, 348 (1989); *Nucl. Phys.* **B337**, 245 (1990).

80. E. V. E. Kovacs, D. K. Sinclair and J. B. Kogut, *Phys. Rev. Lett.* **58**, 751 (1987); *Nucl. Phys.* **B290[FS20]**, 431 (1987).

81. R. Gupta et al. , *Phys. Rev. Lett.* **57**, 2621 (1986).

82. F. Karsch, J. B. Kogut, D. K. Sinclair and H. W. Wyld, *Phys. Lett.* **188B**, 353 (1987).

83. M. Fukugita, H. Mino, M. Okawa, A. Ukawa, *Phys. Rev. Lett.* **65**, 816 (1990); *Phys. Rev.* **D42**, 2936 (1990).

84. S. Gottlieb, in *Proceedings of "Lattice 90", Tallahassee, U.S.A*, Eds. U. M. Heller, A. D. Kennedy and S. Sanielevici, *Nucl. Phys.* **B(PS)20**, 247 (1991).

85. M. P. Grady, D. K. Sinclair and J. B. Kogut, *Phys. Lett.* **200B**, 149 (1988).

86. S. Gottlieb *et al.* , *Phys. Rev.* **D40**, 2389 (1989).

87. A. Vaccarino, in *Proceedings of "Lattice '89", Capri, Italy*, Eds. N. Cabibbo *et al.* , *Nucl. Phys.* **B(PS)17**, 421 (1990).

88. N. Attig, R. V. Gavai, B. Petersson and M. Wolff, *Z. Phys.* **C40**, 471 (1988).

89. M. Fukugita and A. Ukawa, *Phys. Rev.* **D38**, 1971 (1988); M. Fukugita, S. Ohta, Y. Oyanagi and A. Ukawa, *Phys. Rev. Lett.* **58**, 2515 (1987); M. Fukugita, S. Ohta and A. Ukawa, *Phys. Rev. Lett.* **60**, 178 (1988).

90. A. Irbäck, F. Karsch, B. Petersson and H. W. Wyld, *Phys. Lett.* **216B**, 177 (1989).

91. R. Gupta *et al.* , *Phys. Lett.* **201B**, 503 (1988).

92. J. B. Kogut and D. K. Sinclair, *Nucl. Phys.* **B295[FS21]**, 480 (1988).

93. F. Karsch and H. W. Wyld, *Phys. Lett.* 213B, 505 (1988).

94. F. R. Brown *et al.* , *Phys. Rev. Lett.* **65**, 2491 (1990).

95. A. Vaccarino, in *Proceedings of "Lattice 90", Tallahassee, U.S.A*, Eds. U. M. Heller, A. D. Kennedy and S. Sanielevici, *Nucl. Phys.* **B(PS)20**, 263 (1991).

96. S. Gottlieb *et al.* , *Phys. Rev.* **D38**, 2245 (1988).

97. C. Bernard *et al.* , Indiana University Preprint, IUHET-210, 1991.

98. R.V. Gavai, J. Potvin and S. Sanielevici, *Phys. Rev.* **D37**, 1343 (1988).

99. R. V. Gavai and F. Karsch, *Nucl. Phys.* **B261**, 273 (1985).

100. F. Fucito and S. Solomon, *Phys. Rev. Lett.* **55**, 2641 (1985); F. Fucito, S. Solomon and C. Rebbi, *Nucl. Phys.* **B248**, 615 (1984); *Phys. Rev.* **D31**, 1461 (1985).

101. J. B. Kogut and D. K. Sinclair, *Phys. Rev. Lett.* **60**, 1250 (1988).

102. J. B. Kogut and D. K. Sinclair, *Nucl. Phys.* **B344**, 238 (1990).

103. J. B. Kogut, J. Polonyi, H. W. Wyld, and D. K. Sinclair, *Phys. Rev. Lett.* **54**, 1475 (1988).

104. J. B. Kogut and D. K. Sinclair, *Nucl. Phys.* **B295[FS21]**, 465 (1988).

105. N. Christ, in *Proceedings of "Lattice '91", KEK, Tsukuba, Japan*, Ed. M. Fukugita, to appear in *Nucl. Phys.* **B(PS)**.

106. Y. Iwasaki, K. Kanaya, S. Sakai, T. Yoshié, University of Tsukuba Preprint, UTHEP-226, 1991 and in *Proceedings of "Lattice '91", KEK, Tsukuba, Japan*, Ed. M. Fukugita, to appear in *Nucl. Phys.* **B(PS)**.

107. P. Hasenfratz, F. Karsch and I. O. Stamatescu, *Phys. Lett.* **133B**, 221 (1983); T. Çelic, J. Engels and H. Satz, *Phys. Lett.* **133B**, 427 (1983).

108. M. Fukugita, S. Ohta and A. Ukawa, *Phys. Rev. Lett.* **57**, 1974 (1986).

109. Ph. de Forcrand *et al.* , *Phys. Rev. Lett.* **58**, 2011 (1987).

110. Y. Iwasaki and Y. Tsuboi, *Phys. Lett.* **B222**, 269 (1989).

111. A. Ukawa, in *Proceedings of "Lattice '88", FNAL, Batavia, U.S.A*, Eds. A. S. Kronfeld and P. B. Mackenzie, *Nucl. Phys.* **B(PS)9**, 463 (1989).

112. R. Gupta *et al.* , *Phys. Rev.* **D40**, 2072 (1989).

113. K. Bitar, A. D. Kennedy and P. Rossi, *Phys. Lett.* **B234**, 333 (1990).

114. The HEMCGC collaboration, K. M. Bitar *et al.* , *Phys. Rev.* **D43**, 2396 (1991).

115. The MILC collaboration, T. A. DeGrand *et al.* , in *Proceedings of "Lattice '91", KEK, Tsukuba, Japan*, Ed. M. Fukugita, to appear in *Nucl. Phys.* **B(PS)**.

116. Y. Deng, in *Proceedings of "Lattice '88", FNAL, Batavia, U.S.A*, Eds. A. S. Kronfeld and P. B. Mackenzie, *Nucl. Phys.* **B(PS)9**, 334 (1989).

117. J. Engels *et al.* , *Phys. Lett.* **B252**, 625 (1990).

118. J. Hoek, *Nucl. Phys.* **B339**, 732 (1990).

119. F. Karsch and U. Heller, *Nucl. Phys.* **B251[FS13]**, 254 (1985); *Nucl. Phys.* **B258**, 29 (1985).

120. B. Petersson, in *Proceedings of "Quark Matter '90", Menton, France*, Eds. J. P. Blaizot *et al.* , *Nucl. Phys.* **A525**, 237c (1991).

121. R. V. Gavai and A. Gocksch, *Phys. Rev.* **D33**, 614 (1986); K. Redlich and H. Satz, *Phys. Rev.* **D33**, 3747 (1986); T. A. DeGrand and C. E. DeTar, *Phys. Rev.* **D35**, 742 (1987); S. Huang, K. J. M. Moriarty, E. Myers and J. Potvin, *Z. Phys.* **C50**, 221 (1991).

122. M. Gao, in *Proceedings of "Lattice '88", FNAL, Batavia, U.S.A*, Eds. A. S. Kronfeld and P. B. Mackenzie, *Nucl. Phys.* **B(PS)9**, 368 (1989).

123. T. A. DeGrand and C. E. DeTar, *Phys. Rev.* **D34**, 2469 (1986).

124. N. Attig *et al.* , *Phys. Lett.* **B209**, 65 (1988).

125. R. V. Gavai, M. Lev, B. Petersson and H. Satz, *Phys. Lett.* **B203**, 295 (1988).

126. H. Satz, in *Quark-Gluon Plasma*, Advanced Series on Directions in High Energy Physics, Vol. 6, Ed. R. C. Hwa, World Scientific 1990, p. 593.

127. C. DeTar and J. Kogut, *Phys. Rev. Lett.* **59**, 399 (1987); *Phys. Rev.* **D36**, 2828 (1987).

128. S. Gottlieb *et al.* , *Phys. Rev. Lett.* **59**, 1881 (1987).

129. A. Gocksch, P. Rossi and U. Heller, *Phys. Lett.* **B205**, 334 (1988).

130. The MT_c Collaboration, K. D. Born et al. , *Phys. Rev. Lett.* **67**, 302 (1991).

131. C. Bernard et al. , University of Arizona Preprint, AZPH-TH/91-60, 1991.

132. J. Potvin and C. Rebbi, in *Proceedings of "Lattice 90", Tallahassee, U.S.A*, Eds. U. M. Heller, A. D. Kennedy and S. Sanielevici, *Nucl. Phys.* **B(PS)20**, 317 (1991)

133. T. Çelic, J. Engels and H. Satz, *Z. Phys.* **C22**, 301 (1984).

134. Z. Frei and A. Patkós, *Phys. Lett.* **B229**, 102 (1989).

135. L. Kärkkäinen and K. Rummukainen, in *Proceedings of "Lattice 90", Tallahassee, U.S.A*, Eds. U. M. Heller, A. D. Kennedy and S. Sanielevici, *Nucl. Phys.* **B(PS)20**, 309 (1991).

136. J. I. Kapusta, *Nucl. Phys.* **B148**, 461 (1979).

137. P. Hasenfratz and F. Karsch, *Phys. Lett.* **125B**, 308 (1983).

138. N. Bilić and R. V. Gavai, *Z. Phys.* **C23**, 77 (1984)

139. R. V. Gavai, *Phys. Rev.* **D32**, 519 (1985).

140. J. Kogut et al. , *Nucl. Phys.* **B225[FS9]**, 93 (1983).

141. P. E. Gibbs, *Phys. Lett.* **B182**, 369 (1986).

142. I. M. Barbour et al. , *Nucl. Phys.* **B275[FS17]**, 296 (1986).

143. J. Engels and H. Satz, *Phys. Lett.* **159B**, 151 (1985); B. Berg et al. , *Z. Phys.* **C31**, 167 (1986).

144. D. Toussaint, in *Proceedings of "Lattice '89", Capri, Italy*, Eds. N. Cabibbo et al. , *Nucl. Phys.* **B(PS)17**, 248 (1990).

145. E. Dagotto, A. Moreo and U. Wolff, *Phys. Rev. Lett.* **57**, 1292 (1986);*Phys. Lett.* **B186**, 395 (1987).

146. A. Gocksch, *Phys. Rev. Lett.* **61**, 2054 (1988).

147. F. Karsch and K.-H. Mütter, *Nucl. Phys.* **B313**, 541 (1989).

148. I. M. Barbour, in *Proceedings of "Lattice '89", Capri, Italy*, Eds. N. Cabibbo et al. , *Nucl. Phys.* **B(PS)17**, 243 (1990).

149. S. Gottlieb et al. , *Phys. Rev. Lett.* **59**, 2247 (1987); *Phys. Rev.* **D38**, 2888 (1988).

150. L. McLerran, *Phys. Rev.* **D36**, 3291 (1987).

151. R. V. Gavai, J. Potvin and S. Sanielevici, *Phys. Rev.* **D40**, 2743 (1989).

152. T. Banks and E. Rabinovici, *Nucl. Phys.* **B160**, 349 (1979).

153. F. Karsch, E. Seiler and I. O. Stamatescu, *Phys. Lett.* **131B**, 138 (1983);*Phys. Lett.* **157B**, 60 (1985); D. Espriu and J. F. Wheater, *Nucl. Phys.* **B258**, 101 (1985); T. Munehisa and Y. Munehisa, *Z. Phys.* **C32**, 531 (1986).

154. H. G. Evertz, J. Jersák and K. Kanaya, *Nucl. Phys.* **B285[FS19]**, 229 (1987).

155. P. H. Damgaard and U. M. Heller, *Nucl. Phys.* **B294**, 253 (1987).

156. E. Fradkin and S. H. Shenker, *Phys. Rev.* **D19**, 3682 (1979).

157. J. Jersák, DESY/HLRZ preprint, DESY 89-115, HLRZ 89-45, to appear in *Higgs particles – Physics Issues and Searches in High Energy Collisions*, Ed. A. Ali, Plenum Press.

158. W. Langguth and I. Montvay, *Phys. Lett.* **165B**, 135 (1985); J. Jersák, C. B. Lang, T. Neuhaus and G. Vones, *Phys. Rev.* **D32**, 2761 (1985); V. P. Gerdt et al. , *Nucl. Phys.* **B265[FS15]**, 145 (1986); W. Langguth, I. Montvay and P. Weisz, *Nucl. Phys.* **B277**, 11 (1986).

159. R. E. Shrock, in *Proceedings of "Lattice '87"*, *Seillac, France*, Eds. A. Billoire et al. , *Nucl. Phys.* **B(PS)4**, 373 (1988).

160. R. V. Gavai and H. Satz, *Phys. Lett.* **145B**, 248 (1984); G. Curci and R. Tripiccione, *Phys. Lett.* **151B**, 143 (1985); J. Kiskis, *Phys. Rev.* **D35**, 1456 (1987); J. Engels, J. Fingberg and M. Weber, *Nucl. Phys.* **B332**, 737 (1990).

161. K. Jansen and P. Seuferling, *Nucl. Phys.* **B343**, 507 (1990).

162. R. V. Gavai, U. M. Heller, F. Karsch, T. Neuhaus and B. Plache, in *Proceedings of "Lattice '91"*, *KEK, Tsukuba, Japan*, Ed. M. Fukugita, to appear in *Nucl. Phys.* **B(PS)**.

UPPER BOUND ON THE HIGGS MASS

ANNA HASENFRATZ
Department of Physics, University of Arizona
Tucson, AZ-85721

ABSTRACT

The scalar Higgs boson is the most intriguing, still missing part of the electroweak Standard Model. Though its mass is a free parameter of the theory, the consistency of the effective model restricts its value. Recent lattice calculations estimate an upper bound for the Higgs mass around $650 GeV$. This review summarizes the different techniques leading to this value and discusses some of the conceptual problems of the calculations.

1. Introduction

The minimal Standard Model of electroweak interactions describes the low and medium energy data exceptionally well [1]. While the Standard Model is not expected to be valid at arbitrary large energies, experimental data gives no information yet of the energy range where the model breaks down and new interactions come into play. In fact, with present experimental accuracy and energy range there is no signal for the breakdown of the Standard Model at all. Nevertheless, the Standard Model can be only a low energy effective theory, a theory with possibly large, but finite cutoff. In the infinite cut-off limit the Standard Model describes a non-interacting free field theory, it is "trivial" [2].

The success of the Standard Model is mainly due to its many perturbative properties. The $SU(2)$ gauge coupling and the Yukawa couplings (with the possible exception of the yet unknown top-quark-scalar interaction) are small not only at the physical scale, but because of asymptotic freedom, at high energies too. Despite the perturbative success of the Standard Model, one of its main building element, the scalar Higgs particle is still unknown [1]. It has not been detected experimentally nor

---

[1] There are attempts to formulate models without a scalar boson. However, all these models give the Standard Model as low energy effective theory with a scalar particle. In this review we do not consider these alternative high(er) energy theories but rather restrict ourselves to the low energy Standard Model and study its properties as an effective theory [3].

has its mass been predicted theoretically. The present experimental lower bound on the Higgs mass is around $50 GeV$.

The Higgs particle, which is related to the renormalized scalar coupling is a free parameter of the theory. The problem we are considering in this review is the following: is the Higgs mass completely arbitrary within the Standard Model or is it restricted in any way in a physically sensible theory? Alternatively, can the low energy effective Standard Model be consistent with arbitrary renormalized scalar coupling or is λ_R restricted physically? As the scalar coupling is not asymptotically free, even if the renormalized coupling λ_R is small, the running coupling at high energies can be large. The Higgs phenomena is in general non-perturbative.

Already 1-loop perturbation theory suggests that λ_R cannot be arbitrary large [4]. Consider the scalar sector of the Standard Model and denote the bare coupling at the cut-off scale by λ_b. Let us assume for a moment that the 1-loop renormalization group β function is valid even outside the perturbative region. Then the running coupling at some energy E is

$$\lambda(E) = \frac{1}{1/\lambda_b + c \, ln(\Lambda_{cut}/E)},$$ (1)

where $c = 12/8\pi^2$ for the 4-dimensional $O(4)$ scalar model without gauge or fermionic fields. At the scale of the Higgs mass

$$\lambda(m_H) \equiv \lambda_R = \frac{1}{1/\lambda_b + c \, ln(\Lambda_{cut}/m_H)}, \leq \frac{1}{c \, ln(\Lambda_{cut}/m_H)}$$ (2)

the upper bound corresponding to infinite bare coupling $\lambda_b = \infty$. In a physically sensible theory the cut-off should be much larger then the relevant physical energy scales. If we require, for example, $\Lambda_{cut}/m_H \geq 3\pi$, we get $\lambda(m_H) \leq 8\pi^2/12 ln(3\pi)$, which implies for the ratio of the Higgs mass and vacuum condensate F_G

$$R = \frac{m_H}{F_G} = \sqrt{4\lambda(m_H)} \leq 3.4 \qquad or$$
$$m_H \leq 850 GeV.$$ (3)

While the above example illustrates why the Higgs particle mass is bounded in the effective Standard Model, it should not be considered more than an illustration. The obvious shortcomings are

1) The upper bound of the Higgs mass is a non-perturbative problem. It is related to how an infinitely strong bare coupling changes from the cut-off scale to physical energies. Using the 1-loop β function is only a crude approximation.

2) Requiring $\Lambda_{cut}/m_H \geq 3\pi$ is an arbitrary constraint. The cut-off is not a well defined quantity in an effective theory. A finite cut-off nevertheless has measurable, physical consequences at lower energies. Instead of fixing Λ_{cut}/m_H to an arbitrary number, one has to analyze the physical consequences of a finite cut-off. One can require that some specific physical quantity shows for example less then 1% or 5 % cut-off violation effects at the scale of the Higgs mass.

3) Whenever we are working with effective theories, we have to deal with renormalization effects. Regularization schemes can differ by finite inverse cut-off corrections and can show finite cut-off violations at different levels. Even if we require the same physical cut-off violation effects on physical amplitudes at the Higgs mass scale, different regularizations can predict different upper bound values for m_H. How much the upper bound changes within (reasonable) regularization schemes determines the universality of the upper bound calculations.

The above problems were extensively studied in recent years. In the following we will discuss different non-perturbative calculations for the upper bound problem, describe a physical way to characterize cut-off effects and consider the universality problem within certain type of regularization schemes [5].

2. The Model; Definitions and General Considerations

2.1. The role of the gauge fields

In the Standard Model the interactions between the scalar sector and the gauge fields are weak. The SU(2) and U(1) gauge couplings are small, and the effect of the strongly interacting SU(3) sector is indirect, only through the quarks via the Yukawa coupling. For light quarks the Yukawa coupling is small, the interaction is weak.

On the other hand for the upper bound problem we are interested in the strongest possible scalar coupling. We start with infinite bare coupling and study how the running coupling decreases with decreasing energy. It is expected that the small gauge couplings will not influence the running of the scalar coupling. While some non-perturbative effect can not be completely excluded, it is a plausible assumption that the upper bound problem can be studied within the scalar sector taking the gauge fields into account perturbatively.

Without gauge interaction the lowest value of the renormalized coupling $\lambda_R = 0$, the Higgs boson can be massless. The gauge field has an important effect for the lower bound of the Higgs mass though. This problem has been extensively studied in the literature and we do not discuss it in this review [6].

2.2. MC study of coupled $SU(2)$ gauge-scalar system

Monte Carlo studies of the $SU(2)$ gauge-scalar system support the assumption that the gauge field is coupled perturbatively to the scalar sector.

Numerical simulations of the gauge-scalar model are rather difficult [7]. The main problem is the existence of two different mass scales, $m_H \sim 10 m_W$. If the heavy particle is light enough in lattice units that it is not distorted by lattice artifacts $(m_H a \leq 1)$, the light particle is too light and distorted by finite lattice size effects on presently accessible lattices. Nevertheless numerical results from gauge-scalar system simulations are in agreement with other, better controlled calculations. The relation between the W mass and field expectation value follows the perturbative form. The predicted upper bound is about $730 GeV$.

The most important result from the full gauge-scalar calculations is the strong indication that it is indeed sufficient to handle only the scalar part of the model non-perturbatively.

2.3. Connection between the $O(4)$ model and physical quantities

We need to express m_W and m_H in terms of the $O(4)$ model observables.

The action of the continuum $O(4)$ model is

$$S_c = \int d^4 x \left[\frac{1}{2} \sum_{\alpha=1}^{4} (\partial_\mu \phi_\alpha)^2 + \frac{1}{2} m^2 (\sum_{\alpha=1}^{4} \phi_\alpha^2) + \frac{\lambda_c}{4} (\sum_{\alpha=1}^{4} \phi_\alpha^2)^2 - j\phi^0 \right]. \quad (4)$$

In infinite volume the spectrum contains one massive boson and 3 massless Goldstone particles in the broken phase. The basic observables are the vacuum expectation value and the propagators

$$\Sigma = \langle \sigma \rangle, \qquad \left\langle \phi^\alpha(p)\phi^\beta(-p) \right\rangle_c, \qquad (5)$$

where we use $\phi^\alpha = (\sigma, \pi^1, \pi^2, \pi^3)$. We chose the direction of the symmetry breaking to be the first component. The wave function renormalization constant and the scalar mass is defined through the propagators

$$
\begin{aligned}
G_G|_{p^2 \to 0} &= \left\langle \pi^i(p)\pi^j(-p) \right\rangle_{p^2 \to 0} = \delta_{ij} Z_G(p^2 + O(p^4))^{-1}, \\
G_\sigma|_{p^2=\mu^2} &= \langle \sigma(p)\sigma(-p) \rangle_{p^2=\mu^2} = Z_\sigma(\mu^2 + m_H^2(\mu))^{-1}.
\end{aligned}
\qquad (6)
$$

For $\mu = O(m_H)$, $m_H(\mu)$ is close to m_H and the relation can be calculated perturbatively. Note that the σ propagator G_σ is divergent at $p^2 = 0$ due to the Goldstone contribution. The relation of the $O(4)$ observables to m_W is

$$m_W^2 = \frac{1}{4}g_2^2 F_G^2 + O(g_2^4 ln(g_2^2)), \qquad (7)$$

where F_G is the Goldstone decay constant

$$F_G = Z_G^{-1/2}\Sigma. \qquad (8)$$

The derivation of the above formulae can be found in the literature and we refer the reader to [8].

2.4. The lattice regularized $O(4)$ model

Most of the non-perturbative studies used the hypercubic lattice regularized version of the $O(4)$ model. The lattice action is

$$S = -\kappa \sum_{n,\mu,\alpha} \varphi_n^\alpha(\varphi_{n+\mu}^\alpha + \varphi_{n-\mu}^\alpha) + \sum_{n,\alpha} \varphi_n^\alpha \varphi_n^\alpha + \lambda \sum_n (\sum_\alpha \varphi_n^\alpha \varphi_n^\alpha - 1)^2 - J \sum_n \varphi_n^0, \quad (9)$$

where $\kappa, \lambda \geq 0$ are the bare hopping and quartic coupling parameters respectively. The external source allows the proper definition of spontaneous symmetry breaking. Its presence is essential in the use of chiral perturbation theory. The cut-off is proportional to the inverse of the only dimensional parameter, the lattice spacing a. We define the cut-off on the hypercubic lattice as $\Lambda_{cut} = \pi/a$. In the following we denote the dimensionless lattice mass by m_R, $m_R = am_H$.

The continuum normalization is obtained by rescaling

$$\phi^\alpha a = \sqrt{2\kappa}\varphi^\alpha, \quad ja^3 = \frac{J}{\sqrt{2\kappa}}. \qquad (10)$$

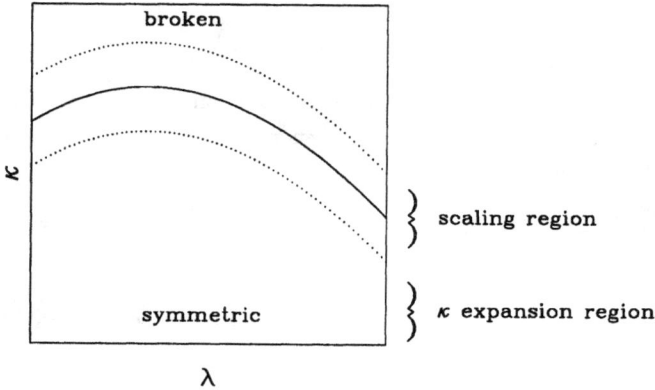

FIGURE 1

The sketch of the phase diagram of the lattice regularized $O(4)$ model

The relation between the continuum and lattice parameters are

$$m^2 a^2 = \frac{1 - 2\lambda}{\kappa} - 8,$$

$$\lambda_c = \frac{\lambda}{\kappa^2} \tag{11}$$

corresponding to the action Eq. (4).

The phase diagram of the lattice regularized model is sketched in Fig.1. The small κ phase is symmetric, separated by a second order phase transition line from the large κ broken phase. At $\lambda = 0$ the critical point is $\kappa_c = 1/8$ or $m^2 a^2 = 0$, the perturbative fixed point. The infinite cutoff limit corresponds to the second order phase transition line where the dimensionless lattice mass $m_R = 0$ therefore $a = 0$. Effective theories with large but finite cut-off are described by the region around the phase transition line. The point (λ, κ) in the phase space where the dimensionless lattice mass is m_R corresponds to an effective theory with

$$\Lambda_{cut} \sim \frac{\pi}{a} = \pi \frac{m_H}{m_R}. \tag{12}$$

Triviality in this language means that no other fixed point but the perturbative Gaussian fixed point at $\lambda = 0, \kappa = 1/8$ exists. Every point of the second order phase transition surface describes the same infinite cut-off theory with $\lambda_R = 0$.

Close to the phase transition line physical observables behave in a universal way. If this scaling behavior is governed by the Gaussian fixed point, the critical exponents can be calculated perturbatively. The scaling formulae for the $O(n)$ model up to two loop level are

$$
\begin{aligned}
\Sigma/a &\sim \tau^{1/2}|ln|\tau||^{\frac{3}{n+8}} , \\
m_R &\sim \tau^{1/2}|ln|\tau||^{-\frac{n+2}{2(n+8)}} ,
\end{aligned}
\tag{13}
$$

where $\tau = \kappa - \kappa_c$. As these scaling laws are derived at the perturbative fixed point, they are obviously valid for small λ. If the theory is trivial, the same scaling formulae should be valid everywhere around the phase transition surface including $\lambda = \infty$. Scaling behavior can be used to check and understand triviality.

In the following we will discuss a few of the different methods recently used to study the broken phase of the scalar $O(4)$ model. We consider only the $\lambda_b = \infty$ case and summarize the results in an "envelope plot" showing the ratio m_H/F_G as the function of $1/m_R = \Lambda_{cut}/\pi m_H$. The symmetric phase of the model has been also extensively studied [9,10]. We do not discuss those calculations here unless they are directly related to the method used in the broken phase. There are several calculations investigating the 1-component ϕ^4 model with similar techniques. The reader is referred to the original literature [10,11].

3. Non-perturbative Studies of the $O(4)$ Model; Analytical Calculations

3.1. Approximate non-perturbative renormalization group equation

The first non-perturbative analytical estimate for the upper bound of the Higgs mass was obtained by solving the exact renormalization group equations in the local potential approximation [12]. An arbitrary $O(n)$ potential $V(\sum_{\alpha=1}^{n} \phi_\alpha^2)$ is allowed but no derivative couplings are considered. The change in the action after a renormalization group transformation $\Lambda_{cut} \to e^{-t}\Lambda_{cut}$ is projected back to the subspace without derivative couplings

$$
\begin{aligned}
S &= \int d^d x \left[\frac{1}{2} \sum_{\alpha=1}^{n} (\partial_\mu \phi_\alpha)^2 + V(\sum_{\alpha=1}^{n} \phi_\alpha^2) \right] \overrightarrow{\Lambda_{cut} \to e^{-t}\Lambda_{cut}} \\
&\longrightarrow S(t) = \int d^d x \left[\frac{1}{2} \sum_{\alpha=1}^{n} (\partial_\mu \phi_\alpha)^2 + V(\sum_{\alpha=1}^{n} \phi_\alpha^2, t) \right].
\end{aligned}
\tag{14}
$$

132

No further approximation is necessary to derive the RG equations for the potential $V(\sum_{\alpha=1}^{n} \phi_\alpha^2, t)$. The function

$$f(x/n^{1/2}, t) = n^{-1/2} \frac{\partial}{\partial x} V(x, t) \tag{15}$$

satisfies the differential equation

$$\frac{\partial f}{\partial t} = \frac{A_d}{2} \left[\frac{1}{n} \frac{f''}{1+f'} + \frac{n-1}{n} \frac{xf'-f}{x^2+xf} \right] + \left(1 - \frac{d}{2}\right) xf' + \left(1 + \frac{d}{2}\right) f, \tag{16}$$

where prime denotes $\partial/\partial x$, A_d is a constant ($1/8\pi^2$ for d=4) and n is the number of real scalar fields (=4).

The non-linear partial differential equation (16) can be solved numerically. From the renormalization group flow the phase diagram, fixed point structure and the critical exponents can be extracted [13]. The method predicts correctly the different fixed points in $2 < d < 4$ and gives reasonable critical exponents in $d = 3$ for the ferromagnetic fixed point. In 4 dimensions it predicts the existence of only one fixed point, the Gaussian one with correct critical exponents.

For the upper bound envelope calculation the RG equation (16) is used to run the system from the strongly coupled region to the vicinity of the Gaussian fixed point where perturbation theory can be used to connect m_H and F_G. The envelope obtained with this method is shown in Fig.2.

The local potential method cannot be improved systematically and it is not possible to estimate the error caused by the approximation. Keeping only one derivative interaction term forces the wave function renormalization constants Z_G and Z_σ to be 1. Other numerical and analytical methods predict that in the 4 dimensional $O(4)$ model both Z_G and Z_σ are close to 1, therefore it is not surprising that the envelope and upper bound obtained in the local potential approximation agrees well with those calculations.

3.2. Strong coupling expansion combined with perturbative RG

This (semi)-analytic technique covers regions with large correlation length that are outside the reach of MC calculations. Together with numerical simulations it gives a complete, controlled description of the model.

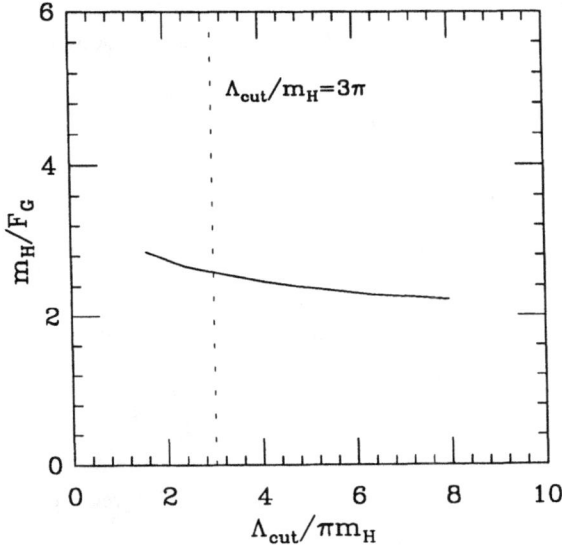

FIGURE 2

The ratio m_H/F_G as the function of the scale change $\Lambda_{cut}/\pi m_H$ for $\lambda = \infty$ as predicted by the approximate non-perturbative RG method.

The analysis is performed at a fixed, arbitrary λ. Hypercubic lattice regularization is considered. The method combines strong coupling (small κ) expansion with weak coupling (perturbative) RG equations (Fig.1.). It has been applied first to the 1-component model and later has been extended to general $O(n)$ theories [10,14].

First consider the symmetric phase. At $\kappa = 0$ the model is trivially solvable. For small κ a high order κ series (high temperature expansion) is calculated for the relevant quantities m_R, λ_R, Z_G. For example, for m_R one has

$$m_R = \frac{1}{\sqrt{\kappa}} \sum_{\nu=0}^{\infty} m_R^{(\nu)}(\lambda)\kappa^{\nu}, \qquad (17)$$

where the coefficients m_R^{ν} have been calculated up to $\nu = 14$ for the $O(n)$ model. The expansion can be controlled from $m_R = \infty$ up to $m_R \approx 0.5$ corresponding to the region from $\kappa = 0$ to approximately $\kappa \leq 0.95\kappa_c$.

Around the critical region weak coupling perturbative renormalization group equations are used. The dependence of the renormalized coupling on the cut-off is described by the Callan-Symanzik β function

$$-\Lambda(\frac{\partial \lambda_R}{\partial \Lambda}) = \beta(m_R, \lambda_R), \qquad (18)$$

or in lattice units

$$m_R(\frac{\partial \lambda_R}{\partial m_R}) = \beta(m_R, \lambda_R). \qquad (19)$$

If we know the β function, the above equation can be integrated from the region $\kappa \approx 0.95\kappa_c$ ($m_R \approx 0.5$) up to the critical line where $m_R = 0$. The initial values for m_R and λ_R are obtained from the high temperature series.

The Callan-Symanzik β function is known up to 3 loops. Assuming that the 3-loop perturbative formula is sufficient to describe the region from $m_R = 0$ to $m_R \approx 0.5$, the combination of the two procedures will cover the whole symmetric phase.

Around the critical region where $m_R \to 0$ the renormalized coupling λ_R goes to zero as well. The dependence of m_R on the coupling is given by the formula

$$m_R = Ce^{-1/\beta_0 \lambda_R}(\beta_0 \lambda_R)^{-\beta_1/\beta_0^2}. \qquad (20)$$

This functional dependence is valid in both the symmetric and broken phase though with different C_s and C_b coefficients. The relation between C_s and C_b can be calculated. If we define the renormalized coupling in the broken phase as

$$\lambda_R = Z_G \frac{3m_H^2}{\Sigma^2}, \qquad (21)$$

we get

$$C_b = e^{1/6}C_s. \qquad (22)$$

The constant C_s can be obtained by comparing the integrated Eq (19) and Eq (20). Using C_b the RG equation is used to predict m_R within the scaling region in the broken phase. Assuming the 3-loop renormalization group equation covers the region $m_R \leq 0.5$ in the broken phase too, the envelope of Fig.3. is obtained for the $O(4)$ model.

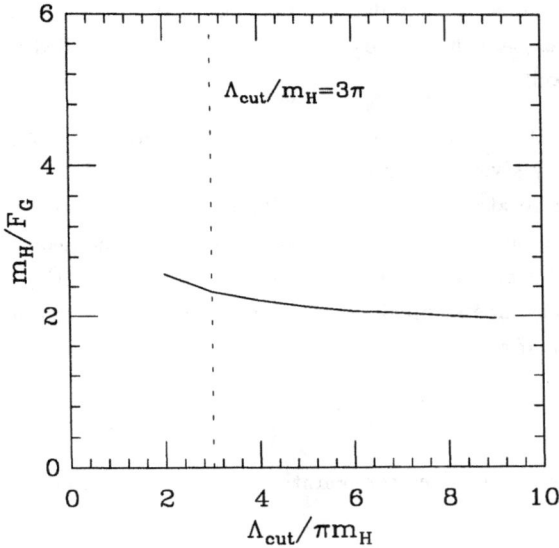

FIGURE 3

The ratio m_H/F_G as the function of the scale change $\Lambda_{cut}/\pi m_H$ for $\lambda = \infty$ as predicted by the method of Ref. [10,14].

The errors of this analysis come from the truncation of the small κ expansion, the truncation of the perturbative expansion and the propagation of those errors during the subsequent steps of the calculation.

4. Non-perturbative Studies of the $O(4)$ Model; Monte-Carlo calculations

4.1. Direct measurements

The first numerical MC calculation for the upper bound problem dates back to 1985 [15]. Since then several extensive numerical simulations for the $O(4)$ model on hypercubic lattices have been published. The methods used were occasionally quite different but the final results are in agreement [16-19].

For the upper bound problem one does not have to work at very large correlation length. It is sufficient to make the cut-off only a few times larger than the scalar mass, which makes numerical simulations feasible. That, and the relative simplicity of the $O(4)$ model explain that these calculations are well controlled statistically and systematically as well.

On a finite lattice there is no spontaneous symmetry breaking. While in the broken phase on any given configuration the spins are ordered, the direction of the symmetry breaking rotates around in the $O(4)$ space due to tunneling between the infinite degenerate ground states. On a finite lattice the spontaneous magnetization averages to zero. The infinite volume limit can be mimiced by rotating the orientation of the spontaneous symmetry breaking to a given direction, for example the first, in every configuration. If we define

$$M^\alpha = \frac{1}{V} \sum_n \phi_n^\alpha, \tag{23}$$

the σ field operator on any given configuration is approximated as

$$\bar\sigma_n = \frac{\phi_n^\alpha M^\alpha}{|M|}. \tag{24}$$

The vacuum expectation value is $\Sigma = \langle\bar\sigma\rangle$, while $\langle\bar\sigma_o\bar\sigma_n\rangle$ gives the longitudinal propagator and m_H. The transversal components are associated with the Goldstone fields. This rotation technique introduces systematical errors in finite volume. The difference between Σ and the measured value of the projection $\langle M^\alpha\phi_x^\alpha/\mid M\mid\rangle$ is of order $V^{-\frac{1}{2}}$ and thus goes to zero as the volume goes to infinity . The volume dependence of the projected correlation function is less understood though [20-22].

We will discuss results at $\lambda = \infty$ only.

The field expectation value Σ and scalar mass m_R obtained using the rotation technique are plotted in Fig.4. The one loop scaling relation is satisfied within errors up to $m_R \sim 0.8$. This supports not only the triviality of the scalar model but shows that there is a relatively wide scaling region where physical quantities are universal.

Finally we need the Goldstone boson wave function renormalization constant. To calculate Z_G we have to investigate the Goldstone propagator. The Goldtsone bosons are massless in the infinite volume limit and light though massive in a finite box. We cannot avoid the problem of light/massless particles associated with the spontaneous symmetry breaking of a continuous symmetry any longer. Chiral

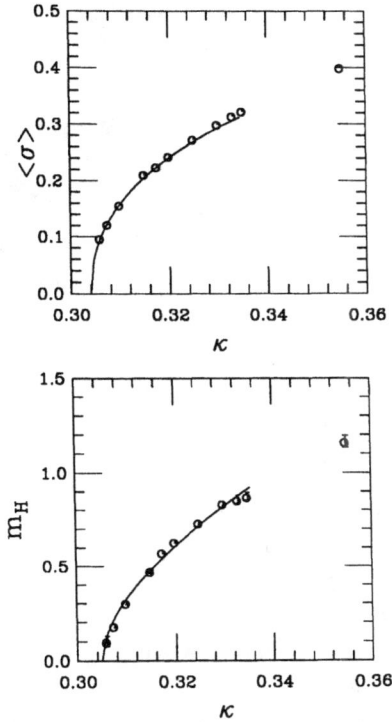

FIGURE 4

The field expectation value and the scalar mass at $\lambda = \infty$. The solid curves correspond to the 1-loop scaling formulae.

perturbation theory offers a well controlled way to describe the effects of Goldstone bosons in finite volume. Its strength is that instead of trying to avoid finite volume effects associated with the light Goldstone particles, it uses the strong dependence on finite volume and finite external source to extract infinite volume, zero external source information. Chiral perturbation theory helps not only in controlling finite size effects but it provides a theoretically correct definition for Σ and Z_G.

4.2. Finite size effects and chiral perturbation theory

The basic physical assumption behind chiral perturbation theory is easy to formulate. The low energy behavior of a system with spontaneously broken continuous symmetry is determined by the massless (in finite volume, light) Goldstone particles and their symmetry properties. The details of the original system are unimportant as long as the massive particles are heavy compared to the Goldstone bosons and the volume is large enough so the finite size effects associated with the massive particles are negligible.

The low energy behavior can be described by an effective Lagrangean with the same symmetry properties as the original system. The couplings of this effective model depend on the low energy infinite volume physical constants of the original theory. This Lagrangean has to be simple enough to allow a systematic perturbative expansion of finite volume/finite external field quantities. This way chiral perturbation theory connects the low energy infinite volume constants to the numerically easier to measure finite volume/finite field quantities.

This theory was first applied to the finite volume, finite temperature dependence of QCD with light quarks [23]. The application for the $O(4)$ problem proved to be an exceptionally good testing ground for the applicability and accuracy of chiral perturbation theory [18]. In recent years the details of chiral perturbation theory up to 2 loop level have been worked out for a variety of statistical mechanics models in addition to the applications for QCD and the Higgs upper bound problem [20,22]. Here we summarize only the most basic formulae as applied to the $O(4)$ model.

In the following we concentrate on the finite volume effects related to the Goldstone bosons and assume that the finite size effects caused by the massive particle are negligible, $m_R \ll 1/L$. Consider the model with a constant, $O(4)$-symmetry breaking external source j (Eq. (4)). For small values of j, the qualitative properties of the system at large volume are controlled by the parameter $\Sigma j V$. If this parameter is small then the expectation value of the field is small and the correlation functions are approximately $O(4)$-symmetric. Holding j fixed and letting the volume grow, we eventually reach the region $\Sigma j V \gg 1$ where the expectation value of the field is approximately the same as at infinite volume. If the volume is large and the source is small, the dependence of the expectation values and correlation functions on V and on j are unambiguously determined by the symmetry properties of the model in terms of the two constants Σ and Z_G (or F_G). In the following, we show that numerical data very clearly exhibit the predicted volume dependence. This way the extrapolation to $V = \infty, j = 0$ can be performed.

At infinite volume, the scale of the $O(4)$ symmetric system is set by the mass of the Higgs particle. In the presence of a small symmetry breaking term j, the model contains a second, independent scale, which may be identified by the mass of the Goldstone bosons, m_G. At leading order in j

$$m_G^2 = \frac{Z_G j}{\Sigma}. \tag{25}$$

The symmetry properties of the model are controlled by the mass of the Goldstone bosons if $m_G << m_H$ and the box is large compared to the Compton wave length m_H^{-1}. For a hypercube of size L^4 that means

$$j << \frac{m_H^2 \Sigma^2}{Z_G} \quad ; \quad L >> m_R^{-1}. \tag{26}$$

If the conditions (26) are met, the expectation values and correlation functions can systematically be expanded in inverse powers of L. The detailed properties of the expansion depend on the ratio of L to the Compton wavelength of the Goldstone bosons [18,20]. In the following we give formulas corresponding to small external source, i.e. for the region

$$m_G L \leq 1. \tag{27}$$

The expectation value of the field is given by a series in $1/L^2$, with coefficients that are non-trivial functions of the product $j L^4$. The first two terms in this series may be written in the form

$$jL^4 \left\langle \varphi_x^0 \right\rangle = u^2 \eta(u)(1 + O(L^{-4})),$$
$$u = \Sigma j L^4 (1 + \frac{r}{F_G^2 L^2}), \tag{28}$$
$$\eta(u) = \frac{1}{u} \frac{I_2(u)}{I_1(u)},$$

where $I_1(u)$ and $I_2(u)$ are the standard Bessel functions of imaginary argument. The value of r depends on the shape of the box. For a symmetric hypercube, $r = 0.2107$.

For illustration Fig.5. shows $\left\langle \varphi^0 \right\rangle$ as a function of the external source j for several values of κ on $8^4, 10^4$ and 12^4 lattices [18]. The curves at any given κ correspond to a one parameter fit which determines Σ in the infinite volume, zero external

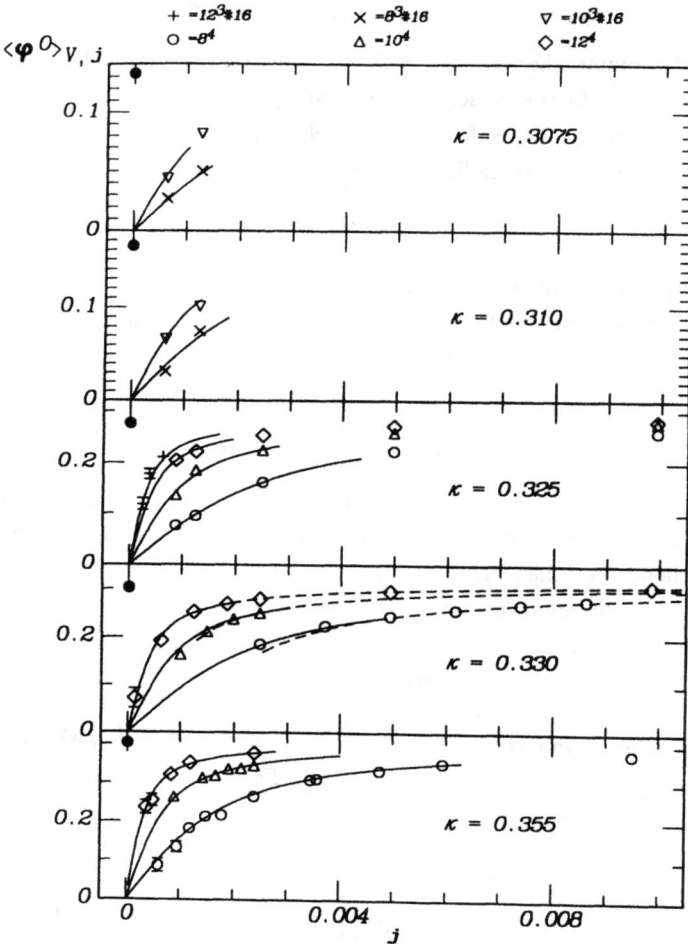

FIGURE 5

The dependence of the field expectation value on the external source on different size lattices.

field limit. The values for Σ in all cases are in complete agreement with the field expectation values obtained using the rotation technique.

The Goldstone wave function renormalization constant Z_G is extracted from

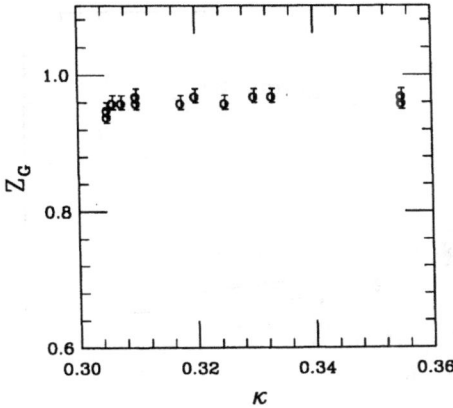

FIGURE 6

The wave function renormalization constant Z_G as a function of κ.

the zero spatial momentum propagator.

$$\frac{1}{L^6} \sum_{\bar{n},\bar{m}} \left\langle \varphi^i_{\bar{n},0} \varphi^k_{\bar{m},t} \right\rangle = G_G(t) \delta^{ik}. \tag{29}$$

Chiral perturbation theory gives the expansion of G_G in powers of $1/L^2$ at fixed t/L and at fixed jL^4

$$G_G(t) = a_G + b_G h(t) + O(1/L^4) \tag{30}$$

where the coefficient b_G is given by

$$b_G = \frac{\Sigma^2}{F_G^2}(1 - \eta(u)). \tag{31}$$

The function $h(t)$ is the spatial integral over the propagator associated with the nonzero modes. On the interval $0 < t < L$, it is given by

$$h(t) = \frac{1}{2L^2}\{(\frac{t}{L} - \frac{1}{2})^2 - \frac{1}{12}\}. \tag{32}$$

Numerical data both at finite and zero external field follows rather precisely the predicted parabolic time dependence. The value obtained for Z_G at several κ values are shown in Fig.6. It is interesting to note that Z_G is finite at the critical point $\kappa_c = 0.304(1)$, in fact it is very close to one in the whole investigated region.

142

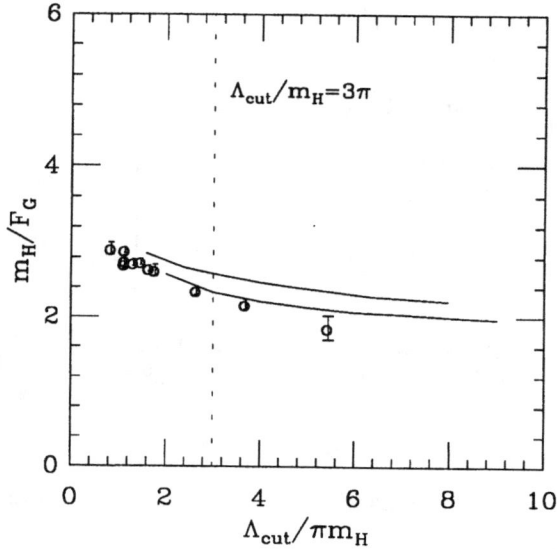

FIGURE 7

Results on the upper bound on hypercubic lattice. The data points are from [16-18], the upper curve is the result of the approximate RG work, the lower curve is from Ref [14].

4.3. Numerical results on the upper bound

Let us collect now the different results on the envelope of m_H/F_G and determine the upper bound for m_H (Fig. 7). The data points from the two large scale collaborations are consistent with each other [16-18]. The analytic results and the approximate RG calculation agrees well with the MC data too. That implies that both the statistical and numerical errors are under control in these works.

The scaling curves for $\langle\sigma\rangle$ and m_R indicates that scaling holds up to $m_H a \approx 0.8$. To be on the safe side we will require $1/m_H a = \Lambda_{cut}/\pi m_H > 3$ which gives the upper bound

$$\frac{m_H}{F_G} \leq 2.6(1),$$
$$m_H \leq 8.2(3)m_W \approx 640(25)GeV.$$

(33)

5. Finite cut-off effects on physical amplitudes

The cut-off in an effective theory does not have a precise definition. Physically it signals the energy scale where the theory loses its meaning or where new interactions from some underlying high energy theory become important. Defining the cut-off in a lattice model as π times the inverse lattice spacing is arbitrary; it would be equally correct to set the cut-off twice as large or three times smaller. In a lattice model with not only nearest neighbor but extended interaction terms it is even less obvious how to define the cut-off. It is equally arbitrary to read off the upper bound for the scalar particle in an $O(4)$ scalar model at $m_H/\Lambda_{cut} = 3\pi$ and not at $m_H/\Lambda_{cut} = 10$ or 3. We need a physically controlled way to define the cut-off of an effective theory to fix these arbitrariness.

The finite cut-off in an effective theory has (in principle) calculable effects on physical quantities at lower energies. To make more precise meaning to the Higgs mass upper bound calculations, we should specify the finite cut-off violation effects on definite physical observables instead of fixing the ratio m_H/Λ_{cut} to some arbitrary number. As finite cut-off effects depend on the regularization scheme, this will make it possible to compare different actions, regularizations as well.

In Ref[24] two physically relevant quantities were suggested: the Goldstone-Goldstone scattering cross section at 90° angle at various center of mass energies and the scalar particle width. The finite cut-off effects on these quantities were estimated in the perturbative, weakly coupled region for two regularization schemes, the F4 and hypercubic nearest neighbor lattice regularizations.

The F4 lattice can be obtained from the hypercubic lattice by removing all sites $n = (n_1, n_2, n_3, n_4)$ with $\sum_\mu n_\mu = odd$. The F4 lattice has a higher rotational symmetry than the hypercubic lattice: in 4 dimensions the lattice propagator has no $O(k^4)$ Lorentz invariance breaking term. One would expect less violation of the $O(4)$ symmetry at low momentum and better scaling properties on the F4 than on the hypercubic lattice.

The details of lattice perturbation theory on F4 lattice and the quite involved 1-loop calculation of the Goldstone scattering amplitude and Higgs decay width both on F4 and hypercubic lattices are discussed in Ref[24,25]. The result on the Goldstone scattering was summarized in Fig.3. of [24]. The finite cut-off violation is expressed through the ratio of the lattice regularized finite cut-off cross section and the continuum infinite cut-off cross section with given value of the center of mass energy W. For example, a 5% cut-off violation with $W/2m_R = 1.0$ corresponds to $m_H/F_G \approx 1$. on the F4 and $m_H/F_G \approx 0.7$ on the hypercubic lattice. One can indeed see that on the F4 lattice the same finite cut-off effects allow a larger ratio of m_H/Λ_{cut} than on the hypercubic lattice; the F4 lattice seems to "behave better".

The cut-off effects on the width of the scalar particle can be expressed in terms of its mass. The calculation gives that the cut-off effects are the same on the width of the scalar particle on the F4 and hypercubic lattices if the masses are related as

$$m_R^{F4} \approx 1.1 m_R^{HC}. \tag{34}$$

In principle similar calculations can be carried out for other regularization schemes making it possible to compare the cut-off effects using well defined physical principles. The calculation are analytically complex and there is no published results other than [24].

One has to note in addition, that the calculation described above is one-loop perturbative. It should be valid as long as the upper bound corresponds to a weakly coupled system, what is the case for both the F4 and hypercubic nearest neighbor actions.

The finite cut-off effects on the Goldstone-Goldstone scattering or on the width of the Higgs particle provide a physically well defined way to compare different regularization schemes. The drawback of this approach is the involved analytic calculation required, and more importantly, that it works only in the perturbative, weakly coupled regime.

A less physical but perhaps more practical way to compare different schemes is to match their cut-off and compare their scaling behavior. At large cut-off all regularization schemes should show scaling behavior. That means not only a universal functional scaling form for the scalar mass, vacuum expectation value, etc. but the same universal value for the ratio m_H/F_G at the same m_H/Λ_{cut} for different schemes with infinite bare quartic coupling. Λ_{cut} is not a well defined physical quantity, but the cut-off of different regularization schemes can be related by using the above universality. The cut-off scales of different regularizations should be matched in such a way that in the scaling region the ratios m_H/F_G agree. One should see deviation from matching when scaling breaks down for at least one of the regularization schemes. One can set a limit on the scaling violation and compare where different schemes reach that limit. Regularization schemes with "poor" scaling will deviate from the universal curve of m_H/F_G at a smaller m_H/Λ_{cut} value while "good" regularization schemes will follow the scaling curve to large ($\sim O(1)$) m_H/Λ_{cut} values. As m_H/F_G is monotonically increasing as the function of m_H/Λ_{cut}, the former, "poor" scaling case will predict a lower upper bound for m_H than the latter, "good" scaling case. It might be possible to construct a regularization scheme which does not violate the set scaling behavior even at $m_H \sim \Lambda_{cut}$. It is a different question whether we can be comfortable with the situation where the physical mass is about the same as the cut-off scale where, in a full theory, one expects new interactions to play important roles.

On a hypercubic lattice both m_H and the vacuum expectation value show scaling consistent within statistical errors with the 2-loop perturbative formulae up to $m_H/\Lambda_{cut} \sim 0.8/\pi$. This indicates that the region $m_H/\Lambda_{cut} \sim 1/3\pi$ is well in the scaling region and reading off the upper bound there is rather conservative. That means that all regularization schemes showing scaling at least up to $m_H/\Lambda_{cut} \sim 1/3\pi$ should give the same upper bound $m_H/F_G \leq 2.6(1)$ when the cut-off 3π times larger than the scalar mass assuming the same cut-off is used for the different schemes.

6. Universality

There is no systematic study yet of the universality of the upper bound problem in the $O(4)$ model though several papers dealt with the problem [26-28]. There is a detailed calculation comparing the hypercubic and F4 lattice nearest neighbor action regularization [27]. It uses the physical criteria of fixed (3% and 0.3%) cut-off violation in the Goldstone-Goldstone scattering amplitude to compare the two schemes. Here we just summarize the result, for the details the reader is referred to the original publication.

It turns out that requiring 3% or 0.3% cut-off effects on the scattering amplitude gives a bound for the scalar mass corresponding to a renormalized coupling still in the perturbative regime even for the F4 lattice, therefore the perturbative calculation of Ref[24] as discussed in the previous chapter can be applied. The results for the upper bound corresponding to 3% cut-off effects on the scattering amplitude are

$$m_H/F_G < 590(60) \quad for F4,$$
$$m_H/F_G < 640(65) \quad for HC,$$

$$(35)$$

and for 0.3% cut-off effects

$$m_H/F_G < 530(60) \quad for F4,$$
$$m_H/F_G < 510(50) \quad for HC.$$

$$(36)$$

Within errors the ratio values are the same for the two different regularizations and very close for the two different scaling violation limits. One can conclude that with the present numerical accuracy the cut-off effects on an F4 and hypercubic lattices with nearest neighbor interactions are not distinguishable if these effects are required to be less than 3%.

A more systematic study of the universality was recently presented in Ref[28]. There the analytically solvable spherical $O(n \to \infty)$ model is considered on F4 lattice with actions containing nearest neighbor and next to nearest neighbor terms. In the 2 dimensional parameter space the action giving the largest ratio m_H/F_G at fixed value of m_H/Λ_{cut} was located. The ratio at this point with $m_H\pi/\Lambda_{cut} = 0.5$ is about 30% higher than that of the nearest neighbor action. The authors argue that similar change can be expected in the $O(4)$ model as well.

While it is well possible to obtain a 30% variation in the upper bound by changing the regularization or action, a world of caution is in order here. In Ref[28] the different actions were *not* compared by using the finite cut-off effects on the Goldstone scattering amplitude or similar physical criteria. The 30% change was obtained by comparing the ratios m_H/F_G at the same value of m_H/Λ_{cut} for the different actions. Here the cut-off is defined as $\Lambda_{cut} = \pi/a$, where a is the nearest neighbor lattice spacing. It is not clear that with a large next to nearest neighbor interaction term the inverse lattice spacing has the same physical meaning as with the original nearest neighbor action. In this case it is essential to use a well defined physical criteria to compare the different actions. If one expects an upper bound outside the perturbative region, perturbative calculation of the finite cut-off violation effects on the Goldstone scattering like in Ref[24] will not be sufficient to compare the different actions. Matching their scale and compare scaling behavior is one possibility in addition to a non-perturbative determination of the cut-off violation effects of the Goldstone scattering or other physical quantity.

7. Summary

The Standard Model of electroweak interactions is necessarily an effective theory with a finite cut-off. In an effective theory the Higgs mass, which is a free parameter of the model is bounded from above by the requirement that all the low energy particles of the cut-off model should be (much) lighter than the cut-off. The upper bound problem is non-perturbative as it corresponds to infinite bare scalar coupling at the cut-off scale.

In recent years several analytical and numerical studies considered the upper bound problem. The results on nearest neighbor hypercubic lattices are all consistent. The numerical calculations are under control especially those using chiral perturbation theory. The upper bound value on a hypercubic lattice is

$$m_H \leq 640(25) GeV \tag{37}$$

when $m_H \leq \Lambda_{cut}/3\pi$ or the Goldstone scattering amplitude has less then 3% cut-off violation.

The upper bound of the scalar particle is regularization dependent. The universality problem has been investigated in depth only for the nearest neighbor F4 and hypercubic lattices in the $O(4)$ model. In these cases the upper bounds turn out to be the same on both lattices within numerical accuracy.

8. References

1) On the present status see for example the plenary talks by J. Ellis and M. Carter at the LP-HEP91 Conference, Geneva, July 1991.

2) K.G. Wilson, Phys. Rev. B4 (1971) 3184;
   K.G. Wilson and J. Kogut, Phys. Rep., 12C (1974) 76;
   M. Aizenman, Phys. Rev. Lett., 47 (1981) 1; Commun. Math. Phys. 86 (1982) 1;
   G.A. Baker and J.M. Kincaid, J. Stat. Phys. 24 (1981) 469;
   D.C.. Brydges, J. Fröhlich and T. Spencer, Commun. Math. Phys., 83 (1982) 123;
   J. Fröhlich, Nucl. Phys. B200 (1982) 281;
   A. D. Sokal, Ann. Inst. H. Poincare A37 (1982) 317;
   M. Aizenman and R. Graham, Nucl. Phys. B225 [FS9] (1983) 261;
   C. Aragaõ de Carvalho, S. Caracciolo and J. Fröhlich, Nucl. Phys. B215 (1983) 209;
   C. B. Lang, Phys. Lett. 155B (1985) 399; Nucl. Phys. B265, FS15 (1986) 630;
   see also the last reference of [6].

3) W. A. Bardeen, C. T. Hill, M. Lindner, Phys. Rev. D41 (1990) 1647;
A. Hasenfratz, P. Hasenfratz, K. Jansen, J. Kuti, Y. Shen, Nucl. Phys. B365 (1991) 79.

4) B. W. Lee, C. Quigg, H. Thacker, Phys. Rev. Lett. 38 (1977) 883.

5) The non-perturbative upper bound problem was first formulated in R. Dashen and H. Neuberger, Phys. Rev. Lett. 50 (1983) 1897;
Since then several summary papers have been published in the yearly Lattice Conferences. See
P. Hasenfratz, Nucl. Phys. B (Proc. Suppl.) 9 (1989) 3;
H. Neuberger, Nucl. Phys. B (Proc. Suppl.) 17 (1990) 17;
J. Shigemitsu, Nucl. Phys. B (Proc. Suppl.) 20 (1991) 515.

6) A. Linde, Pisma Zh. Eksp. Teor. Fiz. 23 (1976); JETP Lett. 23 (1976) 64;
S. Weinberg, Phys. Rev. Lett. 36 (1976) 294;
A. Hasenfratz, P. Hasenfratz, Phys. Rev. D34 (1986) 3160.

7) W. Langhuth and I. Montvay, Z. Phys. C36 (1987) 725;
A. Hasenfratz and T. Neuhaus, Nucl. Phys. B297 (1988) 205;
H. G. Evertz, E. Katznelson, P. G. Lauwers, M. Marcu, Phys. Lett. B221 (1989) 143.

8) See for example the second reference in [5].

9) C. Frick, K. Jansen, J. Jersak, I. Montvay, G. Munster, P. Seuferling, Nucl. Phys. B331 (1990) 515;
M. Lüscher, Commun. Math. Phys. 105 (1986) 153.

10) M. Lüscher and P. Weisz, Nucl.Phys. B290[FS20] (1987) 25; Nucl.Phys. B295[FS21] (1988) 65;

11) J. Kuti, L. Lin, Y. Shen Nucl. Phys. B (Proc. Puppl.) 4 (1988) 397; J. Kuti, Y. Shen Phys. Rev. Lett. 60 (1988) 85.

12) P. Hasenfratz and J. Nager, Z. Phys. C37 (1988) 477.

13) A. Hasenfratz, P. Hasenfratz, Nucl. Phys. B270 FS16 (1986) 685.

14) M. Lüscher, P.Weisz, Nucl. Phys. B318 (1989) 705.

15) M.M. Tsypin Lebedev Physical Inst. Pr. 280 (1985).

16) A. Hasenfratz, T. Neuhaus, K. Jansen, Y. Yoneyama and C. B. Lang, Phys. Lett. 199B (1987) 531;
A. Hasenfratz, K. Jansen, J.Jersak,T. Neuhaus, C. B. Lang and Y. Yoneyama, Nucl. Phys. B317 (1989) 81.

17) J. Kuti,L. Lin, Y. Shen, Nucl. Phys. B (Proc. Suppl.) 9 (1989) 26.

18) A. Hasenfratz, K. Jansen, J.Jersak, C. B. Lang, H. Leutwyler,T. Neuhaus, Z. Phys. C46 (1990) 257; Nucl. Phys. B356 (1991) 332.

19) M. Göckeler , K.Jansen, T. Neuhaus, preprint UCSD/PTH 91-15.

20) H. Leutwyler, Nucl. Phys. B (Proc. Suppl.) 4 (1988) 248;
P. Hasenfratz and H. Leutwyler, Nucl. Phys. B343 (1990) 241.

21) H. Neuberger, Phys. Rev. Lett. 60 (1988) 889; Nucl. Phys. B300 FS22
(1988);
U.M.Heller, H. Neuberger, Phys. Lett. B207 (1988) 189.

22) M. Gockeler, H. Leutwyler, Nucl. Phys. B350 (1991) 228; Phys. Lett. B253
(1991) 193.

23) J. Gasser and H. Leutwyler, Phys. Lett. B184 (1987) 83; Phys. Lett. B188
(1987) 477; Phys. Lett. B189 (1987) 197; Nucl. Phys. B307 (1988) 763;

24) G. Bhanot, K. Bitar, U.M. Heller, H. Neuberger, Nucl. Phys. B343 (1990)
467.

25) H. Neuberger, Phys. Lett. B199 (1987) 536.

26) G. Bhanot, K. Bitar,Phys. Rev. Lett. 61 (1988) 798;
C.B.Lang, Phys. Lett. B229 (1989) 97.

27) G. Bhanot, K. Bitar, U.M. Heller, H. Neuberger, Nucl. Phys. B353 (1991)
551.

28) U.M. Heller, H. Neuberger, P. Vranas, preprint FSU-SCRI-91-94, (1991)

LATTICE YUKAWA MODELS

ROBERT E. SHROCK

Institute for Theoretical Physics
State University of New York
Stony Brook, NY 11794-3840

ABSTRACT

A review is given of lattice Yukawa models. We discuss the phase structures of several different models and the measurements of renormalized Yukawa and scalar couplings. These measurements are consistent with the conclusion that both of these couplings would vanish if one took the ultraviolet cutoff to infinity. The lattice also enables one to simulate effective Yukawa theories which contain a high-energy cutoff marking the limit of their applicability. In this context, the Yukawa studies have shown that a fermion mass generated by a Yukawa interaction cannot be much larger than the mass scale characterizing the symmetry breaking. They have also led to an upper bound for the scalar mass, thereby generalizing previous work on pure scalar models. Studies of different lattice actions suggest that these results are not dependent upon details of the lattice model. We also review some work on lattice Yukawa models with gauge fields. Next, we comment upon efforts to construct chiral gauge theories using lattice regularization, and the generalizations of Yukawa models which arise as global limits of these constructions.

1. INTRODUCTION

The Yukawa theory was originally invented in 1935 to describe the strong force which binds nucleons together to form nuclei.[1] The interaction was taken to have the simple form $\mathcal{L} = g\phi\bar\psi\psi + h.c.$, corresponding to emission or absorption of the spin 0 particle ϕ by the nucleon ψ with coupling strength g. This was later generalized to $g\bar\psi\tau \cdot \pi\psi$, where $\psi = \binom{p}{n}$, $T_i = \tau_i/2$, $i = 1, 2, 3$ are the generators of isospin $SU(2)$, and π denotes the pion isotriplet. This interaction provided an approximate phenomenological description of the (long-range) part of the nucleon-nucleon interaction, which occurs dominantly via one-pion exchange. In this context, the properties of the associated Yukawa quantum field theory were studied.

Since the pion-nucleon coupling is large, $g_{\pi NN} \simeq 14$, it is not possible to use perturbative expansions in $g_{\pi NN}$ to calculate scattering cross sections or bound state energies in this theory. Nevertheless, it is of interest that (in the four-dimensional case of physical interest) this theory is renormalizable, i.e., one can redefine the couplings, masses, and field normalizations consistently order-by-order in perturbation theory so as to absorb the divergences which occur when one calculates Feynman diagrams involving loops. The resultant renormalized couplings are then free parameters. This simple Yukawa theory was later generalized to an SU(3)-invariant coupling involving the baryon and pseudoscalar meson octets. It was also expanded to construct the linear and nonlinear sigma models, which helped further the understanding of spontaneous chiral symmetry breaking and the role of the π and K mesons as nearly Nambu-Goldstone bosons.[2]

Subsequently, the Yukawa interaction reappeared in a different way in particle physics. This time, it formed an important part of the standard $SU(2) \times U(1)_Y$ theory of electroweak interactions, coupling quarks and leptons to a fundamental scalar boson (the Higgs boson) and thereby providing fermion masses and quark mixing.[3,4] Recall the reasons for this. The electroweak theory is a chiral gauge theory, i.e., the left- and right-handed fermions do not transform according to the same representations of the gauge group (equivalently, if one rewrites all fermions as left-handed, Weyl fields, these transform according to complex representations of the gauge group). Consequently, a bare mass term of the form $\bar{\psi}\psi + h.c. = \bar{\psi}_L\psi_R + \bar{\psi}_R\psi_L + h.c.$ would violate the gauge symmetry. Since the gauge symmetry is essential for the renormalizability of the theory, one must avoid such bare mass terms. (This applies for all of the known fermions. For hypothetical right-handed neutrino singlets, one could form bare mass terms.)

How then can one give masses to fermions? One way is to couple them to a scalar Higgs field $\phi = \begin{pmatrix} \phi^+ \\ \phi^0 \end{pmatrix}$ via a Yukawa interaction. Given the fact that the left-handed and right-handed fermions transform, respectively, according to the fundamental, $I = 1/2$ representation and the singlet representation of the weak isospin SU(2) group, it follows that ϕ must have $I = 1/2$ in order for the Yukawa interaction to be gauge-invariant. Further, given the weak hypercharge assignments of the fermions, it also follows that ϕ must be assigned weak hypercharge $Y = 1$ for this Yukawa interaction to be gauge-invariant. In the Lagrangian, the coefficient of the quadratic term in the Higgs potential $V(\phi)$ is chosen so that V has a minimum for $< 0|\phi|0 > \neq 0$. Without loss of generality, one can express the lower component of ϕ, as $\phi^0 = (\xi + i\eta)/\sqrt{2}$ and choose η to gain the vacuum expectation value (vev). Rewriting the theory in terms of a physical Higgs field with zero vev, one finds that the fermions pick up mass terms which, in the simplest case where there is no fermion mixing, are of the form $m_f = 2^{-1/2}y_{ff} < \phi >$, where y_{ff} is the corresponding Yukawa coupling. In the realistic case where one has fermion mixing, i.e. Yukawa interactions which are non-diagonal as well as diagonal in flavor, one must carry out diagonalizations of the resultant fermion mass matrix in order to calculate the fermion masses and the observed weak mixing angles. Via its covariant derivative term, the Higgs field also gives masses to the hitherto (perturbatively) massless gauge bosons. This is known as the Higgs mechanism. The Yukawa couplings thus suffice to give the fermions masses via the spontaneous symmetry breaking of the electroweak symmetry (the usual, but imprecise terminology). They have the key feature that they do this while preserving (perturbative) renormalizability. The standard electroweak model has a single Higgs field, but more complicated models have been considered with multiple Higgs fields.

Before proceeding, it is appropriate to give the reader some warnings. In candor, one should note that the hypothesis of a Yukawa interaction in the standard electroweak theory is at present only a plausible construction, not a verified theory like quantum electrodynamics or even a theory highly supported by experiment like quantum chromodynamics or the gauge sector of the electroweak theory. When the standard model was constructed, much of it was included to describe what were then known experimental facts, such as, for the electroweak sector, the left-handed nature of charged weak currents, and so forth. Subsequently, other features, such as the structure of the weak neutral current, have been verified to high precision, and the charged and neutral vector bosons W and Z have been observed. But there has not been any direct observation of the hypothesized Higgs boson or any direct evidence for the generation of fermion masses and weak mixing via Yukawa interactions. This is not to say that the theory is not not phenomenologically successful; on the contrary, it is in agreement with all well-established data. (The inference of nonzero neutrino masses and mixing which may be suggested by the observed deficiency of solar neutrinos could be accounted for by a straightforward generalization of the standard model, again using Yukawa couplings [and bare mass terms for right-handed Majorana neutrino bilinears] to generate the requisite masses and mixing.)

However, the Higgs and thus also the Yukawa sectors of the standard model do not constitute a complete or satisfactory theory. There are several reasons for this. First, in contrast to the gauge sector, they are quite arbitrary. In the standard model, there is no prediction for the Higgs mass; this is given perturbatively by $m_H^2 = 2\lambda_R v_R^2$, where λ_R and v_R denote the renormalized scalar coupling and vacuum expectation value, respectively. It is known that $v_R = 2^{-1/4} G_F^{-1/2} = 246$ GeV (where G_F is the Fermi weak coupling measured, e.g., in mu decay), but λ_R is unknown. Second, although the Yukawa interaction can describe fermion mass generation phenomenologically, it cannot *predict* the fermion masses or mixing. In the usual perturbative calculational scheme, these are simply arbitrary, free parameters. One chooses values for the various renormalized Yukawa couplings in such a way as to get the observed fermion masses and mixing. That is, one simply gets out what one puts in, at least perturbatively.

What is even more worrisome, one encounters a very basic and mystifying puzzle: given that the scale for fermion masses is set by the electroweak symmetry breaking scale, 246 GeV, why is it that all of the known fermions with the exception of the top quark have masses which are much smaller than this scale? Equivalently stated, why is it that the renormalized Yukawa couplings for these fermions are so tiny? To see how striking this mystery is, one has only to compare the renormalized Yukawa coupling for the electron, $y_e = 3 \times 10^{-6}$ in the standard model (with no lepton mixing, so no off-diagonal Yukawa interactions in the lepton sector) with a coupling which is the paradigm of a small, dimensionless coupling, namely the electromagnetic coupling $e = (4\pi\alpha)^{1/2} = 0.303$. There is also no explanation for why there are three generations of fermions. It is possible that there is no fundamental physical Higgs particle, and that the breaking of electroweak symmetry arises dynamically.[5] In such an approach, it is natural to consider that fermion masses and mixing might not arise from Yukawa interactions but from such terms as four-fermion operators, as in the Nambu-Jona-Lasino model.[6,7] A drawback of the four-fermion operator approach, as noted in Ref. 6, is that it is not perturbatively renormalizable and would thus undermine the property of (perturbative) renormalizability of the standard model. In a different direction, the fermions may be composite, so that their masses would, in principle, be calculable in terms of the dynamics and binding of their constituents. Compositeness could explain why there are multiple generations of fermions, just as the profusion of chemical elements and nuclear isotopes are explained by the composite structure of the atom and nucleus. Indeed, it has recently been pointed out that in this case the conventional wisdom that fermion masses arise via electroweak symmetry breaking may need substantial revision: if fermions are composite, their masses may arise mainly in a manner which is independent of electroweak symmetry breaking.[8] Needless to say, this would mean that even some qualitative expectations which follow from the Yukawa construction in the standard model could be misleading. In particular, if, indeed, fermion masses do arise due to compositeness, in a manner largely independent of electroweak symmetry breaking, one would be able to resolve the apparent mystery alluded to above and to see that it is an artifact of the overly restrictive theoretical assumption that quarks and leptons are fundamental, pointlike particles, and the corollary that their masses must be generated via electroweak symmetry breaking and hence closely related to the scale of this breaking.

Nevertheless, given that Yukawa interactions play a crucial role in the standard electroweak theory, providing both fermion masses and quark mixing, and given the tremendous success of this (admittedly incomplete) theory, it is clearly desirable to understand the properties of Yukawa theories as deeply as possible. The lattice approach is a powerful tool for this purpose, since it deals with the entire functional integral defining the quantum theory and is not limited to perturbative expansions. Within this approach, one can thus study certain questions which are beyond the ability of perturbative approaches. Even before the introduction of fermions, lattice studies had already yielded deeper insights into the Higgs mechanism, especially its nonperturbative aspects.[9] One of the reasons for this was that in a lattice theory with gauge fields (as usually formulated), the functional integration measure over the group is compact. This avoids the infinities which occur in the functional integration measure for the path integral of a continuum theory with gauge fields, and hence also the Faddeev-Popov gauge-fixing term which must be used in the continuum. This work included determinations of the phase structure of gauge-Higgs models based on the simple groups Z_N [10], U(1) [11], SU(2) [12], and the actual electroweak group SU(2) $\times$ U(1) [13], as well as the grand unification group SU(5) [14]. Studies also included measurements of correlation functions and masses for U(1) and SU(2) [15-16] One important general result from pure lattice gauge theory which was also relevant to gauge-Higgs models was the realization that a local gauge symmetry cannot be spontaneously broken, in constrast to the global symmetry. [17] The exact manifest gauge invariance of the lattice formulation was crucial in the proof of this result. However, as has been appreciated better subsequently, it is not straighforward to apply this fact to a deeper understanding of the symmetry realization of the full standard electroweak theory with fermions since the latter is a chiral gauge theory, and it has proved difficult to construct a satisfactory formulation of a chiral gauge theory using lattice techniques. In particular (see further below), a lattice gauge theory can become strictly chiral only in the continuum limit where fermion doublers are completely removed, since these doublers have the effect of rendering the lattice theory vectorlike for finite lattice spacing. Another interesting general result was the demonstration that for a gauge-Higgs model based on a simple gauge group and with a Higgs field which transformed according to the fundamental representation of this group, the confinement and Higgs phases of the lattice theory were analytically connected. [18] Here, in a phase diagram labelled by the plaquette coupling $\beta \propto 1/g^2$ (where g denotes the bare gauge coupling) and the gauge-Higgs coupling κ, the confinement phase was the part of the phase diagram adjacent to the confinement phase of the pure gauge theory which included the strong gauge-coupling limit ($g^2 \to \infty$, i.e. $\beta \to 0$), while the Higgs phase was adjacent to the phase $\beta = \infty$, $\kappa > \kappa_c$ of the frozen-gauge limit where the resultant globally invariant system (spin model) had spontaneous symmetry breaking. This work motivated the development of confining models of weak interactions. [19-20] For the Higgs mechanism in the actual electroweak theory the situation was more complicated owing to the direct product nature of this theory; it was found [13] that the phase adjacent to the strong-gauge coupling limit of this theory (where both of the two bare plaquette couplings $\beta_1 = 1/g_1^2$ and $\beta_2 = 4/g_2^2$ vanished) was separated by a phase boundary from the phase adjacent to the segment $\beta_1 = \beta_2 = \infty$, $\kappa > \kappa_c$ where the frozen-gauge limiting theory with global O(4) invariance had a phase with spontaneous symmetry breaking.

A further related line of research studied the nonperturbative behavior of models with gauge, fermion, and Higgs fields. [21-32] Starting from the simpler case of gauge theories with fermions, one could ask how the properties of these models change when one adds Higgs fields. Alternatively, starting from the case of gauge-Higgs theories, one could ask how the properties of these change when one adds fermions. One might wonder how this could be done without confronting the problem of putting chiral fermions on the lattice. The key was the fact that the SU(2) sector of the standard model, by itself, can be expressed as a vectorlike theory, owing to the properties that (a) SU(2) has only real representations, and

(b) in order for an SU(2) gauge theory to be self-consistent (specifically, to avoid the Witten anomaly associated with $\pi_4(SU(2)) = Z_2$)) if the nonsinglet fermions transform as $I = 1/2$ representations, there must be an even number of them. In nature, requirement (b) is met, indeed, met individually for each generation, since counting quarks plus leptons, there are four SU(2) doublets. Thus, for each pair of left-handed $I = 1/2$ Weyl fermions, one charge-conjugates one of them, obtaining a right-handed fermion, and then combines it with the other left-handed fermion to obtain an $I = 1/2$ Dirac fermion. (This was done explicitly in Ref. 28.) This in turn means that one can at least deal with the SU(2) sector of the standard model while using fermions coupled vectorially to the gauge fields. It was found that there is a phase boundary which separates the confinement phase and Higgs phases. This phase boundary is associated with a chiral phase transition: in the confinement phase, the bilinear fermion condensate $< \bar{\chi}\chi >$ is nonzero, whereas in the Higgs phase it is zero. (In addition, U(1) toy models were studied; in these the condensate also vanishes in the Coulomb phase.) The result for SU(2) disfavored confining models of weak interactions since these required confinement, but no nonzero chiral condensate, since the latter would produce huge dynamically generated masses of order 250 GeV for all of the fermions and would violate electric charge (as one can see by reexpressing the Dirac fields in terms of the original Weyl fields). However, the phase which has confinement also has chiral symmetry breaking, while the Higgs phase with $< \bar{\chi}\chi > = 0$ is no longer analytically connected with the confinement phase so that one can no longer infer that it confines. Much of this work dealt with models without direct couplings between the fermions and Higgs, i.e. with no Yukawa couplings.[21-24,26,27,29-32] Two papers[25,28] also included Yukawa couplings; these will be discussed further below. The determination of the phase structure of the models with gauge, fermion, and Higgs fields was of considerable interest since it elucidated, in a manner not limited to perturbation theory, the effects of the interactions between these three fields. Among other things, the determination of the phase structure indicated which (second-order) phase transitions were relevant to the physical continuum limit of the lattice theory.

Related to this is the important question of the physical values of the various couplings. Here, we shall concentrate on the scalar quartic coupling and Yukawa coupling. As noted above, within a perturbative approach, these are arbitrary (except that the renormalized quartic coupling λ_R must be positive in order that the Hamiltonian be bounded from below), and conventional wisdom has long tacitly accepted this as being a correct description of the couplings in general. However, the renormalization group approach to quantum field theory has shown that couplings may vanish as the ultraviolet (regulator) cutoff is removed, i.e. taken to infinity, so that the resultant theory is actually non-interacting. The term "trivial" has often been used to denote "non-interacting". Although the origin of this term is understandable from the viewpoint of rigorous constructive quantum field theory, it is somewhat unfortunate from the point of view of a more physical approach, since such theories, when reinterpreted as effective theories valid below some cutoff, are not only interacting, but also have an additional predictive feature, namely upper bounds on the couplings which would have vanished had the cutoff been taken strictly to infinity. These, in turn, yield upper bounds on physical scalar and fermion masses, which are of considerable physical interest. An important question here will be the question of whether Yukawa theories (in the physical dimensionality, $d = 4$) have the property of being non-interacting when the ultraviolet cutoff is taken to infinity. Clearly, in order to address this question, it is necessary to have a method of investigation which is capable of nonperturbative calculations. The lattice is eminently suited for this purpose. Of course, the lattice does not retain the full Lorentz invariance of the continuum theory, or, more precisely, the full O(4) Euclidean symmetry of the Wick-rotated continuum theory. However, one understands how to restore this, by defining the continuum theory at a second order phase boundary of the lattice theory, where correlation lengths describing the lattice fields diverge, so that one can take the lattice spacing a to zero, thereby removing traces of discreteness and restoring full O(4) invariance. This is, of course, also necessary in order to have finite masses, since these are defined in terms of correlation lengths according to $ma = 1/\xi$; for m to be finite as $a \to 0$, it is necessary that the correlation length $\xi \to \infty$. The lattice automatically provides an ultraviolet regulator because of the existence of a finite Brillouin zone, which limits the (Euclidean) lattice momenta so that $|p| \lesssim \pi/a$. Thus, one can use a lattice formulation to study the question of the nonperturbative behavior of the couplings in a theory, and this will be a central part of our discussion in this review. It is true that the lattice cutoff is not precisely equivalent to a continuum regulator; the latter does not break the corresponding Euclidean O(4) symmetry. However, this difference disappears in the continuum limit of the lattice theory.

There is a slightly different way to view this. One may regard the standard electroweak theory as an effective theory, which is valid up to some energy scale Λ' but may not be an accurate description of particle physics beyond this. Of course, we know that Λ' is less than the Planck scale of 10^{19} GeV, but how much less we do not know. In this view, the cutoff scale Λ' has direct physical significance, in contrast to the conventional view of the ultraviolet cutoff in a renormalizable quantum field theory, which is usually taken to be just a mathematical device that should eventually be taken to infinity. One should realize that this old view, while perfectly well defined in the context of abstract quantum field theory, contains a tacit assumption which is false, namely that the theory in question is valid up to infinitely large momenta. Thus, a more precise and general view of a given quantum field theory is that it is an effective description which is known to be an accurate representation of the physics up to some mass scale M, and may indeed also apply beyond this scale, but is not necessarily assumed to. In the case of a renormalizable and asymptotically free theory like QCD, the theory is well defined without the need for explicit mention of this possible cutoff. In the case of QED, taken in isolation, one can again get extremely accurate and successful numerical predictions for quantities such as the Lamb shift and the anomalous magnetic moment of charged leptons without taking into account this subtlety. However, as Landau observed[33] (rephrased in modern terminology) this theory is not asymptotically free, and, associated with this is the "zero-charge problem" or triviality property mentioned above for scalar and Yukawa models. Moreover, QED is indeed an example of an effective theory, in the sense that at sufficiently high energies (e.g. in e^+e^- scattering at a center-of-mass energy $\sqrt{s} = 90$ GeV) or, for static quantities like the anomalous magnetic moment of a charged lepton, QED in isolation could not be expected to give correct predictions; one must take account of the fact that it is a part of the full electroweak theory.

Another reason for the interest in lattice studies of Yukawa models is that one of the Yukawa couplings in the standard electroweak theory, namely that for the top quark, is not small compared with unity. For a satisfactory study of the physics of a fermion as heavy as the top quark, it is therefore desirable to use an approach which is not limited to perturbation theory, and here again the lattice is advantageous.

Finally, another motivation for lattice Yukawa studies is that one can investigate a subject which is beyond the ability of continuum perturbation theory, namely the properties of a Yukawa theory with a large bare coupling. In fact, as we shall see, this region is actually amenable to powerful analytic methods which give us much insight into the strong-coupling region of the lattice theory.

Because of its importance, we would to emphasize here the key physical result which one would deduce from the apparent triviality of Yukawa theories, namely upper bounds on fermion masses in the standard model. Given that the scale for such masses is the electroweak symmetry breaking scale, one infers that fermion masses generated by Yukawa interactions could not be substantially larger than this scale. This will be discussed in detail below, and significant complications pertaining to the chiral nature of the electroweak theory will be noted. But here we would like to point out that such a bound can explain a deep feature of the observed fermion mass spectrum: the absence of any (standard model) fermions with masses much larger than the electroweak scale of 246 GeV. First, let us recall the evidence for the latter absence. The LEP (and, with lower accuracy, SLC) measurements of the width of the Z indicate that there are only three weak doublets with associated neutrino weak $I = 1/2, I_3 = 1/2$ eigenstates.[34] Taking into account the requirement of anomaly cancellation in the standard model, this means that there are also only three generations of quarks. Now empirically, within each of the two charge sectors, the quark masses increase rapidly with successive generations. A number of indirect arguments indicate that the top quark mass should not be heavier than about 200 GeV; if it were, it would give too large corrections to the parameter $\rho = (m_w/(m_z \cos\theta_w))^2$ and, where it occurs in loops, to various rare processes such as $K_L^0 \to \mu^+\mu^-$. If there had been a fourth generation, then, barring an abrupt change in the hitherto observed pattern $m_c/m_u \gg 1$, $m_t/m_c \gg 1$, $m_s/m_d \gg 1$, $m_b/m_s \gg 1$, there would also have been at least one fermion, i.e. the next $Q = 2/3$ quark, with a mass which is $>> v$.

This review is organized as follows. In section 2 we shall discuss the general issue of whether or not Yukawa theories are free or interacting when the ultraviolet regulator is removed. Section 3 discusses a particularly simple Yukawa model with a real scalar field, which has been the subject of extensive analytic and numerical studies. Section 4 treats the question of lattice universality by considering several different lattice actions for the Yukawa model with a real scalar field. Section 5 reviews work on Yukawa models with other symmetry groups. In Section 6 we briefly discuss Yukawa models with gauge fields. Section 7 comments on efforts to construct chiral gauge theories with lattice regularization and, in this context, to study certain generalizations of lattice Yukawa models which are obtained as global limits of chiral gauge theories in which the gauge degrees of freedom are frozen out. Some concluding comments and directions for future research are included in Section 8. For related material, the reader should also consult the review of lattice bounds on Higgs masses by A. Hasenfratz in this book.

2. NONPERTURBATIVE DEFINITIONS OF RENORMALIZED

COUPLING CONSTANTS

As mentioned in the introduction, one of the main reasons for interest in lattice studies of $(d = 4$ dimensional) Yukawa theories is the question of whether these theories are free or not. Let us first briefly review the situation concerning a subsector of the Yukawa theory, namely 4D $\lambda\phi^4$ theory. Again, ostensibly, in perturbation theory, the renormalized coupling is an arbitrary (non-negative) parameter which can be chosen at will. However, in the context of the renormalization group, when one calculates the running coupling defined at a momentum scale , one finds

$$\frac{d\lambda}{dt} = A\lambda^2 \tag{2.1}$$

where $A > 0$. Integrating (2.1), one gets

$$\lambda(t_2) = \frac{\lambda(t_1)}{1 - 2A\lambda(t_1)(t_2 - t_1)} \tag{2.2}$$

Now let $t_1 = \ln(\Lambda/\mu)$, and $t_2 = \ln(m/\mu)$, so that $(t_2 - t_1) = -\ln(\Lambda/m)$, where Λ denotes the ultraviolet cutoff, μ denotes an arbitrary mass scale, and m denotes a typical mass in the theory. If one takes $\Lambda \to \infty$,

$$\lambda(m) \to \frac{1}{A\ln(\Lambda/m)} \to 0 \tag{2.3}$$

This suggests that the $\lambda\phi^4$ theory is non-interacting in the limit where one removes the ultraviolet cutoff. Of course, this argument is only heuristic, not rigorous, since the renormalization group equation (2.1) was calculated perturbatively, and only to lowest order. An equivalent and complementary way of viewing the situation is to interchange the values of t_1 and t_2 in (2.1), so that the equation becomes $\lambda(\Lambda) = \lambda(m)/[1 - A\lambda(m)\ln(\Lambda/m)]$. Then if $\lambda(m)$ has some nonzero (positive) value, it follows that as Λ increases through the value $m\exp(1/[A\lambda(m)])$, $\lambda(\Lambda)$ would diverge. This is physically unacceptable and shows that the initial assumption must be false, i.e. $\lambda(m)$ must vanish. Again, this is only heuristic, for the same reason as given above. It was in this form that Landau first presented this argument, in that case, for QED.[33] Moreover, in the QED case, the pole occurs at an energy scale much larger than the mass of the universe, so that the physical signicance was not clear. A large number of studies have been carried out on pure 4D scalar field theories with discrete and continuous symmetry groups.[35-49] Although there is still no rigorous proof of triviality for $d = 4$ (in contrast to $d > 4$, where triviality has been proved[50]) there is analytic and numerical evidence that this theory is, indeed, free if the ultraviolet cutoff is taken to infinity, as the old heuristic arguments suggested. In particular, this has been found to be true of the 4D O(4) scalar field theory which describes the Higgs sector of the standard electroweak model when one removes the fermions and turns off the gauge interactions. Indeed, it is found that for a given value of the bare quartic coupling λ, the effective ("renormalized") scalar coupling λ_R defined at a cutoff Λ is a monotonically decreasing function of Λ. Now in order for the scalar theory to be consistent as an effective theory, it is necessary that the cutoff Λ be larger than the mass of the scalar. If one then requires that the ratio Λ/m_B be larger than some value $\gtrsim 1$, where m_B denotes the (renormalized) scalar mass, then this yields a resultant upper bound on the coupling λ_R. Hence, if one ignores fermions and treats gauge interactions perturbatively, then this property can be used to obtain an estimate of an upper bound on the mass of the physical Higgs boson in the standard electroweak theory.[51-59] (One must incorporate gauge interactions since otherwise one just has a bound for the scalar mass in terms of the scalar vacuum expectation value v; to insert the experimental value of v, one must make use of the relations $m_W^2 = g^2 v^2/4$ and $G_F = g^2/(8m_W^2) = 1/(2v^2)$.) Here we mention a number of recent refinements that have been made to this estimate. First, recall that since the lattice simulates the cutoff Λ by the upper limit on the Euclidean momenta at the edge of the Brillouin zone, $|p| < \pi/a$, it follows that in the continuum limit, $\Lambda \to \infty$. However, in order to obtain the upper limit on λ_R, one does not want Λ too large; instead, one must measure λ_R when Λ/m_B is near to the minimal value which one defines as allowable for the effective theory to be acceptable. (The precise lower limit on Λ/m_B is a convention.) Hence, one must worry about lattice artifacts and their effect on the calculations. In particular, the usual statements about universality of statistical mechanical models at a critical point (i.e. that different lattice actions or Hamiltonians which are defined in the same dimensionality and have the same symmetry group are in the same universality class and have the same critical behavior) do not hold if one is a finite distance away from a critical point. Normally, one expects that if one is reasonably close to the critical point, then nonuniversal effects will be small. However, it is possible to construct actions which will significantly different behavior when one is even just slightly away from the critical point (although one could perhaps view these as perversely chosen). In order to study the role of lattice artifacts, one may perform calculations on different lattice types and compare the results.[55-58] In particular, this was done with a comparison of the usual 4D hypercubic lattice and the so-called F4 lattice. The F4 lattice has the property that it excludes terms of the type $\sum_\mu (p_\mu a)^4$ which violate O(4) invariance and are present for the hypercubic lattice. Based on this study, it was concluded that if one requires less than 0.3 % cutoff effects in $W_L W_L$ scattering (and ignores fermions and puts gauge couplings in perturbatively at the end of the calculation) then the upper bound on the Higgs boson mass is 530 ± 60 GeV with the F4 lattice[56], as compared with 510 ± 50 with the hypercubic lattice.

More recently, however, the old problem of non-universality away from a critical point has come back again and has led to a revision of this estimate. With a different type of action with derivative couplings of the scalar fields, it has been found that the upper bound on the scalar mass can be significantly increased.[59] On the basis of this work, it has been estimated that the upper bound on the Higgs mass might be around 850 GeV rather than the values given previously. Further work will help to clarify the effect of lattice artifacts and non-universality on the lattice bound on the Higgs mass.

It should be mentioned that large-N methods provide an alternative nonperturbative calculational method besides the lattice and have been used to study the issue of triviality of 4D scalar field theory.[60,49]

Having briefly reviewed the situation for pure 4D scalar theories, we next proceed to consider the renormalization group equation for a (4D) Yukawa theory. This has the form, to lowest order,

$$\frac{dy}{dt} = By^3 \tag{2.4}$$

where again $B > 0$. The fact that this coefficient is positive produces behavior which is qualitatively similar to that for the simple 4D $\lambda \phi^4$ theory. Integrating (2.4) gives

$$y(t_2)^2 = \frac{y(t_1)^2}{1 - 2By(t_1)^2(t_2 - t_1)} \tag{2.5}$$

Again, for fixed m, as one removes the ultraviolet cutoff,

$$y(m) \simeq \frac{1}{\left[\ln(\Lambda/m)\right]^{1/2}} \to 0 \tag{2.6}$$

Hence, just as in the simple 4D scalar field theory, one heuristically expects that the physical Yukawa coupling actually vanishes.

It should be noted that in the full Yukawa theory, the renormalization group equation for the scalar sector is more complicated than (2.1); one has[61]

$$\frac{d\lambda}{dt} = A_1\lambda^2 + A_2\lambda y^2 - A_3 y^4 \qquad (2.7)$$

to one-loop order, where $A_i > 0$, $i = 1, 2, 3$ (e.g. Ref. 2). The λy^2 term arises from graphs involving a tree-level quartic scalar interaction with a fermion loop correction to one of the outgoing scalar lines, and the y^4 term arises from the graph in which four scalar lines come in and interact via a fermion loop. Because the coefficient of the A_3 is negative, one is led to inquire whether for sufficiently large y, the sign of $d\lambda/dt$ might switch from positive to negative. Since the actual Yukawa couplings for all of the known fermions except the top quark are so small, one does not expect that the second two terms in (2.7) are important for them. However, since the physical Yukawa coupling for the t quark is of order unity, these additional terms might be signicant in this case. However, in precisely this case, it is not clear that one can trust the perturbative calculation of the coefficients in the renormalization group equations. More generally, the rigorous, constructive field theory definition of the physical couplings involves the actual Green's functions of the theory and is not restricted to any perturbative expansions. As stressed before, in order to investigate the issue of triviality, it is necessary to use a fully nonperturbative calculational method. Let us note here that the term "physical" coupling is equivalent to "renormalized" coupling, again in the constructive field theoretic sense. However, in a nonperturbative approach like that afforded by the lattice, one defines the physical or renormalized coupling in a manner which does not involve any perturbative counterterms.

Before embarking on a discussion of the lattice studies of Yukawa models, let us comment on the physical implications which follow if a theory is free. Naively, one might conclude that one must then reject the Yukawa sector of the standard electroweak theory, since of course one does observe nonzero electroweak interactions. Instead, however, just as with the scalar sector, one reinterprets the theory as being an effective one containing an upper cutoff in energy, and applicable only below this cutoff. The theory with cutoff is then interacting. In fact, one now gains more information about the behavior of the scalar and Yukawa couplings. As will be seen below, numerical simulations are consistent with a monotonic decrease of λ_R and y_R as functions of the correlation length, ξ, which can be defined, e.g., as $1/M_B$, the renormalized boson mass. Since the cutoff $\Lambda = \pi/a$, it thus follows that λ_R and y_R are decreasing functions of Λ/M_B. As in the pure scalar case, one notes that in order for the theory to be an internally consistent effective theory for momenta below Λ, it is necessary that Λ be larger than any of the physical masses in the theory, M_F and M_B. But this then immediately implies upper bounds on λ_R and y_R. Since $m_f \propto y_R v$, the upper bound on y_R implies, in turn, an upper bound on the fermion mass. (Similarly, in a theory with several Yukawa couplings, one would obtain corresponding upper bounds on the full set of fermion masses.) The upper bound on λ_R yields an upper limit on the physical scalar mass, and, putting in gauge interactings perturbatively, this gives an upper limit on the mass of the Higgs boson. The latter result thus generalizes the upper limit derived in a pure 4D scalar model to include fermion effects.

One should note that the $N \to \infty$ limit of a particular continuum 4D Yukawa model was found to be free.[62] There have also been several perturbative efforts to derive upper bounds on fermion masses.[63]

3. STUDY OF A LATTICE YUKAWA MODEL WITH A REAL SCALAR FIELD

3.1 General

One can approach the nonperturbative study of lattice Yukawa models in two different ways. In one direction, one can study simplified models without gauge fields. More realistically, one can include gauge fields as in the standard electroweak theory. Both directions are useful; the models without gauge fields are simpler to analyze and simulate, while the models with gauge fields are closer to the real world but are more complicated to study.

The earliest studies of lattice Yukawa models illustrated both of these two approaches. A study by Shigemitsu in 1987 was the first to use lattice methods to investigate the issue of the nonperturbative behavior of the Yukawa coupling and the question of triviality.[64] This paper addressed this in the simplest context of a Yukawa model with a real scalar field. Again, as an initial study, it employed the quenched approximation, in which the fermion determinant is set equal to a constant. In this approximation, the fermions have no effect on the scalars, and the phase structure is just that of the scalar theory, but one can study how the Yukawa coupling to the scalar sector influences the fermion. Measurements of the renormalized Yukawa coupling y_R were performed, and it was found that y_R was relatively insensitive to the bare Yukawa coupling, as one might expect if $y_R = 0$ when the ultraviolet cutoff was taken to infinity.

Another study in 1987 considered a simple Yukawa model with gauge fields.[25] This work was part of a program dealing with the general nonpertubative behavior of models with gauge, fermion, and Higgs fields. Rather than the actual SU(2) factor group of the electroweak theory, this paper considered a toy model based on the gauge group U(1). After rewriting the SU(2) sector of the standard model in vectorlike form, the fermion content consists of weak isodoublet and isosinglet Dirac fields. As an abelian toy model, Ref. 25 thus used a Higgs field ϕ with U(1) charge $q_\phi = 1$ and two types of (staggered) fermion fields, χ and ξ, with charges $q_\chi = 1$, corresponding to the $I = 1/2$ fermion in SU(2), and $q_\xi = 0$, corresponding to the $I = 0$ fermion in SU(2).

The simplest Yukawa model is that of a fermion field interacting with a real scalar field. In the continuum, this theory is defined by the generating functional (without sources)

$$Z = \int \prod [d\phi_{cn}(x)][d\psi_{cn}(x)][d\bar{\psi}_{cn}(x)]e^{-S} \qquad (3.1a)$$

where the Euclidean action $S = S_B + S_F + S_Y$ is given by

$$S_B = \int d^d x \left[\frac{1}{2}(\partial_\mu \phi_{cn})^2 + \frac{1}{2}m_{cn}^2 \phi_{cn}^2 + \frac{1}{4}\lambda_{cn}\phi_{cn}^4\right] \qquad (3.1b)$$

$$S_F = \int d^d x \left[\bar{\psi}_{cn}\gamma \cdot \partial\psi + m\bar{\psi}_{cn}\psi_{cn}\right] \qquad (3.1c)$$

$$S_Y = y_{cn} \int d^d x \phi_{cn}\bar{\psi}_{cn}\psi_{cn} \qquad (3.1d)$$

Here the fields and couplings have the subscript $cn = continuum$ to distinguish them from the corresponding lattice quantities to be introduced below. The Dirac matrices are defined according to the usual Euclidean conventions:

$$\{\gamma_\mu, \gamma_\nu\} = 2\delta_{\mu\nu} \qquad (3.2)$$

where $\delta_{\mu\nu}$ is the Kronecker delta function. We now discretize this action to put it on a (Euclidean) hypercubic lattice with lattice spacing a and lattice sites n. The lattice spacing will often be taken to be equal to unity for notational convenience.

The discretization of the scalar kinetic term is

$$\frac{1}{2}\int d^d x(\partial_\mu\phi_{cn})^2 \to \frac{1}{2}a^4 \sum_{n,\mu}\left(\frac{\phi_{cn,n+e_\mu a} - \phi_{cn,n}}{a}\right)^2$$

$$= a^2 \sum_{n,\mu}(\phi_{cn,n}^2 - \phi_{cn,n}\phi_{cn,n+e_\mu}) \qquad (3.3)$$

Here $\mu = 1,..,d$ and ae_μ is a lattice vector in the μ'th direction. For realistic models, $d = 4$; however, in some formulas, it will be informative to keep d as a general integer. Note that in order for there to exist a $\gamma_5 = \prod_{\mu=1}^d \gamma_\mu$ with the usual property $\{\gamma_\mu, \gamma_5\} = 0$, it is necessary that d be an even integer.

There are several different discretizations of the fermion action. In the continuum Euclidean formulation, the fermion fields are represented by anticommuting fields

$$\{\psi_{cn}(x), \bar{\psi}_{cn}(y)\} = 0 \qquad (3.4)$$

and this is carried over naturally to the lattice with $\psi_{cn}(x)$ replaced by ψ_n. A discretization of the fermion kinetic term is the symmetric one

$$\int d^4 x \bar{\psi}_{cn}\gamma \cdot \partial\psi_{cn} \to a^4 \sum_{n,\mu}\bar{\psi}_n\gamma_\mu\left(\frac{\psi_{n+e_\mu a} - \psi_{n-e_\mu a}}{2a}\right) \qquad (3.5)$$

However, one finds that with this discretization, the lattice theory produces 2^d different fermion modes. One may simply live with this fact if one is studying properties of the theory which are not rendered unphysical by this fermion replication. This so-called naive action does have the merit that it preserves chiral symmetry; that is, if one sets the bare lattice fermion mass $m = 0$, then the lattice theory is invariant under chiral transformations. As Wilson showed[65], one can remove the replicated modes (often called "doubler" modes) by adding a discretized second-derivative term to the action. This has the advantage of yielding the physical number of fermion modes in the continuum limit, but the disadvantage of breaking chiral symmetry explicitly. A third approach is to note that a spin-diagonalization of the fermion kinetic term is possible, which motivates separating the components of the $2^{[d/2]}$-component continuum fermion fields and placing the individual components on each lattice site. In the continuum limit, one combines the components on each hypercube to construct the Dirac field, so that one reduces the fermion replication to $2^{d-[d/2]}$ or 4 for $d = 4$. This is the staggered fermion formulation[66]; it has the advantage of preserving a continuous remnant of the continuum chiral symmetry but the disadvantage of not removing all of the replicated fermion modes. Henceforth for convenience of notation, unless otherwise indicated, we shall take the lattice spacing $a \equiv 1$. In the staggered fermion formulation, it is natural to represent the Yukawa interaction term as a double sum:

$$y \int d^d x \, \phi_{cn}(x) \bar{\psi}_{cn}(x) \psi_{cn}(x) \rightarrow 2^{-d} \tilde{y} \sum_n \phi_n \sum_{n' \in hc(n)} \bar{\chi}_{n'} \chi_{n'} \qquad (3.6)$$

where $hc(n)$ denotes the hypercube adjacent to the point n on the lattice. Explicitly, the set $n' \in hc(n)$ consists of the points $n' = n + \sum_{\mu=1}^{d} b_\mu e_\mu$ where $b_\mu = 1$ or 0. The reason for this summation over hypercubes is that the continuum Dirac fields are constructed from linear combinations of the one-component χ fields on each hypercube. Since there are 2^d points in the hypercube, the factor 2^{-d} is included for proper normalization. For the mass term, the double sum collapses back to a single sum which can be written simply as $\sum_n \bar{\chi}_n \chi_n$.

In order to make possible numerical simulations of the Yukawa theory with dynamical fermions and measurements of physical couplings, one must choose the lattice action appropriately. A basic complication with such fermionic simulations is that the fermion determinant is not in general positive-definite, so that the integrand of the functional integral cannot be interpreted as a probability measure, i.e. does not lie in the interval $[0, 1]$. Consequently, standard simulation methods cannot be used. To remedy this, one doubles the number of lattice fermion species, which will be labelled with the subscript $f = 1, 2$. This has the effect of replacing the fermion determinant by the square of itself, which, for this theory, is positive semidefinite.

Thus, one lattice formulation of the Yukawa theory with a real scalar field, using staggered fermions, is given by the action $\tilde{S} = \tilde{S}_B + \tilde{S}_F + \tilde{S}_Y$, where

$$\tilde{S}_B = (\frac{\tilde{m}^2}{2} + d) \sum_n \tilde{\phi}_n^2 - \sum_{n,\mu} \tilde{\phi}_n \tilde{\phi}_{n+e_\mu} + \frac{\tilde{\lambda}}{4} \sum_n \tilde{\phi}_n^4 \qquad (3.7a)$$

$$\tilde{S}_F = \frac{1}{2} \sum_{n,\mu} \sum_{f=1}^{2} \bar{\chi}_{n,f} \, \eta_{n,\mu} (\chi_{n+e_\mu,f} - \chi_{n-e_\mu,f}) + \sum_n \sum_{f=1}^{2} m_f \bar{\chi}_{n,f} \chi_{n,f} \qquad (3.7b)$$

$$\tilde{S}_Y = 2^{-d} \tilde{y} \sum_n \tilde{\phi}_n \sum_{n' \in hc(n)} \sum_{f=1}^{2} \bar{\chi}_{n',f} \chi_{n',f} \qquad (3.7c)$$

The bare fermion masses m_f are taken to be zero unless otherwise indicated. The factor $\eta_{n,\mu}$ in the fermion kinetic term arises from the γ matrices and is given by 1 for $\mu = 1$ and by $(-1)^{n_1 + \cdots + n_{\mu-1}}$ for $2 \leq \mu \leq d$. The respective bare parameters are related according to

$$\phi_{cn} = a^{-1} \tilde{\phi} \qquad (3.8a)$$

$$m_{cn}^2 = a^{-2} \tilde{m}^2 \qquad (3.8b)$$

$$\lambda_{cn} = \tilde{\lambda} \qquad (3.8c)$$

$$y_{cn} = \tilde{y} \qquad (3.8d)$$

In the limit $a \to 0$, with the identifications (3.8), the lattice action (3.7) formally reproduces (3.1). The Yukawa coupling in this model is not analogous to that in the standard model, since it involves only one type of fermion (with two flavors), rather than two fermion fields transforming according to different representations of the symmetry group. However, it has the feature that it is invariant under a Z(2) global symmetry.

A number of early lattice studies of the Yukawa model with a real scalar field were carried out with the action (3.7), first for the case of quenched fermions[64] and later for dynamical fermions.[67-68]. The phase diagram of this theory consists of a symmetric paramagnetic phase and a phase in which the global Z(2) symmetry is spontaneously broken via ferromagnetic ordering, with nonzero scalar and bilinear fermion condensates, $< \phi >$ and $< \bar{\chi}\chi >$. The studies with dynamical fermions[67,68] confirmed the original finding in Ref. 64 that the measured values of the renormalized Yukawa coupling $\tilde{y}_R$ coupling were rather insensitive to the value of the bare couplings and lay in a rather restricted range, as would be expected if $\tilde{y}_R$ actually vanished when the cutoff $\propto 1/a \to 0$. Subsequent studies[69-71] have used a different parametrization; we turn to these next.

3.2 Hypercubic Yukawa Coupling

A detailed study of the Yukawa theory with a real scalar field was carried out using the hypercubic discretization of the Yukawa interaction.[69] The lattice theory is defined by

$$Z = \int \prod_{n,f} d\phi_n dx_{n,f} d\bar{x}_{n,f} e^{-S} \tag{3.9a}$$

where $S = S_B + S_F + S_Y$, with

$$S_B = \sum_n \phi_n^2 - 2\kappa \sum_{n,\mu} \phi_n \phi_{n+e_\mu} + \lambda \sum_n (\phi_n^2 - 1)^2 \tag{3.9b}$$

$$S_F = \frac{1}{2} \sum_{n,\mu} \sum_{f=1}^{2} \bar{x}_{n,f} \eta_{n,\mu}(\chi_{n+e_\mu,f} - \chi_{n-e_\mu,f}) + m_f \sum_{n,f} \bar{x}_{n,f} \chi_{n,f} \tag{3.9c}$$

$$S_Y = 2^{-d} y \sum_n \phi_n \sum_{n' \in hc(n);f} \bar{x}_{n',f} \chi_{n',f} \tag{3.9d}$$

In the limit $a \to 0$, this action yields the correct formal continuum (cn) Yukawa theory (with the species replication noted above) if one relates the respective bare parameters as follows:

$$\phi_{cn} = (2\kappa)^{1/2} a^{-1} \phi \tag{3.10a}$$

$$m_{cn}^2 = a^{-2}\left[\frac{(1-2\lambda)}{\kappa} - 2d\right] \tag{3.10b}$$

$$\lambda_{cn} = \kappa^{-2}\lambda \tag{3.10c}$$

$$y_{cn} = (2\kappa)^{-1/2} y \tag{3.10d}$$

There are several motivations for studying the scalar-fermion model using the parametrization (3.9). First, for the purpose of studying the effect of strong Yukawa couplings and for investigating whether these couplings are trivial in the continuum limit, it is necessary to understand the large bare-y region. The reason for this is that it is expected that the renormalized Yukawa coupling $y_{cn,R}$ is a non-decreasing function of the bare y_{cn}. Consequently, in order to gain evidence that $y_{cn,R} = 0$, one would like to be able to investigate arbitrarily large y_{cn}. But with the parametrization (3.7) used earlier[67,68], this was numerically unfeasible; when the local Yukawa interaction was used, the numerical simulation methods were unstable for $\tilde{y} \gtrsim 4$, and when the Yukawa interaction with hypercubic summation was used, the phase boundary at which one would try to construct the continuum limit moved very rapidly to extremely large values of $\bar{m}^2$ as $\tilde{y}$ increased past $\tilde{y} \simeq 5$. In the same way, in order to study the effect of a large bare quartic coupling and to test for possible triviality of this coupling in the combined scalar-fermion theory, it is desirable to study the region of arbitrarily large values of λ_{cn}. But again, this was not feasible with the parametrization (3.7). In contrast, the parametrization (3.9) in terms of (y, κ, λ) allows one to explore a very large region of the space of bare continuum parameters $(y_{cn}, m_{cn}^2, \lambda_{cn})$, including the limits $y_{cn} \to \infty$ and $\lambda_{cn} \to \infty$, as well as the limits $m_{cn}^2 \to \pm\infty$. Given (3.10c) and (3.10d), one can investigate large λ_{cn} and large y_{cn} simply by measuring the properties of the theory at the continuum limit defined where the line of second order phase transitions between the disordered and ferromagnetic phases (to be discussed further below) crosses the $\kappa = 0$ axis.

A second reason for using the parametrization (3.9) is that it is convenient for studying various special limiting cases and thereby gaining insight into the overall phase structure. For example, as the bare quartic coupling $\lambda \to \infty$ in (1.2) (which, for $\kappa > 0$, also means that $\lambda_{cn} \to \infty$), the values that the lattice scalar field ϕ can take are reduced from the real numbers to the discrete set ± 1. In this limit, (3.9b) becomes a simple Ising-type interaction.

For given bare continuum couplings λ_{cn} and m_{cn}^2, (3.10b) yields two solutions for κ:

$$\kappa_\pm = \frac{1}{4\lambda_{cn}}\left[-B \pm (B^2 + 8\lambda_{cn})^{1/2}\right] \qquad (3.11a)$$

where

$$B = (am_{cn})^2 + 2d \qquad (3.11b)$$

Of these solutions, only κ_+ is physically acceptable.[69].

3.3 Phase Structure for Hypercubic Yukawa Coupling

The determination of the phase structure is of interest in its own right and is a necessary first step in studying continuum limits of the lattice theory; this was worked out in Ref. 69. Much of the qualitative and, indeed, quantitative features of the phase structure can be obtained from analytic methods.

For arbitrary y and $m_f = 0$, the theory is invariant under a diagonal U(2) symmetry only involving the fermions and generated by

$$\chi_n \to U\chi_n$$
$$\bar\chi_n \to \bar\chi_n U^\dagger \qquad (3.12)$$

where $U \in U(2)$. For arbitrary y and $m_f = 0$, the theory is also invariant under a Z(2) symmetry involving both scalar and fermion fields and generated by

$$\phi_n \to -\phi_n$$
$$\chi_n \to i(-1)^{\Sigma(n)}\chi_n$$
$$\bar\chi_n \to i(-1)^{\Sigma(n)}\bar\chi_n \qquad (3.13)$$

A nonzero value of the order parameter $< \phi >$, the staggered order parameter $< \phi >_{st}$, or the fermion condensate $< \bar\chi_f\chi_f >$ (with no sum on f) spontaneously breaks the Z(2) symmetry (3.13). These are the order parameters which one monitors for the purpose of mapping out the phase diagram. It is also useful to observe that the theory is symmetric under the simultaneous interchange $\phi_n \to -\phi_n$ and $y \to -y$. Using this symmetry, one can, with no loss of generality, take $y \geq 0$; this will be done henceforth.

For orientation, we first show in Fig. 1 the phase diagram which was determined for the

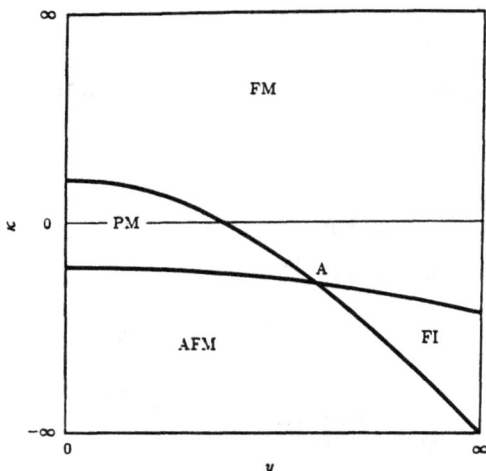

Fig. 1. Phase diagram for the theory (3.9). See text for discussion of phases and of the behavior in the region of point A.

theory (3.9). It consists of (a) a disordered, symmetric, paramagnetic (PM) phase; (b) a phase with ferromagnetic (FM) long-range order; (c) a phase with antiferromagnetic (AFM) long-range order; and (d) a phase with simultaneous FM and AFM long-range order. As was noted in Ref. 69, the behavior in the local neighborhood of point A may involve a meeting of the four respective phases or a more complicated structure. The numerical work (see below) did not suggest the latter possibility, but of course did not completely rule it out. Which of these behaviors occurs was not important for the measurements of renormalized couplings or for the conclusions of this study. Since a goal of Ref. 69 was to study fermion masses generated by Yukawa interactions. and, in particular, heavy fermions with masses of order the electroweak scale (relevant to the top quark), this work investigated the continuum limit defined at the second-order phase boundary between the PM and FM phases, starting from within the FM phase. Recall that on general statistical mechanical grounds, the (uniform, zero-field) susceptibility diverges at a continuous transition associated with the onset of long-range ferromagnetic order, as is the case at the PM-FM phase boundary. (Similarly, the staggered (zero-field) susceptibility diverges at the PM-AFM phase boundary.) Given the expression for the susceptibility in terms of a sum of 2-point correlation functions $< \phi_n \phi_{n'} >$ (fluctuation-dissipation theorem), it follows that this sum must diverge at the transition; in turn, this requires that the associated bosonic correlation length diverge, i.e. $aM_B = 0$. Since the transition is continous and the renormalized fermion mass $aM_F = 0$ in the PM phase, it follows that aM_F continously decreases to zero as one approaches the PM-FM boundary. The phase transition at this boundary thus provides a natural location to define the continuum limit for a continuum Yukawa theory. In passing, we note that if one were to approach this phase bounary from within the PM phase, the goal of studying Yukawa-generated fermion masses could not be realized since the renormalized fermion mass vanishes identically.

Let us now discuss how the phase diagram in Fig. 1 was determined. One can gain much information from considering various special cases of the lattice theory:

3.3.1 $y = 0$

Here the theory reduces to decoupled sectors consisting of a $\lambda\phi^4$ real scalar field theory, and a free massless fermion. The scalar sector is invariant under the global $Z(2)$ symmetry $\phi_n \rightarrow -\phi_n$ while the fermion sector is invariant, for $m_f = 0$, under a certain $U(2)_e \times U(2)_o$ global symmetry. The lattice chiral symmetry of the free fermion, incorporated in (2.1), is, of course, realized explicitly, so that $< \bar{\chi}\chi > = 0$ everywhere. The scalar sector has three phases: a symmetric, disordered, paramagnetic (PM) phase in which the $Z(2)$ global symmetry is realized explicitly, and two phases involving spontaneous symmetry breaking of this symmetry: a ferromagnetic phase in which $< \phi > \neq 0$, and an antiferromagnetic phase in which $< \phi >_{st} \neq 0$. The PM-FM and PM-AFM phase transitions are continuous, with infinite correlation length.

3.3.2 $\kappa = 0$, *Arbitrary* y

Along the line $\kappa = 0$, one can integrate over the scalar field and obtain, as an exact result, a purely fermionic path integral. Before carrying out this integration, one may add the term $H \sum_n \phi_n$ to the action S_B, for the purpose of calculating $< \phi >$. We find

$$Z = v^{N_s} \int \prod_{n,f} d\chi_{n,f} d\bar{\chi}_{n,f} e^{-S'} \tag{3.14a}$$

where N_s denotes the number of sites, $v = e^{-\lambda} J_0(\lambda)$ is an unimportant overall analytic factor, with

$$J_k(\lambda) = \int_0^\infty dx \; x^{\frac{k-1}{2}} e^{-(1-2\lambda)x - \lambda x^2}$$

$$= (2\lambda)^{-(k+1)/4}\Gamma\left(\frac{k+1}{2}\right)e^{\frac{(1-2\lambda)^2}{8\lambda}}D_{-(k+1)/2}\left(\frac{1-2\lambda}{(2\lambda)^{1/2}}\right) \tag{3.14b}$$

and

$$S' = S_F - \sum_n \ln\left[\sum_{\ell=0}^{\infty}\frac{\alpha_n^{2\ell}}{(2\ell)!}R_{2\ell}(\lambda)\right] \tag{3.14c}$$

$$R_k(\lambda) \equiv \frac{J_k(\lambda)}{J_0(\lambda)} \tag{3.14d}$$

$$\alpha_n \equiv H - 2^{-d}y\sum_{n'\in hc(n)}\sum_{f=1}^{2}\bar{\chi}_{n',f}\chi_{n',f} \tag{3.14e}$$

In (3.14b), $D_\nu(z)$ is the parabolic cylinder function of index ν. The formally infinite sum in S' actually truncates at finite order because of the Grassmann nature of the fermion fields. As will be discussed further below, this exact result shows the equivalence, at the quantum level, of a Yukawa theory and a purely fermionic theory with the multifermion couplings specified by (3.14).

To proceed, one uses a mean field type of approximation, as before for the strong gauge-coupling limit of gauge-Higgs theories with fermions.[22,23,25-28] With this method, one replaces fermion operator products higher than quadratic by a quadratic part multiplied by the mean value of the rest. This then enables one to perform the fermionic integrations. Solving for the fermionic condensate (at $H = 0$) using the symmetry between the species $f = 1, 2$, and dropping terms which are negligible in the vicinity of the phase transition, one obtains the mean field consistency condition

$$<\bar{\chi}_f\chi_f> = BI(B) \qquad f = 1,2 \tag{3.15a}$$

where

$$B = 2y^2R_2(\lambda)\left(1 - \frac{3}{2^{d+2}}\right)<\bar{\chi}_f\chi_f> \tag{3.15b}$$

and

$$I(B) = (2\pi)^{-d}\int_{-\pi}^{\pi}\left(\prod_{\mu=1}^{d}dp_\mu\right)\frac{1}{B^2 + \sum_{\mu=1}^{d}\sin^2(p_\mu)} \tag{3.15c}$$

Clearly, at $y = 0$, (3.15) implies that $<\bar{\chi}_f\chi_f> = 0$; furthermore, since the right-hand side can be expanded in a Taylor series in y for small $|y|$, one can infer the same conclusion for a range of y in the neighborhood of $y = 0$. On the other hand, for a range of larger values of y, one can solve (3.15) and obtain a continuous, nonzero solution for $<\bar{\chi}_f\chi_f>$, as will be discussed further below. (Although (3.15) always has $<\bar{\chi}\chi> = 0$ as a solution, if a nonzero solution for $<\bar{\chi}_f\chi_f>$ is allowed, it is the physical solution, since it minimizes the mean field free energy.) It follows that this mean field method implies the existence of a phase transition point at finite y where $< \bar{\chi}_f \chi_f >$ fails to be an analytic function of y. We shall denote this point as $y_{c,MF}$. One can solve (3.15) using the same numerical and approximate analytic methods as were applied earlier for the (zero-temperature) chiral transition in gauge-Higgs systems with fermions.[22-28] These indicate that as y increases through $y_{c,MF}$, $< \bar{\chi}_f \chi_f >$ becomes nonzero, and, furthermore, that the phase transition is continuous. Using a $1/d$ expansion of the right-hand side of (3.15), one obtains the following result for the critical point:

$$y_{c,MF}^2 = \frac{(d/4)}{(1 - \frac{3}{2^{d+2}})R_2(\lambda)} \left[1 - q^{-1} - 2q^{-2} - 7q^{-3} - 35q^{-4} - 215q^{-5} - 1501q^{-6} - O(q^{-7}) \right]$$

(3.16)

where $q = 2d$ is the coordination number of the lattice. The values of $y_{c,MF}$ obtained from (3.16) for $d = 4$ are in excellent agreement with a numerical solution of (3.15). The numerical solution yields the values $y_{c,MF} = 1.24, 1.27, 1.27, 1.22, 0.97,$ and 0.88 at $\lambda = 0, 0.1, 0.5, 1.0, 5.0,$ and ∞.

Using the same mean field approximation to the exact result (3.15), one calculates

$$< \phi > = -2yR_2(\lambda)BI(B)$$
$$= -2yR_2(\lambda) < \bar{\chi}_f \chi_f >$$

(3.17)

(with no sum on f). This shows that $< \phi >$ is nonzero if and only if $< \bar{\chi}_f \chi_f >$ is nonzero, as is from the fact that these order parameters both break the same discrete Z(2) global symmetry.

In order to study the phase structure further, numerical simulations were carried out[69], using the hybrid Monte Carlo algorithm for dynamical fermions. Measurements were carried out for $\lambda = 0, 0.1, 0.5,$ and 1.0. The data gave evidence for the occurrence of a continuous (second-order) phase transition in the thermodynamic limit, in agreement with the mean field prediction. The value of the critical point at $\kappa = 0$ also agreed well with the mean field prediction.

In contrast to the mean field predictions for the critical point, which only become exact in the limit $d \to \infty$, the prediction for the critical exponents are exact for $d \geq d_{u.c.}$, where $d_{u.c.}$ denotes the upper critical dimensionality of the system. (At the upper critical dimensionality, there are characteristically logarithmic corrections to the power law critical singularities, but these do not affect the algebraic values of the critical exponents.) For a bicritical point such as the one here, $d_{u.c.} = 4$, so that one would expect that the mean field values for the critical exponents are exact here (with logarithmic corrections for the actual critical singularities). By expanding the mean field free energy as a function of the order parameter, $< \bar{\chi}_f \chi_f >$, one obtains, for $d - 4 \to 0^+$, the expected Landau-Ginzburg effective free energy, with quadratic and quartic terms, which implies that $< \bar{\chi}_f \chi_f > \sim (y - y_c)^{1/2}$ as $y - y_c \to 0^+$. The mean field result (3.17) implies that $< \phi >$ and $< \bar{\chi}_f \chi_f >$ both have the same critical singularity and critical exponent, so that $< \phi > \sim (y - y_c)^{1/2}$ as $y - y_c \to 0^+$, which corresponds to the usual mean field value of the critical exponent, $\beta = 1/2$. Ref. 69 found that the data on $< \phi >$ and $< \bar{\chi}_f \chi_f >$ was consistent with this value for the critical exponent, although it could not make a very precise comparison.

Further, a generalized mean field method was used to map out the phase boundary linking the transitions at $(y, \kappa) = (0, \kappa_c)$ and $(y_c, 0)$. This provided an analytic understanding of an important feature of the model, viz., the fact that as $|y|$ increases from 0, the Yukawa interaction has the effect of strengthening the tendency toward ferromagnetic ordering, thereby rendering possible long-range ferromagnetic order at progressively smaller and smaller values of κ. For $y = y_c$, this effect is sufficiently strong that the system can order ferromagnetically even in the absence of any direct scalar field coupling $\kappa \sum_{n,\mu} \phi_n \phi_{n+e_\mu}$. Indeed, for $y > y_c$, the Yukawa interaction favors ferromagnetic ordering sufficiently strongly that it is able to overwhelm even a scalar field coupling with $\kappa < 0$ which, by itself, would favor antiferromagnetic ordering.

3.3.3 $\kappa = \infty$

In this limit, one can derive a useful exact result. For technical convenience, assume that $\lambda = \infty$, so that ϕ_n can only take on the values ± 1. Then

$$< \bar{\chi}_f \chi_f > = -y \phi_0 I(y) , \qquad f = 1, 2 \tag{3.18}$$

Clearly, $< \bar{\chi} \chi >$ is an analytic function of y for $\kappa = \infty$. As y increases from 0, $< \bar{\chi} \chi >$ rises linearly at first, then curves over, reaching a maximum at $y \simeq 1.235$, and decreases smoothly, approaching zero like $1/y$ as $y \to \infty$.

3.3.4 $y \to \infty$

Let us define new fermion fields according to $\chi'_{n,f} = y^{1/2} \chi_{n,f}$; then in the path integral, written in terms of the χ' fields, the fermion kinetic terms has a coefficient of $1/y$, while the Yukawa term is independent of y. Consequently, in the limit $y \to \infty$, the fermion kinetic term drops out, and one can perform the fermionic integrations exactly. This yields the result $Z = \int \prod_n d\phi_n e^{-S'_B}$ where

$$S'_B = S_B - \sum_n \ln\left[\left\{ \sum_{n' \in hc(n)} \phi_{n'} \right\}^2\right] \tag{3.19}$$

(If one were to consider N_f species of staggered fermions, then the exponent 2 on the sum over the hypercube in (3.19) would be replaced by N_f.) For finite values of the ϕ_n, the action S'_B is unbounded, since the argument of the logarithm can vanish. An important characteristic of the action (3.19) is that the logarithm term (which is the log of the fermion determinant) always favors ferromagnetic ordering. Its effect is quite strong, since whereas any scalar field configuration yields a finite value for the scalar coupling on a given link, it may yield a negatively infinite value for the logarithm term. In particular, for any scalar field configuration in which the ϕ fields on at least one hypercube are ordered antiferromagnetically, $S'_B = -\infty$, and this field configuration makes zero contribution to the partition function. Hence, a fortiori, the system cannot exist in a completely antiferromagnetically ordered state for $y = \infty$. For $\kappa > 0$, both terms in the action (3.19) thus work in the same direction. A mean field calculation was performed and predicted that even at $\kappa = 0$, the logarithm term is strong enough to produce ferromagnetic long-range order; this agreed with results from numerical simulations.[69] In contrast, however, for $\kappa < 0$, the scalar field coupling favors antiferromagnetic ordering while the logarithm term favors ferromagnetic ordering, so these terms compete against each other to determine the behavior of the system. For any finite y, as $\kappa \to -\infty$, the system approaches perfect antiferromagnetic ordering of the scalar fields. This fact, in conjunction with the fact that such ordering cannot occur at $y = \infty$, implies that the phase boundary of the antiferromagnetically ordered phase extends from ($\kappa = -\kappa_c, y = 0$) downward through the negative-κ half plane, and terminates at the corner ($\kappa = -\infty, y = \infty$). In a later work[70], a large-κ expansion was calculated for $| < \phi > |$ for $y = \infty$ taking $\lambda = \infty$ for technical convenience (this expansion has a finite radius of convergence). The result is (for general d)

$$| < \phi > | = 1 - 2\left(1 - 2^{1-d}\right)^{2^{d+1}} e^{-8\kappa d} - 4d(1 - 2^{1-d})^{2^{d+1}}(1 - 2^{2-d})^{2^d} e^{-8\kappa(2d-1)}$$
$$+ O(e^{-16\kappa d}) \tag{3.20}$$

The decrease in $< \phi >$ as κ decreases from ∞ is much less rapid than in the corresponding bosonic theory without the log term (i.e., d-dim. Ising model), for which

$$| < \phi > | = 1 - 2e^{-8\kappa d} - 4de^{-8\kappa(2d-1)} + O(e^{-16\kappa d}) \tag{3.21}$$

Numerically, for $d = 4$, (3.20) becomes

$$< \phi > = 1 - 0.02788e^{-32\kappa} - 0.002235e^{-56\kappa} + O(e^{-64\kappa}) \tag{3.22}$$

This result shows that the log term from the fermion determinant has the effect of strongly favoring FM long range order as well as inhibiting AFM ordering.

The fermion propagator can also be calculated analytically for large y, using a hopping parameter expansion. Owing to the invariance under translations and $90°$ rotations, the propagator depends only on the distance $r = (\sum_{\mu=1}^d (n_\mu - n'_\mu)^2)^{1/2}$. We thus let $n = 0$, $n' = re_\mu$. To leading order in $1/y$,

$$< \chi_{0,J} \bar{\chi}_{re_\mu, J'} > = (-2)^{-r} (2^{-d} y)^{-(r+1)} \delta_{J,J'} C(r) \tag{3.23a}$$

$$C(r) = < \prod_{t=0}^{r} \left(\sum_{n' \in hc(te_\mu)} \phi_{n'} \right)^{-1} >_{s'} \tag{3.23b}$$

where $< \ldots >_{s'}$ denotes the statistical average in the theory with action (3.19). For fixed r, $|C(r)|$ increases as κ decreases from ∞. Hence, for fixed y, the renormalized fermion mass aM_F decreases as κ decreases from ∞. To evaluate this decrease, one carries out a large-κ expansion of (3.23b) (with $\lambda = \infty$) and finds, to leading order in $1/y$,

$$aM_F = y\left[1 - 2(1 - 2^{-d})(1 - 2^{1-d})2^{2^{d+1}-2}e^{-8\kappa d} + O(e^{-8\kappa(2d-1)})\right] \quad (3.24)$$

Given the exact result (3.18), one sees that the leading term of (4) is actually exact, and is not just restricted to large y. As κ decreases from ∞, aM_F decreases slowly (for $d = 4$, the leading κ-dependent term, $e^{-32\kappa}$, is suppressed by a small coefficient, 0.0341).

3.3.5 *Mapping of Phase Structure for $\kappa < 0$*

In order to supplement these analytic results and to map out the phase structure further for $\kappa < 0$, Ref. 69 used numerical simulations (with dynamical fermions). These showed that as the PM-FM phase boundary crosses the $\kappa = 0$ (horizontal) axis and moves to the negative-κ region, it approaches the PM-AFM boundary. Beyond a certain value of y (e.g. y slightly less than 2 for $\lambda = 1.0$), there is a new phase lying between the FM and AFM phases, which exhibits both FM and AFM order. This was denoted the ferrimagnetic (FI) phase.[69] For example, it was found that at $y = 5$ the FM-FI phase boundary occurs at $\kappa_{c,FM-FI} = -0.2 \pm 0.02$. The simulations were consistent with the hypothesis, but of course could not prove, that the FM, PM, AFM, and FI phases meet exactly at a point; this possibility is shown only for definiteness in Fig. 1. (The behavior in the immediate vicinity of point A was not important to the measurement of renormalized couplings.)

3.3.6 *AFM Phase*

Because the PM phase is analytically connected to the border at $y = 0$, where the fermionic chiral symmetry is necessarily realized explicitly (since the fermions are free), it follows that the renormalized fermion mass $aM_F = 0$ in the PM phase. Interestingly, although this Z(2) symmetry is broken in the AFM phase, one can still prove that $aM_F = 0$ in this phase also.[70] This result provides another way of seeing that the FI-AFM phase boundary extends to the point $(y, \kappa) = (\infty, -\infty)$. Thus, assume the contrary, that at $y = \infty$ this boundary terminates at $\kappa_{FI-AFM,\infty} > -\infty$. Consider a segment at $y = \infty$, $\kappa_{FI-AFM,\infty} > \kappa > -\infty$. Since this segment is part of the AFM phase, $aM_F = 0$. But this contradicts the fact (cf. (3.24)) that at $y = \infty$, $aM_F = \infty$ (end of proof).

These analytic and numerical results thus determined the phase diagram of the Yukawa theory (3.9) to be that shown in Fig. 1.

3.4 Measurement of Renormalized Quantities

As noted before, one studies the continuum limit of the lattice theory at the PM-FM phase boundary, approaching this from within the FM phase, since one wishes to study fermion masses generated by Yukawa interactions, and since one wishes to build in a similarity in this toy model with the actual electroweak theory, where the gauge symmetry is realized in the Higgs mode. Here, by "renormalized" masses and couplings, one means those defined at a second-order phase transition of the lattice theory, where the correlation length(s) is (are) infinite. This definition is not restricted to perturbation theory, in contrast to the standard renormalization theory in continuum quantum field theory, which is defined order-by-order in perturbation theory. Thus, one need not discuss any renormalization counterterms in the lattice action. Since the physical masses and couplings of the continuum theory are defined at the second-order phase transition of the lattice theory, we shall use the terms "physical" and "renormalized" interchangeably in the following.

An important result, proved in Ref. 69, is that the properties of the phase transition and hence also of the continuum limit are the same along the PM-FM boundary in a neighborhood of $\kappa = 0$, including both positive and negative values of κ. This implies that the point on the PM-FM phase boundary where it crosses the $\kappa = 0$ axis is not special; the universality class of the transition there is the same as elsewhere along the boundary in a neighborhood of this point. This in turn shows the equivalence, at the quantum level, of the continuum Yukawa theory and a purely fermionic theory with the multifermion operators specified by the exact result (3.14). By counting powers of the lattice spacing a multiplying these multifermion operators, one sees that the most important such operator is the four-fermion operator. (This fact was also used in the mean field analysis yielding (3.15).) Thus, this shows an equivalence between the four-fermion models[6,7] and Yukawa models.

Since this is a coupled theory, there are really two independent renormalized masses, aM_B and aM_F, and corresponding correlation lengths, ξ_B and ξ_F. Since the PM-FM phase boundary is the continuation into the interior of the phase diagram of the critical point in the pure scalar theory at $y = 0$, it is expected that ξ_B diverges as one approaches this phase boundary from either the symmetric or broken-symmetry phases. In general, in a theory with two or more independent masses, it is not true that they will all diverge at a submanifold in the lattice parameter space where one of them diverges (a counterexample is given in the fourth paper of Ref. 13). However, here, since $M_F = 0$ in the symmetric phase, and since the transition at the PM-FM phase boundary is continuous, we know that M_F must also vanish (i.e., ξ_F diverge) continuously as one approaches the phase boundary from within the broken phase. The simplest statistical mechanical scaling ideas would suggest that, since M_B and M_F are both masses, they should vanish in the same manner as one approaches the critical phase boundary, and consequently the ratio $\rho = M_B/M_F$ should approach a finite (nonzero) constant in this limit. Interestingly, the perturbative (one-loop) renormalization group equations for this theory imply the same conclusion.

The measurements of physical couplings were made using the same hybrid Monte Carlo method that was employed to investigate the phase structure. This utilized an 8^4 lattice, which was large enough for the purposes of the study. For a given set of parameters (y, κ, λ), 120,000 hybrid Monte Carlo iterations were carried out, each of which consisted of 5 molecular dynamics steps followed by a global Metropolis acceptance/rejection update. There are several ways to define a physical (renormalized) Yukawa coupling, y_R. The simplest for numerical purposes is via the relation which holds in the FM phase with spontaneously broken symmetry,

$$y_R = \frac{aM_F}{<\phi_R>} \tag{3.25a}$$

where M_F is the physical (renormalized) fermion mass, and Z_ϕ is the field renormalization constant for the scalar field, defined in the usual manner by

$$\phi_R = Z_\phi^{-1/2} \phi \tag{3.25b}$$

In the continuum Minkowski theory, Z_ϕ would be defined on the physical mass shell. In Euclidean field theory, there is no such mass shell hyperboloid; rather, the condition $p^2 = m_R^2$ defines a 3-sphere in momentum space of no direct physical significance (as is obvious, since it corresponds to a single momentum $|p| = m_R$). Accordingly, the field renormalization constant was defined at zero momentum. The momentum-space fermion propagator is given by

$$\Gamma_F(p) = V^{-1} \sum_{x_1, x_2} e^{ip \cdot (x_1 - x_2)} < \chi(x_2)\bar{\chi}(x_2) > \tag{3.26}$$

The physical fermion mass M_F and fermion field renormalization constant Z_F were obtained from fitting $\Gamma_F(p)$ for timelike $p = (0, 0, 0, p_4)$ to the form

$$\Gamma_F(p_4) = \frac{Z_F}{aM_F + i \sin p_4} \tag{3.27}$$

The renormalized scalar mass aM_B and the scalar field renormalization constant Z_ϕ were determined from an analysis of the spatially Fourier transformed 2-point correlation function. The physical quartic coupling can be defined via the relation

$$\lambda_R \equiv \frac{1}{2} \left(\frac{aM_B}{<\phi_R>} \right)^2 \tag{3.28}$$

applicable in the broken-symmetry phase with $<\phi> \neq 0$. This definition is much easier to use for numerical studies than an alternative definition involving the connected 4-point function.

Most of the measurements were performed for the values $\kappa = 0$ and 0.01, and, independently, $\lambda = 0.1$ and 1.0; y was adjusted to be close to the PM-FM phase boundary, slightly on the ferromagnetic phase. In terms of the "continuum" parametrization (3.7), these points covered the range $9.5 < \tilde{y} < \infty$ and $10^3 < \tilde{\lambda} < \infty$. The results for the renormalized Yukawa and scalar couplings are shown in Figs. 2 and 3, as functions of $(aM_B)^{-1}$. The continuum limit is thus $(aM_B)^{-1} = \infty$.

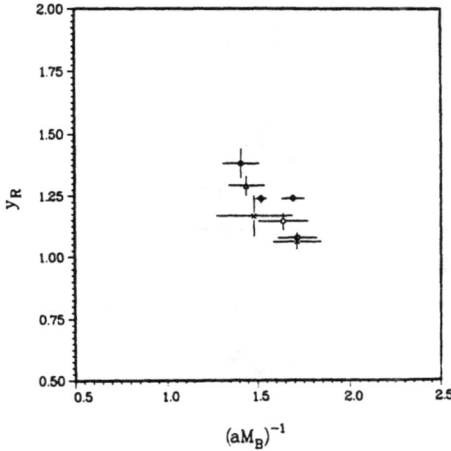

Fig. 2. Measurements of y_R as a function of scalar field correlation length or equivalently $(aM_B)^{-1}$ for $\lambda = 0.1$ ($\bullet$) and $\lambda = 1.0$ ($\circ$), each with $\kappa = 0$ or 0.1. The crosses are the data for $\lambda = 0.1$ with negative κ. See Ref. 69 for further details.

A striking feature of the data is that although the bare couplings $\tilde{y}$ and $\tilde{\lambda}$ range over a very large interval, including infinity, the renormalized couplings lie in a rather small range. For y_R, these values are $\gtrsim 1$, while for λ_R, with our normalization, they range from slightly less than 1 to $\gtrsim 2.5$. This insensitivity of the renormalized couplings to the bare couplings is the type of behavior which one would expect if the renormalized couplings actually vanished for infinite cutoff, $(aM_B)^{-1} = \infty$.

Further, although the data is characterized by a limited range of ξ_B (and ξ_F), it shows that both y_R and λ_R tend to decrease as ξ_B increases. Again, this is the behavior which one would expect if these renormalized couplings vanished as the cutoff goes to infinity. Of course, while these results are consistent with the conclusion that the theory is free, neither they nor even more extensive numerical measurements could ever prove this conclusion. Referring back to eq. (2.8), although the fermion gives rise to a third term with a negative sign, and hence to the possibility that the behavior of the scalar coupling might differ qualitatively in the Yukawa theory from the pure scalar theory, the data of Ref. 69 indicated that, where measured, the qualitative behavior of the scalar coupling is similar to that in the pure scalar theory.

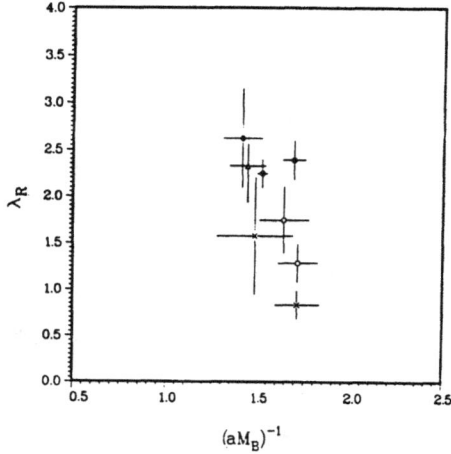

Fig. 3. Measurements of λ_R as a function of $(aM_B)^{-1}$. Symbols are defined as in Fig. 2.

Recalling the discussion in section 2, the data enables one to infer upper bounds on the masses of the physical fermion mass and scalar mass in a continuum theory. The characteristic mass scale here is given by v, the statistical average (vacuum expectation value in the Minkowski theory) of the scalar field ϕ. If one chooses to require that $(aM_B)^{-1}$ must be larger than about 2, so that the physical momentum cutoff $\Lambda = \pi/a > 2\pi M_B$, then the resultant effective renormalized Yukawa coupling $y_R \lesssim 1$, and the renormalized fermion mass $M_F \lesssim v$. The conclusion, then, is that in a Yukawa theory, the mass of the fermion generated by the Yukawa interaction cannot be substantially larger than the symmetry-breaking mass scale. The data also yields a corresponding upper bound on the physical scalar mass. In general, it was found that the presence of the fermion in the full Yukawa theory does not strongly change the behavior of λ_R from that in the pure scalar theory. It would, of course, be desirable to extend these measurements to larger lattices and hence larger correlation lengths. However, one would not expect to observe a substantial decrease in y_R and λ_R with increasing correlation length (cutoff) since the perturbative renormalization group analyses indicate that the asymptotic behaviors would be (cf. eqs. (2.7) and (2.3)) very slow: $y_R \sim (\ln \xi_B)^{-1/2}$ and $\lambda_R \sim (\ln \xi_B)^{-1}$ as $\xi_B \to \infty$. Note that although the behavior observed for these couplings in Figs. 2 and 3 contains pre-asymptotic dependence, one does see that λ_R decreases more rapidly with $(aM_B)^{-1}$ than y_R, as is also expected asymptotically. The data also enable one to compute the ratio M_B/M_F; this is found to be slightly larger than unity and not to vary strongly with ξ_B.

3.5 Other Continuum Limits

In addition to the continuum limit of the lattice theory at the FM-PM phase boundary, one may also investigate the analogous limits along other phase boundaries. In constructing a physically sensible continuum limit, one normally requires not only that the correlation length diverge, but also that the fields, or appropriate functions of the fields, be progressively more and more slowly varying on the scale of the lattice spacing a as one approaches the critical submanifold in the space of lattice parameters (since if this second condition were not met, the continuum fields would not be continuous or differentiable). Along the PM-FM phase boundary, these conditions are clearly satisfied. At $y = 0$, at the critical point separating the PM and AFM phases, although the second condition is not met for the ϕ fields themselves, one can use the symmetry of the theory under the transformations

$$\kappa \to -\kappa \ , \qquad \phi_n \to (-1)^{\Sigma(n)}\phi_n \tag{3.29}$$

to map this critical point to the one between the PM and FM phases and construct the continuum limit in terms of the fields $(-1)^{\Sigma(n)}\phi_n$, which do satisfy the second condition. In the full coupled theory at nonzero y, (3.29) is no longer a symmetry, so this argument does not apply.

The phase boundaries between the FM and FI phase, and between the FI and AFM phases were found to be continuous where measured and hence to provide possible continuum limits. In contrast to the continuum limit at the FM-PM phase boundary, these limits are non-perturbative in the sense that the corresponding phase boundaries do not extend to $y = 0$. It is thus of interest to investigate the behavior at the FM-FI and FI-AFM phase boundaries as different continuum limits of the lattice theory. This was done in Ref. 70. Recall that because the renormalized fermion mass aM_F is nonzero at the FM-FI phase boundary, the continuum limit there is a purely bosonic theory. Again, the quantities $< \bar{\chi}_f \chi_f >$, $< \phi >$, and $< \phi >_{st}$ were measured. At the FM-FI transition, where the staggeredd magnetization turns on as κ decreases through its critical value κ_{FM-FI} at a given y, the data gave an excellent fit to the dependence

$$< \phi >_{st} \propto |\kappa - \kappa_{FM-FI}|^{\theta}(|\kappa| - |\kappa_{FM-FI}|) \tag{3.30}$$

in the vicinity of the transition (where $\theta(x) = 1$ if $x > 0$ and 0 otherwise). Thus, the corresponding critical exponent for the order parameter $< \phi >$ is consistent with $\beta_{<\phi>_{st}} = 1/2$, the Gaussian value. This result is consistent with the conclusion that the continuum theory at the FM-FI phase boundary, which is purely bosonic as discussed above, is non-interacting. Some measurements were also performed at the FI-FM phase boundary.

4. UNIVERSALITY AND THE STUDY OF DIFFERENT LATTICE FORMULATIONS OF A YUKAWA MODEL WITH A REAL SCALAR FIELD

4.1 General

An important issue in any lattice study is that of universality: i.e., which properties of a lattice theory are independent of the details of the lattice action and which are dependent on such details. The latter are often called lattice artifacts. An initial check of universality was made in Ref. 69 between the continuum-type parametrization used in early studies[64,67,68] and the statistical mechanics-type parametrization used there. Measurements were performed for a set of parameters (y, κ, λ) corresponding, via (3.8) and (3.10), to a set $(\tilde{y}, \tilde{m}^2, \tilde{\lambda}) = (5, 25, 1)$ for which measurements had been made in Ref. 68. Good agreement was found.

A further study of universality was made in Ref. 71 by comparing the phase structure of the Yukawa model with a real scalar field and a local Yukawa coupling with that found in Ref. 69 for the same model with a hypercubic coupling. Although for staggered fermions the hypercubic Yukawa interaction (3.6) is the most natural lattice transcription of the continuum Yukawa interaction and was thus used in the lattice actions (3.7c) and (3.9d), the so-called local Yukawa interaction

$$S_{Y,\ell} = -y_\ell \sum_n \phi_n \sum_f \bar{\chi}_{n,f} \chi_{n,f} \qquad (4.1)$$

is also of interest to study. As one approaches a phyiscally reasonable continuum limit, where the correlation lengths diverge, the ϕ fields must vary more and more slowly on the scale of the lattice spacing a in order for the continuum ϕ field to be continuous. Hence, one expects that both the hypercubic and local Yukawa lattice actions should yield the same continuum physics. Note that (4.1) yields the correct formal continuum limit if the bare coupling satisfies

$$y_{cn} = (2\kappa)^{-1/2} y_\ell \qquad (4.2)$$

analogous to (3.10d).

4.2 Phase Structure

The determination of the phase structure of the (4D) lattice Yukawa model with action given by (3.9a-c) together with (4.1) was performed by means similar to those discussed above in section 3. Hence, we shall omit some details, referring the reader to Ref. 71. Note that this theory is again invariant under the same global

Z(2) symmetry (3.13) as was the theory with hypercubic Yukawa interaction. Again, much can be learned from various special cases. At $y_\ell = 0$ the theory is, of course, the same as (3.9) for $y = 0$ and consists of paramagnetic (PM), ferromagnetic (FM), and antiferromagnetic (AFM) phases. For $\kappa = 0$, one can again integrate over the scalar fields and obtain the exact result

$$Z = v^{N_s} \int \prod_{n,f} d\chi_{n,f} d\bar{\chi}_{n,f} e^{-S'_\ell} \tag{4.3a}$$

where N_s denotes the number of sites, $v = e^{-\lambda} J_0$,

$$J_k = (2\lambda)^{-(k+1)/4} \Gamma((k+1)/2) e^{(1-2\lambda)^2/(8\lambda)} D_{-(k+1)/2}((1-2\lambda)(2\lambda)^{-1/2}) \tag{4.3b}$$

$$S'_\ell = S_F - y_\ell^2 R_2 \sum_n (\bar{\chi}_{n,1}\chi_{n,1})(\bar{\chi}_{n,2}\chi_{n,2}) \tag{4.3c}$$

where $D_\nu(z)$ is the parabolic cylinder function and $R_p = J_p/J_0$. As in the theory with a hypercubic Yukawa coupling, this exact result shows an equivalence, at the quantum level, of the Yukawa theory and a purely fermionic theory with the four-fermion coupling, (4.3). In this way, it shows the equivalence of four-fermion models[6,7] and Yukawa theories. Using a mean field approximation as before, one obtains the consistency condition

$$< \bar{\chi}_f \chi_f > = BI(B) , \qquad f = 1,2 \tag{4.4a}$$

(with no sum on f) where

$$B = (y_\ell^2/2) R_2 < \bar{\chi}_f \chi_f > \tag{4.4b}$$

and $I(B)$ is given by (3.15c). Within a neighborhood of $y_\ell = 0$, the only solution to (4.4) is $< \bar{\chi}_f \chi_f > = 0$, but as y_ℓ increases past a critical value $y_{\ell,c1}$ (depending on λ), (4.4) has a nonzero solution for $< \bar{\chi}_f \chi_f >$, which is the physical solution, since it minimizes the mean field free energy. Solving (4.4), one finds that the phase transition at $y_\ell = y_{\ell,c1}$ is continuous. From an analytic $1/d$ expansion (which agrees with a numerical solution), one finds that $y_{\ell,c1;MF}/y_{hc,c;MF} \simeq 1.95$. Using the same methods, one obtains

$$< \phi > = -2y_\ell R_2(\lambda) < \bar{\chi}_f \chi_f > \tag{4.5}$$

(with no sum on f), which is the same as (3.17) with y_{hc} replaced by y_ℓ. The predictions for critical exponents are also the same as for $S_{Y,hc}$. At $\kappa = \infty$, the theory can be solved exactly, and gives (3.18) with $y_{hc} \to y_\ell$.

However, in the region of large bare Yukawa coupling $y_\ell \to \infty$, one finds that the theory with a local Yukawa interaction has a very different phase structure than the one with hypercubic interaction. In this limit, the model with local coupling $S_{Y,loc.}$ reduces to a purely bosonic theory with

$$Z \propto \int \prod_n d\phi_n e^{-S'_{\ell,B}} \tag{4.6a}$$

where

$$S'_{\ell,B} = S_B - \sum_n \ln(\phi_n^2) \qquad (4.6b)$$

(For N_f species of staggered fermions [where N_f is taken to be even to yield a non-negative fermion determinant], $\ln(\phi_n^2)$ would be replaced by $\ln(\phi_n^{N_f})$; .) This theory (4.6), like the bosonic sector at $y = 0$ or $y_\ell = 0$, is invariant under the symmetry (3.29). The finite radius of convergence of a small-$|\kappa|$ expansion implies, as usual in spin models, that there is a symmetric, paramagnetic phase for sufficiently small $|\kappa|$. We denote this phase as PM2. In Ref. 71, a small-$|\kappa|$ series expansion for the susceptibility was calculated to $O(\kappa^7)$, and a d log Padé analysis of this series was performed. This indicated a divergence in the susceptibility at a critical value, $\kappa_{c;y_\ell=\infty}$, associated with a second-order phase transition involving the onset of FM long-range order. Given (3.29), this implies that there is also a phase transition to an AFM ordered phase at $\kappa = -\kappa_{c,y_\ell=\infty}$. The value of the critical point was determined, and it was confirmed that, for a given $\lambda < \infty$, $\kappa_{c,y_\ell=\infty} < \kappa_{c,y_\ell=0}$, as expected since the additional term $-\sum_n \ln(\phi_n^2)$ in $S'_{\ell,B}$ increases the tendency to order (equally for FM and AFM ordering). (For $\lambda = \infty$, this term vanishes, and the theory at $y_\ell = \infty$ is the same as the bosonic theory at $y_\ell = 0$, i.e., the 4D Ising model.) From the d log Padé analysis of the susceptibility series, the critical exponent γ was calculated, and it was found to be consistent with being equal to the Ising value (which is $\gamma = 1$, up to logs), independent of λ, for $\lambda \neq 0$. Thus, the phase structure at large y_ℓ for the model with local Yukawa interaction differs markedly from that for the hypercubic Yukawa interaction, which, at large y, has only FM and FI phases, and neither a symmetric nor an AFM phase.

One can proceed to study the interior of the phase diagram. For large y_ℓ, one uses a $1/y_\ell$ expansion, which has a finite radius of convergence. The phase boundary is most easily calculated for $\lambda = \infty$, for which, to leading order in $1/y_\ell$, the theory has the same form as the original action, with a shifted coupling,

$$2\kappa_{eff} = 2\kappa + N_f/(4y_\ell^2) \qquad (4.7)$$

This implies that the PM2-FM and PM2-AFM phase boundaries both curve downward as y_ℓ decreases from ∞, and that the PM2-FM phase boundary crosses the $\kappa = 0$ axis roughly (for $N_f = 2$) at $y_{\ell,c2} \sim (4\kappa_{c,y_\ell=\infty})^{-1/2}$.

So far, in addition to the ferromagnetic and antiferromagnetic phases, these results show the existence of a symmetric, parametric phase at zero and small y_ℓ and another at large y_ℓ. The question then arises as to whether the two paramagnetic phases are the same (i.e., analytically connected) or distinct. To answer this, the fermion propagator and the resultant fermion mass were calculated at large y_ℓ, again using the $1/y_\ell$ expansion. To leading order in $1/y_\ell$ (and for general d), one finds

$$< \chi_{0,f} \bar{\chi}_{re_\mu,f'} > \, = 2^{-r} y_\ell^{-(r+1)} \delta_{f,f'} < \prod_{t=0}^{r} \phi_{te_\mu}^{-1} >_{S'_{\ell,B}} \qquad (4.8)$$

where $< .. >_{s'_{\ell,B}}$ denotes the statistical average in the bosonic theory (4.6) at $y_\ell = \infty$. (Given the $\delta_{f,f'}$ in (4.8), the dependence on f is obvious and will be suppressed in the notation.) The inverse correlation length is

$$\xi_F^{-1} = - \lim_{r \to \infty} r^{-1} \ln | < \chi_0 \bar{\chi}_{re_\mu} > | \qquad (4.9)$$

with $r + 1$ even (in the PM2 phase, RHS(4.8) $= 0$ for $r + 1$ odd). In the PM2 phase, for small $|\kappa|$ (and arbitrary λ), the resultant renormalized mass is

$$a M_F = \frac{y_\ell}{(2|\kappa|)^{1/2}} \left[1 - O(\kappa^2) \right] - O(y_\ell^{-1}) \qquad (4.10)$$

(In Ref. 71, the equivalent quantity $a m_F \equiv \xi_F^{-1}$ was used as an effective lattice fermion mass; these are related by $a M_F = \sinh(a m_F)$, so that M_F and m_F give the same information in general and are identical in the continuum limit.) The expansion for $a m_F$ corresponding to (4.10) is $a m_F = \ln(2 y_\ell) - (1/2) \ln(2|\kappa|) - O(\kappa^2)$. Although (4.10) applies to leading order in $1/y_\ell$, it is also qualitatively valid for sufficiently large but finite y_ℓ, given the finite radius of convergence of the $1/y_\ell$ expansion. Eq. (4.10) shows analytically that in the PM2 phase, despite the fact that the $Z(2)$ symmetry is realized explicitly, with $< \phi > = < \phi >_{st} = < \bar{\chi}_f \chi_f > = 0$, the fermion mass is nonzero. This contrasts sharply with the behavior in the other $Z(2)$-symmetric phase, PM1, where $a M_F = 0$, as is expected from the fact that this phase is contiguous with the free massless fermion limit at $y_\ell = 0$ and was demonstrated explicitly by dynamical fermion measurements. Given this difference in $a M_F$, it follows that the PM2 phase cannot be analytically connected to the PM1 phase. (This was proved in the usual manner: assume the contrary. Then one could analytically continue the function representing $a M_F$ from the PM1 phase, where it vanishes identically, to the PM2 phase, which would contradict the result that $a M_F \neq 0$ in the PM2 phase.) Recall that in the continuum, chiral symmetry can, *a priori*, be realized explicitly in two ways: (a) massless physical fermions or (b) massive, parity-doubled fermions. Although the $Z(2)$ symmetry is not precisely equivalent to usual chiral symmetry, since it also involves a transformation of the scalar field, it is similar in forbidding a fermion mass term in the action. Evidently, the phases PM1 and PM2 exhibit realizations of the $Z(2)$ symmetry which are reminiscent of (a) and (b).

The phase structure inferred from these results is shown in Fig. 4. As was noted in Ref. 71, the behavior in the neighborhood of point B may involve a meeting of the four respective phases (the possibility shown, for simplicity, in Fig. 4), or a more complicated structure. Since this was not important to the universality study of Ref. 71, the issue of the behavior near point B was not further explored there. However, it was subsequently investigated, and some hints of a possible intermediate ferrimagnetic phase in the region B were found (see below and Ref. 75). The presence or absence of such an intermediate FI does not affect the conclusions concerning universality derived from the comparison of the hypercubic and local Yukawa actions.

For the Ising limit $\lambda = \infty$ (where $\phi_n \equiv \sigma_n = \pm 1$), one can prove the following rigorous theorem: aM_F, calculated from eq. (4.8) (applicable for large y_t), is a

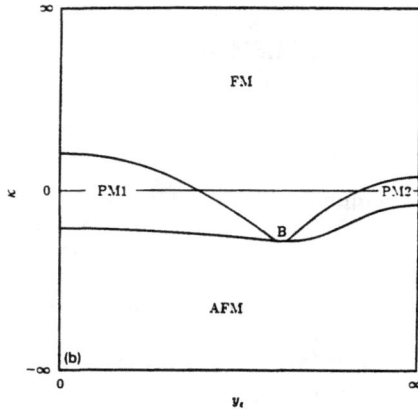

Fig. 4. Phase diagram for the theory (3.9). The figure shows one of several possibilities for the behavior in the region of point B. See text for discussion of phases and of the behavior near point B.

monotonically decreasing function of $|\kappa|$. This follows because the Griffiths-Kelly-Sherman (GKS) inequality[72] implies that for $\kappa \geq 0$, $< \prod_{t=0}^{r} \sigma_{te_\mu}^{-1} >_{s'_{t,B}}$ is a monotonically increasing function of κ. Combining this with the symmetry (3.29) shows that $| < \prod_{t=0}^{r} \sigma_{te_\mu}^{-1} >_{s'_{t,B}} |$ is a monotonically increasing function of $|\kappa|$, which yields the theorem. The GKS inequality has been extended to real scalar lattice field theory[72] ($\lambda < \infty$) and implies that for $\kappa \geq 0$, correlation functions such as $< \prod_j \phi_j >_{s'_{t,B}}$ are monotonically increasing functions of κ, where j denotes some set of points in Z^d. Although the specific application of the generalized GKS inequality to correlation functions involving inverse powers of ϕ has not, to our knowledge, appeared in the literature, the explicit calculation of $< \prod_{t=0}^{r} \phi_{te_\mu}^{-1} >_{s'_{t,B}}$, e.g., for small κ, shows that it is also a monotonically increasing function of κ for arbitrary λ. This, with (3.29), then implies that $| < \prod_{t=0}^{r} \phi_{te_\mu}^{-1} >_{s'_{t,B}} |$ is a monotonically increasing function of $|\kappa|$ for arbitrary λ. Eq. (4.10) agrees with this general theorem and shows that in phase PM2, for arbitrary λ, aM_F decreases as $|\kappa|$ increases and that as one crosses the PM2-FM or PM2-AFM phase boundaries, aM_F does not vanish. In contrast, given that the PM2-FM and PM2-AFM transitions are second-order at $y = \infty$, and the finite radius of convergence of the $1/y_\ell$ expansion, the PM2-FM and PM2-AFM boundaries should also be second order for a range of finite y_ℓ, with divergent correlation length ξ_B. Hence, as one approaches the PM2-FM phase boundary, $M_F/M_B \to \infty$, the fermions freeze out, and the resultant continuum theory is purely bosonic.

The general theorem implies, in particular, that for $\lambda = \infty$, as one approaches the PM2-FM boundary from within the FM phase, aM_F increases. An explicit calculation (for $\lambda = \infty$) at large $|\kappa|$ agrees with this, yielding

$$| < \prod_{t=0}^{r} \phi_{ic_\mu}^{-1} >_{S'_{t,B}} | \simeq 1 - 2(r+1)e^{-4q|\kappa|} \tag{4.11}$$

where $q = 2d$ is the coordination number of the lattice. Hence, for large $|\kappa|$ (and large y_ℓ), $am_F \simeq \ln(2y_\ell) + 2e^{-4q|\kappa|}$, or equivalently,

$$aM_F \simeq y_\ell \left[1 + 2e^{-4q|\kappa|} \right] - O(y_\ell^{-1}) \tag{4.12}$$

The property that aM_F increases as one approaches the FM-PM2 phase boundary from within the FM phase is the opposite of what was found for the approach to the (similarly bosonic) continuum limit of the model with a hypercubic Yukawa coupling at the FM-FI phase boundary, also at large y. The behavior of am_F at large y for the model with local coupling agrees with that found in a quenched numerical simulation of a related model with naive fermions[73].

The comparative study of the Yukawa theory with hypercubic and local Yukawa interactions in Refs. 69 and 71 thus showed that the phase structures were qualitatively the same for small Yukawa coupling: a symmetric, paramagnetic phase for small $|\kappa|$ together with broken-symmetry ferromagnetic and antiferromagnetic phases at sufficiently large positive and negative κ, respectively. In particular, this study suggested that the behavior at the PM-FM phase boundary in the model with hypercubic Yukawa interaction appears to be universal, and not dependent upon details of the specific lattice action. The same universality then applies to the the measurements of renormalized masses and couplings in Ref. 69 and the consistency of those results with the conclusion that in the infinite cutoff limit, the scalar and Yukawa couplings would vanish.

In this context, it is of interest to inquire to what extent constructive field theorists have been able to prove results pertaining to triviality of 4D Yukawa theories. Of course, given that the mathematical proofs of the triviality of the pure $\lambda\phi^4$ field theory apply only for $d > 4$, not the physical dimension $d = 4$, and given that fermions and the associated anticommuting variables in functional integrals only make the problem more difficult, one would not expect that there is any mathematical proof of triviality of 4D Yukawa theories. Although much work has been done for Yukawa models in $d = 2$ and some in $d = 3$, very little progress has been made on the issue of triviality in $d = 4$.[74] One study of a 4D Yukawa model[74] began

186

by integrating out the fermions, to obtain a purely bosonic theory. This is some-what reminiscent of the integration of the fermions to obtain a bosonic theory at infinite bare Yukawa coupling in Refs. 69 and 71. However, the fermions are not explicit dynamical variables in this limit; the susceptibility series expansion and d log Padé analysis performed in Ref. 71 to show that the universality class of the resultant bosonic theory is consistent with being Gaussian took advantage of this fact. Thus, it appears that progress on investigating the renormalized couplings in the 4D Yukawa theory, as defined via lattice regularization, will continue to rely on numerical simulations (and the associated work of determining the lattice phase structure and locations of phase boundaries).

In passing, we note that the results of Ref. 71 also showed the existence of a second paramagnetic phase also in the phase diagram for the lattice theory with local Yukawa interaction and so-called continuum parametrization. This theory thus has FM, PM1, and PM2 phases (the first two had been found earlier[68]).

Some further information on the model with a local Yukawa coupling was obtained in Ref. 75, which performed an extended calculation of the effective action at strong Yukawa coupling. To order $1/y_{ell}^4$, for $\lambda = \infty$, it was found that $S_{eff} = S'_{\ell,B} + S_2 + S_4$ where $S'_{\ell,B}$ is given by (4.6) and

$$S_4 = S_{4,p} + S_{4,2\ell} + S_{4,cr} \tag{4.13}$$

where

$$S_{4,p} = -2^{-3}y_\ell^{-4}N_f \sum_{n,\mu,\nu} \sigma_{plaq.} \tag{4.14}$$

$$S_{4,2\ell} = 2^{-5}y_\ell^{-4}N_f \sum_{n,\mu} \sigma_n \sigma_{n+2e_\mu} \tag{4.15}$$

$$S_{4,cr} = 2^{-5}y_\ell^{-4}N_f \sum_{n,\mu,\nu;\mu\neq\nu} \sigma_n \sigma_{n+e_\mu+e_\nu} \tag{4.16}$$

The subscripts denote the types of graphs which yield the various terms (p = plaque-tte, 2ℓ = 2-link straight path, cr = corner). Evidently, the various $O(y_\ell^{-4})$ terms favor different types of ordering. The $S_{4,p}$ term favors FM and AFM ordering (equally), but $S_{4,2\ell}$ and $S_{4,cr}$ disfavor both FM and AFM ordering. Hence, at $O(y_\ell^{-4})$ and thus also to higher order in $1/y_\ell$, the effective theory is frustrated, i.e., field configurations which minimize the contributions to the action from some terms do not minimize the contributions from other terms. This provided an analytic suggestion that there might exist a new phase at intermediate y_ℓ and κ which is characterized by a type of ordering more complicated than simple FM or AFM. Such an intermediate phase would, in general, be of ferrimagnetic (FI) type.

Another result given in Ref. 75 concerned the universality class of the FM-PM2 transition. Since the $1/y_\ell$ expansion has a finite radius of convergence, it follows that the universality class of the FM-PM2 transition is the same for an interval of y_t from $y_t = \infty$ to a finite values of y_t (depending on κ) as it is at $y_t = \infty$. Recall that the universality class of the FM-PM2 transition in the latter theory had been studied and determined to be consistent with that of the 4D Ising model, which is Gaussian (up to logs).[71] (A similar comment applies to the PM2-AFM phase boundary.) Further, given the strong evidence that that the 4D $\lambda\phi^4$ theory is noninteracting, one can conclude that the same is true of the bosonic continuum theory defined on the FM-PM2 phase boundary for large y_t.

4.3 Other Actions

Work has also been done on Yukawa models with a real scalar field using naive fermions. Here, the Yukawa interaction is taken simply as

$$S_Y = -y \sum_{n,f} \phi_n \bar{\psi}_{n,f} \psi_{nf},$$ (4.17)

where $\psi_{n,f}$ are naive lattice Dirac fermions. An early study was performed with quenched fermions.[74] In the quenched approximation, since the fermions are not dynamical, the phase structure of the full Yukawa theory cannot be determined; in this approximation, the phase structure is given by that of the pure scalar theory. However, one can study the behavior of the quenched fermion quantities. It was found that y_R decreases with increasing bosonic correlation length ξ_B at small bare y, consistent with the conclusion that the continuum theory has $y_R = 0$. In contrast, as discussed above, for large bare y, aM_F and y_R increased with ξ_B.

One can also comment on modifications of the hypercubic Yukawa coupling used in Ref. 69. For this purpose, we first recall the equivalence which was established in Ref. 69 and 71 between the Yukawa theory with, respectively, a hypercubic and a local Yukawa interaction, and purely fermionic theories with multifermion couplings, of which the four-fermion coupling was the most important near the continuum limit. (This equivalence was also discussed subsequently in Ref. 76.) Given this equivalence, one can make use of the results from a lattice study[77] of a 2D model with a four-fermion coupling. Since the symmetry group is Z(2) in both cases, one expects qualitatively similar phase structure for the 2D and 4D models. Ref. 77 considered defining the hypercubic sum over alternate hypercubes rather than all hypercubes. In this formulation, one loses the shift symmetry of the staggered fermion action, which, in turn, causes problems with the continuum limit of the theory.[66] Indeed, it was for this reason that the sum over all hypercubes was used in Ref. 69. In passing, we note[78] that if, despite this problem, one were nevertheless to try to use the hypercubic Yukawa interaction with summation only over alternate hypercubes, one would obtain a different phase structure at strong Yukawa coupling, including a second symmetric phase (see also Ref. 79)).

Some examples were given in Ref. 80 to demonstrate the large variation in lattice phase structure for different lattice Yukawa actions, all of which yield the

188

same formal continuum limit. Thus, consider the Yukawa theory with a real scalar field, staggered fermions, and a local Yukawa interaction, but with (4.1) replaced by

$$S_{Y,\ell} = -y \sum_{n,f} \phi_n [1 + (y\phi_n)^2]^{-1} \bar{\chi}_{n,f} \chi_{n,f} \tag{4.18}$$

One easily sees that this model yields the same formal continuum limit as (4.1) since it only differs by irrelevant operators. However, it exhibits a fundamentally different phase structure. Now $S_{Y,\ell}$ in (4.18) is invariant under

$$y\phi_n \to (y\phi_n)^{-1} \tag{4.19}$$

In particular, as $y \to \infty$, the Yukawa interaction vanishes. Using analytic methods, one infers that the phase structure of this model consists of an FM, a single PM, and an AFM phase, in contrast to the case with the model with Yukawa interactions (4.1) (and to the model with naive fermions and usual local fermion coupling). As one approaches the FM-PM phase boundary from within the FM phase, the renormalized fermion mass $aM_F \to 0$ everywhere, unlike the respective behaviors at the FM-PM2 and FM-FI phase boundaries of the models with $S_{Y,\ell}$ and $S_{Y,hc}$. The interaction (4.18) is only one of an infinite class, with the property that the term vanishes as $y \to \infty$; for example, one could also consider the more general class of Yukawa interactions

$$S_{Y,\ell} = -y \sum_{n,f} \phi_n [1 + (y\phi_n)^{2p}]^{-q} \bar{\chi}_{n,f} \chi_{n,f} \tag{4.20}$$

where p and q are integers with $p \geq 1$ and $q \geq 1$. The interaction (4.18) is the special case with $p = q = 1$. Similarly, one can replace the usual Yukawa interaction for naive fermions by the analogue of (4.18) with the staggered fermion bilinear replaced by the corresponding naive fermion bilinear $\bar{\psi}_{n,f} \psi_{n,f}$ and one would again find results similar to those in the staggered fermion case: the phase structure consists of a single PM phase together with FM and AFM phases.

An analogous replacement can also be made in the case of the Yukawa model with a hypercubic Yukawa interaction; here one replaces (3.9d) by

$$S_{Y,hc'} = 2^{-4} y \sum_n \phi_n [1 + (y\phi_n)^{2p}]^{-q} \sum_{n' \in hc(n);f} \bar{\chi}_{n,f} \chi_{n,f} \tag{4.21}$$

As before, for $p = q = 1$, (4.21) is symmetric under $y\phi_n \to (y\phi_n)^{-1}$, and hence one can deduce that the phase structure for this model consists of an FM, AFM, and single PM phase.

In general, then, a major result from the studies of the Yukawa theory with a real scalar field using different actions is that the phase structure at small bare

Yukawa coupling is qualitatively similar, while the behavior at large bare y is dependent upon details of the lattice action.

5. YUKAWA MODELS WITH CONTINUOUS GLOBAL SYMMETRY GROUPS

5.1 U(1), U(1)$_L \times$ U(1)$_R$

Several studies have been carried out of Yukawa models invariant under continuous global symmetry groups, the simplest being U(1). Using dynamical fermion simulations (with a hybrid Monte Carlo algorithm), Ref. 81 studied the phase structure of a Yukawa model with a complex scalar field and two species of naive fermions. In the models with naive fermions, one can actually perform independent U(1) transformations on left- and right-handed fields, with the scalar field having an appropriate transformation property to keep the action globally invariant; thus the full symmetry group is U(1)$_L \times$ U(1)$_R$. A similar comment applies to Yukawa theories with other global symmetry groups, like Z(2)$_L \times$ Z(2)$_R$ and SU(2)$_L \times$ SU(2)$_R$. Analytic methods were used in Ref. 82 to investigate the phase structure of a U(1) model with a complex scalar field, staggered fermions, and a certain local Yukawa coupling. A U(1) Yukawa model with two species of naive fermions was studied with analytic and numerical simulations using the hybrid Monte Carlo algorithm in Ref. 83. The inferred phase structure is qualitatively similar to that for the Yukawa theory with a real scalar field and local Yukawa coupling; it consists of distinct symmetric phases at weak and strong Yukawa coupling together with a ferromagnetic and antiferromagnetic phases and perhaps a ferrimagnetic phase. Of course, in the broken-symmetry phases, since the symmetry group is continuous, there is a goldstone boson mode, in contrast to the situation with a theory having a discrete symmetry group.

Recently, a U(1) $\times$ U(1) Yukawa model with staggered fermions and a hypercubic Yukawa interaction has been studied in detail.[84] The phase structure was found to be qualitatively the same as that of the theory with a real scalar field, as shown in Fig. 1, i.e. to consist of a PM, FM, and FI phase. As in Ref. 69, measurements were carried out of fermion and scalar masses and renormalized couplings, again using the hybrid Monte Carlo algorithm for dynamical fermions. The difficulties associated with the goldstone mode were addressed in this work. Two different methods were utilized to study the spontaneous symmetry breaking which would occur in the limit of infinite lattice volume. First, a nonzero bare fermion mass am_f was used, and an extrapolation was made to $am_f = 0$. Second, as in Ref. 69, the theory with bare fermion mass was simulated directly. To circumvent the problem of goldstone rotations washing out a nonzero order parameter, a global rotation on fields was carried out before taking measurements. These two methods were found to give generally consistent results. As in Ref. 69, the bare values of lattice parameters (y, κ, λ) corresponded to a very large range of $\tilde{y}$ and $\tilde{\lambda}$ in the continuum-type parametrization. Again as in Ref. 69, the renormalized couplings y_R and λ_R were found to lie in a relatively small range. In particular, the renormalized Yukawa coupling was never much larger than the value obtained from the tree-level partial wave unitarity bound in the corresponding continuum theory. Thus, for both the Z(2) and U(1) × U(1) Yukawa theories, the measurements are consistent with the hypothesis that, at the continuum limit defined at the PM-FM phase boundary (from within the FM phase), the continuum theory is consistent with being free.

5.2 $SU(2)_L \times SU(2)_R$

A study of a Yukawa model with $SU(2)_L \times SU(2)_R$ global symmetry was initated in Ref. 85 using quenched naive fermions and a radially frozen scalar field. The renormalized fermion mass and Yukawa coupling were measured and it was found that at small bare y the results were consistent with the hypothesis that $y_R = 0$ in the continuum limit, while at large bare y the fermion mass grew with increasing correlation length. This was qualitatively the same behavior that had been observed in the theory with a real scalar field and local Yukawa interaction. Analytic methods were applied to this theory in Ref. 86. Numerical studies of the distribution of eigenvalues of the fermion determinant have also shed light on Yukawa models; such a study was carried out for $Z(2)_L \times Z(2)_R$ and $SU(2)_L \times SU(2)_R$ Yukawa theories in Ref. 87. For another review of these models, see Ref. 88.

6. LATTICE YUKAWA MODELS WITH GAUGE FIELDS

Since Yukawa interactions are relevant to the physical world as part of the full electroweak gauge theory, it is clearly worthwhile and important to study lattice Yukawa models with gauge fields included. The first such work was carried out in 1987 and considered a simple U(1) model.[25] A subsequent study was carried out for SU(2). [28] These used analytic methods to investigate the phase structure and nature of the chiral phase transition. We describe them in turn. It would be interesting to explore such models further in the future.

6.1 U(1)

Ref. 25 considered a U(1) lattice gauge model with a Higgs field ϕ having charge $q_\phi = 1$ and two different fermion fields, χ and ξ, with charges $q_\chi = 1$ and $q_\xi = 0$. One may think of this as a simplified, abelian model of the SU(2) part of the standard electroweak theory, with χ corresponding to the $I = \frac{1}{2}$ fermions, ξ corresponding to the $I = 0$ fermions, and ϕ to the $I = \frac{1}{2}$ Higgs. Since (as will be discussed further below) the SU(2) sector of the standard electroweak sector can be written as a vectorlike theory, one does not need to deal with the problems of a lattice chiral gauge theory for this group. Thus, given that the U(1) model was used as a simplification of SU(2), vectorially coupled fermions were also adopted.

This model is defined by the discretized path integral

$$Z = \int \prod_{n,\mu} dU_{n,\mu} d\phi_n d\phi_n^\dagger d\chi_n d\bar{\chi}_n d\xi_n d\bar{\xi}_n e^{-S} \qquad (6.1a)$$

where $S = S_G + S_H + S_F + S_Y$, with

$$S_G = \beta_g \sum_{plaq.} [1 - P] \qquad (6.1b)$$

$$S_H = -2\kappa \sum_{n,\mu} Re\{\phi_n^\dagger U_{n,\mu} \phi_{n+e_\mu}\} \qquad (6.1c)$$

$$S_F = \frac{1}{2} \sum_{n,\mu} \eta_{n,\mu} [(\bar{\chi}_n U_{n,\mu} \chi_{n+e_\mu} - \bar{\chi}_{n+e_\mu} U_{n,\mu}^\dagger \chi_n) + (\bar{\xi}_n \xi_{n+e_\mu} - \bar{\xi}_{n+e_\mu} \xi_n)]$$

$$+ \sum_n (m_\chi \bar{\chi}_n \chi_n + m_\xi \bar{\xi}_n \xi_n) \qquad (6.1d)$$

$$S_Y = \sum_n (h\bar{\chi}_n \xi_n \phi_n + h^* \bar{\xi}_n \chi_n \phi_n^\dagger) \qquad (6.1e)$$

$\beta_g = 1/g^2$, and $P = Re\{U_{plaq.}\}$. (The relation between the notation used in Ref. 25 and elsewhere in this review is $\beta_h \equiv \kappa$ and $h \equiv y$.) The Higgs fields satisfy $\phi_n^\dagger \phi_n = 1$. (Of course, this "fixed-length" condition on the lattice Higgs fields does not imply that the continuum Higgs fields are of fixed length.) With no loss of generality, one takes $\beta_h \geq 0$. The fields χ_n and ξ_n are staggered fermions. By means of a global U(1) transformation on ϕ, one can render h real and non-negative. For $h = 0$, the theory degenerates into two noncommunicating sectors, one of which describes the U(1) gauge-Higgs model with the charged fermion χ and the other of which describes a free fermion ξ. The former sector (with $m_\chi = 0$) undergoes a chiral transition, which completely separates two generic regions of the phase diagram: in the confinement phase, chiral symmetry is spontaneously broken by the formation of the condensate $< \bar{\chi}\chi >$ while in the remaining region, the chiral symmetry is realized explicitly, with $< \bar{\chi}\chi >= 0$. Analytic work at $\beta_g = 0$ predicted that this transition was continuous (second-order) there[22]; numerical simulations[23] agreed with this result and indicated that the chiral transition also appeared to be continuous, where probed, for $\beta_g > 0$. For technical convenience, Ref. 25 considered the limit of strong gauge coupling, $\beta_g = 0$. Because a small-β_g expansion away from a point where the theory is analytic has (at least) a finite radius of convergence, the properties of the $\beta_g = 0$ special case also describe the interior of the phase diagram, in at least a strip adjacent to the $\beta_g = 0$ side.

Exact integrations are first performed over Higgs and gauge fields. This yields a path integral with a fermionic action which is bilinear, except for a term

$$S_4 = \frac{1}{4}(1 - r^2) \sum_{n,\mu} \bar{\chi}_n \chi_n \bar{\chi}_{n+e_\mu} \chi_{n+e_\mu} \qquad (6.2a)$$

where

$$r \equiv I_1(2\kappa)/I_0(2\kappa) \tag{6.2b}$$

with $I_\nu(x)$ the modified Bessel function. The variable r increases from 0 to 1 as κ increases from 0 to ∞ and serves as an equivalent (positive) variable. For $r = 1$, the gauge field is frozen out, and the theory can be exactly re-expressed as a theory of two free fermions $f_{1,2}$ with masses

$$m_{f_{1,2}} = (1/2)[m_\chi + m_\xi \pm \{(m_\chi - m_\xi)^2 + 4h^2\}^{1/2}] \tag{6.3}$$

Next, a mean field type approximation is used to replace the quartic fermion term S_4 by $(d/2) < \bar{\chi}\chi > \sum_n \bar{\chi}_n\chi_n$. (The reliability of this method had been previously demonstrated for gauge-Higgs models with fermions.[22,23]) In a spin model, such a mean field method would reduce the calculation essentially to a single-site integration or summation. Here one still has a coupled, infinite set of fermionic integrations to perform, but since the fermionic action is now bilinear, these can in fact be done exactly. Using this result, one then calculates the fermion condensates. For simplicity, one sets $m_\chi = m_\xi = 0$ so that the Yukawa interaction is the only source of explicit chiral symmetry breaking. A physical motivation for this is that in the standard electroweak theory, fermion masses arise entirely from the Yukawa couplings.

One obtains the mean field consistency condition

$$< \bar{\chi}\chi > = A \int (dp)sF \tag{6.4a}$$

where

$$A = (d/2)(1 - r^2) < \bar{\chi}\chi > \tag{6.4b}$$

$$\int (dp) \equiv (2\pi)^{-d} \int_{-\pi}^{\pi} d^d p \tag{6.4c}$$

$$s = \sum_{\mu=1}^{d} \sin^2 p_\mu \tag{6.4d}$$

$$F = [(h^2 + rs)^2 + A^2s]^{-1} \tag{6.4e}$$

(Note that (6.4a) is symmetric under $< \bar{\chi}\chi > \to - < \bar{\chi}\chi >$; by convention, one chooses $< \bar{\chi}\chi > \geq 0$.) For $r = 0$ and $h < h_c$, where

$$h_c = (d/2)^{1/2} \tag{6.5}$$

(6.4a) has a nonzero solution for $< \bar{\chi}\chi >$. Furthermore, because of the analyticity in r of a small-r expansion of the right-hand side of (6.4a), this equation also has a nonzero solution for $< \bar{\chi}\chi >$ in a neighborhood of $r = 0$. However, for $r = 1$ (i.e., $\kappa = \infty$) where the gauge field is frozen out and the model reduces to free fermions,

(6.4a) clearly only has the solution $< \bar{\chi}\chi > = 0$; this is also true in a neighborhood $r \lesssim 1$. It follows that $< \bar{\chi}\chi >$ must vanish non-analytically at a a critical point $r = r_c$ (depending on d and h^2) and remain zero for $r_c < r \leq 1$.

The mean field prediction for the phase transition point r_c was determined as a function of the Yukawa coupling h by numerical and analytic solution of (6.4). In the large-d limit, an exact solution of (6.4) was calculated: in terms of the scaled condensate

$$\bar{M}_\chi \equiv (d/2)^{1/2} < \bar{\chi}\chi > \tag{6.6}$$

one found $r_c = 1/\sqrt{2}$, independent of h (for finite h), and

$$M_\chi^2 = \frac{(1 - 2r^2)}{(1 - r^2)^2} \tag{6.7}$$

for $r < r_c$. Since experience in statistical mechanics indicates that the mean field approximation should be exact in the large-d limit, (6.7) should also be an exact solution for the scaled condensate. (Note that in this limit, there is a phase transition for any finite h since $h_c = \infty$.) The mean field methods predict that the chiral transition remains continuous for nonzero Yukawa coupling h, as it was[22,23] for $h = 0$. The fact that, where it is nonzero, $< \bar{\chi}\chi >$ is a decreasing function of r and thus κ is understood as follows: the spontaneous chiral symmetry breaking (χSB) is due to the gauge-fermion interaction, and this is effectively weakened as r increases because the gauge field configurations are more restricted (and, indeed, rendered trivial at $r = 1$).

Interestingly, through its Yukawa interaction with the charged fermion, the fermion ξ, which is neutral with respect to the U(1) gauge interaction, picks up a condensate. This is described by the mean field consistency equation

$$< \bar{\xi}\xi > = -h^2 A \int (dp) F \tag{6.8}$$

This was solved for $< \bar{\xi}\xi >$ in Ref. 25. Thus, both in the absence of any explicit χSB, and in the presence of sufficiently weak explicit χSB by the Yukawa interaction (6.1e), the theory is predicted to have a chiral phase transition.

6.2 SU(2)

The SU(2) Yukawa model studied in Ref. 28 has an $I = 1/2$ Higgs field ϕ, an $I = 1/2$ fermion field χ, and $I = 0$ fermion fields ξ and ζ. This particle content is modelled directly on that of the SU(2) sector of the standard electroweak theory, after it is rewritten in vectorlike form. The theory is defined by

$$Z = \int \prod_{n,\mu} dU_{n,\mu} d\phi_n d\phi_n^\dagger d\chi_n d\bar{\chi}_n d\xi_n d\bar{\xi}_n d\zeta_n d\bar{\zeta}_n e^{-S} \tag{6.9a}$$

where $S = S_G + S_H + S_F + S_Y$, with

$$S_G = \beta_g \sum_{plaq.} [1 - P] \tag{6.9b}$$

$$S_H = -2\kappa \sum_{n,\mu} Re\{\phi_n^\dagger U_{n,\mu} \phi_{n+e_\mu}\} \tag{6.9c}$$

$$S_F = \frac{1}{2} \sum_{n,\mu} \eta_{n,\mu} \Big[(\bar{\chi}_n U_{n,\mu} \chi_{n+e_\mu} - \bar{\chi}_{n+e_\mu} U_{n,\mu}^\dagger \chi_n)$$

$$+ (\bar{\xi}_n \xi_{n+e_\mu} - \bar{\xi}_{n+e_\mu} \xi_n) + (\bar{\zeta}_n \zeta_{n+e_\mu} - \bar{\zeta}_{n+e_\mu} \zeta_n) \Big] \tag{6.9d}$$

$$S_Y = \sum_n \Big[h_1 \bar{\xi}_n \phi_n^\dagger \chi_n + h_1^* \bar{\chi}_n \phi_n \xi_n + h_2 \bar{\zeta}_n \epsilon(\chi_n \phi_n) + h_2^* \epsilon(\bar{\chi}_n \phi_n^\dagger) \zeta_n \Big] \tag{6.9e}$$

with $\beta_g = 4/g^2$ and $P = (1/2)Tr\{U_{plaq.}\}$. (SU(2) indices are suppressed in the notation; in particular, the compact notation $\epsilon(CD)$ denotes the SU(2)-singlet quantity $\epsilon_{ij} C^i D^j$, where ϵ_{ij} is the totally antisymmetric tensor density for SU(2) and C and D transform according to the fundamental representation of this group. Again, for technical convenience, one uses radially frozen Higgs fields with $\phi_n^\dagger \phi_n = 1$. One can rescale the SU(2)- singlet fields ξ and ζ to absorb possible phases in the Yukawa couplings h_1, and h_2, respectively, and thereby render these couplings real (and nonnegative). The two different types of Yukawa interactions reflect the Yukawa terms in the standard SU(2) × U(1) model.

As in the study of the U(1) model[25], an exact integration was performed over gauge and Higgs fields in the strong gauge coupling limit $\beta_g = 0$. A mean field method was then applied to analyze the resulting theory. This predicted that if

$$\frac{1}{h_1^4} + \frac{1}{h_2^4} > \frac{8}{d^2} \tag{3.10}$$

then the theory has a spontaneous chiral symmetry breaking phase transition at finite κ. Consistency conditions were calculated for the condensates and were solved to obtain the mean field prediction for the critical point. As in the U(1) case, the critical exponent for the order parameters (fermion condensates) was predicted to be the usual mean field value, $\beta_{MF} = 1/2$. Analytic $1/d$ expansions for the critical point were calculated, and exact solutions were given for the condensates and critical point in the large-d limit. Ref. 28 also gave the explicit formulas for reexpressing a chiral SU(2) gauge theory in vectorlike form. This study, like that of the U(1) model, illustrated the rich interplay of spontaneous chiral symmetry breaking by the gauge-fermion interaction (as in QCD), as modulated by the gauge-Higgs interaction, together with explicit chiral symmetry breaking by the Yukawa interactions.

7. CHIRAL GAUGE THEORIES AND THEIR GLOBAL LIMITS

7.1 General

The subject of this review, lattice Yukawa models, has an overlap with the effort to construct (global limits of) chiral gauge theories on the lattice. We shall therefore include a brief discussion of this topic here. However, it should be mentioned that these efforts have not yet reached a fully satisfactory state.

Although the standard electroweak model has been quite successful phenomenologically, it leaves a number of fundamental questions unanswered, including the nature of the symmetry breaking and in particular, why the chiral gauge symmetry $SU(2) \times U(1)_Y$ is realized in a "spontaneously broken" manner whereas the vectorial gauge invariances of $SU(3)_{color}$ and $U(1)_{e.m.}$ are realized symmetrically, (in confined and deconfined modes, respectively). The Higgs sector of the standard electroweak model also suffers from the necessity of fine-tuning to counteract the quadratic shift in the Higgs mass squared, which renders it naturally of order the largest mass scale in the full theory. Indeed, it may be that there is no Higgs boson, and that the electroweak symmetry breaking is due to some other physics. It is thus of fundamental importance to understand the properties of chiral gauge theories. This is crucial if we are to achieve a satisfactory understanding of the electroweak symmetry breaking observed in nature.

Moreover, it is now known that the mass of the top quark is not small compared with the masses of the W and Z bosons. Within the framework of the standard model, this corresponds to a physical Yukawa coupling of order unity. Since this coupling is not small, it is desirable in studying the properties of the electroweak theory to use a method of analysis which is capable of nonperturbative calculations. The lattice discretization of the path integral provides a powerful tool for this purpose, in addition to constituting a mathematically rigorous definition of this theory, and a regulator for ultraviolet divergence.

In principle, the lattice approach also has the capability of studying pure chiral gauge theories without any fundamental Higgs fields. Quite early on, it was suggested[89] that a pure chiral gauge theory with just fermions would, for sufficiently strong coupling, produce a bilinear fermion condensate, which would dynamically break the gauge symmetry. Extending earlier work[90], it was also conjectured that such a theory might confine nonsinglets and form massless gauge-singlet composite fermions[91]. However, in the absence of reliable nonperturbative calculational methods, it has not, so far, been possible to explore these suggestions and settle the issue of the spectrum of a chiral gauge theory. (Some initial nonperturbative work has been done with continuum large-N methods.[92])

A basic obstacle to formulating a chiral gauge theory (with or without Higgs fields) on the lattice is the presence of fermion doubling: each lattice fermion field yields $2^d = 16$ actual fermion modes, half of one chirality and half of the other, so that the theory is non-chiral (e.g., Ref. 93.). If one were to add a QCD-like Wilson term to remove doubled fermion modes in the limit of zero lattice spacing, $a \to 0$, this would break the chiral gauge invariance, since this operator connects left- and right-handed fermion fields, which have different transformation properties under the chiral gauge group.

7.2 Approaches to the Construction of Lattice Chiral Gauge Theories

The problem of fermion doublers has been addressed in two different contexts. One was that of a pure chiral gauge theory with fermions but no explicit Higgs fields. [94,95] The approach of these works was actually aimed at the use of the lattice as a regulator for an anomalous chiral gauge theory and was partly motivated by suggestions in the context of continuum field theory[96,97] that it was possible to achieve a consistent quantization of an anomalous chiral gauge theory by adding a kind of Wess-Zumino term to the action. In practice, this involved inserting an integration over a new auxiliary gauge degree of freedom in the measure of the generating functional for the theory. The usual discretized $\bar{\psi} D^2 \psi$ Wilson term was treated essentially as a gauge-fixed form of a chirally gauge-invariant term involving the auxiliary gauge degree of freedom.

A very different proposal for constructing a pure chiral gauge theory on the lattice was suggested in Ref. 98. Here, one makes use of several single-site and multisite four-fermion operators involving additional fermions. Among the motivations for this work is an important property of a chiral gauge theory (without bare fermion mass terms which would explicitly break the gauge symmetry): such a theory is invariant, at the classical level, under a group of chiral symmetries. At the quantum level, a certain subset of these are broken by instantons, leaving a remaining set of exact global chiral symmetries. These may or may not be broken spontaneously; some insight into whether this happens and what remaining exact chiral symmetries might survive is provided by the global anomaly matching conditions of 't Hooft.[89] These, in turn, have played an important role in efforts to construct composite models of fermions, in which one hypothesizes a preonic chiral gauge group with massless preons and tries to investigate whether the theory can yield massless composite fermions. (The more ambitious goal of getting realistic fermions with masses which are nonzero but much smaller than the compositeness scale remains an outstanding challenge to theorists.) In the context of the standard model, both the global axial U(1) of QCD with massless quarks and the global symmetries of baryon and lepton number are broken in this manner by instantons.[99] An objective addressed in Ref. 98 was to construct a lattice action which would break the subset of the global chiral symmetries which were broken in the continuum by instantons and to preserve the remaining subset. The problem is that, essentially because of the finite lattice spacing and the related fermion doubling with resultant zero axial fermionic charge, the initial lattice theory does not have any anomalies. The idea of Ref. 98 is to break the subset of global symmetries which are anomalous in the continuum theory by the device of adding certain four-fermion operators which do this explicitly. It is further hoped that the additional fermions involved in the various four-fermion operators would succeed in decoupling the doublers for the original fermions. The resultant lattice action is quite complicated. It has not so far been used for any numerical work, for the following reason. Since it is not possible to represent anticommuting variables directly in numerical treatments, one always integrates out the fermions first. However, this is only possible for an action which is quadratic in the fermion fields. Hence, a numerical study of this formulation would necessarily begin by rewriting each of the four-fermion operators in terms of a Yukawa-type interaction with a corresponding auxiliary scalar. In practice, this would yield an action somewhat similar to that of Ref. 94 or a related approach[100,101] to be discussed below. However, rather than just one scalar field (defined on sites), it would involve several such fields defined on sites and also certain fields defined on links. Some analytic work has been done with this proposal.[98]

The second context in which the doubling problem was addressed was that of the standard model with its Higgs doublet[100,101]; here, one modified the Wilson term by a coupling to the lattice Higgs field ϕ in such a way as to make it invariant under local chiral gauge transformations. A common idea underlying vboth approaches was that this operator could succeed in decoupling the fermion doublers and thereby render possible a chiral gauge theory, but would otherwise be an irrelevant operator (as, by power counting, it is). However, in a perturbative analysis, one finds that the fermion doublers appear not to be removed at the second-order transition where the continuum limit is defined (e.g., Ref. 102). If this were the whole story, it would mean that such a construction would fail to decouple the fermion doublers. Consequently, a number of studies were carried numerically and analytically, for both the pure chiral gauge theory with fermions, and the theory with Higgs fields, whether the lattice constructions actually do succeed in removing the fermion doublers in the continuum limit. We shall discuss some of this work further below.

Another approach to constructing chiral gauge-Higgs theories on the lattice made use of what were called mirror fermions.[103] The idea was that these would succeed in decoupling the doubler modes for the original fermions. Several analytic and numerical investigations of this proposal were carried out and indicated that such decoupling could occur. It appears that neither the lattice constructions[94,95,100,101] using a gauge-invariant second-derivative involving a scalar coupling to the fermions (sometimes called a "Wilson-Yukawa" term), nor the construction involving mirror fermions[103] satisfactorily reproduces the global chiral symmetry structure of a general continuum chiral gauge theory, in particular, the instanton-induced breaking of a subset of the global chiral symmetries. In certain cases, this is not a problem: for example, since the homotopy group $\pi_3(U(1))$ is trivial, there are no instantons in a (vectorial or chiral) U(1) gauge theory, and hence for this case, there is no instanton-induced breaking of global chiral symmetries.

A still different proposal for lattice chiral gauge theories uses terms which explicitly break the chiral gauge invariance; it envisions tuning bare parameters in such a way as to restore the exact chiral gauge invariance in the continuum limit.[104]

7.3 Some Calculations with Lattice Chiral Models

Here we shall give some further details which illustrate the approaches using a gauge-invariant second derivative term involving a scalar. These will not be comprehensive, and are only included to give a flavor of the ideas. Because the lattice formulations of a pure chiral gauge theory with fermions and of a chiral gauge-Higgs theory differ in the presence or absence of a certain term in the action, our discussion will subsume both.

In undertaking a lattice study of the $SU(2) \times U(1)_Y$ lattice electroweak model, one may, for simplicity, begin with the $SU(2)$ and $U(1)_Y$ factor groups separately. Since $SU(2)$ has only real representations, one can re-express the $SU(2)$ sector of the standard model, involving $2N$ left-handed $I = 1/2$ Weyl fields (four for each generation), as a vectorlike theory with N Dirac fields.[105,22]; this was done explicitly in Ref. 28. As was noted before, this fact made it possible to avoid dealing with chiral fermions in a number of previous studies of this sector.[26,28-30] Indeed, an $SU(2)$ chiral gauge theory with an odd number of left-handed $I = 1/2$ fermions is mathematically inconsistent because of an anomaly[106] associated with $\pi_4(SU(2)) = Z(2)$. Thus, the $SU(2)$ factor group by itself cannot be used as a simplified laboratory in which to explore chiral gauge theories on the lattice. If, however, one does not try to study a full chiral gauge theory but instead only its global limit where the gauge degrees of freedom are absent, then one can consider an $SU(2)$ theory with an odd number of left-handed $I = 1/2$ fermion fields (e.g. one). Actually, in this case, the theory involves a global $SU(2)_L \times SU(2)_R$ invariance. Several studies have been carried out with this global model.[107-109] These have investigated the phase diagram and have demonstrated the decoupling of doublers and tuning of physical fermion masses.

In contrast to $SU(2)$, the $U(1)$ group does provide a useful simplified context in which to study chiral gauge theories on the lattice, since it has complex representations. In particular, the hypercharge $U(1)_Y$ factor group in the standard model embodies two properties of the full electroweak gauge theory: (a) it cannot be expressed in vectorlike form; (b) the right-handed fermions are nonsinglets under $U(1)_Y$, as they are under $SU(2) \times U(1)_Y$ (whereas the right-handed fermions are singlets under weak $SU(2)$). The hypercharge factor group was studied in Refs. 110-111. The theory may be defined by the Euclidean generating functional (with sources omitted)

$$Z = \int \prod_{n,\mu,j} [dU_{n,\mu}][d\psi_{n,j}][d\bar\psi_{n,j}][dg_n]e^{-S} \tag{7.1a}$$

where

$$S = S_G + S_F + S_g \tag{7.1b}$$

with

$$S_G = -\beta_G \sum_{plaq.} \mathcal{R}e\{U_{plaq.}\} \tag{7.1c}$$

$$S_F = S_{F0} + S_W + S_h \tag{7.1d}$$

$$S_{F0} = \frac{1}{2}\sum_{n,\mu,j}\bar\psi_{n,j}\gamma_\mu\big[P_L(U_{n,\mu}^{Y_{j,L}}\psi_{n+e_\mu,j} - U_{n,-\mu}^{Y_{j,L}}\psi_{n-e_\mu,j})+$$

$$P_R(U_{n,\mu}^{Y_{j,R}}\psi_{n+e_\mu,j} - U_{n,-\mu}^{Y_{j,R}}\psi_{n-e_\mu,j})\big] \tag{7.1e}$$

$$S_{WY} = -\frac{1}{2}\sum_{n,\mu,j} r_j\bar\psi_{n,j}\Big[P_L g_n^{Y_{j,R}}(g_{n+e_\mu}^{-Y_{j,L}}\psi_{n+e_\mu,j} + g_{n-e_\mu}^{-Y_{j,L}}\psi_{n-e_\mu,j} - 2g_n^{-Y_{j,L}}\psi_{n,j})$$

$$+ (L \leftrightarrow R)\Big] \tag{7.1f}$$

$$S_h = \sum_{n,j} h_j\bar\psi_{n,j}\big[P_L g_n^{Y_{j,R}-Y_{j,L}} + P_R g_n^{Y_{j,L}-Y_{j,R}}\big]\psi_{n,j} \tag{7.1g}$$

$$S_g = -2\kappa\sum_{n,\mu}\mathcal{R}e\{(g_n^\dagger U_{n,\mu}g_{n+e_\mu}\} \tag{7.1h}$$

Here, $U_{n,-\mu} \equiv U_{n-e_\mu,\mu}^\dagger$, j denotes the type of fermion, P_χ, $\chi = L, R$, are chiral projection operators, and $Y_{j,\chi}$ denotes the hypercharge of $\psi_{\chi,j} \equiv P_\chi\psi_j$. (Ref. 111 allowed for more general forms of S_G and S_g; (7.1) is sufficient for our purposes here.) In (7.1) S_{WY} is the second-derivative term designed to decouple the fermion doublers, and S_Y is a Yukawa term. The field g_n is an element of the group $U(1)_Y$. In the formulation of Ref. 94, g_n is an auxiliary integration variable so that the kinetic term S_g is absent in the bare action. In this case, the integration over $[dg_n]$ in the measure is equivalent to the integration over all gauge transformations since $\int \prod_n[dg_n] = 1$. In the approach of Refs. 100-101, g_n plays the role of a fixed-length Higgs-type field. Since one uses the naturally compactified $U(1)_Y$ on the lattice, the $Y_{j,x}$ are necessarily integral, in contrast to the continuum (cn) noncompact $U(1)_Y$. The theory (7.1) is invariant under the local $U(1)_Y$ gauge transformation

$$U_{n,\mu} \to \eta_n U_{n,\mu}\eta_{n+e_\mu}^\dagger \tag{7.2a}$$

$$g_n \to \eta_n g_n \tag{7.2b}$$

$$\psi_{n,j,\chi} \to \eta_n^{Y_{j,\chi}}\psi_{n,j,\chi} \tag{7.2c}$$

$$\bar\psi_{n,j,\chi} \to \bar\psi_{n,j,\chi}\eta_n^{-Y_{j,\chi}} \tag{7.2d}$$

where $\chi = L, R$ and $\eta_n \in U(1)_Y$.

To see that the term S_{WY} fails to remove fermion doublers perturbatively, consider, for simplicity, the special case where $\beta_G = \infty$, so that the local chiral $U(1)_Y$ gauge symmetry is reduced to a global symmetry. Then (suppressing the flavor index j) the bare masses of the fundamental mode of ψ (denoted f) and the lowest-lying doubler mode (denoted d) are given perturbatively by $am_{f,bare} = h < g >$ and $am_{d,bare} = (h + 2r) < g >$, respectively. Hence,

$$m_{d,bare} - m_{f,bare} = 2ra^{-1} < g > = 2r(2\kappa)^{-1/2}v_{cn} \tag{7.3}$$

where v_{cn} is the symmetry breaking mass scale. Since this mass scale is finite, (7.3) shows that in this perturbative region the doublers are not removed from the spectrum. The key to circumventing this problem is to take advantage of the fact that in this type of model there is also a nonperturbative region analogous to the FM-PM2 phase boundary in the Yukawa theory found in Ref. 71. Because one has an additional parameter available for tuning, namely r, it turns out that one can achieve the decoupling of fermion doublers while tuning the renormalized mass of the fundamental lattice fermion mode am_f so that it vanishes as one approaches the continuum limit. This, in turn, makes possible a finite, acceptable mass m_f for the physical fermion. Using a hopping parameter expansion, Ref. 111 demonstrated that this decoupling and tuning could be carried out (for the full gauge theory). It should perhaps be emphasized that these constructions would only yield a chiral gauge theory in the continuum limit; for finite, even if large, correlation length, the theory is not truly chiral since the doublers are not completely removed. The decoupling and tuning were also demonstrated via quenched numerical simulations of the global limit of the $U(1)_Y$ hypercharge theory in Refs. 110-111. For the special case of the neutrino, decoupling and tuning were studied via the same hopping parameter method in Ref. 112 and via use of a symmetry in the action[113]. Further related papers are Refs. 114-115. Some discussion of renormalized couplings is given in Ref. 111; to explore this question further is important, but requires both a satisfactory construction of chiral lattice theories and adequate methods for the numerical simulation of such theories (see further below). Decoupling and tuning of physical fermion masses have also been discussed for a lattice formulation of the full $SU(2) \times U(1)_Y$ electroweak theory.[112,116] For a recent review, see Ref. 117.

One interesting point is that the continuum theories produced by (7.1) with $\kappa = 0$ (so that the g_n fields are auxiliary variables) and with $\kappa \neq 0$ are probably in the same universality class[111], just as was the case with the Yukawa model studied in Ref. 69 (and with similar Yukawa models). If indeed this is the case, it would provide a deep insight into the properties of chiral gauge theories: since (7.1) with $\kappa = 0$ may be regarded as the lattice regularization of a pure chiral gauge theory without Higgs fields, this equivalence of the quantum theories would suggest that a pure chiral gauge theory behaves like a chiral gauge-Higgs theory (i.e., a theory where the kinetic term for the scalar field is present so that it is a dynamical field rather than an auxiliary integration variable).[112,111]

So far, simulations of chiral lattice models have been carried out only in the case where the gauge degrees of freedom are frozen out and the symmetry reduces to a global chiral symmetry. For U(1) this has been done primarily for technical convenience, whereas, as discussed above, it was crucial for the work on SU(2) since with unfrozen gauge fields, one would have had to include an even number of left-handed fermions, and hence the theory would be vectorlike, not chiral. The global limit does have physical interest, because the most tractable continuum limit of the lattice electroweak theory occurs where the SU(2) and U(1)$_Y$ bare gauge couplings both approach zero[13]. (In this respect at least, the situation would be similar to that in lattice QCD, where the continuum limit occurs at zero bare gauge coupling.)

7.4 Some Problems

There are a number of serious outstanding problems with work on constructing chiral gauge theories on the lattice. We list some of them here. As we have already noted above, although the approach using four-fermion operators[98] addresses the matter of global chiral symmetries in a promising way, it is so complicated that no lattice simulations have been done with it to date, and they would be very cumbersome.

The approaches using a gauge-invariant second-derivative term involving coupling with a scalar field do not appear to deal adequately with instanton-induced breaking of global symmetries (which occurs for the full electroweak group, although not for a U(1) chiral gauge theory). After an initial study in Ref. 98, this has recently been the subject of several discussions.[118-121] Another serious problem is that in these approaches one makes crucial use of nonperturbative behavior to remove the fermion doublers. But it is then very difficult to make contact with the successful continuum electroweak perturbation theory. Another possible problem is that the action may not satisfy reflection positivity, which is a sufficient (but not necessary) condition for a hermitian continuum theory. In contrast, for certain ranges of lattice parameters, it has been shown that the mirror fermion approach does satisfy reflection positivity.[103]. See also Ref. 121 and certain claims in Ref. 122.

A particularly disappointing fact concerns numerical simulations. For QCD, such simulations have played a very important role in exploiting the potential of the lattice to gain nonperturbative information about the theory. Ideally, one would hope that lattice constructions of chiral gauge theories might make possible similar numerical simulations. But there is a fundamental difference between these two types of theories. For both vectorial and chiral theories, the fermion determinant $\det(A)$ is not positive semidefinite (where the fermion action is of the form $\bar{\psi} A \psi$ in a compact notation). There are no known reliable numerical simulation techniques which can deal with such a quantity, since it cannot be represented as a probability. For dynamical fermion simulations of vectorial theories, one usually circumvents this by actually simulating a theory with fermion determinant $\det(A^\dagger A)$. However, this

202

method is not possible for a chiral theory since it automatically renders the theory vectorial. This means that reliable numerical simulations can only be done in the quenched approximation. (Some exploratory work has been done with the complex Langevin algorithm[124], but the theorems which imply convergence of the regular Langevin algorithm do not apply to this case.) One will recall that early simulations of QCD were also done with quenched fermions and, in general, the results were not drastically changed when dynamical fermion simulations were later performed. But again, the situation is qualitatively worse here: in the quenched approximation, one does not include fermion loops. Consequently such an approximation is not sensitive to whether a chiral gauge theory has or does not have anomalies in gauged currents. Normally, in formulating a chiral gauge theory, one requires that it be anomaly-free, i.e. that it have no anomalies in gauged currents, since this would spoil perturbative renormalizability. Indeed, it is a profound property of the standard $SU(3) \times SU(2) \times U(1)_Y$ gauge theory that it is anomaly-free. Since the anomaly-free property of a chiral gauge theory, and the standard model in particular, is so important and fundamental, the insensitivity of the quenched approximation to it is a very serious shortcoming. Of course, if one intends only to study globally invariant chiral models with no gauge fields, the quenched approximation may be useful. But, unfortunately, since reliable numerical simulations are restricted to the quenched approximation, they are very unlikely to yield useful new information about chiral gauge theories in general or the standard model in particular.

8. CONCLUSIONS

Lattice studies of Yukawa models have produced many valuable results. This work has realized the potential of the lattice approach for obtaining information which cannot be obtained via the usual perturbation theory. We now have a good understanding of much of the phase structure of such lattice models. Numerical simulations have yielded data which is consistent with the hypothesis that conclusion that both the renormalized Yukawa and scalar couplings would vanish if one were to take the ultraviolet cutoff to infinity. The lattice also enables one to simulate effective Yukawa theories which contain a high-energy cutoff marking the limit of their applicability. In this context, the Yukawa studies have shown that a fermion mass generated by a Yukawa interaction cannot be much larger than the mass scale characterizing the symmetry breaking. They have also led to an upper bound for the scalar mass, thereby generalizing previous work on pure scalar models. Some work has been done on Yukawa models with gauge fields. Further analytic and numerical work on simple Yukawa models and their generalizations to include gauge fields will definitely be worthwhile.

Much effort has also gone into the construction of chiral gauge theories on the lattice; the global limits of these constitute a kind of generalization of Yukawa theories in which the action contains one or more terms involving fermions on different sites coupled with scalar fields. This is an area where very difficult problems still remain.

We hope that this review has given readers some insight into what has so far been achieved in studies of lattice Yukawa models. Obviously in the space available, it has not been possible to discuss all of the research which has been done, and we have naturally concentrated on those areas with which the author had greater familiarity. We hope that for those not already working on such studies, the review has excited their interest sufficiently that they may wish to begin research in this area, since there is still much to be learned.

ACKNOWLEDGMENTS

The author would like to thank S. Aoki, A. Hasenfratz, I-H. Lee, and J. Shigemitsu for many valuable discussions. He is indebted to many other colleagues in lattice field theory for stimulating conversations. He also grateful to M. Creutz for the invitation to write this review and for a number of helpful comments. This research was partially supported by the NSF grant PHY-89-08495.

204

REFERENCES

1. H. Yukawa, Proc. Phys. Math. Soc. Japan **17** (1935) 48; ibid., **19** (1937) 712. For discussions of this history of this development, see, e.g. H. Yukawa, *Tabibito (The Traveller)*, transl. L. M. Brown and R. Yoshida (World Scientific, Singapore, 1979); S. Hayakawa, in L. M. Brown and L. Hoddeson, eds., *The Birth of Particle Physics* (Cambridge University Press, Cambridge, 1983), pp. 82-107; A. Pais, *Inward Bound; of Matter and Forces in the Physical World* (Oxford University Press, Oxford, 1986), pp. 429-436.

2. For a review, see, e.g., B. W. Lee, *Chiral Dynamics* (Gordon and Breach, New York, 1972).

3. S. Weinberg, Phys. Rev. Lett. **19** (1967) 1264; A. Salam, in *Elementary Particle Theory: Relativistic Groups and Analyticity*, ed. N. Svartholm (Almqvist and Wiksell, Stockholm, 1968), p. 367; S. L. Glashow, Nucl. Phys. **B22** (1961) 579.

4. For reviews, see, e.g., E. Abers and B. W. Lee, Phys. Rept. **9C** (1973) 1; H. Fritzsch and P. Minkowski, Phys. Rept. **73** (1981) 67; C. Itzykson and J.-B. Zuber, *Quantum Field Theory* (McGraw-Hill, New York, 1980); T.-P. Cheng and L.-F. Li, *Gauge Theory of Elementary Particle Physics* (Oxford, New York, 1984); P. Frampton, *Gauge Field Theories* (Benjamin, New York, 1987).

5. A (not fully realistic) example is technicolor (= hypercolor); see S. Weinberg, Phys. Rev. **D13** (1976) 974; **D19** (1979) 1277; L. Susskind, Phys. Rev. **D20** (1979) 2619. For a review of this and similar work, see E. Farhi and R. Jackiw, eds., *Dynamical Gauge Symmetry Breaking* (World Scientific, Singapore, 1982).

6. Y. Nambu and G. Jona-Lasinio, Phys. Rev. **122** (1961) 345.

7. For some recent work in this direction, see Y. Nambu, Chicago preprint 89-08; V. Miransky, M. Tanabashi, and K. Yamawaki, Mod. Phys. Lett. **A4** (1989) 1043; Phys. Lett. **B221** (1989); *Proceedings of the 1989 Workshop on Dynamical Symmetry Breaking*, Nagoya, eds. T. Muta and K. Yamawaki (Nagoya, 1990); and *1990 International Workshop on Strong Coupling Gauge Theories and Beyond*, eds. T. Muta and K. Yamawaki (World Scientific, Singapore, 1991).

8. R. E. Shrock, Phys. Lett. **B273** (1991) 493.

9. For reviews of work on lattice gauge-Higgs models, see C. Creutz, L. Jacobs, and C. Rebbi, Phys. Rept. **95** (1983) 201; R. E. Shrock, in *Proceedings of the 1987 International Symposium on Field Theory on the Lattice, Lattice-87*, Nucl. Phys. B (Proc. Suppl.) 4 (1988) 373; J. Jersák, in the *Proceedings of the 8'th INFN Eloisotron Project Workshop*, 1989 (Plenum, New York, 1990).

10. M. Creutz, Phys. Rev. **D21** (1980) 1006; Phys. Rev. Lett. **43** (1979) 553.

11. K. C. Bowler et al., Phys. Lett. **B104** (1981) 481; D. J. E. Callaway and L. Carson, Phys. Rev. **D25** (1982) 531.

12. C. B. Lang, C. Rebbi, and M. Virasoro, Phys. Lett. **B104** (1981) 294.

13. R. E. Shrock, Phys. Lett. **B162** (1985); Nucl. Phys. **B267** (1986); Phys. Rev. Lett. **56** (1986) 2124; Nucl. Phys. **B278** (1986) 380; Phys. Lett. **B180** (1986) 269.

14. K. Olynyk and J. Shigemitsu, Phys. Rev. Lett. **54** (1985) 2403; Phys. Lett. **B154** (1985) 278.

15. I. Montvay, Nucl. Phys. **B269** (1986) 170; W. Langguth, I. Montvay, and P. Weisz, Nucl. Phys. **B277** (1986) 11.

16. J. Jersák, C. B. Lang, T. Neuhaus, and G. Vones, Phys. Rev. **D32** (1985) 2761; H. Evertz, K. Jansen, J. Jersák, C. B. Lang, and T. Neuhaus, Nucl. Phys. **B285** (1987) 590.

17. S. Elitzur, Phys. Rev. **D12** (1975) 3978.

18. E. Fradkin and S. H. Shenker, Phys. Rev. **D19** (1979) 3682.

19. L. Abbot and E. Farhi, Phys. Lett. **B101** (1981) 69; Nucl. Phys. **B189** (1981) 547.

20. M. Claudson, E. Farhi, and R. L. Jaffee, Phys. Rev. **D34** (1986) 873.

21. I-H. Lee and J. Shigemitsu, Phys. Lett. **B178** (1986) 93.

22. I-H. Lee and R. E. Shrock, Phys. Rev. Lett. **59** (1987) 14.

23. I-H. Lee and R. E. Shrock, Nucl. Phys. **B190** (1987) 275.

24. I-H. Lee and R. E. Shrock, Phys. Lett. **B196** (1987) 82.

25. I-H. Lee and R. E. Shrock, Phys. Lett. **B199** (1987) 541.

26. I-H. Lee and R. E. Shrock, Phys. Lett. **B201** (1988) 497.

27. I-H. Lee and R. E. Shrock, Nucl. Phys. **B305** (1988) 286.

28. I-H. Lee and R. E. Shrock, Nucl. Phys. **B305** (1988) 305.

29. A. De and J. Shigemitsu, Nucl. Phys. **B307** (1988) 376.

30. S. Aoki, I-H. Lee, and R. E. Shrock, Phys. Lett. **B207** (1988).

31. E. Dagotto and J. Kogut, Phys. Lett. **B208** (1988) 475.

32. S. Aoki, I-H. Lee, and R. E. Shrock, Phys. Lett. **B219** (1989).

33. L. D. Landau, "On the Quantum Theory of Fields", in "Niels Bohr and the Development of Physics", eds. W. Pauli, V. Weisskopf, and L. Rosenfeld (McGraw-Hill, New York, 1955); L. D. Landau, A. Abrikosov, and I. Khalatnikov, Dokl. Akad. Nauk **95** (1954) 773.

34. Particle Data Group, *Review of Particle Properties*, Phys. Lett. **239** (1990) 1.

35. K. Wilson, Phys. Rev. **B4** (1971) 3184; K. Wilson and J. Kogut, Phys. Rept. **12C** (1974) 75.

36. G. Baker and J. Kincaid, J. Stat. Phys. **24** (1981) 469.

37. J. Fröhlich, Nucl. Phys. **B200** (1982) 281.

38. C. Aragao de Carvalho, S. Caracciolo, and J. Fröhlich, Nucl. Phys. **B215** (1983) 209.

39. B. Freedman, P. Smolensky, and D. Weingarten, Phys. Lett. **B113** (1982) 481.

40. C. B. Lang, Phys. Lett. **B155** (1985) 399; Nucl. Phys. **B265** (1986) 630.

41. M. Lüscher and P. Weisz, Nucl. Phys. **B290** (1987) 25.

42. M. Lüscher and P. Weisz, Nucl. Phys. **B295** (1988) 65.

43. M. Lüscher and P. Weisz, Phys. Lett. **B212** (1988) 472.

44. M. Lüscher and P. Weisz, Nucl. Phys. **B318** (1989) 705.

45. J. Kuti and Y. Shen, Phys. Rev. Lett. **60** (1988) 85.

46. J. Kuti, L. Lin, and Y. Shen, Phys. Rev. Lett. **61** (1988) 678.

47. A. Hasenfratz, K. Jansen, C. B. Lang, T. Neuhaus, and H. Yoneyama, Phys. Lett. **B199** (1987) 531.

48. A. Hasenfratz, K. Jansen, J. Jersák, C. B. Lang, T. Neuhaus, H. Yoneyama, Nucl. Phys. **B317** (1989) 81.

49. D. J. E. Callaway, Phys. Rep. **167** (1989) 241.

50. M. Aizenmann, Commun. Math. Phys. **86** (1982) 1.

51. R. Dashen and H. Neuberger, Phys. Rev. Lett. **50** (1983) 1897.

52. A. Hasenfratz and P. Hasenfratz, Phys. Rev. **D34** (1986) 3160.

53. A. Hasenfratz and T. Neuhaus, Nucl. Phys. **B297** (1988) 205.

54. P. Hasenfratz and J. Nager, Z. Phys. **C37** (1988) 477.

55. G. Bhanot and K. Bitar, Phys. Rev. Lett. **61** (1988) 798.

56. G. Bhanot, K. Bitar, U. Heller, and H. Neuberger, Nucl. Phys. **B343** (1990) 467; ibid., **B353** (1991) 551.

57. For reviews, see P. Hasenfratz, in *Lattice 88*, Nucl. Phys. B (Proc. Suppl.) 9 (1989) 3; H. Neuberger, in *Lattice 89*, Nucl. Phys. B (Proc. Suppl.) 17 (1990) 17.

58. R. Mawhinney and R. S. Willey, to be published; R. S. Willey, private communication.

59. H. Neuberger, U. Heller, and P. Vranos, "How to Put a Heavier Higgs on the Lattice", Rutgers preprint RUTP-91-31.

60. M. Einhorn, Nucl. Phys. **B246** (1984) 75.

61. See, e.g., T. P. Cheng, E. Eichten, and L.-F. Li, Phys. Rev. **D9** (1974) 2259.

62. M. Einhorn and H. Goldberg, Phys. Rev. Lett. **57** (1986) 2115.

63. L. Maiani, G. Parisi, and R. Petronzio, Nucl. Phys. **B136** (1978); M. Lindner, Z. Phys. **C31** (1986) 298; R. Flores and M. Sher, Phys. Rev. **D27** (1983) 1679; M. J. Duncan, R. Philippe, and M. Sher, Phys. Lett. **B153** (1985) 165.

64. J. Shigemitsu, Phys. Lett. **B189** (1987) 164.

65. K. Wilson, in *New Phenomena in Subnuclear Physics* (Erice Summer School), ed. A. Zichichi (Plenum, New York, 1976), p. 69.

66. T. Banks, L. Susskind, and J. Kogut, Phys. Rev. **D13** (1976) 1043; L. Susskind, Phys. Rev. **D16** (1977) 3031; J. Kogut, M. Stone, G. W. Wyld, S. H. Shenker, J. Shigemitsu, and D. K. Sinclair, Nucl. Phys. **B225** (1983) 326.

67. J. Polonyi and J. Shigemitsu, Phys. Rev. **D38** (1988) 3231; J. Shigemistu, in *Proceedings of the International Symposium on Lattice Field Theory, Lattice-88*, Nucl. Phys. B (Proc. Suppl.) 9 (1989) 96.

68. J. Shigemitsu, Phys. Lett. **B226** (1989) 364.

69. I-H. Lee, J. Shigemitsu, and R. E. Shrock, Nucl. Phys. **B330** (1990) 225.

70. S. Aoki, I-H. Lee, D. Mustaki, J. Shigemitsu, and R. E. Shrock, Phys. Lett. **B244** (1990) 301.

71. I-H. Lee, J. Shigemitsu, and R. E. Shrock, Nucl. Phys. **B335** (1990) 265.

72. R. B. Griffiths, J. Math. Phys. **8** (1967) 478,484; D. G. Kelly and S. Sherman, J. Math. Phys. **9** (1968) 466; J. Ginibre, Commun. Math. Phys. **16** (1970) 310.

73. A. Hasenfratz and T. Neuhaus, Phys. Lett. 220B (1989) 435.

74. J. Dimock, J. Glimm, A. Kupiainen, private communications (1989-1990); J. Dimock, Commun. Math. Phys. **109** (1987) 379.

75. A. Abada and R. E. Shrock, Phys. Rev. **D43** (Rapid Commun.) (1991) R304.

76. A. Hasenfratz, P. Hasenfratz, K. Jansen, J. Kuti, and Y. Shen, "The equivalence of the top quark condensate and the elementary Higgs field", preprint UCSD/PTH-91-06/AZPH-TH/91-12.

77. Y. Cohen, S. Elitzur, and E. Rabinovici, Phys. Lett. **B104** (1981) 289; Nucl. Phys. **B220** (1983) 102.

78. R. Shrock, in *Lattice-89*, Nucl. Phys. B (Proc. Suppl.) 17 (1990) 480.

79. J. Berlin, A. Hasenfratz, U. Heller, and M. Klomfass, Phys. Lett. **B249** (1990) 485.

80. R. E. Shrock, in *Lattice-90*, Nucl. Phys. B (Proc. Suppl.) 20 (1991) 585.

81. D. Stephenson and A. Thornton, Phys. Lett. **B212** (1988) 479; A. Thornton, ibid., **B214** (1988) 577; **B221** (1989) 151.

82. A. Hasenfratz, W. Liu, and T. Neuhaus, Phys. Lett. **B236** (1990) 339.

83. M. A. Stephanov and M. M. Tsypin, Phys. Lett. **B242** (1990) 432; M. A. Stephanov, Phys. Lett. **B266** (1991) 447.

84. S. Aoki, J. Shigemitsu, and J. Sloan, Nucl. Phys., in press.

85. W. Bock, A. De, K. Jansen, J. Jersák, and T. Neuhaus, Phys. Lett. **B231** (1989) 283.

86. M. A. Stephanov and M. M. Tsypin, Phys. Lett. **B261** (1991) 109.

87. I. Barbour, W. Bock, C. Davies, A. K. De, D. Henty, J. Smit, and T. Trappenberg, HLRZ-90-42.

88. J. Shigemitsu, in *Lattice-90*, Nucl. Phys. B (Proc. Suppl.) 20 (1991) 515.

89. H. Georgi, Nucl. Phys. (1980) ; S. Raby, S. Dimopoulos, and L. Susskind, Nucl. Phys. **B169** (1980) 373.

90. G. 't Hooft, in *Recent Developments in Gauge Theories* (1979 Cargèse Summer School) (Plenum, 1980), p. 135.

91. S. Dimopoulos, S. Raby, and L. Susskind, Nucl. Phys. **B173** (1980) 208.

92. E. Eichten, R. Peccei, J. Preskill, and D. Zeppenfeld, Nucl. Phys. **B268** (1986) 161.

93. H. B. Nielsen and M. Ninomiya, Nucl. Phys. **B185** (1981) 20; **B193** (1981) 173, erratum: **B193** (1981) 173; Phys. Lett. **105** (1981) 219.

94. S. Aoki, in *Lattice-87*, Nucl. Phys. B (Proc. Suppl.) 4 (1988) 479; Phys. Rev. Lett. **60** (1988) 2109; Phys. Rev. **D38** (1988) 618; *Lattice-88*, Nucl. Phys. (Proc. Suppl.) 9 (1989) 584; Phys. Rev. **D40** (1989) 2729.

95. K. Funakubo and R. Kashiwa, Phys. Rev. Lett. **60** (1988) 2133.

96. L. Faddeev and S. Shatashvili, Phys. Lett. B167 (1986) 225.

97. K. Harada and I. Tsutsui, Phys. Lett. B183 (1987) 311.

98. E. Eichten and J. Preskill, Nucl. Phys. **B268** (1986) 179.

99. G. 't Hooft, Phys. Rev. Lett. **37** (1976) 8; Phys. Rev. **D14** (1976) 3432.

100. P. D. V. Swift, Phys. Lett. **B145** (1984) 256.

101. J. Smit, Acta Phys. Polon. **B17** (1986) 531; *Lattice-87*, Nucl. Phys. B (Proc. Suppl.) 4 (1988) 451; *Lattice-89*, Nucl. Phys. B (Proc. Suppl.) 17 (1990) 3.

102. S. J. Hands and D. B. Carpenter, Nucl. Phys. **B266** (1986) 285.

103. I. Montvay, Phys. Lett. **199B** (1987) 89; Nucl. Phys. **B305** (1988) 389; K. Farakos, G. Koutsombas, L. Lin, J. Ma, I. Montvay, and G. Münster, Nucl. Phys. **B350** (1991) 474. For another different approach, see S. V. Zenkin, Phys. Lett. **B252** (1990) 636.

104. A. Borrelli, L. Maiani, G. C. Rossi, R. Sisto, and M. Testa, Phys. Lett. **B221** (1989); Nucl. Phys. **B333** (1990) 335; L. Maiani, G. C. Rossi, and M. Testa, Phys. Lett. **B261** (1991) 479.

105. H. Georgi, as cited in Ref. 20.

106. E. Witten, Phys. Lett. **117B** (1982) 324.

107. W. Bock, A. K. De, K. Jansen, J. Jersák, and T. Neuhaus, Phys. Lett. **B231** (1989) 283.

108. W. Bock, A. K. De, K. Jansen, J. Jersák, T. Neuhaus, and J. Smit, Phys. Lett. **B232** (1989) 486; Nucl. Phys. **B344** (1990) 207.

109. W. Bock and A. K. De, Phys. Lett. **B245** (1990) 207; W. Bock, A. K. De, C. Frick, K. Jansen, and T. Trappenberg, preprint HLRZ-91-21.

110. S. Aoki, I-H. Lee, J. Shigemitsu, and R. E. Shrock, Phys. Lett. **B243** (1990) 403.

111. S. Aoki, I-H. Lee, and R. E. Shrock, Nucl. Phys. **B355** (1991) 383.

112. S. Aoki, I-H. Lee, and S.-S. Xue, BNL Report 42494 (Feb. 1989); Phys. Lett. **229B** (1989) 79; I-H. Lee, in *Lattice-89*, Nucl. Phys. B Proc. Suppl. 17 (1990) 457.

113. M. Golterman and D. Petcher, Phys. Lett. **B225** (1989) 159.

114. M. Golterman and D. Petcher, Phys. Lett. **B247** (1990) 370; Nucl. Phys. **B359** (1991) 91.

115. S. Aoki, I-H. Lee, and R. E. Shrock, preprint ITP-SB-91-25.

116. S. Aoki, I-H. Lee, and R. E. Shrock, Phys. Rev. **D45** (1992) (Rapid Commun.) R13.

117. I. Montvay, in *Lattice-91*, Nucl. Phys. B (Proc. Suppl.), in press.

118. T. Banks, preprint RU-91-13.

119. M. J. Dugan and A. V. Manohar, Phys. Lett. **B265** (1991) 137.

120. S. Aoki, Tsukuba preprint UTHEP-224, to be published in the *Proceedings of the Brookhaven Workshop on QCD at Finite Temperature (1991)*.

121. S. Aoki, Tsukuba preprint UTHEP-230, to appear in *Lattice-91*, Nucl. Phys. B (Proc. Suppl.), in press.

122. M. Golterman, D. Petcher, and J. Smit, preprint WUSL-HEP-91-GP1; W. Bock, A. K. De, and J. Smit, ITFA-HLRZ preprint.

124. S. Sanielevici, H. Gausterer, M. Golterman, and D. Petcher, in *Lattice-90*, Nucl. Phys. B (Conf. Suppl.) 20 (1991) 581.

BOSONIC ALGORITHMS

ALAN D. SOKAL

Department of Physics
New York University
4 Washington Place
New York NY 10003, USA
SOKAL@ACF3.NYU.EDU

The goal of this chapter is to give an introduction to current research on Monte Carlo methods in quantum field theory and statistical mechanics, with an emphasis on:

1) the conceptual foundations of the method, including the possible dangers and misuses, and the correct use of statistical error analysis; and

2) new Monte Carlo algorithms for problems in critical phenomena and quantum field theory, aimed at reducing or eliminating the "critical slowing-down" found in conventional algorithms.

I hope that this chapter will be useful both as a "how-to" manual and annotated bibliography for practitioners of Monte Carlo, and as a "consumer protection guide" for physicists who want to be able to evaluate the reliability of published Monte Carlo work.

Here is a summary: Section 1 reviews the basic principles of dynamic Monte Carlo methods. Section 2 discusses the practical problems of statistical data analysis. Section 3 surveys the various "local" algorithms for simulating lattice spin and gauge models: Metropolis, heat-bath, overrelaxation, Langevin and hybrid. Section 4 surveys recent work on "collective-mode" algorithms aimed at reducing the critical slowing-down: Fourier acceleration, multi-grid, Swendsen-Wang and embedding algorithms. Section 5 lists some important unsolved problems.

This chapter is based in large part on my 1989 Lausanne lectures [1]; further information and references can be found there, as well as in the LATTICE 'nn reviews of Adler [2], Wolff [3] and myself [4].

I assume that the reader is familiar with the Euclidean formulation of quantum field theory, which maps quantum field theory (at least for bosons) into classical equilibrium statistical mechanics, and which transforms the problem of the continuum limit (= removal of the ultraviolet cutoff) into the problem of critical behavior. I shall thus use interchangeably the languages of statistical mechanics and field theory.

1 Dynamic Monte Carlo Methods: Fundamental Principles

1.1 General Theory

Correlation functions in classical equilibrium statistical mechanics or Euclidean quantum field theory are expectation values with respect to the Boltzmann-Gibbs probability measure $\mu_{eq}(\varphi) = Z^{-1}e^{-H(\varphi)}$, where φ denotes a generic field configuration ($\varphi \in S \equiv$ configuration space) and $H(\varphi)$ is the Hamiltonian (with inverse temperature β absorbed) or Euclidean action. The idea of the Monte Carlo method is to compute numerically such expectation values $\langle F \rangle_{\mu_{eq}} \equiv \int d\varphi\, \mu_{eq}(\varphi)\, F(\varphi)$ by generating a long sequence $\varphi^{(1)}, \ldots, \varphi^{(n)}$ of random samples from μ_{eq} and then using the sample mean $\overline{F}^{(n)} \equiv n^{-1} \sum_{i=1}^{n} F(\varphi^{(i)})$ as an estimate of the theoretical mean $\langle F \rangle_{\mu_{eq}}$. So how can we generate such random samples?

Monte Carlo methods can be classified as *static* or *dynamic*. Static methods are those that generate a sequence of *statistically independent* samples from the desired probability distribution μ_{eq}. These techniques are widely used in Monte Carlo numerical integration in spaces of not-too-high dimension [5]. But they are unfeasible for most applications in statistical physics and quantum field theory, in which μ_{eq} describes a very large number of coupled degrees of freedom.

The idea of *dynamic* Monte Carlo methods is to invent a *stochastic process* with state space S having μ_{eq} as its unique equilibrium distribution. We then simulate this stochastic process, starting from an arbitrary initial configuration; once the system has reached equilibrium, we measure time averages, which converge (as the run time tends to infinity) to μ_{eq}-averages. In physical terms, we are inventing a *stochastic time evolution* for the given system. It must be emphasized, however, that this time evolution *need not correspond to any real "physical" dynamics*: rather, the dynamics is simply a numerical algorithm, and it is to be chosen, like all numerical algorithms, on the basis of its computational efficiency.

In practice, the stochastic process is always taken to be a *Markov process*.[1] That is, we invent (somehow or other) a transition probability matrix $P(\varphi \rightarrow \varphi')$ that is ergodic ("from any state one can reach any other") and that leaves μ_{eq} invariant, i.e.

$$\int d\varphi\, \mu_{eq}(\varphi)\, P(\varphi \rightarrow \varphi') = \mu_{eq}(\varphi'). \tag{1}$$

Then the general theory of Markov chains [6, 7, 8, 9] guarantees that:

a) μ_{eq} is the *unique* invariant measure for P.

b) The Markov chain converges as $t \rightarrow \infty$ to the equilibrium distribution μ_{eq}, irrespective of the initial configuration $\varphi^{(0)}$. More precisely, for any observable F,

---

[1] The books of Kemeny and Snell [6] and Iosifescu [7] are excellent references on the theory of Markov chains with *finite* state space. At a somewhat higher mathematical level, the books of Chung [8] and Nummelin [9] deal with the cases of *countable* and *general* state space, respectively.

the sample mean $\overline{F}^{(n)}$ converges (with probability 1) as $n \to \infty$ to the theoretical mean $\langle F \rangle_{\mu_{eq}}$ (strong law of large numbers), and does so with fluctuations of size $\sim n^{-1/2}$ (central limit theorem).

So far, so good! But while this is a *valid* Monte Carlo algorithm for generating samples from μ_{eq}, it may or may not be an *efficient* one. The key difficulty is that the successive configurations $\varphi^{(1)}, \varphi^{(2)}, \ldots$ of the Markov process are *correlated*, perhaps very strongly. The Markov chain is characterized by an "autocorrelation time" τ: roughly speaking, this is the time required for the process to lose memory of its current configuration. Thus, a run of length n provides only $\sim n/\tau$ "effectively independent" samples; as a result, one expects the estimates produced from the dynamic Monte Carlo simulation to have a variance $\sim \tau$ times larger than the same run length in the corresponding static Monte Carlo (independent sampling).

To make this all precise, let $F = F(\varphi)$ be an observable, and consider the *stationary* Markov process (i.e. start the system in the equilibrium distribution μ_{eq}, or equivalently, "equilibrate" it for a very long time prior to observing the system). Then $\{F_t\} \equiv \{F(\varphi^{(t)})\}$ is a stationary stochastic process with mean

$$\mu_F \equiv \langle F_t \rangle = \langle F \rangle_{\mu_{eq}} \equiv \int d\varphi \, \mu_{eq}(\varphi) \, F(\varphi) \tag{2}$$

and *unnormalized autocorrelation function*[2]

$$\begin{aligned} C_{FF}(t) &\equiv \langle F_s F_{s+t} \rangle - \mu_F^2 \\ &= \int d\varphi \, \mu_{eq}(\varphi) \, F(\varphi) \, P^{|t|}(\varphi \to \varphi') \, F(\varphi') - \mu_F^2 \, . \end{aligned} \tag{3}$$

Define also the *normalized autocorrelation function*

$$\rho_{FF}(t) \equiv C_{FF}(t)/C_{FF}(0) \, . \tag{4}$$

Typically $\rho_{FF}(t)$ decays exponentially ($\sim e^{-|t|/\tau}$) for large t; we therefore define the *exponential autocorrelation times*

$$\tau_{exp,F} \equiv \limsup_{t \to \infty} \frac{t}{-\log|\rho_{FF}(t)|} \tag{5}$$

$$\tau_{exp} \equiv \sup_F \tau_{exp,F} \tag{6}$$

Thus, $\tau_{exp,F}$ is the relaxation time of the slowest mode which couples to F, and τ_{exp} is the relaxation time of the slowest mode in the system. (For most observables, $\tau_{exp,F} = \tau_{exp}$; the exception is when F is "orthogonal" to the slowest mode, usually due to some symmetry.) If the state space S is finite, then τ_{exp} is guaranteed to be finite; but if the state space is infinite, τ_{exp} could be $+\infty$. Also, τ_{exp} can equivalently

---

[2]In the statistics literature, this is called the *autocovariance function*.

be defined in terms of the spectrum of the transition probability matrix P considered as an operator on the Hilbert space $L^2(\mu_{eq})$: see [10, 1].

On the other hand, the statistical error in estimates of $\langle F \rangle_{\mu_{eq}}$ is controlled by the *integrated autocorrelation time*[3]

$$\tau_{int,F} \equiv \tfrac{1}{2} \sum_{t=-\infty}^{\infty} \rho_{FF}(t) \tag{7a}$$

$$= \tfrac{1}{2} + \sum_{t=1}^{\infty} \rho_{FF}(t) \tag{7b}$$

in the sense that the sample mean $\overline{F}^{(n)} \equiv n^{-1} \sum_{t=1}^{n} F_t$ has variance[4]

$$\text{var}(\overline{F}^{(n)}) = \frac{1}{n^2} \sum_{r,s=1}^{n} C_{FF}(r-s) \tag{8a}$$

$$= \frac{1}{n} \sum_{t=-(n-1)}^{n-1} \left(1 - \frac{|t|}{n}\right) C_{FF}(t) \tag{8b}$$

$$\approx \frac{1}{n}(2\tau_{int,F}) C_{FF}(0) \quad \text{for } n \gg \tau \tag{8c}$$

In other words, a run of length n contains $n/2\tau_{int,F}$ "effectively independent" samples. (This is sometimes expressed by saying that the "statistical inefficiency" of dynamic Monte Carlo, relative to static Monte Carlo, is $2\tau_{int,F}$.)

In summary, the exponential and integrated autocorrelation times play different roles in Monte Carlo simulation: τ_{exp} is very natural from the point of view of the theory of dynamic critical phenomena, while $\tau_{int,F}$ is of practical importance. Often it is assumed (either tacitly or explicitly) that τ_{exp} and $\tau_{int,F}$ are of the same order of magnitude, at least for "reasonable" observables F; but in fact this is *not* true in general, as will be explained in Section 3.3.

Returning to the general theory, we note that one convenient way of satisfying the stationarity condition (1) is to satisfy the following *stronger* condition, called *detailed balance*[5]: for each pair $\varphi, \varphi' \in S$,

$$\mu_{eq}(\varphi) P(\varphi \to \varphi') = \mu_{eq}(\varphi') P(\varphi' \to \varphi) . \tag{9}$$

---

[3]The factor of $\tfrac{1}{2}$ is purely a matter of convention; it is inserted so that $\tau_{int,F} \approx \tau_{exp,F}$ if $\rho_{FF}(t) \approx e^{-|t|/\tau}$ with $\tau \gg 1$. Some authors use a different convention, omitting the $\tfrac{1}{2}$ in either (7a) or (7b).

[4]The variance of a random variable A is $\text{var}(A) \equiv \langle A^2 \rangle - \langle A \rangle^2$. More generally, the covariance of two random variables A and B is $\text{cov}(A, B) \equiv \langle AB \rangle - \langle A \rangle \langle B \rangle$.

[5]In the mathematical literature, a Markov chain satisfying the detailed-balance condition is called *reversible*. For the physical significance of this term, see Kemeny and Snell [6, section 5.3] or Iosifescu [7, section 4.5].

215

[Integrating (9) over φ, we recover (1).] Detailed balance is equivalent to the *self-adjointness* of P as on operator on the space $L^2(\mu_{eq})$. Thus, if detailed balance holds, the spectrum of P is *real* and lies in an interval $[\lambda_{min}, \lambda_{max}] \subset [-1, 1]$; and the autocorrelation function $\rho_{FF}(t)$ has a spectral representation

$$\rho_{FF}(t) = \int_{\lambda_{min}}^{\lambda_{max}} \lambda^{|t|} \, d\sigma_{FF}(\lambda) \tag{10}$$

with a *nonnegative* spectral weight $d\sigma_{FF}(\lambda)$. ¿From this it is easy to prove [10, 1] that

$$\tau_{int,F} \le \frac{1}{2} \left(\frac{1 + e^{-1/\tau_{exp}}}{1 - e^{-1/\tau_{exp}}} \right) \approx \tau_{exp} . \tag{11}$$

Finally, let us make a remark about transition probabilities P that are "built up out of" other transition probabilities $P_1, P_2, \ldots, P_n$:

a) If $P_1, P_2, \ldots, P_n$ satisfy the stationarity condition (resp. the detailed-balance condition) for μ_{eq}, then so does any convex combination $P = \sum_{i=1}^n \lambda_i P_i$. Here $\lambda_i \ge 0$ and $\sum_{i=1}^n \lambda_i = 1$.

b) If $P_1, P_2, \ldots, P_n$ satisfy the stationarity condition for μ_{eq}, then so does the product $P = P_1 P_2 \cdots P_n$. (Note, however, that P does *not* in general satisfy the detailed-balance condition, even if the individual P_i do.[6])

Algorithmically, the convex combination amounts to choosing *randomly*, with probabilities $\{\lambda_i\}$, from among the "elementary operations" P_i. (It is crucial here that the λ_i are *constants*, independent of the current configuration of the system; only in this case does P leave μ_{eq} stationary in general.) Similarly, the product corresponds to performing *sequentially* the operations $P_1, P_2, \ldots, P_n$.

We now discuss two general methods for constructing transition matrices P that leave invariant a given measure μ_{eq}: *partial resampling* (which underlies the heat-bath, multi-grid, Swendsen-Wang and embedding algorithms, among others), and the *Metropolis-Rosenbluth-Rosenbluth-Teller-Teller-Hastings procedure*.

1.2 Partial Resampling

Let the variables $\{\varphi\}$ of the system be divided somehow into two subsets, call them $\{\psi\}$ and $\{\theta\}$. For fixed values of the $\{\theta\}$ variables, μ_{eq} induces a conditional probability distribution of $\{\psi\}$ given $\{\theta\}$, call it $P^{\mu_{eq}}(\{\psi\} | \{\theta\})$. Indeed, if μ_{eq} is given by the usual Boltzmann-Gibbs form with Hamiltonian $H(\varphi) \equiv H(\psi, \theta)$, then $P^{\mu_{eq}}(\{\psi\} | \{\theta\})$ is given simply by the same Hamiltonian but with the $\{\theta\}$ variables considered as fixed.

---

[6]Recall that if A and B are self-adjoint operators, then AB is self-adjoint *if and only if* A and B commute.

Now, any algorithm for updating $\{\psi\}$ with $\{\theta\}$ fixed that leaves invariant all of the distributions $P^{\mu_{eq}}(\,\cdot\mid\{\theta\})$ will also leave invariant μ_{eq}. One possibility is to use an *independent resampling* of $\{\psi\}$: we throw away the old values $\{\psi\}$, and take $\{\psi'\}$ to be a new random variable chosen from the probability distribution $P^{\mu_{eq}}(\,\cdot\mid\{\theta\})$, independent of the old values. This is feasible in practice if $\{\psi\}$ is of not-too-high dimension. For example, if $\{\psi\}$ are the variables associated with a single lattice site or link, this defines the *heat-bath updating* of that site or link. The full heat-bath algorithm then consists of updating successively all sites or links, in either random or sequential order (see Section 3.1). More generally, $\{\psi\}$ might be the variables associated with a block of sites or links, in which case we can speak of *block heat-bath updating*.

On the other hand, if $\{\psi\}$ is a large set of variables, independent resampling is probably unfeasible. Nevertheless, we are free to use *any* updating that leaves invariant the appropriate conditional distributions, even though $\{\psi'\}$ is not independent of $\{\psi\}$. Of course, in this generality "partial resampling" includes *all* dynamic Monte Carlo algorithms — we could just take $\{\psi\}$ to be the entire system — but it is in many cases conceptually useful to focus on some subset of variables. The partial-resampling idea will be at the heart of the multi-grid Monte Carlo method (Section 4.2), the Swendsen-Wang algorithm (Section 4.4) and the embedding algorithms (Section 4.5).

Partial resampling has a nice geometric interpretation: the configuration space is partitioned into "fibers" (or "leaves") — namely the sets of configurations with fixed $\{\theta\}$ — and the algorithm moves around the current fiber, using any updating procedure that leaves invariant the conditional probability distribution of μ_{eq} restricted to that fiber. Of course, one must combine this with other moves, or with a different fibering, in order to make the algorithm ergodic.

1.3   Metropolis-Hastings Procedure

A very general method for constructing transition matrices satisfying detailed balance for a given distribution μ_{eq} was introduced in 1953 by Metropolis *et al.* [11], with a slight extension two decades later by Hastings [12]. The idea is the following: Let $P^{(0)}(\varphi \to \varphi')$ be an *arbitrary* transition matrix on S. We call $P^{(0)}$ the *proposal matrix*; we shall use it to generate *proposed* moves $\varphi \to \varphi'$ that will then be accepted or rejected with probabilities $a_{\varphi\varphi'}$ and $1 - a_{\varphi\varphi'}$, respectively. If a proposed move is rejected, then we make a "null transition" $\varphi \to \varphi$. Therefore, the transition matrix $P(\varphi \to \varphi')$ of the full algorithm is

$$P(\varphi \to \varphi') \;=\; P^{(0)}(\varphi \to \varphi')\, a_{\varphi\varphi'} \;+\; \delta(\varphi,\varphi') \int P^{(0)}(\varphi \to \varphi'')\,(1 - a_{\varphi\varphi''})\, d\varphi'' \;, \quad (12)$$

where of course we must have $0 \le a_{\varphi\varphi'} \le 1$ for all φ, φ'. It is easy to see that P satisfies detailed balance for μ_{eq} if and only if

$$\frac{a_{\varphi\varphi'}}{a_{\varphi'\varphi}} \;=\; \frac{\mu_{eq}(\varphi')\, P^{(0)}(\varphi' \to \varphi)}{\mu_{eq}(\varphi)\, P^{(0)}(\varphi \to \varphi')} \qquad\qquad (13)$$

for all pairs $\varphi \neq \varphi'$. But this is easily arranged: just set

$$a_{\varphi\varphi'} = F\left(\frac{\mu_{eq}(\varphi')\,P^{(0)}(\varphi' \to \varphi)}{\mu_{eq}(\varphi)\,P^{(0)}(\varphi \to \varphi')}\right) ,\qquad(14)$$

where $F\colon [0, +\infty] \to [0, 1]$ is any function satisfying

$$\frac{F(z)}{F(1/z)} = z \qquad \text{for all } z. \qquad(15)$$

The choice suggested by Metropolis *et al.* is

$$F(z) = \min(z, 1) ; \qquad(16)$$

this is the *maximal* function satisfying (15). Another choice sometimes used is

$$F(z) = \frac{z}{1+z} . \qquad(17)$$

Of course, it is still necessary to check that P is ergodic; this is usually done on a case-by-case basis.

Note that if the proposal matrix $P^{(0)}$ happens to *already* satisfy detailed balance for μ_{eq}, then we have $\mu_{eq}(\varphi')\,P^{(0)}(\varphi' \to \varphi)/\mu_{eq}(\varphi)\,P^{(0)}(\varphi \to \varphi') = 1$, so that $a_{\varphi\varphi'} = 1$ (if we use the Metropolis choice of F) and $P = P^{(0)}$. On the other hand, no matter what $P^{(0)}$ is, we obtain a matrix P that satisfies detailed balance for μ_{eq}. So the Metropolis-Hastings procedure can be thought of as a prescription for minimally modifying a given transition matrix $P^{(0)}$ so that it satisfies detailed balance for μ_{eq}.

A very important point is that the desired equilibrium measure μ_{eq} occurs in (14) only in the *ratio* $\mu_{eq}(\varphi')/\mu_{eq}(\varphi)$. In statistical mechanics we have $\mu_{eq}(\varphi) = Z^{-1}e^{-H(\varphi)}$, and hence

$$\frac{\mu_{eq}(\varphi')}{\mu_{eq}(\varphi)} = e^{-[H(\varphi')-H(\varphi)]} . \qquad(18)$$

Note that the partition function Z has disappeared from this expression; this is crucial, as Z is almost never explicitly computable!

Many textbooks and articles describe the Metropolis-Hastings procedure only in the special case in which the proposal matrix $P^{(0)}$ is *symmetric*, namely $P^{(0)}(\varphi \to \varphi') = P^{(0)}(\varphi' \to \varphi)$.[7] In this case (14) reduces to

$$a_{\varphi\varphi'} = F\left(\frac{\mu_{eq}(\varphi')}{\mu_{eq}(\varphi)}\right) = F\left(e^{-[H(\varphi')-H(\varphi)]}\right) . \qquad(19)$$

Using the Metropolis acceptance probability $F(z) = \min(z, 1)$, we obtain the following rules for acceptance or rejection:

[7]The original article of Metropolis *et al.* [11] assumed this symmetry. The more general procedure was pointed out by Hastings [12].

- If $\Delta H \equiv H(\varphi') - H(\varphi) \leq 0$, then we accept the proposal *always* (i.e. with probability 1).

- If $\Delta H > 0$, then we accept the proposal with probability $e^{-\Delta H}$ (< 1). That is, we choose a random number r uniformly distributed on $[0, 1]$, and we accept the proposal if $r \leq e^{-\Delta H}$.

But there is nothing special about $P^{(0)}$ being symmetric; *any* proposal matrix $P^{(0)}$ is perfectly legitimate, and the Metropolis-Hastings procedure is defined quite generally by (14).

Let us emphasize once more that the Metropolis-Hastings procedure is a *general technique*; it produces an infinite family of different algorithms depending on the choice of the proposal matrix $P^{(0)}$. In the literature the term "Metropolis algorithm" is often used to denote the algorithm resulting from some *particular* commonly-used choice of $P^{(0)}$, but it is important not to be misled.

1.4 Stochastic Linear Iterations for the Gaussian Model

As explained in Section 1.1, the performance of a dynamic Monte Carlo algorithm is controlled by its autocorrelation times τ_{exp} and $\tau_{int,F}$. It is useful to study a class of algorithms in which these autocorrelation times can be computed explicitly. Here we analyze such a class, namely the stochastic linear iterations for Gaussian models [13, Section 8]. This class includes, among others, the single-site heat-bath algorithm (with deterministic sweep of the sites), the stochastic overrelaxation algorithm and the multi-grid Monte Carlo algorithm — all, of course, in the Gaussian case only. We show that the behavior of the stochastic algorithm is completely determined by the behavior of the corresponding deterministic algorithm for solving linear equations. In particular, we show that the exponential autocorrelation time τ_{exp} of the stochastic linear iteration is *equal* to the relaxation time of the corresponding linear iteration.

Consider any quadratic Hamiltonian

$$H(\varphi) = \tfrac{1}{2}(\varphi, A\varphi) - (f, \varphi), \tag{20}$$

where A is a symmetric positive-definite matrix. The corresponding Gaussian measure

$$d\mu_{eq}(\varphi) = \text{const} \times e^{-\frac{1}{2}(\varphi, A\varphi) + (f, \varphi)} \, d\varphi \tag{21}$$

has mean $A^{-1}f$ and covariance matrix A^{-1}. Next consider any first-order stationary linear stochastic iteration of the form

$$\varphi^{(n+1)} = M\varphi^{(n)} + Nf + Q\xi^{(n)}, \tag{22}$$

where M, N and Q are fixed matrices and the $\xi^{(n)}$ are independent Gaussian random vectors with mean zero and covariance matrix C. The iteration (22) has a unique stationary distribution if and only if the spectral radius $\rho(M) \equiv \lim_{n \to \infty} \|M^n\|^{1/n}$

is < 1; and in this case the stationary distribution is the desired Gaussian measure (21) for all f if and only if

$$N = (I - M)A^{-1} \qquad (23a)$$
$$QCQ^T = A^{-1} - MA^{-1}M^T \qquad (23b)$$

(here T denotes transpose).

The reader familiar with numerical linear algebra [14] will recognize the close analogy with deterministic iterative methods for solving the linear system $A\varphi = f$, or equivalently for minimizing the functional $H(\varphi)$. Indeed, such methods are of the form (22) with Q set equal to zero. This is precisely the "zero-temperature limit" of the stochastic problem, in which H is replaced by βH with $\beta \to +\infty$: then the Gaussian measure (21) approaches a delta function concentrated at the unique minimum of H (namely, the solution of the linear equation $A\varphi = f$), and the "noise" term disappears ($Q \to 0$). In other words:

(a) The linear deterministic problem is the zero-temperature limit of the Gaussian stochastic problem; and the first-order stationary linear deterministic iteration is the zero-temperature limit of the first-order stationary linear stochastic iteration. Therefore, any stochastic linear iteration for generating samples from the Gaussian measure (21) gives rise to a deterministic linear iteration for solving the linear equation $A\varphi = f$, simply by setting $Q = 0$.

(b) Conversely, the stochastic problem and iteration are the nonzero-temperature generalizations of the deterministic ones. In principle this means that a deterministic linear iteration for solving $A\varphi = f$ can be generalized to a stochastic linear iteration for generating samples from (21), if and only if the matrix $A^{-1} - MA^{-1}M^T$ is positive-semidefinite: just choose a matrix Q satisfying (23b). In practice, however, such an algorithm is computationally tractable only if the matrix Q has additional nice properties such as sparsity (or triangularity with a sparse inverse).

We shall see in subsequent sections how the heat-bath, overrelaxation, Langevin and multi-grid algorithms fit into this framework.

It is straightforward to analyze the dynamic behavior of the stochastic linear iteration (22). Using (22) and (23) to express $\varphi^{(n)}$ in terms of the independent random variables $\varphi^{(0)}, \xi^{(0)}, \xi^{(1)}, \ldots, \xi^{(n-1)}$, we find after a bit of manipulation that

$$\langle \varphi^{(n)} \rangle = M^n \langle \varphi^{(0)} \rangle + (I - M^n)A^{-1}f \qquad (24)$$

and

$$\mathrm{cov}(\varphi^{(s)}, \varphi^{(t)}) = M^s \, \mathrm{cov}(\varphi^{(0)}, \varphi^{(0)})(M^T)^t + \begin{cases} [A^{-1} - M^s A^{-1}(M^T)^s](M^T)^{t-s} & \text{if } s \leq t \\ M^{s-t}[A^{-1} - M^t A^{-1}(M^T)^t] & \text{if } s \geq t \end{cases} \qquad (25)$$

(here "cov" denotes covariance). Now let us either start the stochastic process in equilibrium

$$\langle \varphi^{(0)} \rangle = A^{-1} f \tag{26a}$$

$$\mathrm{cov}(\varphi^{(0)}, \varphi^{(0)}) = A^{-1} \tag{26b}$$

or else let it relax to equilibrium by taking $s, t \to +\infty$ with $s - t$ fixed. Either way, we conclude that in equilibrium (22) defines a Gaussian stationary stochastic process with mean $A^{-1} f$ and autocovariance matrix

$$\mathrm{cov}(\varphi^{(s)}, \varphi^{(t)}) = \begin{cases} A^{-1}(M^T)^{t-s} & \text{if } s \leq t \\ M^{s-t} A^{-1} & \text{if } s \geq t \end{cases} \tag{27}$$

Moreover, since the stochastic process is Gaussian, all higher-order time-dependent correlation functions are determined in terms of the mean and autocovariance. Thus, the matrix M determines the autocorrelation functions of the Monte Carlo algorithm.

Another way to state these relationships is to recall [15, 16] that the Hilbert space $L^2(\mu_{eq})$ is isomorphic to the bosonic Fock space $\mathcal{F}(U)$ built on the "energy Hilbert space" (U, A): the "n-particle states" are the homogeneous Wick polynomials of degree n in the shifted field $\tilde{\varphi} = \varphi - A^{-1} f$. (If U is one-dimensional, these are just the Hermite polynomials.) Then the transition probability $P(\varphi^{(n)} \to \varphi^{(n+1)})$ induces on the Fock space an operator

$$P = \Gamma(M^T) \equiv I \oplus M^T \oplus (M^T \otimes M^T) \oplus \cdots \tag{28}$$

that is the second quantization of the operator M^T on the energy Hilbert space (see [13, Section 8] for details). It follows from (28) that

$$\| \Gamma(M)^n \restriction \mathbf{1}^{\perp} \|_{L^2(\mu_{eq})} = \| M^n \|_{(U,A)} \tag{29}$$

(here $\restriction$ denotes restriction) and hence that

$$\rho(\Gamma(M) \restriction \mathbf{1}^{\perp}) = \rho(M) . \tag{30}$$

Moreover, P is self-adjoint on $L^2(\mu_{eq})$ [i.e. satisfies detailed balance] if and only if M is self-adjoint with respect to the energy inner product, i.e.

$$MA = AM^T ; \tag{31}$$

and in this case

$$\rho(\Gamma(M) \restriction \mathbf{1}^{\perp}) = \| \Gamma(M) \restriction \mathbf{1}^{\perp} \|_{L^2(\mu_{eq})} = \rho(M) = \| M \|_{(U,A)} . \tag{32}$$

In summary, we have shown that the dynamic behavior of any stochastic linear iteration is completely determined by the behavior of the corresponding deterministic linear iteration. In particular, the exponential autocorrelation time τ_{exp} (slowest decay rate of any autocorrelation function) is given by

$$\exp(-1/\tau_{exp}) = \rho(M) , \tag{33}$$

and this decay rate is achieved by at least one observable which is linear in the field φ. In other words, the (worst-case) convergence rate of the Monte Carlo algorithm is precisely *equal* to the (worst-case) convergence rate of the corresponding deterministic iteration.

2 Statistical Analysis of Dynamic Monte Carlo Data

Many published Monte Carlo studies contain statements like:

> We ran for a total of 100000 iterations, discarding the first 50000 iterations (for equilibration) and then taking measurements once every 100 iterations.

It is important to note that unless further information is given — namely, the autocorrelation time of the algorithm — *such statements have no value whatsoever.*

Is a run of 100000 iterations long enough? Are 50000 iterations sufficient for equilibration? That depends on how big the autocorrelation time is. The purpose of this section is to give some practical advice for choosing the parameters of a dynamic Monte Carlo simulation, and to give an introduction to the statistical theory that puts this advice on a sound mathematical footing.

There are two fundamental — and quite distinct — issues in dynamic Monte Carlo simulation:

- *Initialization bias.* If the Markov chain is started in a distribution α that is not equal to the stationary distribution μ_{eq}, then there is an "initial transient" in which the data do not reflect the desired equilibrium distribution μ_{eq}. This results in a *systematic error* (bias).

- *Autocorrelation in equilibrium.* The Markov chain, once it reaches equilibrium, provides *correlated* samples from μ_{eq}. This correlation causes the *statistical error* (variance) to be a factor $2\tau_{int,F}$ larger than in independent sampling.

Let us discuss these issues in turn.

Initialization bias. Often the Markov chain is started in some chosen configuration φ; then $\alpha = \delta_\varphi$. For example, in an Ising model, φ might be the configuration with "all spins up"; this is sometimes called an *ordered* or *cold* start. Alternatively, the Markov chain might be started in a random configuration chosen according to some simple probability distribution α. For example, in an Ising model, we might initialize the spins randomly and independently, with equal probabilities of up and down; this is sometimes called a *random* or *hot* start. In all these cases, the initial distribution α is clearly *not* equal to the equilibrium distribution μ_{eq}. Therefore, the system is initially "out of equilibrium". The general theory guarantees that the system approaches equilibrium as $t \to \infty$, but we need to know something about the *rate* of convergence to equilibrium.

Using the exponential autocorrelation time τ_{exp}, we can set an *upper bound* on the amount of time we have to wait before equilibrium is "for all practical purposes" attained. For example, if we wait a time $20\tau_{exp}$, then the deviation from equilibrium (in the L^2 sense) will be at most e^{-20} ($\approx 2 \times 10^{-9}$) times the initial deviation from equilibrium. There are two difficulties with this bound. Firstly, it is usually impossible to apply in practice, since we almost never know τ_{exp} (or a rigorous upper bound for it). Secondly, even if we can apply it, it may be overly conservative; indeed, there exist perfectly reasonable algorithms in which $\tau_{exp} = +\infty$ [17].

Lacking rigorous knowledge of the autocorrelation time τ_{exp}, we should try to estimate it both *theoretically* and *empirically*. To make a heuristic theoretical estimate of τ_{exp}, we attempt to understand the physical mechanism(s) causing slow convergence to equilibrium; but it is always possible that we have overlooked one or more such mechanisms, and have therefore grossly underestimated τ_{exp}. To make a rough empirical estimate of τ_{exp}, we measure the autocorrelation function $C_{FF}(t)$ for a suitably large set of observables F [see below]; but there is always the danger that our chosen set of observables has failed to include one that has strong enough overlap with the slowest mode, again leading to a gross underestimate of τ_{exp}.

On the other hand, the actual rate of convergence to equilibrium from a given initial distribution α may be much faster than the worst-case estimate given by τ_{exp}. So it is usual to determine *empirically* when "equilibrium" has been achieved, by plotting selected observables as a function of time and noting when the initial transient appears to end. More sophisticated statistical tests for initialization bias can also be employed [18].

In all empirical methods of determining when "equilibrium" has been achieved, a serious danger is the possibility of *metastability*. That is, it could *appear* that equilibrium has been achieved, when in fact the system has only settled down to a long-lived (metastable) region of configuration space that may be very *far* from equilibrium. The only sure-fire protection against metastability is a *proof* of an upper bound on τ_{exp} (or more generally, on the deviation from equilibrium as a function of the elapsed time t). The next-best protection is a convincing heuristic argument that metastability is unlikely (i.e. that τ_{exp} is not too large); but as mentioned before, even if one rules out several potential physical mechanisms for metastability, it is very difficult to be certain that one has not overlooked others. If one cannot rule out metastability on theoretical grounds, then it is helpful at least to have an idea of what the possible metastable regions look like; then one can perform several runs with different initial conditions typical of each of the possible metastable regions, and test whether the answers are consistent. For example, near a first-order phase transition, most Monte Carlo methods suffer from metastability associated with transitions between configurations typical of the distinct pure phases. We can try initial conditions typical of each of these phases (e.g. for many models, a "hot" start and a "cold" start). Consistency between these runs does not *guarantee* that metastability is absent, but it does give increased confidence. Plots of observables as a function of time are also useful indicators of possible metastability.

But when all is said and done, no purely empirical estimate of τ from a run of length n can be guaranteed to be even approximately correct. What we *can* say is that if $\tau_{estimated} \ll n$, then *either* $\tau \approx \tau_{estimated}$ or else $\tau \gtrsim n$.

Once we know (or guess) the time needed to attain "equilibrium", what do we do with it? The answer is clear: we discard the data from the initial transient, up to some time n_{disc}, and include only the subsequent data in our averages. In principle, this is (asymptotically) unnecessary, because the systematic errors from this initial transient will be of order τ/n, while the statistical errors will be of order $(\tau/n)^{1/2}$. But in practice, the coefficient of τ/n in the systematic error may be fairly large, if the initial distribution is very far from equilibrium. By throwing away the data from the initial transient, we lose nothing, and avoid a potentially large systematic error.

Autocorrelation in equilibrium. As explained in Section 1.1, the variance of the sample mean $\overline{F}$ in a dynamic Monte Carlo method is a factor $2\tau_{int,F}$ higher than it would be in independent sampling. Otherwise put, a run of length n contains only $n/2\tau_{int,F}$ "effectively independent data points".

This has several implications for Monte Carlo work. On the one hand, it means that the the *computational efficiency* of the algorithm is determined principally by its autocorrelation time. More precisely, if one wishes to compare two alternative Monte Carlo algorithms for the same problem, then *the better algorithm is the one that has the smaller autocorrelation time, when time is measured in units of computer (CPU) time.* [In general there may arise tradeoffs between "physical" autocorrelation time (i.e. τ measured in *iterations*) and computational complexity *per iteration*.] So accurate measurements of the autocorrelation time are essential to evaluating the computational efficiency of competing algorithms.

On the other hand, even for a fixed algorithm, knowledge of $\tau_{int,F}$ is essential for determining run lengths — is a run of 100000 sweeps long enough? — and for setting error bars on estimates of $\langle F \rangle$. Roughly speaking, error bars will be of order $(\tau/n)^{1/2}$; so if we want 1% accuracy, then we need a run of length $\approx 10000\tau$, and so on. Above all, there is a basic self-consistency requirement: the run length n must be $\gg$ than the estimates of τ produced by that same run, otherwise *none* of the results from that run should be believed. On the other hand, while self-consistency is a *necessary* condition for the trustworthiness of Monte Carlo data, it is not a *sufficient* condition; there is always the danger of metastability.

Already we can draw a conclusion about the relative importance of initialization bias and autocorrelation as difficulties in dynamic Monte Carlo work. Let us *assume* that the time for initial convergence to equilibrium is comparable to (or at least not too much larger than) the equilibrium autocorrelation time $\tau_{int,F}$ (for the observables F of interest) — this is often but not always the case. Then initialization bias is a relatively trivial problem compared to autocorrelation in equilibrium. To eliminate initialization bias, it suffices to discard $\approx 20\tau$ of the data at the beginning of the run; but to achieve a reasonably small statistical error, it is necessary to make a run of length $\approx 1000\tau$ or more. So the data that must be discarded at the beginning, n_{disc}, is a negligible fraction of the total run length n. This estimate also shows that the exact value of n_{disc} is not particularly delicate: anything between $\approx 20\tau$ and $\approx n/5$ will eliminate essentially all initialization bias while paying less than a 10% price in the final error bars.

The hypothetical paper quoted at the beginning of this section should, therefore, be immediately considered suspect: either the run length 100000 is too short, or the discard interval 50000 is too conservative.

In the remainder of this section I would like to discuss in more detail the statistical analysis of dynamic Monte Carlo data (assumed to be already "in equilibrium"), with emphasis on how to estimate the autocorrelation time $\tau_{int,F}$ and how to compute valid error bars. What is involved here is a branch of mathematical statistics called *time-series analysis*. An excellent exposition can be found in the books of of Priestley [19] and Anderson [20].

Let $\{F_t\}$ be a real-valued stationary stochastic process with mean

$$\mu \equiv \langle F_t \rangle , \tag{34}$$

unnormalized autocorrelation function

$$C(t) \equiv \langle F_s\, F_{s+t} \rangle - \mu^2 , \tag{35}$$

normalized autocorrelation function

$$\rho(t) \equiv C(t)/C(0) , \tag{36}$$

and integrated autocorrelation time

$$\tau_{int} = \tfrac{1}{2} \sum_{t=-\infty}^{\infty} \rho(t) . \tag{37}$$

Our goal is to estimate μ, $C(t)$, $\rho(t)$ and τ_{int} based on a finite (but large) sample $F_1, \ldots, F_n$ from this stochastic process.

The "natural" estimator of μ is the sample mean

$$\overline{F}^{(n)} \equiv \frac{1}{n} \sum_{i=1}^{n} F_i . \tag{38}$$

This estimator is unbiased (i.e. $\langle \overline{F}^{(n)} \rangle = \mu$) and has variance

$$\mathrm{var}(\overline{F}^{(n)}) = \frac{1}{n} \sum_{t=-(n-1)}^{n-1} \left(1 - \frac{|t|}{n} \right) C(t) \tag{39a}$$

$$\approx \frac{1}{n} (2\tau_{int}) C(0) \quad \text{for } n \gg \tau \tag{39b}$$

Thus, even if we are interested only in the static quantity μ, it is necessary to estimate the dynamic quantity τ_{int} in order to determine valid error bars for μ.

The "natural" estimator of $C(t)$ is

$$\widehat{C}(t) \equiv \frac{1}{n - |t|} \sum_{i=1}^{n-|t|} (F_i - \mu)(F_{i+|t|} - \mu) \tag{40}$$

if the mean μ is known, and

$$\widehat{\widehat{C}}(t) \equiv \frac{1}{n - |t|} \sum_{i=1}^{n-|t|} (F_i - \overline{F}^{(n)})(F_{i+|t|} - \overline{F}^{(n)}) \tag{41}$$

if the mean μ is unknown. We emphasize the conceptual distinction between the autocorrelation function $C(t)$, which for each t is a *number*, and the estimator $\widehat{C}(t)$ or $\widehat{\widehat{C}}(t)$, which for each t is a *random variable*. As will become clear, this distinction is also of *practical* importance. $\widehat{C}(t)$ is an unbiased estimator of $C(t)$, and $\widehat{\widehat{C}}(t)$ is almost unbiased (the bias is of order $1/n$) [20, p. 463]. Their variances and covariances are [20, pp. 464–471] [19, pp. 324–328]

$$\mathrm{var}(\widehat{C}(t)) = \frac{1}{n} \sum_{m=-\infty}^{\infty} \left[C(m)^2 + C(m+t)C(m-t) + \kappa(t, m, m+t) \right]$$
$$+ o\left(\frac{1}{n}\right) \tag{42}$$

$$\mathrm{cov}(\widehat{C}(t), \widehat{C}(u)) = \frac{1}{n} \sum_{m=-\infty}^{\infty} [C(m)C(m+u-t) + C(m+u)C(m-t)$$
$$+ \kappa(t, m, m+u)] + o\left(\frac{1}{n}\right) \tag{43}$$

where $t, u \geq 0$ and κ is the connected 4-point autocorrelation function

$$\kappa(r, s, t) \equiv \langle (F_i - \mu)(F_{i+r} - \mu)(F_{i+s} - \mu)(F_{i+t} - \mu) \rangle$$
$$-C(r)C(t-s) - C(s)C(t-r) - C(t)C(s-r). \tag{44}$$

To leading order in $1/n$, the behavior of $\widehat{\widehat{C}}$ is identical to that of $\widehat{C}$.

The "natural" estimator of $\rho(t)$ is

$$\widehat{\rho}(t) \equiv \widehat{C}(t)/\widehat{C}(0) \tag{45}$$

if the mean μ is known, and

$$\widehat{\widehat{\rho}}(t) \equiv \widehat{\widehat{C}}(t)/\widehat{\widehat{C}}(0) \tag{46}$$

if the mean μ is unknown. The variances and covariances of $\widehat{\rho}(t)$ and $\widehat{\widehat{\rho}}(t)$ can be computed (for large n) from (43); we omit the detailed formulae.

The "natural" estimator of τ_{int} would seem to be

$$\widehat{\tau}_{int} \stackrel{?}{=} \frac{1}{2} \sum_{t=-(n-1)}^{n-1} \widehat{\rho}(t) \tag{47}$$

(or the analogous thing with $\widehat{\widehat{\rho}}$), *but this is wrong!* The estimator defined in (47) has a variance that does not go to zero as the sample size n goes to infinity [19, pp. 420–431], so it is clearly a very bad estimator of τ_{int}. Roughly speaking, this is because the sample autocorrelations $\widehat{\rho}(t)$ for $|t| \gg \tau$ contain much "noise" but little "signal"; and there are so many of them (order n) that the noise adds up to a total variance of order 1. (For a more detailed discussion, see [19, pp. 432–437].) The solution is to cut off the sum in (47) using a "window" $\lambda(t)$ which is ≈ 1 for $|t| \lesssim \tau$ but ≈ 0 for $|t| \gg \tau$:

$$\widehat{\tau}_{int} \equiv \tfrac{1}{2} \sum_{t=-(n-1)}^{n-1} \lambda(t)\,\widehat{\rho}(t)\,. \tag{48}$$

This retains most of the "signal" but discards most of the "noise". A good choice is the rectangular window

$$\lambda(t) = \begin{cases} 1 & \text{if } |t| \le M \\ 0 & \text{if } |t| > M \end{cases} \tag{49}$$

where M is a suitably chosen cutoff. This cutoff introduces a bias

$$\text{bias}(\widehat{\tau}_{int}) = -\tfrac{1}{2} \sum_{|t|>M} \rho(t) + o\left(\frac{1}{n}\right)\,. \tag{50}$$

On the other hand, the variance of $\widehat{\tau}_{int}$ can be computed from (43); after some algebra, one obtains

$$\text{var}(\widehat{\tau}_{int}) \approx \frac{2(2M+1)}{n}\,\tau_{int}^2\,, \tag{51}$$

where we have made the approximation $\tau \ll M \ll n$.[8] The choice of M is thus a tradeoff between bias and variance: the bias can be made small by taking M large enough so that $\rho(t)$ is negligible for $|t| > M$ (e.g. $M = $ a few times τ usually suffices), while the variance is kept small by taking M to be no larger than necessary consistent with this constraint. We have found the following "automatic windowing" algorithm [21] to be convenient: choose M to be the smallest integer such that $M \ge c\widehat{\tau}_{int}(M)$. If $\rho(t)$ were roughly a pure exponential, then it would suffice to take $c \approx 4$ (since $e^{-4} < 2\%$). However, in many cases $\rho(t)$ is expected to have an asymptotic or pre-asymptotic decay slower than exponential, so it is usually prudent to take c at least 6; and in some situations (i.e. when $\rho(t)$ has extremely slow decay) it is necessary to take c as large as 10 or 20.

We have found this automatic windowing procedure to work well in practice, *provided* that a sufficient quantity of data is available ($n \gtrsim 1000\tau$). However, at present we have very little understanding of the conditions under which this windowing algorithm may produce biased estimates of τ_{int} or of its own error bars. Further theoretical and experimental study of the windowing algorithm — e.g. experiments on various exactly-known stochastic processes, with various run lengths — would be highly desirable.

An alternative approach [22] is to use the estimated autocorrelations $\widehat{\rho}(t)$ for $|t| \le M$ to produce an *extrapolation* to $|t| > M$ which is then summed. By using

---

[8]We have also assumed that the only strong peaks in the Fourier transform of $C(t)$ are at zero frequency. This assumption is valid if $C(t) \ge 0$, but could fail if there are strong *anti*correlations.

these extrapolated values (rather than simply zero!), this estimator could in principle have a smaller bias than the simple windowing method. But it is also susceptible to serious errors arising from the extrapolation of noisy data. Further theoretical and experimental study would again be desirable.

3   Local Algorithms

3.1   Single-Site Heat-Bath Algorithm

The single-site (or single-link) heat-bath algorithm is the special case of partial resampling (Section 1.2) in which the $\{\psi\}$ variables are those belonging to a single lattice site (or link) and the resamplings are independent. Consider, for example, the Ising model: on each site i of some finite d-dimensional lattice, we place a random variable σ_i taking the values ± 1. The Hamiltonian is

$$H(\sigma) = -\sum_{\langle ij \rangle} \sigma_i \sigma_j , \qquad (52)$$

where the sum runs over all nearest-neighbor pairs. The corresponding Gibbs measure is

$$\mu_{eq}(\sigma) = Z^{-1} \exp[-\beta H(\sigma)] . \qquad (53)$$

Now focus on a single site i; the conditional probability distribution of σ_i, given all the other spins $\{\sigma_j\}_{j \neq i}$, is

$$P^{\mu_{eq}}(\sigma_i \,|\, \{\sigma_j\}_{j \neq i}) = \text{const}\,(\{\sigma_j\}_{j \neq i}) \times \exp\left[\beta \sigma_i \sum_{j \text{ n.n. of } i} \sigma_j\right] . \qquad (54)$$

(Note that this conditional distribution is precisely that of a single Ising spin σ_i in an "effective magnetic field" produced by the fixed neighboring spins σ_j.) The single-site heat-bath algorithm updates σ_i by choosing a new spin value σ_i', *independent of the old value* σ_i, from the conditional distribution (54); all the other spins $\{\sigma_j\}_{j \neq i}$ remain unchanged. This defines a transition matrix P_i in which only the spin at site i is touched. The full "single-site heat-bath algorithm" involves sweeping through the entire lattice in either a random or periodic fashion, i.e. either

$$P = \frac{1}{V} \sum_i P_i \qquad \text{(random site updating)} \qquad (55)$$

or

$$P = P_{i_1} P_{i_2} \cdots P_{i_V} \qquad \text{(sequential site updating)} \qquad (56)$$

(here V is the volume). In the former case, the transition matrix P satisfies detailed balance for μ_{eq}. In the latter case, P does *not* in general satisfy detailed balance for μ_{eq}, but it does satisfy stationarity for μ_{eq}, which is all that really matters. It is easy to see that P is ergodic, since both values of σ_i' have nonzero probability.

Analogous algorithms can be developed for more complicated models, e.g. $P(\varphi)$ models, nonlinear σ-models and lattice gauge theories. In each case, we focus on a single field variable (holding all the other variables fixed), and give it a new value, *independent of the old value*, chosen from the appropriate conditional distribution. Of course, the feasibility of this algorithm depends on our ability to construct an efficient subroutine for generating the required single-site (or single-link) random variables, which are typically those of a single spin in an "effective magnetic field". This is trivially done for finite-state spin or gauge models and for Gaussian theories; it has also been done for φ^4 theories [13], N-vector models [23], and $U(1)$ [24, 25, 26], $SU(2)$ [23, 27] and $SU(3)$ [28] spin or gauge models. For these more complicated models, the heat-bath algorithm is often less efficient in practice than a carefully-tuned Metropolis algorithm (Section 3.2); but because it is a well-defined algorithm with no free parameters, it serves as a clear standard of comparison, which is useful in the development of more sophisticated algorithms.

Note that in a spin model with nearest-neighbor interactions, the heat-bath updates of *non-neighboring* sites may be carried out without mutual interference (the corresponding operators P_i commute). In particular, one may update simultaneously the entire even sublattice, and then the entire odd sublattice. This *checkerboard* (or *red-black*) updating is well suited to vectorization and parallelization. Similar but slightly more complicated schemes can be arranged for lattice gauge theories.

Note, finally, that the single-site heat-bath algorithm for generating random samples from $e^{-\beta H(\varphi)}$ is the precise stochastic analogue of the Gauss-Seidel algorithm for minimizing $H(\varphi)$. The Gauss-Seidel algorithm chooses φ_i' to minimize the conditional Hamiltonian $H(\cdot, \{\varphi_j\}_{j \neq i})$ with $\{\varphi_j\}_{j \neq i}$ fixed, while the heat-bath algorithm chooses φ_i' to be a random sample from $e^{-\beta H(\cdot, \{\varphi_j\}_{j \neq i})}$. As $\beta \to +\infty$ the stochastic problem reduces to the deterministic one, and the heat-bath algorithm reduces to Gauss-Seidel.

3.2 Single-Site Metropolis Algorithm

The single-site (or single-link) Metropolis algorithm is the special case of the Metropolis-Hastings procedure (Section 1.3) in which the proposed transitions consist of some change to a single site (or link) variable. For the Ising model (52)/(53), the obvious proposal is to flip the spin σ_i:

$$P_i^{(0)}(\{\sigma\} \to \{\sigma'\}) = \begin{cases} 1 & \text{if } \sigma_i' = -\sigma_i \text{ and } \sigma_j' = \sigma_j \text{ for all } j \neq i \\ 0 & \text{otherwise} \end{cases} \tag{57}$$

Here $P_i^{(0)}$ is symmetric, so the acceptance probability [using the Metropolis choice (16)] is

$$a_i(\{\sigma\} \to \{\sigma'\}) = \min(e^{-\beta \Delta E}, 1), \tag{58}$$

where

$$\Delta E \equiv E(\{\sigma'\}) - E(\{\sigma\}) = 2\sigma_i \sum_{j \text{ n.n. of } i} \sigma_j. \tag{59}$$

So ΔE is easily computed by comparing the status of σ_i and its neighbors.[9]

This defines a transition matrix P_i in which only the spin at site i is touched. The full "single-spin-flip Metropolis algorithm" involves sweeping through the entire lattice in either a random or periodic fashion. It used to be thought obvious that this algorithm is ergodic for all $\beta \neq 0$ — and this is even stated explicitly in my own lecture notes [1] — but Gidas [29] has recently shown by explicit counterexample that this belief is *false*! [The trouble is that the Metropolis acceptance function flips the spin with probability 1 whenever $\Delta E = 0$. This pathology could be avoided by using a less radical acceptance function such as $F(z) = z/(1+z)$.]

For continuous-spin models, the Metropolis proposal usually consists of some random change to the spin φ_i. For example, in a model of real-valued "spins" φ_i (such as a φ^4 model), one might propose the move $\varphi \to \varphi_i' \equiv \varphi_i + \Delta_i$, where Δ_i is uniformly distributed in the interval $[-\delta, \delta]$. Here δ is a free parameter of the algorithm, and it must be chosen carefully: if δ is too large, nearly all proposed moves will be rejected, and the autocorrelation time of the algorithm will be large ($\gtrsim 1/f$, where f is the acceptance fraction); while if δ is too small, the system will make a small-step random walk through the configuration space, and the autocorrelation time will again be large ($\sim 1/\delta^2$). A commonly-used rule of thumb is to choose δ so that the acceptance rate is around 50%; although this rule seems reasonable to me, I do not know of any theoretical justification for it. A better criterion is to do test runs with several different values of δ and measure the autocorrelation times as described in Section 2; the best value of δ is the one that gives the smallest autocorrelation time $\tau_{int,F}$ for the observables of interest.

An interesting variant on the Metropolis algorithm is the "quasi-heat-bath" method of Fredenhagen and Marcu [30].

3.3   Interlude: Critical Slowing-Down

We have now defined a rather large class of dynamic Monte Carlo algorithms: the single-site heat-bath algorithm, the single-spin-flip Metropolis algorithm, and so on. How well do these algorithms perform?

Away from phase transitions, they perform rather well. However, near a phase transition, the autocorrelation time grows rapidly. In particular, near a critical point (second-order phase transition), the autocorrelation time typically diverges as

$$\tau \sim \min(L, \xi)^z , \qquad (60)$$

where L is the linear size of the system, ξ is the correlation length of an infinite-volume system at the same temperature, and z is a *dynamic critical exponent*. This phenomenon is called *critical slowing-down*; it severely hampers the study of critical phenomena by Monte Carlo methods. Most of the remainder of this chapter will

---

[9]The alert reader will note that *in the Ising case* the single-spin-flip Metropolis algorithm with the choice $F(z) = z/(1+z)$ of acceptance function is equivalent to the single-site heat-bath algorithm. But this correspondence does not hold for more complicated models.

be devoted, therefore, to describing recent progress in inventing new Monte Carlo algorithms with radically reduced critical slowing-down.

The critical slowing-down of the conventional algorithms arises fundamentally from the fact that their updates are *local*: in a single step of the algorithm, "information" is transmitted from a given site only to its nearest neighbors. Crudely one might guess that this "information" executes a random walk around the lattice. In order for the system to evolve to an "essentially new" configuration, the "information" has to travel a distance of order ξ, the (static) correlation length. One would guess, therefore, that $\tau \sim \xi^2$ near criticality, i.e. that the dynamic critical exponent z equals 2.[10] This guess is correct for the Gaussian model (free field).[11] For other models, we have a situation analogous to theory of static critical phenomena: the dynamic critical exponent is a nontrivial number that characterizes a rather large class of algorithms (a so-called "dynamic universality class"). In any case, for most models of interest, the dynamic critical exponent for local algorithms is close to 2 (usually somewhat higher) [32]. Accurate measurements of dynamic critical exponents are, however, very difficult — even more difficult than measurements of static critical exponents — and require enormous quantities of Monte Carlo data: run lengths of $\approx 10000\tau$, when τ is itself getting large!

We can now make a rough estimate of the computer time needed to study the Ising model near its critical point, or quantum chromodynamics near the continuum limit. Each sweep of the lattice takes a time of order L^d, where d is the spatial (or space-"time") dimensionality of the model. And we need $\approx 2\tau$ sweeps in order to get *one* "effectively independent" sample. So this means a computer time of order $L^d \xi^z \gtrsim \xi^{d+z}$.[12] For high-precision statistics one might want 10^6 "independent" samples. The reader is invited to plug in $\xi = 100$, $d = 4$ (or $d = 3$ if you're a condensed-matter physicist) and get depressed. It should be emphasized that the factor ξ^d is inherent in *all* Monte Carlo algorithms for spin models and field theories (but not for self-avoiding walks, see [1, 35, 36]). The factor ξ^z could, however, conceivably be reduced or eliminated by a more clever algorithm, as will be discussed in Section 4.

(Let me remark that near a first-order transition the slowing-down is even more severe: typically $\tau \sim \exp(cL^{d-1})$, as required for tunneling through very improbable configurations involving interfaces. However, a very recent idea of Berg and Neuhaus [37] may succeed in reducing this to a power of L.)

The definition (60) of the dynamic critical exponent z is, however, a bit too sloppy in one important respect: as discussed in Section 1.1, there are several distinct notions of autocorrelation time τ, notably τ_{exp} and $\tau_{int,F}$ — so might there exist

---

[10]This argument can alternatively be phrased in terms of the diffusion of "domain walls", i.e. of the boundaries separating regions in which the field φ takes radically different average values.

[11]Indeed, for the Gaussian model this random-walk picture can be made rigorous: see [31] combined with [13, Section 8].

[12]Clearly one must take $L \gtrsim \xi$ in order to avoid severe finite-size effects. Typically one approaches the critical point with $L \approx c\xi$, where $c \approx 2-4$, and then uses finite-size scaling [33, 34] to extrapolate to the infinite-volume limit.

distinct dynamic critical exponents z_{exp} and $z_{int,F}$? It seems to be generally believed that, at least for "reasonable" dynamics (such as the traditional local ones), τ_{exp} and $\tau_{int,F}$ are of the *same* order of magnitude in the critical region, i.e. they diverge with the *same* dynamic critical exponent z. This belief is implicit in articles which refer simply to "the" dynamic critical exponent z; and it is made explicit in some of my own papers of a few years back (such as [38]). It now seems to me quite obvious that τ_{exp} and $\tau_{int,F}$ *need not be of the same order of magnitude*, i.e. they *need not scale with the same dynamic critical exponent* z. So we do need to define distinct dynamic critical exponents z_{exp} and $z_{int,F}$:[13]

$$\tau_{exp,F} \sim \tau_{exp} \sim \min(L,\xi)^{z_{exp}} \tag{61}$$

$$\tau_{int,F} \sim \min(L,\xi)^{z_{int,F}} \tag{62}$$

Nearly always one has $\tau_{int,F} \lesssim \tau_{exp,F}$ — in particular, this is *provable* if detailed balance holds, see (11) — hence $z_{int,F} \leq z_{exp}$. But $z_{int,F}$ *can be strictly smaller than* z_{exp}! One known example is the pivot algorithm for the ordinary random walk, which is exactly soluble [21, Section 3.3]. But I claim that in fact $z_{int,F} < z_{exp}$ should be regarded as the *typical* behavior!

To see this, consider the following analogies between dynamic and static critical phenomena:

Dynamic	Static
time $\longleftrightarrow$	space
$\tau_{exp} \longleftrightarrow$	ξ
$\tau_{int,F} \longleftrightarrow$	$\begin{cases} \text{susceptibility } \chi \text{ (if } F = \mathcal{M}) \\ \text{specific heat } C_h \text{ (if } F = \mathcal{E}) \\ \text{etc.} \end{cases}$

Now we know perfectly well that the susceptibility and correlation length have different critical exponents ($\gamma \neq \nu$); so shouldn't one expect that the integrated and exponential autocorrelation times do likewise ($z_{int,F} \neq z_{exp}$)? Indeed, there is a scaling law relating γ to ν and the exponent η describing the decay of correlations at criticality; and one expects a similar scaling law for dynamic correlations:

$$\rho_{FF}(t) \sim t^{-p_F} f(t/\tau_{exp}) \qquad \longleftrightarrow \qquad G_{FF}(x) \sim x^{-(d-2+\eta_F)} f(x/\xi)$$
$$\Downarrow \qquad\qquad\qquad\qquad\qquad\qquad \Downarrow$$
$$z_{int,F} = (1-p_F)z_{exp} \qquad\qquad\qquad \gamma_F = (2-\eta_F)\nu$$

---

[13]Even for observables F that are orthogonal to the slowest mode, one typically expects that $\tau_{exp,F} \sim \tau_{exp}$. For example, in any linear stochastic iteration for a Gaussian model, the observables $:\varphi^n:$ have an exponential autocorrelation time which is exactly $1/n$ times that of the slowest mode [13, Section VIII]. So it is not necessary to define distinct exponents $z_{exp,F}$; they will all equal z_{exp}.

So one expects $z_{int,F} < z_{exp}$ except in the special case when $p_F = 0$. This latter does occur in some simple examples (e.g. in most algorithms for the Gaussian model), but I see no reason for it to occur even in the single-site Metropolis or heat-bath dynamics for the two-dimensional Ising model! It is thus crucial, when reporting estimates of dynamic critical exponents, to distinguish between $z_{int,F}$ and z_{exp} (and to specify F!).

It is worth noting that the phenomenon of critical slowing-down is not confined to Monte Carlo simulations: very similar difficulties were encountered more than 50 years ago by numerical analysts concerned with the numerical solution of partial differential equations (PDEs). For example, the single-site heat-bath algorithm is the stochastic analogue of the Gauss-Seidel algorithm in numerical analysis, and the dynamic critical behavior of the two algorithms is qualitatively similar (in the Gaussian case, it is quantitatively the same). Similarly, several of the algorithms invented by numerical analysts to reduce critical slowing-down in solving elliptic PDEs — successive overrelaxation (SOR) [14], Fourier preconditioning [39] and multi-grid (MG) [40, 41] — have stochastic analogues: stochastic overrelaxation, Fourier acceleration and multi-grid Monte Carlo (MGMC). These algorithms are discussed in Sections 3.4, 4.1 and 4.2, respectively.

3.4   Stochastic Overrelaxation

An excellent survey of overrelaxation algorithms has been given by Adler [2], so I shall limit myself to sketching the main ideas.

Let us start with a problem of classical numerical analysis: solving a linear system $A\varphi = f$, where A is a given symmetric positive-definite matrix and f is a given vector. This is equivalent to the problem of minimizing the Hamiltonian

$$H(\varphi) = \tfrac{1}{2}(\varphi, A\varphi) - (f, \varphi). \tag{63}$$

The Gauss-Seidel algorithm attacks this problem by sweeping through the sites i in some chosen order, at each stage replacing φ_i by that new value φ'_i which minimizes H when the other fields $\{\varphi_j\}_{j \neq i}$ are held fixed at their current values. In other words, if we write

$$H(\varphi_i, \{\varphi_j\}_{j \neq i}) = \tfrac{1}{2}a_i(\varphi_i - b_i)^2 + c_i, \tag{64}$$

where $a_i = A_{ii} > 0$ and b_i, c_i are easily computable functions of $\{\varphi_j\}_{j \neq i}$, then the Gauss-Seidel algorithm involves replacing φ_i by $\varphi'_i = b_i$. This algorithm has severe critical slowing-down: for $A = -\Delta + m^2$, the convergence time ($\equiv$ number of sweeps needed to reduce the error by a factor e) grows as m^{-2} [14]. That is, the dynamic critical exponent is $z = 2$, as predicted by the heuristic random-walk argument.

The numerical analysts have, however, invented a better scheme: instead of replacing φ_i by the value b_i that minimizes $H(\cdot, \{\varphi_j\}_{j \neq i})$, let us intentionally *overshoot* this value by a factor $\omega > 1$. That is, let us take

$$\varphi'_i = \omega b_i + (1 - \omega)\varphi_i. \tag{65}$$

Note that if $\omega < 2$, the energy still decreases, although not as much as it would with $\omega = 1$. This scheme is called *successive overrelaxation (SOR)* [14].

It is far from obvious that this short-run asceticism (choosing an apparently "sub-optimal" value of φ_i') will lead to long-run bounty (faster convergence to the minimum of H). But this turns out to be the case, as a result of very subtle coherent effects. By either Fourier [42] or matrix [14] methods, one can show that the optimal choice of ω behaves as

$$\omega_{opt} \approx \frac{2}{1 + Cm} \tag{66}$$

when $m \to 0$ (here C is a computable constant), and that with this choice the convergence time grows only as m^{-1}. That is, the dynamic critical exponent is $z = 1$.[14]

Both the Gauss-Seidel and SOR algorithms have stochastic analogues. The stochastic analogue of Gauss-Seidel is the single-site heat-bath algorithm: the new value φ_i' is a random sample from $e^{-\beta H(\cdot, \{\varphi_j\}_{j \neq i})}$, i.e. it is a Gaussian random variable of mean b_i and variance $1/\beta a_i$. Similarly, the stochastic analogue of SOR is *stochastic SOR* [43]: the new value φ_i' is a Gaussian random variable of mean $\omega b_i + (1 - \omega)\varphi_i$ and variance $\omega(2 - \omega)/\beta a_i$. A simple computation shows that this update satisfies detailed balance with respect to the distribution $e^{-\beta H(\cdot, \{\varphi_j\}_{j \neq i})}$. The general analysis of stochastic linear iterations for the Gaussian model (Section 1.4) then shows that the autocorrelation time τ_{exp} of stochastic SOR is exactly equal to the convergence time of deterministic SOR, i.e. the dynamic critical exponent is $z = 1$.[15]

The stochastic SOR algorithm has an obvious extension to Hamiltonians that are "multiquadratic", i.e. H is quadratic in each φ_i when the $\{\varphi_j\}_{j \neq i}$ are considered fixed. For non-multiquadratic models there is no *direct* extension of stochastic SOR; rather, a variety of schemes have been proposed (notably for σ-models and lattice gauge theories), each of which attempts to mimic the *essential physics* of Gaussian stochastic SOR. I refer the reader to Adler [2] for a thorough review. Numerical experiments have yielded $z \approx 1$ for a one-dimensional $O(4)$ σ-model [44], $z \approx 1.2$ for a two-dimensional XY model [45], and $z \approx 1.1$ for a two-dimensional $O(3)$ σ-model [46]. So it seems possible to achieve z near to, though perhaps not quite equal to, the free-field value of 1. More detailed studies of the dynamic critical behavior of stochastic SOR (as a function of β, L and ω) on a variety of non-Gaussian models would be welcome.

---

[14]The reader may wonder how the SOR algorithm escapes the heuristic random-walk argument leading to $z = 2$ (Section 3.3). The answer, which arises when one looks at the rigorous version of the random-walk argument [31], is that the putative "random walk" corresponding to the SOR algorithm would have negative probabilities.

[15]It is crucial here that the updates are carried out sequentially in a fixed order [cf. (56)]. If instead the sites are selected randomly [cf. (55)], the coherence is destroyed, and it is easy to show that $z = 2$.

234

3.5 Langevin Algorithm

The Langevin algorithm is discussed in detail in Creutz' chapter in this book, so I shall limit myself to a brief introduction.

Consider, for starters, a field φ taking values in a linear configuration space — for example, a Gaussian or φ^4 model — with Hamiltonian $H(\varphi)$. Now consider the first-order stochastic differential equation

$$\frac{d\varphi_i}{dt} = -\frac{1}{2}\sum_j C_{ij}\frac{\partial H}{\partial \varphi_j} + \xi_i \,, \tag{67}$$

where i denotes both spatial and internal indices, C is a fixed symmetric positive-definite matrix, and ξ is Gaussian noise with mean zero and covariance

$$\langle \xi_i(t)\xi_j(t')\rangle = C_{ij}\delta(t-t') \,. \tag{68}$$

(For now the reader may assume that $C_{ij} = \delta_{ij}$; but the added generality will be useful later when we discuss Fourier acceleration.) The equation (67)–(68) defines a kind of "Aristotelian dynamics with noise"; among Monte Carlists it is called the *Langevin equation*.[16] Because the noise ξ is independent for distinct times t, the stochastic process $\{\varphi(t)\}$ is a *Markov process* (of diffusion type). It is not difficult to show that the transition probability kernel $P_t(\varphi \to \varphi')$ satisfies detailed balance with respect to the Gibbs measure $\mu_{eq}(\varphi) = Z^{-1}e^{-H(\varphi)}$; in particular, the Gibbs measure is an equilibrium distribution for the Markov process. Under mild conditions on H, it is the unique equilibrium distribution, and $\lim_{t\to+\infty} P_t(\varphi \to \varphi') = \mu_{eq}(\varphi')$ for all initial conditions φ.

For computer simulation, the stochastic differential equation (67) must be replaced by a finite-difference (discrete-time) approximation. The simplest such approximation is the "forward Euler" discretization

$$\varphi_i^{(n+1)} = \varphi_i^{(n)} - \frac{\epsilon}{2}\sum_j C_{ij}\frac{\partial H}{\partial \varphi_j}(\varphi^{(n)}) + \sqrt{\epsilon}\,\xi_i^{(n)} \,, \tag{69}$$

where ϵ is the time-step size (so that $t = n\epsilon$) and ξ is Gaussian noise with mean zero and covariance

$$\langle \xi_i^{(m)}\xi_j^{(n)}\rangle = C_{ij}\delta_{mn} \,. \tag{70}$$

The stochastic difference equation (69)–(70) defines a Markov chain whose one-step transition probability is

$$P_{Euler}(\varphi \to \varphi') = \text{const} \times \exp\left[-\frac{1}{2\epsilon}\sum_{i,j}\alpha_i(C^{-1})_{ij}\alpha_j\right] \,, \tag{71}$$

[16]This appellation seems to be historically incorrect: as far as I can make out, Langevin wrote a *second-order* stochastic differential equation for the particle positions, namely "Newtonian dynamics with friction and noise". Of course, in the absence of an external potential, this is equivalent to an equation of the form (67) for the *velocities*.

where

$$\alpha_i \equiv \varphi_i' - \varphi_i + \frac{\epsilon}{2} \sum_j C_{ij} \frac{\partial H}{\partial \varphi_j}(\varphi^{(n)}) . \tag{72}$$

Unfortunately, the equilibrium distribution of this Markov chain is *not* μ_{eq}, but some other measure $\mu_{Euler}(\varphi) = \text{const} \times \exp[-H_{eff,Euler}(\varphi)]$. For small ϵ, perturbation theory yields [47][17]

$$H_{eff,Euler}(\varphi) = H(\varphi) + \frac{\epsilon}{8} \sum_{i,j} C_{ij} \left[2\frac{\partial^2 H}{\partial\varphi_i \partial\varphi_j} - \frac{\partial H}{\partial\varphi_i}\frac{\partial H}{\partial\varphi_j} \right] + O(\epsilon^2) . \tag{73}$$

Thus, if our goal is to produce random samples from the Gibbs measure μ_{eq}, use of (69)–(70) leads to a systematic error of order ϵ. By a higher-order discretization (second-order stochastic Runge-Kutta algorithm), this systematic error can be reduced to order ϵ^2 [48].

The Langevin dynamics (67)–(68) and its discrete-time approximations can also be defined for fields living in a nonlinear manifold, as arise in a nonlinear σ-model or lattice gauge theory. The subtlety here is that the noise covariance matrix C_{ij} may now depend on the field configuration φ (it is a kind of "metric tensor" in the manifold), and it is necessary therefore to distinguish between the Itô and Stratonovich interpretations of the stochastic differential equation [49, 50, 51].

Early workers using the Langevin dynamics took the approach of running the simulation with several small step sizes ϵ and then attempting an extrapolation to $\epsilon = 0$. In this way they hoped to eliminate, or at least reduce, the systematic error associated with the finite step-size. The trouble is that the error bars (for fixed CPU time) grow as $\epsilon^{-1/2}$ as $\epsilon \to 0$, so that the extrapolation becomes very noisy.

Then, around 1986, it was recognized [52] that there is a very simple solution to the problem of systematic error: we use P_{Euler} (or a higher-order discretization) as the *proposal matrix* $P^{(0)}$ in the Metropolis-Hastings procedure, and then accept or reject the proposed move in accordance with the Metropolis formula (14)/(16). Since P_{Euler} *almost* satisfies detailed balance for the Gibbs measure μ_{eq}, one may expect that the acceptance fraction will be near 1 if the time-step size ϵ is sufficiently small. A more careful discussion [53] shows that for a (noncritical) lattice system of volume $V = L^d$, the time-step size ϵ must be taken $\lesssim V^{-1/3}$ to have a reasonable acceptance rate; this yields an autocorrelation time of order $1/\epsilon \sim V^{1/3}$ in units of sweeps, or $V^{4/3}$ in CPU units.

The principal drawback of the Langevin algorithm (in either its original or corrected version) is the same as that of other local algorithms: severe critical slowing-down. With $C_{ij} = \delta_{ij}$ (the so-called "unaccelerated" algorithm), the heuristic argument in Section 3.3 suggests that the dynamic critical exponent z is approximately 2. For the free field (Gaussian model), a direct calculation using the theory of Section 1.4 confirms that $z = 2$ (see Section 4.1). Some non-Gaussian models can be analyzed by dynamic renormalization-group methods [54].

---

[17]In general, (71) does *not* satisfy detailed balance for μ_{Euler}, i.e. the Markov chain is not reversible.

236

3.6   Hybrid Algorithm

The hybrid algorithm is discussed in detail in Creutz' chapter in this book as well as in [55], so I shall limit myself to a brief introduction.

Recall that the slow convergence of the Metropolis and Langevin algorithms arises because they perform a random walk in configuration space, leading to a dynamic critical exponent $z \approx 2$. If we could perform instead a purposeful (deterministic) walk through configuration space, then we might expect to obtain $z \approx 1$. We do this by introducing "canonical momenta" π_i and seeking to simulate the probability distribution

$$\mu_{eq}(\varphi, \pi) = \mathcal{Z}^{-1} e^{-\mathcal{H}(\varphi, \pi)} d\varphi \, d\pi \qquad (74)$$

defined by the "Hamiltonian"

$$\mathcal{H}(\varphi, \pi) = \tfrac{1}{2} \sum_{i,j} \pi_i C_{ij} \pi_j + H(\varphi), \qquad (75)$$

where C is a fixed symmetric positive-definite matrix. (For now the reader may assume that $C_{ij} = \delta_{ij}$; but the added generality will be useful later when we discuss Fourier acceleration.) Note that in this distribution φ and π are independent random variables; nevertheless, we make the apparently perverse choice to update them in a way that couples them, namely Hamilton's equations:

$$\frac{d\varphi_i}{dt} = \frac{\partial \mathcal{H}}{\partial \pi_i} = \sum_j C_{ij} \pi_j \qquad (76a)$$

$$\frac{d\pi_i}{dt} = -\frac{\partial \mathcal{H}}{\partial \varphi_i} = -\frac{\partial H}{\partial \varphi_i} \qquad (76b)$$

This *molecular dynamics* (or *microcanonical*) updating leaves invariant the "Hamiltonian" $\mathcal{H}(\varphi, \pi)$ as well as the phase-space volume element $d\varphi \, d\pi$, so it preserves μ_{eq}. The trouble is that it is not ergodic (at best it could be ergodic on surfaces of constant $\mathcal{H}$). So every once in a while we must supplement the molecular-dynamics moves by *momentum refreshments* in which new π's, independent of the old ones, are chosen from the Gaussian distribution $e^{-\frac{1}{2}(\pi, C\pi)}$. The combined dynamics is called the *hybrid dynamics* [56].

For computer simulation, Hamilton's equations (76) must be replaced by a finite-difference (discrete-time) approximation. As in the Langevin algorithm, this leads to systematic errors arising from the finite time-step size ϵ. The solution is the same [57]: use some discrete-time approximation to Hamilton's equations as a *proposal matrix* in the Metropolis-Hastings procedure, then accept or reject in the usual way. More precisely, we want the discrete-time evolution $f: (\varphi, \pi) \mapsto (\varphi', \pi')$ to satisfy:

- *Reversibility.*   If $f(\varphi, \pi) = (\varphi', \pi')$, then $f(\varphi', -\pi') = (\varphi, -\pi)$.

- *Preservation of phase-space volumes.* $\det \dfrac{\partial(\varphi', \pi')}{\partial(\varphi, \pi)} = \pm 1$.

Then the proposed move is accepted with probability[18]

$$a\big((\varphi,\pi)\to(\varphi',\pi')\big) \;=\; \min\Big(1,\, e^{-[\mathcal{H}(\varphi',\pi')-\mathcal{H}(\varphi,\pi)]}\Big)\,. \tag{77}$$

As $\epsilon\to 0$ the discrete-time evolution becomes energy-conserving ($\Delta\mathcal{H}=0$), and the acceptance rate tends to 1.

The hybrid dynamics with accept-reject test is often called the Hybrid Monte Carlo (HMC) algorithm. But this name is misleading, since *all* of the algorithms at issue here — with or without the accept-reject test — are Monte Carlo algorithms. A more accurate name is the *exact* (or *corrected*) hybrid algorithm.

There exist reversible volume-preserving schemes for Hamilton's equations that are accurate to arbitrarily high order in the time-step ϵ [58]. The simplest such scheme is the *leapfrog* method: using this scheme, the time-step size ϵ must be taken $\lesssim V^{-1/4}$ to have a reasonable acceptance rate [59]; this yields an autocorrelation time of order $1/\epsilon \sim V^{1/4}$ in units of sweeps, or $V^{5/4}$ in CPU units. Higher-order schemes can reduce the exponent $5/4$ arbitrarily close to 1.

The dynamic critical behavior of the hybrid algorithm can be studied analytically in the Gaussian case. Suppose that $H(\varphi)=\frac{1}{2}(\varphi,A\varphi)$, where A is a symmetric positive-definite matrix having eigenvalues $\omega_{min}^2\equiv\omega_1^2\le\omega_2^2\le\ldots\le\omega_N^2\equiv\omega_{max}^2$. [For the usual free field, $A=-\Delta+m^2$ so that $\omega_{min}=m$ and $\omega_{max}\sim 1$.] Suppose, next, that $C_{ij}=\delta_{ij}$. And suppose, finally, that the momentum refreshings occur at exponentially-distributed *random* time intervals of mean T_0. Then a straightforward calculation [60] shows that the autocorrelation time is

$$\tau_{exp} \;=\; \begin{cases} \dfrac{2T_0}{1-\sqrt{1-(2\omega_{min}T_0)^2}} & \text{for } T_0\le 1/2\omega_{min} \\[2ex] 2T_0 & \text{for } T_0\ge 1/2\omega_{min} \end{cases} \tag{78}$$

In particular, the optimal mean refreshing interval is $T_{0,opt}=1/2\omega_{min}$, and this choice gives $\tau_{exp}=1/\omega_{min}=2T_0$. On the other hand, for stability of the discretized algorithm we must take $\epsilon\lesssim 1/\omega_{max}$. Therefore, each molecular-dynamics trajectory ($\sim$ autocorrelation time) involves $\sim T_0/\epsilon \sim \omega_{max}/\omega_{min}$ lattice sweeps; so the autocorrelation time (in sweep units) for the optimally-tuned hybrid algorithm is $\sim \omega_{max}/\omega_{min} \sim m^{-1}$. That is, the dynamic critical exponent is $z=1$, as originally hoped.

For non-Gaussian theories, little is known about the dynamic critical behavior of the hybrid algorithm. Based on the free-field results, it is reasonable to conjecture that the optimal mean refreshing interval is of order $\xi^{\approx 1}$ (here ξ is the spatial correlation

---

[18]Here we have applied the Metropolis-Hastings procedure not to f itself, but to f followed by a momentum reversal $\pi\mapsto-\pi$; and we have used the fact that $\mathcal{H}(\varphi,\pi)=\mathcal{H}(\varphi,-\pi)$. Of course, this momentum reversal is irrelevant since it will be followed immediately by a momentum refreshing. (Alternatively, the molecular-dynamics/accept-reject move can be followed immediately by an "extra" momentum reversal. But this "extra" momentum reversal must then be carried out irrespective of whether the proposed molecular-dynamics move was accepted or rejected.)

length), and that this choice leads to a dynamic critical exponent $z \approx 1$. An initial study of the two-dimensional XY model [61] is consistent with these conjectures; but much more study is needed.

4 Collective-Mode Algorithms

In Section 3.3 we learned that the simple local algorithms are unfortunately extremely inefficient in the vicinity of a critical point: they suffer from severe critical slowing-down. How can this difficulty be overcome?

Our heuristic analysis of the *physics* of critical slowing-down told us that the slow modes are the long-wavelength modes, if the updating is purely local. (This is also borne out by the exact solution of the heat-bath or Langevin dynamics in the free-field case.) The natural solution is therefore to speed up those modes by some sort of *collective-mode* (nonlocal) updating. It is necessary, then, to *identify physically* the appropriate collective modes, and to devise an *efficient computational algorithm* for speeding up those modes. These two goals are unfortunately in conflict: it is very difficult to devise collective-mode algorithms that are not so nonlocal that their increased computational complexity per iteration outweighs the reduction in critical slowing-down. (For example, in $d = 4$, an algorithm that eliminates critical slowing-down but has computational complexity $O(V^2)$ is as bad as an $O(V)$ algorithm whose dynamic critical exponent is $z = 4$ — i.e. much *worse* than the conventional algorithms.)

Specific implementations of the collective-mode idea are thus highly model-dependent: one has to use one's knowledge of the *physics* of a given model to build into the algorithm the collective modes that *the system wants* — and do so at not too great a computational cost. Four classes of collective-mode algorithms have been invented so far, and found to be advantageous in at least some models:

- Fourier acceleration [62, 47]

- Multi-grid Monte Carlo (MGMC) [63, 13, 24, 23]

- Auxiliary-variable (cluster) algorithms [64, 65]

- Embedding algorithms [65, 66]

Fourier acceleration and MGMC are very similar in spirit (though quite different technically). Their performance is thus probably qualitatively similar, in the sense that they probably work well for the same models and work badly for the same models. We discuss these methods in Sections 4.1 and 4.2, and compare them briefly in Section 4.3. The remaining sections are devoted to auxiliary-variable and embedding algorithms.

4.1 Fourier Acceleration

Consider, for starters, the Langevin dynamics (67) in the Gaussian case $H(\varphi) = \frac{1}{2}(\varphi, A\varphi)$. This linear stochastic differential equation can be solved exactly [imitating

(24)/(25)]; the autocorrelation time is $\tau_{exp} = 2/\lambda_{min}$, where λ_{min} is the smallest eigenvalue of the matrix CA (or what is equivalent, the symmetric matrix $C^{1/2}AC^{1/2}$). On the other hand, for stability we must take the time-step size $\epsilon \lesssim 1/\lambda_{max}$. Thus, each autocorrelation time involves $\sim \tau_{exp}/\epsilon \sim \lambda_{max}/\lambda_{min}$ lattice sweeps. For $A = -\Delta + m^2$ and $C = I$, this means $\sim m^{-2}$ lattice sweeps, which is the familiar dynamic critical exponent $z = 2$. On the other hand, it is now obvious how to eliminate this critical slowing-down: take $C = A^{-1}$, so that we have $\tau_{exp}/\epsilon \sim \lambda_{max}/\lambda_{min} \sim 1$. In physical terms, by taking $C = A^{-1}$ we have made a momentum-dependent change of time scales so that all Fourier modes of the field φ equilibrate at roughly the same rate (namely $\tau \sim 1$). This idea is called *Fourier acceleration* [62, 47]. An analogous trick works for the hybrid algorithm [67]. Note that any choice $C = F(\nabla)$ can be implemented efficiently (in periodic boundary conditions) using the fast Fourier transform (FFT); each iteration takes a time of order $V \log V$.

What about non-Gaussian models? It is reasonable to conjecture that a good choice for the acceleration kernel is $C = (-\Delta + \widetilde{m}^2)^{-\kappa}$ with $\widetilde{m} \sim \min(\xi, L)^{-1}$ and $\kappa \approx 1$ (or possibly $\kappa \approx 1 - \eta/2$ where η is the static critical exponent governing the spin-spin correlation function at criticality). Preliminary studies of the two-dimensional XY [47, 67, 68] and $SU(3)$ chiral models [69] indicate that Fourier acceleration does succeed in reducing significantly the autocorrelation time of the Langevin and hybrid algorithms. But a serious measurement of the dynamic critical exponent z for a Fourier-accelerated Langevin or hybrid algorithm has yet to be performed. (Any such measurement should use the exact versions of the algorithms.)

Fourier acceleration applies without change to *Abelian* gauge theories: the point is that the *change* in an Abelian gauge field is a gauge-invariant quantity, so its Fourier modes have a physical meaning [70]. For a non-Abelian theory, by contrast, Fourier acceleration can only work in a fixed smooth gauge, such as Landau gauge [71]. Again, detailed numerical studies are badly needed.

4.2   Multi-Grid Algorithms

As mentioned previously, critical slowing-down was already encountered long ago by numerical analysts concerned with the numerical solution of partial differential equations. An ingenious solution, now called the multi-grid (MG) method, was proposed in 1964 by the Soviet numerical analyst Fedorenko [72]: the idea is to consider, in addition to the original ("fine-grid") problem, a sequence of auxiliary "coarse-grid" problems that approximate the behavior of the original problem for excitations at successively longer length scales (a sort of "coarse-graining" procedure). The local updates of the traditional algorithms are then supplemented by coarse-grid updates. To a present-day physicist, this philosophy is remarkably reminiscent of the renormalization group — so it is all the more remarkable that it was invented two years before the work of Kadanoff [73] and seven years before the work of Wilson [74]! After a decade of dormancy, multi-grid was revived in the mid-1970's [75], and was shown to be an extremely efficient computational method. In the early 1980's, multi-grid methods became an active area of research in numerical analysis, and were applied to a wide variety of problems in *classical* physics [40, 76]. Then, in 1986, it was shown [63] how a stochastic generalization of the multi-grid method — multi-grid Monte Carlo (MGMC) — can be applied to problems in *statistical*, and hence also Euclidean *quantum*, physics.

In this section we begin by giving a brief introduction to the deterministic multi-grid method; we then explain the stochastic analogue.[19]

Before entering into details, let us emphasize that although the multi-grid method and the block-spin renormalization group (RG) are based on very similar *philosophies* — dealing with a single length scale at a time — they are in fact *very different*. In particular, the conditional coarse-grid Hamiltonian employed in the MGMC method is *not* the same as the renormalized Hamiltonian given by a block-spin RG transformation. The RG transformation computes the *marginal*, not the conditional, distribution of the block means — that is, it *integrates* over the complementary degrees of freedom, while the MGMC method *fixes* these degrees of freedom at their current (random) values. The conditional Hamiltonian employed in MGMC is given by an explicit finite expression, while the marginal (RG) Hamiltonian cannot be computed in closed form. The failure to appreciate these distinctions has unfortunately led to much confusion in the literature.[20]

4.2.1 Deterministic Multi-Grid

Suppose we want to solve the lattice Poisson equation $-\Delta\varphi = f$ in a region $\Omega \subset \mathbf{Z}^d$ with zero Dirichlet data:

$$(-\Delta\varphi)_x \equiv 2d\varphi_x - \sum_{x': |x-x'|=1} \varphi_{x'} = f_x \tag{79}$$

for $x \in \Omega$, with $\varphi_x \equiv 0$ for $x \notin \Omega$. One simple iterative algorithm arises by solving (79) repeatedly for φ_x:

$$\varphi_x^{(n+1)} = \frac{1}{2d} \left[\sum_{x': |x-x'|=1} \varphi_{x'}^{(n)} + f_x \right]. \tag{80}$$

(80) is called the *Jacobi iteration*. It is convenient to consider also a slight generalization of (80): let $0 < \omega \le 1$, and define

$$\varphi_x^{(n+1)} = (1-\omega)\varphi_x^{(n)} + \frac{\omega}{2d} \left[\sum_{x': |x-x'|=1} \varphi_{x'}^{(n)} + f_x \right]. \tag{81}$$

(81) is called the *damped Jacobi iteration* with damping parameter ω.

---

[19]For an excellent introduction to the deterministic multi-grid method, see Briggs [41]; more advanced topics are covered in the book of Hackbusch [40]. Both MG and MGMC are discussed in detail in [13].

[20]For further discussion, see [13, Section 10.1].

If the domain Ω is a square $\{1,\ldots,L\} \times \{1,\ldots,L\}$, we can solve exactly for the eigenvectors and eigenvalues of the damped Jacobi iteration: they are

$$\varphi_x^{(p)} = \sin p_1 x_1 \sin p_2 x_2 \tag{82}$$

$$\lambda_p = (1-\omega) + \frac{\omega}{2}(\cos p_1 + \cos p_2) \tag{83}$$

where $p_1, p_2 = \frac{\pi}{L+1}, \frac{2\pi}{L+1}, \ldots, \frac{L\pi}{L+1}$. The convergence rate is governed by the eigenvalue of largest magnitude, namely

$$\lambda_{\frac{\pi}{L+1}, \frac{\pi}{L+1}} = 1 - \omega\left[1 - \cos\frac{\pi}{L+1}\right]$$
$$= 1 - O(L^{-2}). \tag{84}$$

It follows that $O(L^2)$ iterations are needed for the damped Jacobi iteration to converge adequately.

This behavior is in precise agreement with the heuristic predictions of Section 3.3: the long-wavelength modes ($p \ll 1$) are very slow ($\lambda_p \approx 1 - cp^2$). On the other hand, it is important to note that the *only* slow modes in the damped Jacobi iteration are the long-wavelength modes (provided that ω is not near 1): as long as, say, $\max(p_1, p_2) \geq \frac{\pi}{2}$, we have $0 \leq \lambda_p \leq \frac{3}{4}$ (for $\omega = \frac{1}{2}$), independent of L. It follows that the short-wavelength components of the error $e^{(n)} \equiv \varphi^{(n)} - \varphi$ can be effectively killed by a few (say, five or ten) damped Jacobi iterations. The remaining error has primarily long-wavelength components, and so is slowly varying in x-space. But a slowly varying function can be well represented on a coarser grid: if, for example, we were told $e_x^{(n)}$ only at *even* values of x, we could nevertheless reconstruct with high accuracy the function $e_x^{(n)}$ at *all* x by, say, linear interpolation. This suggests an improved algorithm for solving (79): perform a few damped Jacobi iterations on the original grid, until the (unknown) error is smooth in x-space; then set up an auxiliary coarse-grid problem whose solution will be approximately this error (this problem will turn out to be a Poisson equation on the coarser grid); perform a few damped Jacobi iterations on the coarser grid; and then transfer (interpolate) the result back to the original (fine) grid and add it in to the current approximate solution.

There are two advantages to performing the damped Jacobi iterations on the coarse grid. Firstly, the iterations take less work, because there are fewer lattice points on the coarse grid (2^{-d} times as many for a factor-of-2 coarsening in d dimensions). Secondly, with respect to the coarse grid the long-wavelength modes no longer have such long wavelength: *their wavelength has been halved* (i.e. their wavenumber has been doubled). This suggests that those modes with, say, $\max(p_1, p_2) \geq \frac{\pi}{4}$ can be effectively killed by a few damped Jacobi iterations on the coarse grid. And then we can transfer the remaining (smooth) error to a yet coarser grid, and so on recursively. These are the essential ideas of the multi-grid method.

Let us now give a precise definition of the multi-grid algorithm. For simplicity we shall restrict attention to problems defined in variational form[21]: thus, the

---

[21]In fact, the multi-grid method can be applied to the solution of linear or nonlinear systems of

goal is to minimize a real-valued function ("Hamiltonian") $H(\varphi)$, where φ runs over some N-dimensional real vector space U. We shall treat quadratic and non-quadratic Hamiltonians on an equal footing. In order to specify the algorithm we must specify the following ingredients:

1) A sequence of *coarse-grid spaces* $U_M \equiv U$, U_{M-1}, U_{M-2}, ..., U_0. Here $\dim U_l \equiv N_l$ and $N = N_M > N_{M-1} > N_{M-2} > \cdots > N_0$.

2) *Prolongation* (or "interpolation") *operators* $p_{l,l-1}$: $U_{l-1} \to U_l$ for $1 \le l \le M$.

3) *Basic* (or "smoothing") *iterations* S_l: $U_l \times \mathcal{H}_l \to U_l$ for $0 \le l \le M$. Here $\mathcal{H}_l$ is a space of "possible Hamiltonians" defined on U_l; we discuss this in more detail below. The role of S_l is to take an approximate minimizer φ_l' of the Hamiltonian H_l and compute a new (hopefully better) approximate minimizer $\varphi_l'' = S_l(\varphi_l', H_l)$. [For the present we can imagine that S_l consists of a few iterations of damped Jacobi for the Hamiltonian H_l.] Most generally, we shall use two smoothing iterations, S_l^{pre} and S_l^{post}; they may be the same, but need not be.

4) *Cycle control parameters* (integers) $\gamma_l \ge 1$ for $1 \le l \le M$, which control the number of times that the coarse grids are visited.

We discuss these ingredients in more detail below.

The multi-grid algorithm is then defined recursively as follows:

> **procedure** $mgm(l, \varphi, H_l)$
> **comment** This algorithm takes an approximate minimizer φ of the Hamiltonian H_l, and overwrites it with a better approximate minimizer.
>
> $\varphi \leftarrow S_l^{pre}(\varphi, H_l)$
> **if** $l > 0$ **then**
>   compute $H_{l-1}(\,\cdot\,) \equiv H_l(\varphi + p_{l,l-1}\,\cdot\,)$
> $\psi \leftarrow 0$
>   **for** $j = 1$ **until** γ_l **do** $mgm(l-1, \psi, H_{l-1})$
> $\varphi \leftarrow \varphi + p_{l,l-1}\psi$
> **endif**
> $\varphi \leftarrow S_l^{post}(\varphi, H_l)$
> **end**

Here is what is going on: We wish to minimize the Hamiltonian H_l, and are given as input an approximate minimizer φ. The algorithm consists of three steps:

---

equations, whether or not these equations come from a variational principle. See, for example, [40] and [13, Section 2].

1) *Pre-smoothing*. We apply the basic iteration (e.g. a few sweeps of damped Jacobi) to the given approximate minimizer. This produces a better approximate minimizer in which the high-frequency (short-wavelength) components of the error have been reduced significantly. Therefore, the error, although still large, is *smooth* in x-space (whence the name "smoothing iteration").

2) *Coarse-grid correction*. We want to move rapidly towards the minimizer φ^* of H_l, using coarse-grid updates. Because of the pre-smoothing, the error $\varphi - \varphi^*$ is a smooth function in x-space, so it should be well approximated by fields in the range of the prolongation operator $p_{l,l-1}$. We will therefore carry out a coarse-grid update in which φ is replaced by $\varphi + p_{l,l-1}\psi$, where ψ lies in the coarse-grid subspace U_{l-1}. A sensible goal is to attempt to choose ψ so as to minimize H_l; that is, we attempt to minimize

$$H_{l-1}(\psi) \equiv H_l(\varphi + p_{l,l-1}\psi) \,. \tag{85}$$

To carry out this approximate minimization, we use a few (γ_l) iterations of the best algorithm we know — namely, multi-grid itself! And we start at the best approximate minimizer we know, namely $\psi = 0$! The goal of this coarse-grid correction step is to reduce significantly the low-frequency components of the error in φ (hopefully without creating large new high-frequency error components).

3) *Post-smoothing*. We apply, for good measure, a few more sweeps of the basic smoother. (This would protect against any high-frequency error components which may inadvertently have been created by the coarse-grid correction step.)

The foregoing constitutes, of course, a single step of the multi-grid algorithm. In practice this step would be repeated several times, as in any other iteration, until the error has been reduced to an acceptably small value. The advantage of multi-grid over the traditional (e.g. damped Jacobi) iterative methods is that, with a suitable choice of the ingredients $p_{l,l-1}$, $\mathcal{S}_l$ and so on, only a few (maybe five or ten) iterations are needed to reduce the error to a small value, *independent of the lattice size L*. This contrasts favorably with the behavior (84) of the damped Jacobi method, in which $O(L^2)$ iterations are needed.

The multi-grid algorithm is thus a general framework; the user has considerable freedom in choosing the specific ingredients, which must be adapted to the specific problem. We now discuss briefly each of these ingredients; more details can be found in Chapter 3 of the book of Hackbusch [40].

Coarse grids. Most commonly one uses a uniform factor-of-2 coarsening between each grid Ω_l and the next coarser grid Ω_{l-1}. The coarse-grid points could be either a subset of the fine-grid points (Fig. 1) or a subset of the dual lattice (Fig. 2). These schemes have obvious generalizations to higher-dimensional cubic lattices. Coarsenings by a larger factor (e.g. 3) could also be considered, but are generally disadvantageous. Note that each of the above schemes works also for periodic boundary conditions provided that the linear size L_l of the grid Ω_l is *even*. For this reason it is most convenient to take the linear size $L \equiv L_M$ of the original (finest) grid $\Omega \equiv \Omega_M$ to be a power of 2, or at least a power of 2 times a small integer.

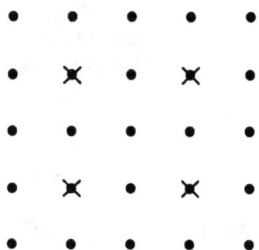

Figure 1: Standard coarsening (factor-of-2) in dimension $d = 2$. Dots are fine-grid sites and crosses are coarse-grid sites.

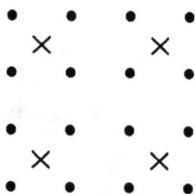

Figure 2: Staggered coarsening (factor-of-2) in dimension $d = 2$.

˘ *Prolongation operators.* For a coarse grid as in Fig. 2, a natural choice of prolongation operator is *piecewise-constant injection*:

$$(p_{l,l-1}\varphi_{l-1})_{x_1\pm\frac{1}{2},x_2\pm\frac{1}{2}} = (\varphi_{l-1})_{x_1,x_2} \qquad \text{for all } x \in \Omega_{l-1} \qquad (86)$$

(illustrated here for $d = 2$). It can be represented in an obvious shorthand notation by the stencil

$$\begin{bmatrix} 1 & 1 \\ 1 & 1 \end{bmatrix}. \qquad (87)$$

For a coarse grid as in Fig. 1, a natural choice is *piecewise-linear interpolation*, one example of which is the *nine-point prolongation*

$$\begin{bmatrix} \frac{1}{4} & \frac{1}{2} & \frac{1}{4} \\ \frac{1}{2} & 1 & \frac{1}{2} \\ \frac{1}{4} & \frac{1}{2} & \frac{1}{4} \end{bmatrix}. \qquad (88)$$

Higher-order interpolations (e.g. quadratic or cubic) can also be considered. All these prolongation operators can easily be generalized to higher dimensions.

We have ignored here some important subtleties concerning the treatment of the boundaries in defining the prolongation operator. Fortunately we shall not have to worry much about this problem, since most applications in quantum field theory and statistical mechanics use periodic boundary conditions.

Coarse-grid Hamiltonians. What does the coarse-grid Hamiltonian H_{l-1} look like? If the fine-grid Hamiltonian H_l is quadratic,

$$H_l(\varphi) = \tfrac{1}{2}(\varphi, A_l\varphi) - (f_l, \varphi), \qquad (89)$$

then so is the coarse-grid Hamiltonian H_{l-1}:

$$\begin{aligned} H_{l-1}(\psi) &\equiv H_l(\varphi + p_{l,l-1}\psi) & (90) \\ &= \tfrac{1}{2}(\psi, A_{l-1}\psi) - (d, \psi) + \text{const}, & (91) \end{aligned}$$

where

$$\begin{aligned} A_{l-1} &\equiv p_{l,l-1}^* A_l\, p_{l,l-1} & (92) \\ d &\equiv p_{l,l-1}^*(f - A_l\varphi) & (93) \end{aligned}$$

The coarse-grid problem is thus also a linear equation whose right-hand side is just the "coarse-graining" of the residual $r \equiv f - A_l\varphi$; this coarse-graining is performed using the adjoint of the interpolation operator $p_{l,l-1}$. The exact form of the coarse-grid operator A_{l-1} depends on the fine-grid operator A_l and on the choice of interpolation operator $p_{l,l-1}$. For example, if A_l is the nearest-neighbor Laplacian and $p_{l,l-1}$ is piecewise-constant injection, then it is easily checked that A_{l-1} is also a nearest-neighbor Laplacian (multiplied by an extra 2^{d-1}). On the other hand, if $p_{l,l-1}$ is piecewise-linear interpolation, then A_{l-1} will have nearest-neighbor *and next-nearest-neighbor* terms (but nothing worse than that).

Clearly, the point is to choose classes of Hamiltonians $\mathcal{H}_l$ with the property that if $H_l \in \mathcal{H}_l$ and $\varphi \in U_l$, then the coarse-grid Hamiltonian H_{l-1} defined by (85) necessarily lies in $\mathcal{H}_{l-1}$. In particular, it is convenient (though not in principle necessary) to choose all the Hamiltonians to have the same "functional form"; this functional form must be one which is stable under the coarsening operation (85). For example, suppose that the Hamiltonian H_l is a φ^4 theory with nearest-neighbor gradient term and possibly site-dependent coefficients:

$$H_l(\varphi) = \frac{\alpha}{2} \sum_{\langle xx' \rangle} (\varphi_x - \varphi_{x'})^2 + \sum_x V_x(\varphi_x), \tag{94}$$

where

$$V_x(\varphi_x) = \lambda \varphi_x^4 + \kappa_x \varphi_x^3 + A_x \varphi_x^2 + h_x \varphi_x. \tag{95}$$

Suppose, further, that the prolongation operator $p_{l,l-1}$ is piecewise-constant injection (86). Then the coarse-grid Hamiltonian $H_{l-1}(\psi) \equiv H_l(\varphi + p_{l,l-1}\psi)$ can easily be computed: it is

$$H_{l-1}(\psi) = \frac{\alpha'}{2} \sum_{\langle yy' \rangle} (\psi_y - \psi_{y'})^2 + \sum_y V_y'(\psi_y) + \text{const}, \tag{96}$$

where

$$V_y'(\psi_y) = \lambda' \psi_y^4 + \kappa_y' \psi_y^3 + A_y' \psi_y^2 + h_y' \psi_y \tag{97}$$

and

$$\alpha' = 2^{d-1}\alpha \tag{98}$$

$$\lambda' = 2^d \lambda \tag{99}$$

$$\kappa_y' = \sum_{x \in B_y} (4\lambda \varphi_x + \kappa_x) \tag{100}$$

$$A_y' = \sum_{x \in B_y} (6\lambda \varphi_x^2 + 3\kappa_x \varphi_x + A_x) \tag{101}$$

$$h_y' = \sum_{x \in B_y} (4\lambda \varphi_x^3 + 3\kappa_x \varphi_x^2 + 2A_x \varphi_x + h_x) + \alpha \sum_{\substack{x \in B_y \\ x' \notin B_y \\ |x - x'| = 1}} (\varphi_x - \varphi_{x'}) \tag{102}$$

Here B_y is the block consisting of those 2^d sites of grid Ω_l which are affected by interpolation from the coarse-grid site $y \in \Omega_{l-1}$ (see Figure 2). Note that the coarse-grid Hamiltonian H_{l-1} has the same functional form as the "fine-grid" Hamiltonian H_l: it is specified by the coefficients α', λ', κ_y', A_y' and h_y'. The step "*compute H_{l-1}*" therefore means to compute these coefficients. Note also the importance of allowing in (94) for φ^3 and φ terms and for site-dependent coefficients: even if these are not present in the original Hamiltonian $H \equiv H_M$, they will be generated on coarser grids. Finally, we emphasize that the coarse-grid Hamiltonian H_{l-1} depends implicitly on the current value of the fine-lattice field $\varphi \in U_l$; although our notation suppresses this dependence, it should be kept in mind.

Basic (smoothing) iterations. We have already discussed the damped Jacobi iteration as one possible smoother. More commonly one uses instead the Gauss-Seidel iteration: it is easier to program than Jacobi, requires only half the storage space, and can be shown to perform slightly better. (The only reason we introduced damped Jacobi at all is that it is easier to understand and to analyze.)

Thus, S_l^{pre} and S_l^{post} will consist, respectively, of m_1 and m_2 iterations of the Gauss-Seidel algorithm. The balance between pre-smoothing and post-smoothing is usually not very crucial; only the total $m_1 + m_2$ seems to matter much. Indeed, one (but not both!) of m_1 or m_2 could be zero, i.e. either the pre-smoothing or the post-smoothing could be omitted entirely. Increasing m_1 and m_2 improves the convergence rate of the multi-grid iteration, but at the expense of increased computational labor per iteration. The optimal tradeoff seems to be achieved in most cases with $m_1 + m_2$ between about 2 and 4. The coarsest grid Ω_0 is a special case: it usually has so few grid points (perhaps only one!) that S_0 can be an exact solver.

Cycle control parameters and computational labor. Usually the parameters γ_l are all taken to be equal, i.e. $\gamma_l = \gamma \geq 1$ for $1 \leq l \leq M$. Then one iteration of the multi-grid algorithm at level M comprises one visit to grid M, γ visits to grid $M-1$, γ^2 visits to grid $M-2$, and so on. Thus, γ determines the degree of emphasis placed on the coarse-grid updates. ($\gamma = 0$ would correspond to the pure Gauss-Seidel iteration on the finest grid alone.)

We can now estimate the computational labor required for one iteration of the multi-grid algorithm. Each visit to a given grid involves $m_1 + m_2$ Gauss-Seidel sweeps on that grid, plus some computation of the coarse-grid Hamiltonian and the prolongation. The work involved is proportional to the number of lattice points on that grid. Let W_l be the work required for these operations on grid l. Then, for grids defined by a factor-of-2 coarsening in d dimensions, we have

$$W_l \approx 2^{-d(M-l)} W_M , \qquad (103)$$

so that the total work for one multi-grid iteration is

$$
\begin{aligned}
work(MG) &= \sum_{l=M}^{0} \gamma^{M-l} W_l \\
&\approx W_M \sum_{l=M}^{0} (\gamma 2^{-d})^{M-l} \\
&\leq W_M (1 - \gamma 2^{-d})^{-1} \qquad \text{if } \gamma < 2^d .
\end{aligned}
\qquad (104)
$$

Thus, provided that $\gamma < 2^d$, the work required for one entire multi-grid iteration is no more than $(1 - \gamma 2^{-d})^{-1}$ times the work required for $m_1 + m_2$ Gauss-Seidel iterations (plus a little auxiliary computation) on the finest grid alone — *irrespective of the total number of levels*. The most common choices are $\gamma = 1$ (which is called the V-cycle) and $\gamma = 2$ (the W-cycle).

Convergence proofs. For certain classes of Hamiltonians H — primarily quadratic ones — and suitable choices of the coarse grids, prolongations, smoothing iterations and cycle control parameters, it can be proven rigorously[22] that the multi-grid iteration matrices M_l satisfy a *uniform* bound

$$\|M_l\| \le C < 1, \tag{105}$$

valid *irrespective of the total number of levels*. Thus, a fixed number of multi-grid iterations (maybe five or ten) are sufficient to reduce the error to a small value, *independent of the lattice size L*. In other words, *critical slowing-down has been completely eliminated*.

The rigorous convergence proofs are somewhat arcane, so we cannot describe them here in any detail, but certain general features are worth noting. The convergence proofs are most straightforward when linear or higher-order interpolation and restriction are used, and $\gamma > 1$ (e.g. the W-cycle). When either low-order interpolation (e.g. piecewise-constant) or $\gamma = 1$ (the V-cycle) is used, the convergence proofs become much more delicate. Indeed, if *both* piecewise-constant interpolation and a V-cycle are used, then the uniform bound (105) has *not* yet been proven, and it is most likely *false*! To some extent these features may be artifacts of the current methods of proof, but we suspect that they do also reflect real properties of the multi-grid method, and so the convergence proofs may serve as guidance for practice. For example, in our work we have used piecewise-constant interpolation (so as to preserve the simple nearest-neighbor coupling on the coarse grids), and thus for safety we stick to the W-cycle. There is in any case much room for further research, both theoretical and experimental.

To recapitulate, the extraordinary efficiency of the multi-grid method arises from the combination of two key features:

1) The convergence estimate (105). This means that only $O(1)$ iterations are needed, independent of the lattice size L.

2) The work estimate (104). This means that each iteration requires only a computational labor of order L^d (the fine-grid lattice volume).

It follows that the complete solution of the minimization problem, to any specified accuracy ε, requires a computational labor of order L^d.

---

[22]For a detailed exposition of multi-grid convergence proofs, see [40, Chapters 6–8, 10, 11], [77] and the references cited therein. The additional work needed to handle the piecewise-constant interpolation can be found in [78].

Unigrid point of view. Let us look again at the multi-grid algorithm from the variational standpoint. One natural class of iterative algorithms for minimizing H are the so-called *directional methods*: let $p_0, p_1, \ldots$ be a sequence of "direction vectors" in U, and define $\varphi^{(n+1)}$ to be that vector of the form $\varphi^{(n)} + \lambda p_n$ which minimizes H. The algorithm thus travels "downhill" from $\varphi^{(n)}$ along the line $\varphi^{(n)} + \lambda p_n$ until reaching the minimum of H, then switches to direction p_{n+1} starting from this new point $\varphi^{(n+1)}$, and so on. For a suitable choice of the direction vectors $p_0, p_1, \ldots$, this method can be proven to converge to the global minimum of H [79, pp. 513–520].

Now, some iterative algorithms for minimizing $H(\varphi)$ can be recognized as special cases of the directional method. For example, the Gauss-Seidel iteration is a directional method in which the direction vectors are chosen to be unit vectors $e_1, e_2, \ldots, e_N$ (i.e. vectors which take the value 1 at a single grid point and zero at all others), where $N = \dim U$. [One step of the Gauss-Seidel iteration corresponds to N steps of the directional method.] Similarly, it is not hard to see [80] that the multi-grid iteration with the variational choices of restriction and coarse-grid operators, and with Gauss-Seidel smoothing at each level, is itself a directional method: some of the direction vectors are the unit vectors $e_1^{(M)}, e_2^{(M)}, \ldots, e_{N_M}^{(M)}$ of the fine-grid space, but other direction vectors are the images in the fine-grid space of the unit vectors of the coarse-grid spaces, i.e. they are $p_{M,l} e_1^{(l)}, p_{M,l} e_2^{(l)}, \ldots, p_{M,l} e_{N_l}^{(l)}$. The exact order in which these direction vectors are interleaved depends on the parameters m_1, m_2 and γ which define the cycling structure of the multi-grid algorithm. For example, if $m_1 = 1$, $m_2 = 0$ and $\gamma = 1$, the order of the direction vectors is $\{M\}$, $\{M-1\}$, $\ldots$, $\{0\}$, where $\{l\}$ denotes the sequence $p_{M,l} e_1^{(l)}, p_{M,l} e_2^{(l)}, \ldots, p_{M,l} e_{N_l}^{(l)}$. If $m_1 = 0$, $m_2 = 1$ and $\gamma = 1$, the order is $\{0\}, \{1\}, \ldots, \{M\}$. The reader is invited to work out other cases.

Thus, the multi-grid algorithm (for problems defined in variational form) is a directional method in which the direction vectors include both "single-site modes" $\{M\}$ and also "collective modes" $\{M-1\}, \{M-2\}, \ldots, \{0\}$ on all length scales. For example, if $p_{l,l-1}$ is piecewise-constant injection, then the direction vectors are characteristic functions χ_B (i.e. functions which are 1 on the block $B \subset \Omega$ and zero outside B), where the sets B are successively single sites, cubes of side 2, cubes of side 4, and so on. Similarly, if $p_{l,l-1}$ is linear interpolation, then the direction vectors are triangular waves of various widths.

The multi-grid algorithm has thus an alternative interpretation as a collective-mode algorithm working solely in the fine-grid space U. We emphasize that this "unigrid" viewpoint [80] is mathematically fully equivalent to the recursive definition given earlier. But it gives, we think, an important additional insight into what the multi-grid algorithm is really doing.

For example, for the simple model problem (Poisson equation in a square), we know that the "correct" collective modes are sine waves, in the sense that these modes diagonalize the Laplacian, so that in this basis the Jacobi or Gauss-Seidel algorithm would give the exact solution in a *single* iteration. On the other hand, the multi-grid method uses square-wave (or triangular-wave) updates, which are not exactly the "correct" collective modes. Nevertheless, the multi-grid convergence proofs [40, 77, 78] assure us that they are "close enough": the norm of the multi-grid iteration matrix M_l is bounded away from 1, uniformly in the lattice size, so that an accurate solution is reached in a *very few* MG iterations (in particular, critical slowing-down is completely eliminated). This viewpoint also explains why MG convergence is more delicate for piecewise-constant interpolation than for piecewise-linear: the point is that a sine wave (or other slowly varying function) can be approximated to arbitrary accuracy (in energy norm) by piecewise-linear functions but *not* by piecewise-constant functions.

We remark that McCormick and Ruge [80] have advocated the "unigrid" idea not just as an alternate point of view on the multi-grid algorithm, but as an alternate *computational procedure*. To be sure, the unigrid method is somewhat simpler to program, and this could have pedagogical advantages. But one of the key properties of the multi-grid method, namely the $O(L^d)$ computational labor per iteration, is sacrificed in the unigrid scheme. Instead of (103)–(104) one has

$$W_l \approx W_M \qquad (106)$$

and hence

$$
\begin{aligned}
work(UG) \quad &\approx \quad W_M \sum_{l=M}^{0} \gamma^{M-l} \\
&\sim \quad \begin{cases} M W_M & \text{if } \gamma = 1 \\ \gamma^M W_M & \text{if } \gamma > 1 \end{cases} \qquad (107)
\end{aligned}
$$

Since $M \approx \log_2 L$ and $W_M \sim L^d$, we obtain

$$
work(UG) \quad \sim \quad \begin{cases} L^d \log L & \text{if } \gamma = 1 \\ L^{d+\log_2 \gamma} & \text{if } \gamma > 1 \end{cases} \qquad (108)
$$

For a V-cycle the additional factor of $\log L$ is perhaps not terribly harmful, but for a W-cycle the additional factor of L is a severe drawback (though not as severe as the $O(L^2)$ critical slowing-down of the traditional algorithms). Thus, we do *not* advocate the use of unigrid as a computational method if there is a viable multi-grid alternative. Unigrid could, however, be of interest in cases where true multi-grid is unfeasible, as may occur for non-Abelian lattice gauge theories.

Multi-grid algorithms can also be devised for some models in which state space is a nonlinear manifold, such as nonlinear σ-models and lattice gauge theories [13, Sections 3–5]. The simplest case is the XY model: both the fine-grid and coarse-grid field variables are *angles*, and the interpolation operator is piecewise-constant (with angles added modulo 2π). Thus, a coarse-grid variable ψ_y specifies the angle by which the 2^d spins in the block B_y are to be simultaneously rotated. A similar strategy can be employed for nonlinear σ-models taking values in a group G (the so-called "principal chiral models"): the coarse-grid variable ψ_y simultaneously left-multiplies the 2^d spins in the block B_y. For nonlinear σ-models taking values in a nonlinear manifold M on which a group G acts [e.g. the n-vector model with $M = S_{n-1}$ and $G = SO(n)$], the coarse-grid-correction moves are still simultaneous rotation; this means that while the fine-grid fields lie in M, the coarse-grid fields all lie in G. Similar ideas can be applied to lattice gauge theories; the key requirement is to respect the geometric (parallel-transport) properties of the theory. Unfortunately, the resulting algorithms appear to be practical only in the abelian case. (In the non-abelian case, the coarse-grid Hamiltonian becomes too complicated.) Much more work needs to be done on devising good interpolation operators for non-abelian lattice gauge theories.

4.2.2 Multi-Grid Monte Carlo

Classical equilibrium statistical mechanics is a natural generalization of classical statics (for problems posed in variational form): in the latter we seek to minimize a Hamiltonian $H(\varphi)$, while in the former we seek to generate random samples from the Boltzmann-Gibbs probability distribution $e^{-\beta H(\varphi)}$. The statistical-mechanical problem reduces to the deterministic one in the zero-temperature limit $\beta \to +\infty$.

Likewise, many (but not all) of the deterministic iterative algorithms for minimizing $H(\varphi)$ can be generalized to stochastic iterative algorithms — that is, dynamic Monte Carlo methods — for generating random samples from $e^{-\beta H(\varphi)}$. For example, the stochastic generalization of the Gauss-Seidel algorithm is the single-site heat-bath algorithm; and the stochastic generalization of multi-grid is multi-grid Monte Carlo.

Recall that in the Gauss-Seidel algorithm, the grid points are swept in some order, and at each stage the Hamiltonian is minimized as a function of a single variable φ_x, with all other variables $\{\varphi_y\}_{y\neq x}$ being held fixed. The single-site heat-bath algorithm has the same general structure, but the new value φ'_x is chosen randomly from the conditional distribution of $e^{-\beta H(\varphi)}$ given $\{\varphi_y\}_{y\neq x}$, i.e. from the one-dimensional probability distribution

$$P(\varphi'_x)\, d\varphi'_x = \text{const} \times \exp\left[-\beta H(\varphi'_x, \{\varphi_y\}_{y\neq x})\right] d\varphi'_x \qquad (109)$$

(where the normalizing constant depends on $\{\varphi_y\}_{y\neq x}$). It is not difficult to see that this operation leaves invariant the Gibbs distribution $e^{-\beta H(\varphi)}$. As $\beta \to +\infty$ it reduces to the Gauss-Seidel algorithm.

It is useful to visualize geometrically the action of the Gauss-Seidel and heat-bath algorithms within the space U of all possible field configurations. Starting at the current field configuration φ, the Gauss-Seidel and heat-bath algorithms propose to move the system along the line in U consisting of configurations of the form $\varphi' = \varphi + t\delta_x$ ($-\infty < t < \infty$), where δ_x denotes the configuration which is 1 at site x and zero elsewhere. In the Gauss-Seidel algorithm, t is chosen so as to minimize the Hamiltonian restricted to the given line; while in the heat-bath algorithm, t is chosen randomly from the the conditional distribution of $e^{-\beta H(\varphi)}$ restricted to the given line, namely the one-dimensional distribution with probability density $P_{cond}(t) \sim \exp[-H_{cond}(t)] \equiv \exp[-H(\varphi + t\delta_x)]$.

The method of partial resampling (Section 1.2) generalizes the heat-bath algorithm in two ways:

1) The "fibers" used by the algorithm need not be lines, but can be higher-dimensional linear or even nonlinear manifolds.

2) The new configuration φ' need not be chosen *independently* of the old configuration φ (as in the heat-bath algorithm); rather, it can be selected by any updating procedure which leaves invariant the conditional probability distribution of $e^{-\beta H(\varphi)}$ restricted to the fiber.

The multi-grid Monte Carlo (MGMC) algorithm is a partial-resampling algorithm in which the "fibers" are the sets of field configurations that can be obtained one from another by a coarse-grid-correction step, i.e. the sets of fields $\varphi + p_{l,l-1}\psi$ with φ fixed and ψ varying over U_{l-1}. These fibers form a family of parallel affine subspaces in U_l, of dimension $N_{l-1} = \dim U_{l-1}$.

The ingredients of the MGMC algorithm are identical to those of the deterministic MG algorithm, with one exception: the deterministic smoothing iteration S_l is replaced by a stochastic smoothing iteration (for example, single-site heat-bath). That is, $S_l(\cdot, H_l)$ is a stochastic updating procedure $\varphi_l \to \varphi_l'$ that leaves invariant the Gibbs distribution $e^{-\beta H_l}$. The MGMC algorithm is then defined as follows:

> **procedure** $mgmc(l, \varphi, H_l)$
> **comment** This algorithm updates the field φ in such a way as to
> leave invariant the probability distribution $e^{-\beta H_l}$.
>
> $\varphi \leftarrow S_l^{pre}(\varphi, H_l)$
> **if** $l > 0$ **then**
>   **compute** $H_{l-1}(\cdot) \equiv H_l(\varphi + p_{l,l-1} \cdot)$
> $\psi \leftarrow 0$
>   **for** $j = 1$ **until** γ_l **do** $mgmc(l-1, \psi, H_{l-1})$
> $\varphi \leftarrow \varphi + p_{l,l-1}\psi$
> **endif**
> $\varphi \leftarrow S_l^{post}(\varphi, H_l)$
> **end**

The alert reader will note that this algorithm is *identical* to the deterministic MG algorithm presented earlier; only the meaning of S_l is different.

The validity of the MGMC algorithm is proven inductively, starting at level 0 and working upwards. That is, if $mgmc(l-1, \cdot, H_{l-1})$ is a stochastic updating procedure that leaves invariant the probability distribution $e^{-\beta H_{l-1}}$, then $mgmc(l, \cdot, H_l)$ leaves invariant $e^{-\beta H_l}$. Note that the coarse-grid-correction step of the MGMC algorithm differs from the heat-bath algorithm in that the new configuration φ' is *not* chosen independently of the old configuration φ; to do so would be impractical, since the fiber has such high dimension. Rather, φ' (or what is equivalent, ψ) is chosen by a *valid updating procedure* — namely, MGMC itself!

The MGMC algorithm has also an alternate interpretation — the *unigrid* viewpoint — in which the fibers are one-dimensional and the resamplings are independent.

More precisely, the fibers are lines of the form $\varphi' = \varphi + t\chi_B$ ($-\infty < t < \infty$), where χ_B denotes the function which is 1 for sites belonging to the block B and zero elsewhere. The sets B are taken successively to be single sites, cubes of side 2, cubes of side 4, and so on. (If linear interpolation were used, then the "direction vectors" χ_B would be replaced by triangular waves of various widths.) Just as the deterministic unigrid algorithm chooses t so as to minimize the "conditional Hamiltonian" $H_{cond}(t) \equiv H(\varphi + t\chi_B)$, so the stochastic unigrid algorithm chooses t randomly from the one-dimensional distribution with probability density $P_{cond}(t) \sim \exp[-H_{cond}(t)]$. Conceptually this algorithm is no more complicated than the single-site heat-bath algorithm. But physically it is of course very different, as the direction vectors χ_B represent *collective modes* on all length scales.

We emphasize that the stochastic unigrid algorithm is *mathematically and physically equivalent* to the multi-grid Monte Carlo algorithm described above. But it is useful, we believe, to be able to look at MGMC from either of the two points of view: independent resamplings in one-dimensional fibers, or non-independent resamplings (defined recursively) in higher-dimensional (coarse-grid) fibers. On the other hand, the two algorithms are not *computationally* equivalent. One MGMC sweep requires a CPU time of order volume (provided that $\gamma < 2^d$), while the time for a unigrid sweep grows faster than the volume [cf. the work estimates (104) and (108)]. Therefore, we advocate unigrid only as a conceptual device, not as a computational algorithm.

How well does MGMC perform? The answer is highly model-dependent:

• For the *Gaussian model*, it can be proven rigorously [63, 13, 78] that τ is *bounded* as criticality is approached (empirically $\tau \approx 1 - 2$); therefore, critical slowing-down is *completely eliminated*. The proof is a simple Fock-space argument (Section 1.4) combined with the convergence proof for deterministic MG.

• For the φ^4 *model*, numerical experiments [63] show that τ diverges with the *same* dynamic critical exponent as in the heat-bath algorithm; the gain in efficiency thus approaches a *constant* factor $F(\lambda)$ near the critical point. This behavior can be understood [13, Section 9.1] as due to the double-well nature of the φ^4 potential, which makes MGMC ineffective on large blocks. Thus, the correct collective modes at long length scales are nonlinear excitations *not* well modelled by $\varphi \to \varphi + t\chi_B$. (See Section 4.5 for an algorithm that appears to model these excitations well, at least for λ not too small.)

• For the $d = 2$ *XY model*, the numerical data [24, 81] show a more complicated behavior: As the critical temperature is approached from above, τ diverges with a dynamic critical exponent $z_{int,\mathcal{M}^2} = 1.4 \pm 0.3$ for the MGMC algorithm (in either V-cycle or W-cycle), compared to $z_{int,\mathcal{M}^2} = 2.1 \pm 0.3$ for the heat-bath algorithm. Thus, critical slowing-down is significantly reduced but is still very far from being eliminated. On the other hand, below the critical temperature, τ is very small ($\approx 1 - 2$), uniformly in L and β (at least for the W-cycle); critical slowing-down appears to be completely eliminated. This very different behavior in the two phases can be understood physically: in the low-temperature phase the main excitations are spin waves, which are well handled by MGMC (as in the Gaussian model); but near the critical temperature the important excitations are widely separated vortex-antivortex pairs, which are apparently not easily created by the MGMC updates.

• For the $O(4)$ nonlinear σ-model in two dimensions, which is asymptotically free, the numerical data [23] show a very strong reduction, but not the total elimination, of critical slowing-down. For a W-cycle it is found that $z_{int,\mathcal{M}^2} = 0.60 \pm 0.07$. We are able [23, Section 4] to understand heuristically why $z \neq 0$: the critical slowing-down arises from the very slow (1/log) approach to free-field behavior in this non-superrenormalizable asymptotically-free model. Unfortunately, we are at present unable to predict theoretically, even very roughly, the value of z.

• For the CP^3 σ-model in two dimensions, a preliminary study of a unigrid algorithm with piecewise-linear updating and a V-cycle found $z_{int,\mathcal{M}^2} \approx 0.2 - 0.3$; while for the $SU(3)$ chiral model, confusing results were obtained [82]. Further studies are in progress.

4.3 Interlude: Fourier Acceleration vs. Multi-Grid

It is convenient to pause at this point to make a brief comparison of Fourier acceleration with multi-grid.

Fourier acceleration and multigrid are philosophically and physically very similar (though their technical details are quite different). Both are based on an intuition from the free-field (Gaussian) model, and both can be proven to eliminate completely the critical slowing-down in this model.[23] Both offer the system collective-mode updates of *fixed shape* (sine waves in Fourier acceleration, typically square or triangular waves in multigrid) on *all length scales*, and allow the system to choose the amplitude. (These algorithms make sense, therefore, only for systems of continuous-valued spins.) Both algorithms are expected to work well (i.e. have $z \approx 0$) for systems that are in some sense near-Gaussian, such as asymptotically free continuous-spin models or the low-temperature (spin-wave) phase of the two-dimensional XY model. Both algorithms are expected to work badly (i.e. have $z \approx 2$) for systems in which the dominant large-scale collective modes have discrete elements, such as spin-flips in the one-component φ^4 model or vortices in the two-dimensional XY model near the Kosterlitz-Thouless transition. In summary, the performance of Fourier acceleration and MGMC is probably very similar, in the sense that they probably work well for the same models and work badly for the same models; it is even conceivable that in many models they are in the same dynamic universality class.

The available studies of MGMC, discussed at the end of Section 4.2, are consistent with this general picture. Correspondingly detailed studies of the dynamic critical

---

[23]This is not, of course, such a great feat: the critical slowing-down in the Gaussian model can also be eliminated by throwing away the computer and solving the model analytically. I consider the Gaussian model, rather, as a constraint: if an algorithm does not work well for the Gaussian model, then it is unlikely to work well for near-Gaussian models (e.g. asymptotically free theories) either.

behavior of the Fourier-accelerated Langevin and hybrid algorithms have not yet been done — but they ought to be!

4.4 Auxiliary-Variable (Cluster) Algorithms

A very different type of collective-mode algorithm was proposed three years ago by Swendsen and Wang [64] for Potts spin models. Since then, there has been an explosion of work trying to understand why this algorithm works so well and why it does not work even better, and trying to improve or generalize it. The basic idea behind all algorithms of Swendsen-Wang type is to augment the given model by means of *auxiliary variables*, and then to simulate this augmented model. (Algorithms of this type are sometimes called "cluster algorithms", but this term is too narrow, because the relevant objects are not always clusters — see below.)

Consider the Hamiltonian for the ferromagnetic q-state Potts model [83]:

$$H(\sigma) = -\sum_{\langle ij \rangle} J_{ij} \left(\delta_{\sigma_i, \sigma_j} - 1 \right), \tag{110}$$

where $J_{ij} \geq 0$ for all i, j. The partition function is then

$$\begin{aligned} Z &= \sum_{\{\sigma\}} \exp\left[\sum_{\langle ij \rangle} J_{ij} \left(\delta_{\sigma_i, \sigma_j} - 1 \right) \right] \\ &= \sum_{\{\sigma\}} \prod_{\langle ij \rangle} \left[(1 - p_{ij}) + p_{ij} \delta_{\sigma_i, \sigma_j} \right] \end{aligned} \tag{111}$$

where we have defined $p_{ij} = 1 - \exp(-J_{ij})$. We now employ the deep identity

$$a + b = \sum_{n=0}^{1} \left[a \delta_{n,0} + b \delta_{n,1} \right], \tag{112}$$

where a and b are real numbers.[24] That is, we introduce on each bond $\langle ij \rangle$ an auxiliary variable n_{ij} taking the values 0 and 1, and obtain

$$Z = \sum_{\{\sigma\}} \sum_{\{n\}} \prod_{\langle ij \rangle} \left[(1 - p_{ij}) \delta_{n_{ij},0} + p_{ij} \delta_{n_{ij},1} \delta_{\sigma_i, \sigma_j} \right]. \tag{113}$$

Let us now take seriously the $\{n\}$ as dynamical variables: we can think of n_{ij} as an *occupation variable* for the bond $\langle ij \rangle$ (1 = occupied, 0 = empty). We therefore define the *Fortuin-Kasteleyn-Swendsen-Wang (FKSW) model* to be a joint model having q-state Potts spins σ_i at the sites and occupation variables n_{ij} on the bonds, with the joint probability distribution implied by (113). Finally, let us see what happens if we sum over the $\{\sigma\}$ at fixed $\{n\}$. Each occupied bond $\langle ij \rangle$ imposes a constraint that the

---

[24]This identity is valid in an arbitrary abelian semigroup, but such generality will not be needed here.

spins σ_i and σ_j must be in the same state, but otherwise the spins are unconstrained. We therefore group the sites into connected clusters (two sites are in the same cluster if they can be joined by a path of occupied bonds); then all the spins within a cluster must be in the same state (all q values are equally probable), and distinct clusters are independent. It follows that

$$Z = \sum_{\{n\}} \left(\prod_{\langle ij \rangle : \, n_{ij}=1} p_{ij} \right) \left(\prod_{\langle ij \rangle : \, n_{ij}=0} (1 - p_{ij}) \right) q^{\mathcal{C}(n)} , \qquad (114)$$

where $\mathcal{C}(n)$ is the number of connected clusters (including one-site clusters) in the graph whose edges are the bonds having $n_{ij} = 1$. The corresponding probability distribution is called the *random-cluster model with parameter q* [84]: it is a generalized bond-percolation model with non-local correlations coming from the factor $q^{\mathcal{C}(n)}$, and for $q = 1$ it reduces to ordinary bond percolation.

We have thus verified the following facts about the FKSW model:

a) $Z_{Potts} = Z_{FKSW} = Z_{RC}$.

b) The marginal distribution of μ_{FKSW} on the Potts variables $\{\sigma\}$ (integrating out the $\{n\}$) is precisely the Potts model $\mu_{Potts}(\sigma)$.

c) The marginal distribution of μ_{FKSW} on the bond occupation variables $\{n\}$ (integrating out the $\{\sigma\}$) is precisely the random-cluster model $\mu_{RC}(n)$.

The conditional distributions of μ_{FKSW} are also simple:

d) The conditional distribution of the $\{n\}$ given the $\{\sigma\}$ is as follows: independently for each bond $\langle ij \rangle$, one sets $n_{ij} = 0$ in case $\sigma_i \neq \sigma_j$, and sets $n_{ij} = 0, 1$ with probability $1 - p_{ij}, p_{ij}$, respectively, in case $\sigma_i = \sigma_j$.

e) The conditional distribution of the $\{\sigma\}$ given the $\{n\}$ is as follows: independently for each connected cluster, one sets all the spins σ_i in the cluster to the same value, chosen equiprobably from $\{1, 2, \ldots, q\}$.

The Swendsen-Wang (SW) algorithm [64] simulates the joint model (113) by alternately applying the conditional distributions (d) and (e) — that is, by alternately generating new bond occupation variables (independent of the old ones) given the spins, and new spin variables (independent of the old ones) given the bonds. [This is yet another example of partial resampling.] Each of these operations can be carried out in a computer time of order volume: for generating the bond variables this is trivial, and for generating the spin variables it relies on an efficient (linear-time) algorithm for computing the connected clusters.[25] It is easy to see that the SW algorithm

---

[25]Determining the connected components of an undirected graph is a classic problem of computer science. The depth-first-search and breadth-first-search algorithms [88] have a running time of order V, while the Fischer-Galler-Hoshen-Kopelman algorithm [89] has an observed mean running time

is ergodic and leaves invariant the distribution (113).[26]

It is certainly plausible that the SW algorithm might have less critical slowing-down than the conventional (single-spin-update) algorithms: the reason is that a local move in one set of variables can have highly nonlocal effects in the other. For example, setting $n_b = 0$ on a single bond may disconnect a cluster, causing a big subset of the spins in that cluster to be flipped simultaneously. In some sense, therefore, the SW algorithm is a collective-mode algorithm in which the collective modes are *chosen by the system* rather than imposed from the outside as in Fourier acceleration or multigrid. (The miracle is that this is done in a way that preserves the correct Gibbs measure.)

How well does the SW algorithm perform? Table 1 shows some data [98] on a two-dimensional Ising model at the bulk critical temperature; for comparison we give also data on the single-site Metropolis algorithm [99]. These data are consistent with $\tau_{SW} \sim L^{\approx 0.3}$ [64] or alternatively with $\tau_{SW} \sim \log L$ [103] (it seems difficult to distinguish except by using *extremely* large lattices, e.g. L up to 10000 or more). By contrast, the Metropolis algorithm has $z \approx 2.13$ [99]. For $L = 1024$, this translates into a factor-of-200000 advantage for SW over Metropolis. Even granting that one iteration of the Swendsen-Wang algorithm may be a factor of $\sim 10 - 100$ more costly in CPU time than one iteration of a conventional algorithm (the exact factor depends on the efficiency of the cluster-finding subroutine), the SW algorithm wins already for $L \gtrsim 25$.

For other Potts models, the performance of the SW algorithm is less spectacular than for the two-dimensional Ising model, but it is still very impressive. In Table 2 we give the current best estimates of the dynamic critical exponent z_{SW} for q-state Potts models in d dimensions, as a function of q and d. (For the SW algorithm the decay of the energy-energy autocorrelation function appears to be very close to exponential, hence $z_{int,\mathcal{E}} \approx z_{exp}$.) All these exponents are much lower than the $z \gtrsim 2$ observed in the single-spin-flip algorithms.

Although the SW algorithm performs extraordinarily well, we understand very little about *why* these exponents take the values they do. Some cases are easy. If $q = 1$, then all spins are in the same state (the *only* state!), and all bonds are thrown

---

of order V in percolation-type problems [90]. (There exist sophisticated variants of the FGHK algorithm [91] which improve its asymptotic worst-case behavior, but these variants appear not to be advantageous in practice.) In recent years much attention has been devoted to finding algorithms that run efficiently on vector or parallel computers. The Shiloach-Vishkin algorithm [92] runs on a vector machine or on a SIMD parallel machine with *shared* memory. Several algorithms have been proposed for a SIMD parallel machine with *local* memory and a hypercubic communications grid [93, 94, 95].

[26] *Historical remark:* The random-cluster model was introduced in 1969 by Fortuin and Kasteleyn [84]; they derived the identity $Z_{Potts} = Z_{RC}$ and some corresponding identities for correlation functions. These relations were rediscovered several times during the subsequent two decades [96]. Surprisingly, however, no one seems to have noticed the *joint* probability distribution (113) that underlay all these identities; this was discovered implicitly by Swendsen and Wang [64], and was made explicit by Edwards and Sokal [97].

258

L	χ	$\tau_{int,\mathcal{E}}(SW)$	$\tau_{exp,\mathcal{M}}(Metr)$
16	139.58 (0.04)	3.258 (0.005)	286 (4)
32	470.12 (0.20)	4.016 (0.005)	1258 (28)
64	1581.4 (0.5)	4.899 (0.010)	5380 (140)
128	5319.2 (2.4)	5.874 (0.016)	23950 (480)
256	17900 (7)	6.928 (0.030)	*104500* (est.)
512	60185 (28)	8.144 (0.055)	*458000* (est.)
1024	202219 (117)	9.37 (0.48)	*2008000* (est.)

Table 1: Data for two-dimensional Ising model at criticality. Susceptibility χ and Swendsen-Wang autocorrelation time $\tau_{int,\mathcal{E}}$ ($\approx \tau_{exp,\mathcal{E}}$ since $\mathcal{E}$ = energy $\approx$ slowest mode) are from [98]; see also [100, 101]. Metropolis autocorrelation time $\tau_{exp,\mathcal{M}}$ ($\mathcal{M}$ = magnetization $\approx$ slowest mode) is from [99]; data in italics are extrapolations. Metropolis autocorrelation time τ_{exp} for observables in the even subspace (e.g. $|\mathcal{M}|$ or $\mathcal{M}^2$) is about a factor of 3 smaller [102]. Standard error is shown in parentheses.

	$q=1$	$q=2$	$q=3$	$q=4$
$d=1$	0	0	0	0
$d=2$	0	0 × log (?) ≈ 0.3 (?)	0.55 ± 0.03	≈ 1 (exact?)
$d=3$	0	≈ 0.34 (?) ≈ 0.75 (?)	—	—
$d=4$	0	≈ 1 (exact?)	—	—

Table 2: Current best estimates of the dynamic critical exponent $z_{int,\mathcal{E}}$ ($\approx z_{exp}$) for the Swendsen-Wang algorithm. References: $d=2$, $q=2$ [64, 22, 103, 104, 98]; $d=2$, $q=3$ [105, 104]; $d=2$, $q=4$ [105]; $d=3$, $q=2$ [106, 22, 64, 98]; $d=4$, $q=2$ [107, 98]. Error bar is a 95% confidence interval.

independently, so the autocorrelation time is zero. (Here the SW algorithm just reduces to the standard *static* algorithm for independent bond percolation.) If $d = 1$ (more generally, if the lattice is a *tree*), the SW dynamics is exactly soluble: the behavior of each bond is independent of each other bond, and $\tau_{exp} \to -1/\log(1 - 1/q) < \infty$ as $\beta \to +\infty$. But the remainder of our understanding is very murky. Two principal insights have been obtained so far:

a) A calculation yielding $z_{SW} = 1$ in a mean-field (Curie-Weiss) Ising model [108]. This suggests (but of course does not prove) that $z_{SW} = 1$ for Ising models ($q = 2$) in dimension $d \geq 4$.

b) A rigorous proof that $z_{SW} \geq \alpha/\nu$ [105]; physically this is due to the slow convergence of energy-like observables. This bound, while valid for all d and q, is extremely far from sharp for the Ising models in dimensions 3 and higher. But it is reasonably good for the 3- and 4-state Potts models in two dimensions, and in the latter case it may even be sharp.

But much remains to be understood!

Numerous modifications and generalizations of the SW algorithm have been proposed. For Potts and related models we have:

- Single-cluster variant [65]

- Duality-improved SW algorithm ($d = 2$ only) [109]

- Multi-scale SW algorithms [110, 111]

- SW algorithm for Potts lattice gauge theories [112, 113]

- Algorithm for fully frustrated Ising models [114]

For non-Potts models, several generalizations have been proposed [97, 115], but the most promising ideas at present seem to be the *embedding algorithms* discussed in the next section.

In the single-cluster (1C) variant of the SW algorithm [65], one builds only a *single* cluster (starting at a randomly chosen site) and flips it — as opposed to the standard SW algorithm, which enumerates *all* the clusters in the lattice. Clearly, one step of the single-cluster SW algorithm makes less change in the system than one step of the standard SW algorithm, but it also takes much less work; what matters is the dynamic critical exponent z *measured in CPU-time units*. One advantage of the single-cluster algorithm is that the probability of choosing a cluster is proportional to its size (since we pick a random *site*), so the work is concentrated preferentially on larger clusters. Another advantage is that the successive clusters are less strongly correlated. So it would not be surprising if $z_{1C,CPU}$ were smaller than z_{SW}. The measurements thus far [22, 104, 116] paint a confusing picture: $z_{1C,CPU}$ and z_{SW} seem to be roughly the same in $d = 2$; $z_{1C,CPU}$ seems to be slightly smaller in $d = 3$ (but this is far from certain); and $z_{1C,CPU}$ seems to be significantly smaller in $d = 4$. Better measurements are definitely needed.

A second generalization, which works only in *two* dimensions, augments the SW algorithm by transformations to the dual lattice [109]. Preliminary data show the complete elimination of critical slowing-down for the energy $\mathcal{E}$ — but *not* for other observables, such as $(\mathcal{E} - \langle \mathcal{E} \rangle)^2$ — in the "ultra-scaling region" $|\beta - \beta_c| \ll L^{-1/\nu}$.

The idea of the multi-scale SW algorithm [110] is to carry out only a partial FKSW transformation, but then to apply this concept recursively in a multigrid style. This is a very appealing idea, and it was originally claimed [110] that the critical slowing-down is completely eliminated. However, it has now been proven rigorously [111] that this is not the case, at least for the version of the algorithm originally proposed; in fact, the same lower bound $z \geq \alpha/\nu$ that was proven for the standard SW algorithm holds also for the multi-scale SW algorithm. It is an open question whether there exists a multi-scale SW algorithm that has a dynamic critical exponent strictly smaller than that of standard SW.

The SW algorithm can be generalized in a straightforward manner to Potts lattice gauge theories (more precisely, lattice gauge theories with a *finite abelian* gauge group G and *Potts (δ-function) action*). The auxiliary variables $\{n\}$ live now on *plaquettes*, and the SW update requires generating a random gauge field subject to the constraint of zero curvature on each occupied plaquette. Doing this efficiently (i.e. in time of order volume) seems to be a difficult problem in computational algebraic topology (an almost nonexistent field [117]); it has been done in $d = 3$ by a clever use of duality [112]. [I emphasize that "clusters" play no role in this algorithm: clusters (= connected components) are zeroth cohomology, whereas what is relevant here is *first* cohomology mod G.] Experiments on the three-dimensional Z_2 gauge theory [112, 113] yield a dynamic critical exponent $z \approx 0.6 - 0.7$.

The SW algorithm can easily be generalized to Ising models with both ferromagnetic and antiferromagnetic couplings, but this generalized algorithm does *not* work well [64]. The trouble is that the ferromagnetic and antiferromagnetic bonds work against each other in making a phase transition, but work together in making the SW bonds percolate; therefore, the SW bonds begin to percolate well above the critical temperature, and near criticality almost all the lattice belongs to a single huge cluster. (Flipping a huge cluster is equivalent to flipping its complement, which consists of many very small clusters.) An interesting way of handling frustrated Ising models was recently proposed by Kandel *et al.* [114]: their idea is to consider all the bonds in a plaquette as a single unit, and to apply a cleverly chosen FKSW transformation to this entity. Preliminary results suggest that this method works well when *every* plaquette is frustrated, but not in cases of partial frustration. Devising an efficient algorithm for simulating frustrated spin models is thus a very important — and apparently very difficult — open problem.

Finally, it is worth indicating the general idea behind all algorithms of Swendsen-Wang type. Consider an arbitrary statistical-mechanical model with variables $\{\varphi\}$, and let $W(\{\varphi\})$ be its Boltzmann weight. The idea is then to introduce auxiliary variables $\{n\}$ according to

$$W(\{\varphi\}) = \sum_{\{n\}} W^{\{n\}}(\{\varphi\}) . \qquad (115)$$

Here the $\{n\}$ are any kind of variables you like, discrete or continuous (in the latter case the sum would be an integral), and you are free to decompose W into partial Boltzmann weights $W^{\{n\}}$ any way you please. Usually $W = \prod_b W_b$ where the "bonds" b are sites, links or plaquettes, and correspondingly $W^{\{n\}} = \prod_b W_{n_b}$; but this is not mandatory. One then simulates the joint model

$$W_{joint}(\{\varphi\}, \{n\}) = W^{\{n\}}(\{\varphi\}) \qquad (116)$$

by any legal algorithm (usually by alternately applying the conditional distributions given $\{\varphi\}$ and $\{n\}$). Such an algorithm may or may not reduce the critical slowing-down; that depends on the *physics* behind the decomposition (115). But this formalism provides, in any event, an easy way to check the validity of proposed SW-type algorithms.

4.5 Embedding Algorithms

Thanks to Swendsen and Wang, we now have a fantastically good algorithm for simulating ferromagnetic Ising (and Potts) models. Can we extend our success to non-Potts models such as nonlinear σ-models and lattice gauge theories?

The most promising methods at present seem to be the *embedding algorithms*: the idea is to "embed" Ising variables $\{\varepsilon\}$ "inside" the original model, and then simulate the induced Ising model using the ordinary SW algorithm (or the single-cluster variant).

For one-component spins, this embedding is the obvious decomposition into magnitude and sign [118]. Let the Hamiltonian be

$$H(\varphi) = -\beta \sum_{\langle xy \rangle} \varphi_x \varphi_y + \sum_x V(\varphi_x) , \qquad (117)$$

where $\beta \geq 0$ and $V(\varphi) = V(-\varphi)$. We write

$$\varphi_x = \varepsilon_x |\varphi_x| , \qquad (118)$$

where $\varepsilon_x \equiv \text{sgn}(\varphi_x) = \pm 1$. For *fixed* values of the magnitudes $\{|\varphi|\}$, the conditional probability distribution of the $\{\varepsilon\}$ is given by an Ising model with *ferromagnetic* (though space-dependent) couplings $J_{xy} \equiv \beta |\varphi_x||\varphi_y|$. Therefore, the $\{\varepsilon\}$ model can be updated efficiently by the Swendsen-Wang algorithm (or its single-cluster variant). Heat-bath or MGMC sweeps must also be performed, in order to update the magnitudes.

Wolff's embedding algorithm [65, 66] for $O(N)$-invariant spin models ($N \geq 2$), independently invented by Hasenbusch [119], is equally simple. Let the Hamiltonian be

$$H(\boldsymbol{\sigma}) = -\beta \sum_{\langle xy \rangle} \boldsymbol{\sigma}_x \cdot \boldsymbol{\sigma}_y + \sum_x V(|\boldsymbol{\sigma}_x|) , \qquad (119)$$

with $\beta \geq 0$. Now fix a unit vector $\mathbf{r} \in \mathbf{R}^N$, and write

$$\boldsymbol{\sigma}_x = \boldsymbol{\sigma}_x^\perp + \varepsilon_x |\boldsymbol{\sigma}_x \cdot \mathbf{r}| \mathbf{r} , \qquad (120)$$

where $\boldsymbol{\sigma}_x^\perp \equiv \boldsymbol{\sigma}_x - (\boldsymbol{\sigma}_x \cdot \mathbf{r})\mathbf{r}$ and $\boldsymbol{\sigma}_x^\| \equiv (\boldsymbol{\sigma}_x \cdot \mathbf{r})\mathbf{r}$ are the components of $\boldsymbol{\sigma}_x$ perpendicular and parallel to $\mathbf{r}$, and $\varepsilon_x \equiv \mathrm{sgn}(\boldsymbol{\sigma}_x \cdot \mathbf{r}) = \pm 1$. (Flipping ε_x corresponds to reflecting $\boldsymbol{\sigma}_x$ in the hyperplane perpendicular to $\mathbf{r}$.) Therefore, for *fixed* values of the $\{\boldsymbol{\sigma}^\perp\}$ and $\{|\boldsymbol{\sigma} \cdot \mathbf{r}|\}$, the probability distribution of the $\{\varepsilon\}$ is given by an Ising model with *ferromagnetic* couplings $J_{xy} \equiv \beta |\boldsymbol{\sigma}_x \cdot \mathbf{r}||\boldsymbol{\sigma}_y \cdot \mathbf{r}|$. The algorithm is then: Choose at random a unit vector $\mathbf{r}$; fix the $\{\boldsymbol{\sigma}^\perp\}$ and $\{|\boldsymbol{\sigma} \cdot \mathbf{r}|\}$ at their current values, and update the $\{\varepsilon\}$ by either SW or 1CSW. For fixed-length spins, no other moves are required: the random choice of $\mathbf{r}$ suffices to make the algorithm ergodic.

A third example was provided recently by Evertz, Hasenbusch, Marcu, Pinn and Solomon [120]: it concerns solid-on-solid (SOS) models such as the discrete Gaussian model. An SOS model has integer-valued fields n_x living on lattice sites, and the Hamiltonian is of the form

$$H(n) = \sum_{\langle xy \rangle} V(|n_x - n_y|) . \qquad (121)$$

The embedding is as follows: Choose cleverly a "reflection level" $M \in \mathbf{Z}$ or $\mathbf{Z} + \frac{1}{2}$ (this is the subtle part), and write

$$n_x = M + \varepsilon_x |n_x - M| . \qquad (122)$$

One then fixes the $\{|n_x - M|\}$ and updates the $\{\varepsilon\}$ by SW or 1CSW. This embedding is closely analogous to the Wolff embedding (120) for the XY model, if one identifies an SOS height with an XY angle.

A fourth example was provided by Ben-Av, Evertz, Marcu and Solomon [121]: it concerns the $SU(2)$ lattice gauge theory at finite temperature, but only at $N_t = 1$!

The general idea behind all these algorithms is the following: "Foliate" the configuration space of the original model into "leaves" isomorphic to the configuration space of the "embedded" model. [In the above examples the embedded model is an Ising model, but the idea is much more general. For example, one might consider embeddings of XY spins in a higher σ-model, $U(1)$ spins in an $SU(N)$ gauge theory, etc.] One then moves around the current leaf, using any legitimate Monte Carlo algorithm for simulating the conditional probability distribution restricted to that leaf (i.e. the induced Hamiltonian for the embedded model). Of course, one must combine this with other moves, or with a different foliation, in order to make the algorithm ergodic. (The reader will recognize this as yet another instance of partial resampling.)

The performance of an embedding algorithm is determined by the combined effect of two *completely distinct* issues:

i) How well the embedding captures the important large-scale collective modes of the original model.

ii) How well some particular algorithm (e.g. standard SW or single-cluster SW) succeeds in updating the embedded model.

I wish to emphasize the importance of studying these questions *separately*. If the physically relevant large-scale collective motions of the original model cannot be obtained by motions *within* a leaf, then the embedding algorithm will have severe critical slowing-down *no matter what* method is used to update the embedded variables. On the other hand, if the embedding algorithm with a *particular* choice of updating method for the embedded variables shows severe critical slowing-down, this does *not* necessarily mean that the embedding works badly: the poor performance might be due to slow decorrelation in the inner updating subroutine, and could possibly be remedied by switching to a better algorithm for updating the embedded model. (This is particularly likely to occur if the induced Hamiltonian for the embedded model exhibits *frustration*.) It is crucial to distinguish these two issues, if we wish to obtain *physical insight*.

How can we disentangle these two effects? To study question (i), we can investigate the *idealized embedding algorithm* defined by *independent* resampling on each leaf. (In MGMC this is called the "idealized two-grid algorithm".) To approximate this in practice, one makes N_{hit} hits of the best available method for simulating the embedded model, and extrapolates to $N_{hit} = \infty$. (A familiar analogue is approximating the single-site heat-bath algorithm by multi-hit Metropolis.) I emphasize that this is *not* claimed to be an efficient algorithm (usually $N_{hit} = 1$ is optimal for fixed CPU time). Rather, it is a *test procedure* for gaining physical insight into the embedding; it is expensive but indispensable. To study question (ii), one can investigate the autocorrelation behavior of particular algorithms for the embedded model, using the induced Hamiltonians generated from a few "typical" configurations of the original model.

Let's make these ideas concrete by looking at the Wolff algorithm for the N-vector model. At first thought it may seem strange (and somehow "unphysical") to try to find Ising-like (i.e. discrete) variables in a model with a *continuous* symmetry group. However, upon reflection (pardon the pun) one sees what is going on [122]: if the spin configuration is slowly varying (e.g. a long-wavelength spin wave), then the induced Ising Hamiltonian tends to decouple along the surfaces where J_{xy} is small, hence where $\sigma \cdot \mathbf{r} \approx 0$. The regions where $\sigma \cdot \mathbf{r} > 0$ and $\sigma \cdot \mathbf{r} < 0$ then get flipped *independently*, and this corresponds to a long-wavelength collective mode (Figure 3). So it is quite plausible that the *idealized* Wolff algorithm could have very small (or even zero) critical slowing-down in models where the important large-scale collective modes are spin waves. (An additional argument [122] is needed to explain how the Wolff embedding deals with vortices in the two-dimensional XY model.) To see why the *practical* Wolff algorithm using SW or 1CSW updates also works well, it suffices to note that the induced Ising Hamiltonian is *ferromagnetic*, and that for such an Ising model SW and 1CSW work well.

264

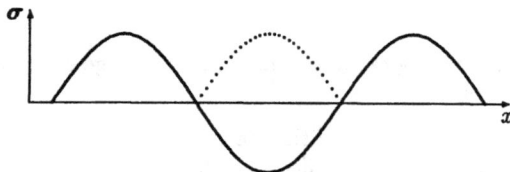

Figure 3: Action of the Wolff algorithm on a long-wavelength spin wave. For simplicity, both spin space (σ) and physical space (x) are depicted as one-dimensional.

Numerical tests of the Wolff algorithm confirm these predictions. For the two-dimensional models with $N = 2, 3, 4$, the data show $z \lesssim 0.1$, both in the idealized algorithm and in the practical algorithm with SW [122] or 1CSW [66] updates. For the three-dimensional XY model, a simulation using standard SW updates ($N_{hit} = 1$) found $z_{int,\mathcal{E}} \approx 0.46$ [123], while one using single-cluster updates found $z_{int,\mathcal{E},CPU} \approx 0.25$ [124]. But these latter exponents may well be due to critical slowing-down in the inner SW or 1CSW subroutine; a study of the *idealized* Wolff algorithm for this model would be very useful.

In view of the extraordinary success of the Wolff algorithm for spin models, it is tempting to try to extend it to lattice gauge theories with continuous gauge group [for example, $U(1)$, $SU(N)$ or $SO(N)$]. Gauge theories differ from N-vector models in two ways:

a) The field takes values in a *group* rather than a sphere. [$U(1)$ and $SU(2)$ are spheres, but higher Lie groups are not.]

b) The field is a 1-form rather than a 0-form, i.e. it lives on *links* rather than sites. Correspondingly, the energy is the *curl* of the field rather than its gradient, and it lives on *plaquettes* rather than links. As a result, the theory has a *local gauge invariance* rather than just a global symmetry.

The deep physical difference between gauge and spin models is, of course, item (b). The fact of gauge invariance, and the transverseness of physical excitations in a gauge theory, will impose severe constraints on the as-yet-unknown analogue of the embedding (120) [if indeed such an analogue exists].[27] So far no progress has been made in this direction (though some insight might possibly be gleaned from the Swendsen-Wang algorithm for Potts lattice gauge theories [112, 113]).

Instead, Caracciolo, Edwards, Pelissetto and Sokal [125] have addressed the less profound, but still highly nontrivial, problem (a). They asked whether the embedding (120) can be generalized to nonlinear σ-models taking values in *manifolds other than*

---

[27]The same issue arises in devising multi-grid algorithms for gauge theories [13, Section V].

spheres — such as $SU(N)$ for $N \geq 3$ — and, if so, what is the dynamic critical behavior of the corresponding idealized Wolff algorithm. Their approach is as follows: First they ask what are the fundamental properties of the embedding (120) that cause the Wolff algorithm to work so well. Then they ask whether embeddings having these properties exist also in other Riemannian manifolds M; this is a question in differential geometry to which they are able to give a fairly complete answer. Finally, they perform a numerical study to test (in one case) whether this theoretical reasoning is correct. The conclusion of this analysis is quite surprising: roughly speaking, they claim that a generalized Wolff algorithm can work well (i.e. have $z \ll 2$) *only* if the manifold M is either a sphere, a product of spheres, or the quotient of such a space by a discrete group (for example, real projective space RP^{N-1}). If correct, this conclusion is quite disappointing, and lends renewed impetus to other classes of collective-mode algorithms such as multi-grid Monte Carlo and Fourier acceleration.

5   Some Open Problems

We conclude by listing some of the principal open problems:

1. Study, both theoretically and experimentally, the various windowing algorithms for estimating τ_{int} (Section 2).

2. Investigate theoretically the conditions under which $z_{int,F}$ should or should not be equal to z_{exp} (Section 3.3).

3. Make detailed studies of the dynamic critical behavior of the various stochastic overrelaxation algorithms, on a variety of non-Gaussian models (Section 3.4).

4. Same, for the hybrid algorithm (Section 3.6).

5. Same, for the Fourier-accelerated Langevin and hybrid algorithms (Section 4.1).

6. Determine in what gauges, if any, Fourier acceleration can work well for a non-Abelian gauge theory (Section 4.1).

7. Clarify, both theoretically and experimentally, the behavior of multi-grid Monte Carlo in asymptotically free spin models (Section 4.2). In particular, clarify the role (if any) played by the choice of interpolation (piecewise-constant or piecewise-linear) and cycle (V or W).

8. Devise good multi-grid Monte Carlo algorithms for both Abelian and non-Abelian gauge theories, and study their dynamic critical behavior (Section 4.2).

9. Understand theoretically the dependence of z_{SW} on q and d (Section 4.4). In particular, understand why the rigorous lower bound $z_{SW} \geq \alpha/\nu$ is close to sharp (or sharp?) for $d = 2$, $q = 2, 4$, but so poor for $d = 3, 4$ Ising.

10. Clarify, both theoretically and experimentally, the dynamic critical behavior of the single-cluster variant of the SW algorithm (Section 4.4). Is $z_{1C,CPU} < z_{SW}$? If so, why?

11. Devise a good algorithm for generating a random gauge field subject to the constraint of zero curvature on a specified set of plaquettes (Section 4.4).

12. Devise an efficient algorithm for simulating frustrated spin models (Section 4.4).

13. Clarify, both theoretically and experimentally, the dynamic critical behavior of Wolff-type embedding algorithms (in both the idealized and practical versions) for nonlinear σ-models taking values in a Riemannian manifold M (Section 4.5).

14. Devise a good Wolff-type embedding algorithm for Abelian or non-Abelian lattice gauge theory, and study its dynamic critical behavior (Section 4.5).

Acknowledgments

Many of the ideas reported here have grown out of joint work with my colleagues Sergio Caracciolo, Robert Edwards, Sabino Ferreira, Jonathan Goodman, Xiao-Jian Li, Andrea Pelissetto and Dan Zwanziger. I thank them for many pleasant and fruitful collaborations. I have also learned much from discussions with Richard Brower, Michael Creutz, Alan Horowitz, Mal Kalos, Neal Madras, Claudio Parrinello, Robert Swendsen, Ulli Wolff and too many others to thank here.

The author's research is supported in part by the U.S. National Science Foundation grant DMS-8911273, the U.S. Department of Energy contract DE-FG02-90ER40581, the NATO Collaborative Research Grant CRG 910251, and the Pittsburgh Supercomputing Center grant PHY890035P.

References

[1] A.D. Sokal, *Monte Carlo Methods in Statistical Mechanics: Foundations and New Algorithms*, Cours de Troisième Cycle de la Physique en Suisse Romande (Lausanne, June 1989).

[2] S.L. Adler, Nucl. Phys. B (Proc. Suppl.) **9**, 437 (1989).

[3] U. Wolff, Nucl. Phys. B (Proc. Suppl.) **17**, 93 (1990).

[4] A.D. Sokal, Nucl. Phys. B (Proc. Suppl.) **20**, 55 (1991).

[5] J.M. Hammersley and D.C. Handscomb, *Monte Carlo Methods* (Methuen, London, 1964), Chapter 5.

[6] J. G. Kemeny and J. L. Snell, *Finite Markov Chains* (Springer, New York, 1976).

[7] M. Iosifescu, *Finite Markov Processes and Their Applications* (Wiley, Chichester, 1980).

[8] K. L. Chung, *Markov Chains with Stationary Transition Probabilities*, 2nd ed. (Springer, New York, 1967).

[9] E. Nummelin, *General Irreducible Markov Chains and Non-Negative Operators* (Cambridge Univ. Press, Cambridge, 1984).

[10] A. D. Sokal and L. E. Thomas, J. Stat. Phys. **54**, 797 (1989).

[11] N. Metropolis, A.W. Rosenbluth, M.N. Rosenbluth, A.H. Teller and E. Teller, J. Chem. Phys. **21**, 1087 (1953).

[12] W.K. Hastings, Biometrika **57**, 97 (1970).

[13] J. Goodman and A.D. Sokal, Phys. Rev. **D40**, 2035 (1989).

[14] R.S. Varga, *Matrix Iterative Analysis* (Prentice-Hall, Englewood Cliffs, N.J., 1962).

[15] E. Nelson, in *Constructive Quantum Field Theory*, ed. G. Velo and A. Wightman (Lecture Notes in Physics #25) (Springer, Berlin, 1973).

[16] B. Simon, *The $P(\varphi)_2$ Euclidean (Quantum) Field Theory* (Princeton Univ. Press, Princeton, N.J., 1974), Chapter I.

[17] A. D. Sokal and L. E. Thomas, J. Stat. Phys. **51**, 907 (1988).

[18] D. Goldsman, Ph.D. thesis, School of Operations Research and Industrial Engineering, Cornell University (1984); L. Schruben, Oper. Res. **30**, 569 (1982) and **31**, 1090 (1983).

[19] M.B. Priestley, *Spectral Analysis and Time Series*, 2 vols. (Academic, London, 1981), Chapters 5-7.

[20] T.W. Anderson, *The Statistical Analysis of Time Series* (Wiley, New York, 1971).

[21] N. Madras and A.D. Sokal, J. Stat. Phys. **50**, 109 (1988).

[22] U. Wolff, Phys. Lett. **B228**, 379 (1989).

[23] R.G. Edwards, S.J. Ferreira, J. Goodman and A.D. Sokal, Multi-Grid Monte Carlo. III. Two-Dimensional $O(4)$ Nonlinear σ-Model, SCRI preprint (1991).

[24] R.G. Edwards, J. Goodman and A.D. Sokal, Nucl. Phys. **B354**, 289 (1991).

[25] D.J. Best and N.I. Fisher, Appl. Statist. [= J. Roy. Statist. Soc., Series C] **28**, 152 (1979).

[26] L. Devroye, *Non-Uniform Random Variate Generation* (Springer, New York, 1986).

[27] M. Creutz, Phys. Rev. **D21**, 2308 (1980); K. Fabricius and O. Haan, Phys. Lett. **B143**, 459 (1984); A.D. Kennedy and B.J. Pendleton, Phys. Lett. **B156**, 393 (1985); P. de Forcrand, J. Stat. Phys. **43**, 1077 (1986).

[28] E. Pietarinen, Nucl. Phys. **B190**[FS3], 349 (1981).

[29] B. Gidas, private communication.

[30] K. Fredenhagen and M. Marcu, Phys. Lett. **B193**, 486 (1987).

[31] J. Goodman and N. Madras, Courant Institute preprint (1987).

[32] G.F. Mazenko and O.T. Valls, Phys. Rev. **B24**, 1419 (1981); C. Kalle, J. Phys. **A17**, L801 (1984); J.K. Williams, J. Phys. **A18**, 49 (1985); R.B. Pearson, J.L. Richardson and D. Toussaint, Phys. Rev. **B31**, 4472 (1985); S. Wansleben and D.P. Landau, J. Appl. Phys. **61**, 3968 (1987).

[33] M.N. Barber, in *Phase Transitions and Critical Phenomena*, vol. 8, ed. C. Domb and J.L. Lebowitz (Academic Press, London, 1983).

[34] K. Binder, J. Comput. Phys. **59**, 1 (1985).

[35] K. Kremer and K. Binder, Comput. Phys. Reports **7**, 259 (1988).

[36] N. Madras and G. Slade, *The Self-Avoiding Walk*, book in preparation.

[37] B.A. Berg and T. Neuhaus, Phys. Lett. **B267**, 249 (1991) and Phys. Rev. Lett. **68**, 9 (1992); B.A. Berg, U. Hansmann and T. Neuhaus, SCRI preprint TH–91/125 (1991).

[38] S. Caracciolo and A.D. Sokal, J. Phys. **A19**, L797 (1986).

[39] G.G. Batrouni, A. Hansen and M. Nelkin, Phys. Rev. Lett. **57**, 1336 (1986); G.G. Batrouni and A. Hansen, J. Stat. Phys. **52**, 747 (1988); P. Rossi, C.T.H. Davies and G.P. Lepage, Nucl. Phys. **B297**, 287 (1988); P. Rossi and C.T.H. Davies, Phys. Lett. **B202**, 547 (1988); C.T.H. Davies *et al.*, Phys. Rev. **D37**, 1581 (1988); G. Katz *et al.*, Phys. Rev. **D37**, 1589 (1988).

[40] W. Hackbusch, *Multi-Grid Methods and Applications* (Springer, Berlin, 1985).

[41] W.L. Briggs, *A Multigrid Tutorial* (SIAM, Philadelphia, 1987).

[42] H. Neuberger, Phys. Rev. Lett. **59**, 1877 (1987) and Phys. Lett. **B207**, 461 (1988); R.J. LeVeque and L.N. Trefethen, IMA J. Numer. Anal. **8**, 273 (1988); L.M. Adams, R.J. LeVeque and D.M. Young, SIAM J. Numer. Anal. **25**, 1156 (1988).

[43] S.L. Adler, Phys. Rev. **D23**, 2091 (1981); C. Whitmer, Phys. Rev. **D29**, 306 (1984).

[44] U.M. Heller and H. Neuberger, Phys. Rev. **D39**, 616 (1989).

[45] R. Gupta, J. DeLapp, G.G. Batrouni, G.C. Fox, C.F. Baillie and J. Apostolakis, Phys. Rev. Lett. **61**, 1996 (1988).

[46] J. Apostolakis, C.F. Baillie and G.C. Fox, Phys. Rev. **D43**, 2687 (1991).

[47] G.G. Batrouni, G.R. Katz, A.S. Kronfeld, G.P. Lepage, B. Svetitsky and K.G. Wilson, Phys. Rev. **D32**, 2736 (1985).

[48] E. Helfand, Bell System Tech. J. **58**, 2289 (1979); H.S. Greenside and E. Helfand, Bell System Tech. J. **60**, 1927 (1981); W. Rümelin, SIAM J. Numer. Anal. **19**, 604 (1982); J.R. Klauder and W.P. Petersen, SIAM J. Numer. Anal. **22**, 1153 (1985); R.L. Honeycutt, Phys. Rev. **A45**, 600 (1992); P.E. Kloeden and E. Platen, J. Stat. Phys. **66**, 283 (1992).

[49] N.G. van Kampen, J. Stat. Phys. **24**, 175 (1981).

[50] L. Arnold, *Stochastic Differential Equations: Theory and Applications* (Wiley, New York, 1974); N. Ikeda and S. Watanabe, *Stochastic Differential Equations and Diffusion Processes* (North-Holland, Amsterdam–New York–Oxford, 1981); K.D. Elworthy, *Stochastic Differential Equations on Manifolds* (Cambridge University Press, 1982); Ya.I. Belopolskaya and Yu.L. Dalecky, *Stochastic Equations and Differential Geometry* (Kluwer Academic Publishers, Dordrecht, 1990).

[51] I.T. Drummond, S. Duane and R.R. Horgan, Nucl. Phys. **B220**[FS8], 119 (1983); M.B. Halpern, Nucl. Phys. **B228**, 173 (1983); M. Namiki, I. Ohba and K. Okano, Prog. Theor. Phys. **72**, 350 (1984); S. Caracciolo, H.-C. Ren and Y.-S. Wu, Nucl. Phys. **B260**, 381 (1985); M. Claudson and M.B. Halpern, Ann. Phys. **166**, 33 (1986); G.G. Batrouni, H. Kawai and P. Rossi, J. Math. Phys. **27**, 1646 (1986); J. Zinn-Justin, Nucl. Phys. **B275**[FS17], 135 (1986); H. Gausterer and S. Sanielevici, Comput. Phys. Commun. **52**, 43 (1988).

[52] P.J. Rossky, J.D. Doll and H.L. Friedman, J. Chem. Phys. **69**, 4628 (1978); R.T. Scalettar, D.J. Scalapino and R.L. Sugar, Phys. Rev. **B34**, 7911 (1986).

[53] M. Creutz, Phys. Rev. **D38**, 1228 (1988).

[54] J. Zinn-Justin, Nucl. Phys. **B275**[FS17], 135 (1986); S. Pugnetti, Nucl. Phys. **B300**, 143 (1988); K. Okano and L. Schülke, Phys. Lett. **B201**, 108 (1988).

[55] A.D. Kennedy, in *Probabilistic Methods in Quantum Field Theory and Quantum Gravity* (1989 Cargèse lectures), ed. P.H. Damgaard, H. Hüffel and A. Rosenblum (Plenum, New York, 1990), pp. 209–223.

[56] S. Duane, Nucl. Phys. **B257**[FS14], 652 (1985); S. Duane and J.B. Kogut, Phys. Rev. Lett. **55**, 2774 (1985); S. Duane and J.B. Kogut, Nucl. Phys. **B275**, 398 (1986).

[57] S. Duane, A.D. Kennedy, B.J. Pendleton and D. Roweth, Phys. Lett. **B195**, 216 (1987).

[58] A.D. Kennedy, Nucl. Phys. B (Proc. Suppl.) **9**, 457 (1989); M. Creutz and A. Gocksch, Phys. Rev. Lett. **63**, 9 (1989); M. Campostrini and P. Rossi, Nucl. Phys. **B329**, 753 (1990).

[59] M. Creutz, Phys. Rev. **D38**, 1228 (1988); R. Gupta, G.W. Kilcup and S.R. Sharpe, Phys. Rev. **D38**, 1278 (1988); H. Gausterer and M. Salmhofer, Phys. Rev. **D40**, 2723 (1989); S. Gupta, A. Irbäck, F. Karsch and B. Petersson, Phys. Lett. **B242**, 437 (1990); A.D. Kennedy and B. Pendleton, Nucl. Phys. B (Proc. Suppl.) **20**, 118 (1991).

[60] S. Duane, Nucl. Phys. **B257**[FS14], 652 (1985); G.M. Buendía, J. Phys. **A22**, 5065 (1989); A.D. Kennedy and B. Pendleton, Nucl. Phys. B (Proc. Suppl.) **20**, 118 (1991).

[61] S. Gupta, Dynamical critical properties of the hybrid Monte Carlo algorithm, CERN preprint CERN–TH.6178/91 (1991).

[62] G. Parisi, in *Progress in Gauge Field Theory* (1983 Cargèse lectures), ed. G. 't Hooft *et al.* (Plenum, New York, 1984).

[63] J. Goodman and A.D. Sokal, Phys. Rev. Lett. **56**, 1015 (1986).

[64] R.H. Swendsen and J.-S. Wang, Phys. Rev. Lett. **58**, 86 (1987).

[65] U. Wolff, Phys. Rev. Lett. **62**, 361 (1989).

[66] U. Wolff, Nucl. Phys. **B322**, 759 (1989); Phys. Lett. **B222**, 473 (1989); Nucl. Phys. **B334**, 581 (1990).

[67] J.B. Kogut, Nucl. Phys. **B275** [FS17], 1 (1986).

[68] E. Dagotto and J.B. Kogut, Phys. Rev. Lett. **58**, 299 (1987).

[69] E. Dagotto and J.B. Kogut, Nucl. Phys. **B290** [FS20], 451 (1987).

[70] S. Duane, R. Kenway, B.J. Pendleton and D. Roweth, Phys. Lett. **176B**, 143 (1986).

[71] C. Davies, G. Batrouni, G. Katz, A. Kronfeld, P. Lepage, P. Rossi, B. Svetitsky and K. Wilson, J. Stat. Phys. **43**, 1073 (1986).

[72] R.P. Fedorenko, Zh. Vychisl. i Mat. Fiz. **4**, 559 (1964) [USSR Comput. Math. and Math. Phys. **4**, 227 (1964)].

[73] L.P. Kadanoff, Physics **2**, 263 (1966).

[74] K.G. Wilson, Phys. Rev. **B4**, 3174, 3184 (1971).

[75] A. Brandt, in *Proceedings of the Third International Conference on Numerical Methods in Fluid Mechanics* (Paris, July 1972), ed. H. Cabannes and R. Temam (Lecture Notes in Physics #18) (Springer, Berlin, 1973); R.A. Nicolaides, J. Comput. Phys. **19**, 418 (1975) and Math. Comp. **31**, 892 (1977); A. Brandt, Math. Comp. **31**, 333 (1977); W. Hackbusch, in *Numerical Treatment of Differential Equations* (Oberwolfach, July 1976), ed. R. Bulirsch, R.D. Griegorieff and J. Schröder (Lecture Notes in Mathematics #631) (Springer, Berlin, 1978).

[76] W. Hackbusch and U. Trottenberg, editors, *Multigrid Methods* (Lecture Notes in Mathematics #960) (Springer, Berlin, 1982); Proceedings of the First Copper Mountain Conference on Multigrid Methods, ed. S. McCormick and U. Trottenberg, Appl. Math. Comput. **13**, no. 3-4, 213-470 (1983); D.J. Paddon and H. Holstein, editors, *Multigrid Methods for Integral and Differential Equations* (Clarendon Press, Oxford, 1985); Proceedings of the Second Copper Mountain Conference on Multigrid Methods, ed. S. McCormick, Appl. Math. Comput. **19**, no. 1-4, 1-372 (1986); W. Hackbusch and U. Trottenberg, editors, *Multigrid Methods II* (Lecture Notes in Mathematics #1228) (Springer, Berlin, 1986); S.F. McCormick, editor, *Multigrid Methods* (SIAM, Philadelphia, 1987); *Proceedings of the Third Copper Mountain Conference on Multigrid Methods*, ed. S. McCormick (Dekker, New York, 1988).

[77] J. Mandel, S. McCormick and R. Bank, in *Multigrid Methods*, ed. S.F. McCormick (SIAM, Philadelphia, 1987), Chapter 5.

[78] J. Goodman and A.D. Sokal, unpublished.

[79] J.M. Ortega and W.C. Rheinboldt, *Iterative Solution of Nonlinear Equations in Several Variables* (Academic Press, New York–London, 1970).

[80] S.F. McCormick and J. Ruge, Math. Comp. **41**, 43 (1983).

[81] A. Hulsebos, J. Smit and J.C. Vink, Nucl. Phys. **B356**, 775 (1991).

[82] M. Hasenbusch and S. Meyer, Phys. Rev. Lett. **68**, 435 (1992).

[83] F.Y. Wu, Rev. Mod. Phys. **54**, 235 (1982); **55**, 315 (E) (1983).

272

[84] P.W. Kasteleyn and C.M. Fortuin, J. Phys. Soc. Japan **26** (Suppl.), 11 (1969); C.M. Fortuin and P.W. Kasteleyn, Physica **57**, 536 (1972); C.M. Fortuin, Physica **58**, 393 (1972); C.M. Fortuin, Physica **59**, 545 (1972).

[85] K. Krickeberg, *Probability Theory* (Addison-Wesley, Reading, Mass., 1965), Sections 4.2 and 4.5.

[86] R.H. Schonmann, J. Stat. Phys. **52**, 61 (1988).

[87] A.D. Sokal, unpublished.

[88] E.M. Reingold, J. Nievergelt and N. Deo, *Combinatorial Algorithms: Theory and Practice* (Prentice-Hall, Englewood Cliffs, N.J., 1977), Chapter 8; A. Gibbons, *Algorithmic Graph Theory* (Cambridge University Press, 1985), Chapter 1; S. Even, *Graph Algorithms* (Computer Science Press, Potomac, Maryland, 1979), Chapter 3.

[89] B.A. Galler and M.J. Fischer, Commun. ACM **7**, 301, 506 (1964); D.E. Knuth, *The Art of Computer Programming*, vol. 1, 2nd ed., (Addison-Wesley, Reading, Massachusetts, 1973), pp. 353–355, 360, 572; J. Hoshen and R. Kopelman, Phys. Rev. **B14**, 3438 (1976).

[90] K. Binder and D. Stauffer, in *Applications of the Monte Carlo Method in Statistical Physics*, ed. K. Binder (Springer-Verlag, Berlin, 1984), section 8.4; D. Stauffer, *Introduction to Percolation Theory* (Taylor & Francis, London, 1985).

[91] R.E. Tarjan, *Data Structures and Network Algorithms* (Philadelphia: SIAM, 1983), Chapter 2; R.E. Tarjan and J. van Leeuwen, J. Assoc. Comput. Mach. **31**, 245 (1984).

[92] Y. Shiloach and U. Vishkin, J. Algorithms **3**, 57 (1982).

[93] C.F. Baillie and P.D. Coddington, Concurrency: Practice and Experience **3**, 129 (1991).

[94] R.C. Brower, P. Tamayo and B. York, J. Stat. Phys. **63**, 73 (1991).

[95] J. Apostolakis, P. Coddington and E. Marinari, Europhys. Lett. **17**, 189 (1992).

[96] C.-K. Hu, Phys. Rev. **B29**, 5103 (1984); T.A. Larsson, J. Phys. **A19**, 2383 (1986) and **A20**, 2239 (1987).

[97] R.G. Edwards and A.D. Sokal, Phys. Rev. **D38**, 2009 (1988).

[98] P.D. Coddington, private communication; see also P.D. Coddington and C.F. Baillie, Empirical relations between static and dynamic exponents for Ising model cluster algorithms, Syracuse University preprint SCCS-141 (1991).

[99] N. Ito, M. Taiji and M. Suzuki, J. Phys. Soc. Japan **56**, 4218 (1987).

[100] A.D. Sokal, in *Computer Simulation Studies in Condensed Matter Physics: Recent Developments*, ed. D.P. Landau, K.K. Mon and H.-B. Schüttler (Springer, Berlin-Heidelberg, 1988).

[101] X.-J. Li, Cluster algorithms in Monte Carlo simulations, Ph.D. dissertation, New York University (January 1992).

[102] S. Tang and D.P. Landau, Phys. Rev. **B36**, 567 (1987).

[103] D.W. Heermann and A.N. Burkitt, Physica **A162**, 210 (1990).

[104] C.F. Baillie and P.D. Coddington, Phys. Rev. **B43**, 10617 (1991).

[105] X.-J. Li and A.D. Sokal, Phys. Rev. Lett. **63**, 827 (1989).

[106] J.-S. Wang, Physica **A164**, 240 (1990).

[107] W. Klein, T. Ray and P. Tamayo, Phys. Rev. Lett. **62**, 163 (1989).

[108] T.S. Ray, P. Tamayo and W. Klein, Phys. Rev. **A39**, 5949 (1989).

[109] R.G. Edwards and A.D. Sokal, Swendsen-Wang algorithm augmented by duality transformations for the two-dimensional Potts model, in preparation.

[110] D. Kandel, E. Domany, D. Ron, A. Brandt and E. Loh, Phys. Rev. Lett. **60**, 1591 (1988); D. Kandel, E. Domany and A. Brandt, Phys. Rev. **B40**, 330 (1989).

[111] X.-J. Li and A.D. Sokal, Phys. Rev. Lett. **67**, 1482 (1991).

[112] R. Ben-Av, D. Kandel, E. Katznelson, P.G. Lauwers and S. Solomon, J. Stat. Phys. **58**, 125 (1990).

[113] R.C. Brower and S. Huang, Phys. Rev. **B41**, 708 (1990).

[114] D. Kandel, R. Ben-Av and E. Domany, Phys. Rev. Lett. **65**, 941 (1990).

[115] F. Niedermayer, Phys. Rev. Lett. **61**, 2026 (1988).

[116] P. Tamayo, R.C. Brower and W. Klein, J. Stat. Phys. **58**, 1083 (1990).

[117] J. Saludes i Closa, Butll. Soc. Catalana Cièn. **3**, 127 (1984); I.K. Svetlichnova, Vestnik Mosk. Univ. Mat. **39**, no. 3, 79 (1984) [= Moscow Univ. Math. Bull. **39**, no. 5, 100 (1984)].

[118] R.C. Brower and P. Tamayo, Phys. Rev. Lett. **62**, 1087 (1989).

[119] M. Hasenbusch, Nucl. Phys. **B333**, 581 (1990).

[120] H.G. Evertz, M. Hasenbusch, M. Marcu, K. Pinn and S. Solomon, Nucl. Phys. B (Proc. Suppl.) **20**, 80 (1991) and Phys. Lett. **B254**, 185 (1991).

[121] H.G. Evertz, R. Ben-Av, M. Marcu and S. Solomon, Nucl. Phys. B (Proc. Suppl.) **20**, 85 (1991).

[122] R.G. Edwards and A.D. Sokal, Phys. Rev. **D40**, 1374 (1989).

[123] M. Hasenbusch and S. Meyer, Phys. Lett. **B241**, 238 (1990).

[124] W. Janke, Phys. Lett. **A148**, 306 (1990).

[125] S. Caracciolo, R.G. Edwards, A. Pelissetto and A.D. Sokal, Nucl. Phys. B (Proc. Suppl.) **20**, 72 (1991) and paper in preparation.

ALGORITHMS FOR SIMULATING FERMIONS

MICHAEL CREUTZ

Physics Department, Brookhaven National Laboratory
Upton, NY 11973, USA

ABSTRACT

I review some of the approaches to including the effects of dynamical quark loops in Monte Carlo simulations of quantum field theories. The discussion begins with a brief introduction to anticommuting variables and ends with higher order improvements to the hybrid Monte Carlo scheme. Pseudofermion, Langevin, and various combinations thereof are compared.

1. Introduction

Fermions provide some of the biggest challenges to the field of lattice gauge theory. One difficulty appears already at the level of formulating an appropriate action. Here we have the notorious doubling of species appearing in the simplest schemes incorporating chiral symmetry. The two popular schemes for handling this are the Kogut-Susskind formulation where each site carries only a single component of the Dirac spinors, and the Wilson projection operator approach where chiral symmetry is abandoned with the hope that it will be recovered in the continuum limit. Both of these schemes are mentioned in other chapters of this book. Here I will be quite generic and assume we have an acceptable lattice transcription of the Dirac equation.

The purpose of this chapter is to discuss the other primary difficulty with fermions: the extreme computational difficulties that appear on adding them to Monte Carlo simulations. While several large scale simulations are ongoing, even the most meager results strain the most advanced computational resources available. This chapter discusses the algorithms currently in use, several of which are quite clever. Nevertheless, there remains a certain awkwardness in known approaches that hints that new and better techniques will evolve.

For this discussion I will be quite generic and assume we are interested in a path integral of form

$$Z = \int (dA)(d\psi)(d\psi^*) \, \exp(-S_G(A) - \psi^* M(A)\psi). \tag{1}$$

Here the gauge fields are formally denoted A and fermionic fields ψ and ψ^*. As I will be concentrating on fermionic details, I will ignore the technicality that the gauge fields are group elements. All details of the fermionic formulation are hidden in the matrix $M(A)$. While I may call A a gauge field, the algorithms are general, and have potential applications in other field theories and condensed matter physics.

The numerical difficulties with fermionic fields stem from their being anticommuting quantities. Thus it is not immediately straightforward to place them on a computer, which likes to manipulate numbers. Indeed, the Boltzmann factor is formally an operator in a Grassmann space, and cannot be directly interpreted as a probability for Monte Carlo purposes. All algorithms in current use eliminate the fermions at the outset by a formal analytic integration. This is possible because most actions in practice are, or can easily be made, quadratic in the fermionic fields. The fermion integrals are then over generalized gaussians. Unfortunately, the resulting expressions involve the determinant of a large, albeit sparse, matrix. This determinant introduces nonlocal couplings between the bosonic degrees of freedom, making the path integrals over the remaining fields rather time consuming. Nevertheless, various tricks have been developed to minimize the pain with fermions, and in other chapters of this book you can find numerous results from such calculations. This chapter reviews some of these tricks.

2. Anticommuting variables

I begin with a brief review of Grassmann variables [1]. I start with a set $\{\psi_i\}$ of anticommuting variables

$$[\psi_i, \psi_j]_+ \equiv \psi_i \psi_j + \psi_j \psi_i = 0. \tag{2}$$

To keep the formulae simple I combine spatial dependence, internal symmetry, and spinor indices into a single index i.

These anticommutation relations can be realized via matrices. Had I just two variables, I could represent the ψ_i using two sets of independent Pauli matrices $\{\sigma\}$ and $\{\tau\}$. In fairly standard notation I could satisfy Eq. (2) by writing $\psi_1 = \sigma^+$ and $\psi_2 = \sigma^z \tau^+$. Going on to more variables I would need to increase the number of independent matrices used in this direct product. As the number of independent Grassmann variables increases, the overall matrix dimension necessary to represent the algebra increases exponentially. Because of this growth, using explicit representations for the Grassmann variables has not, at least so far, been particularly useful for computational purposes. For this reason I will proceed more formally and just treat the ψ_i abstractly.

Generalizing complex conjugation to include these variables, I adopt the convention that corresponding to each ψ_i I have another independent Grassmann variable ψ_i^*. Furthermore, I postulate

$$(\psi_i^*)^* = \psi_i$$
$$(\psi_1 \ldots \psi_n)^* = \psi_n^* \ldots \psi_1^*. \tag{3}$$

If I consider just a single variable ψ, then ψ^2 must vanish, and a general function $f(\psi)$ can be expanded with two terms

$$f(\psi) = f_0 + \psi f_1. \tag{4}$$

Similarly, with n variables, a general function can be expanded as a finite polynomial with 2^n terms.

To define integration over an anticommuting variable, I demand the properties of linearity and invariance under a translation of variables. These are summarized in the axioms

$$\int d\psi(f(\psi)\alpha + g(\psi)\beta) = \left(\int d\psi f(\psi)\right)\alpha + \left(\int d\psi g(\psi)\right)\beta \tag{5a}$$

$$\int d\psi f(\psi) = \int d\psi f(\psi + \psi'). \tag{5b}$$

This is sufficient to imply that for the function in Eq. (4)

$$\int d\psi f(\psi) = K f_1 \tag{6}$$

where the normalization K is undetermined. I adopt the convention $K = i$ so that

$$\int d\psi \; \psi = i$$
$$\int d\psi \; 1 = 0 \tag{7}$$
$$\int d\psi^* d\psi \; \psi^* \psi = 1.$$

Note that under multiplicative rescaling a Grassmann integral behaves as

$$\int d\psi f(\psi a) = \left(\int d\psi f(\psi)\right) a. \tag{8}$$

This can be written in the heuristic form $d(\psi a) = (d\psi)/a$.

For integration over several anticommuting variables, I have

$$\int d\psi_1 \ldots d\psi_n \psi_1 \ldots \psi_n = i^n (-1)^{n(n-1)/2}. \tag{9}$$

The analog of Eq. (8) in this case is

$$\int (d\psi) f(M\psi) = |M| \int d\psi f(\psi) \tag{10}$$

where M is an arbitrary matrix, $|M|$ its determinant, and $(d\psi)$ denotes $d\psi_1 \ldots d\psi_n$. Note that eq. (10) immediately implies the Matthews-Salam [2] formula for a fermionic Gaussian integral

$$\int (d\psi d\psi^*) \, e^{-\psi^* M \psi} = |M| \tag{11}$$

278

where $(d\psi d\psi^*) = d\psi_1 d\psi_1^* \ldots d\psi_n d\psi_n^*$.

Eq. (11) provides an easy way out of the difficulty that our partition function is not an ordinary integral. Indeed, I explicitly integrate out the fermions to convert Eq. (1) to

$$Z = \int (dA) \, |M| \, e^{-S_a}. \tag{12}$$

This is now an integral over ordinary numbers and therefore in principle amenable to Monte Carlo attack.

For the remainder of this chapter I assume that the fermions have been formulated such that $|M|$ is positive and thus the integrand in Eq. (12) can be regarded as proportional to a probability measure. If this is not so, one can always double the number of fermionic species, using $M^\dagger$ for the extra ones, thus replacing M by $MM^\dagger$. The case where M is not positive is not yet well understood, and indeed in some such cases the path integral may not be well defined. These problems are probably closely connected with the difficulties of placing chiral fermions on the lattice. This issue is quite important; with a chemical potential present to give a background fermion density, the determinant is not in general positive. Indeed, it is not yet known how to simulate this physically important problem.

Direct Monte Carlo study of the partition function in Eq. (12) is still not practical because of the large size of the matrix M. In our compact notation, this is a square matrix of dimension equal to the number of lattice sites times the number of Dirac components times the number of internal symmetry degrees of freedom. Thus, it is typically a tens of thousands by tens of thousands matrix, precluding any direct attempt to calculate its determinant. It is, however, generally an extremely sparse matrix because most popular actions do not directly couple distant sites. All the Monte Carlo algorithms used in practice for fermions make essential use of this fact.

3. Exact local algorithms

Many current simulations make additional approximations beyond the lattice cutoff and the statistical errors inherent in Monte Carlo simulations. Such approximations are in principle not required, but save substantial computer time. Here I first discuss a few of the approaches that do not make such additional assumptions. These are in a sense "exact" because with enough Monte Carlo statistics they will give the correct correlations for the partition function. They are still approximations in the sense of having a finite lattice spacing and finite lattice size, but these approximations are already present in the bosonic simulations. Later I consider some approximate variations which are considerably faster. Finally I will show how to make some of those approximate approaches "exact" again.

Some time ago Weingarten and Petcher [3] presented a simple "exact" algorithm. They observed that by introducing auxiliary set of complex scalar fields ϕ one can rewrite Eq. (12) in the form

$$Z = \int (dA)(d\phi^* \, d\phi) \exp(-S_G - \phi^* M^{-1} \phi). \qquad (13)$$

Thus a successful fermionic simulation would be possible if one could obtain configurations of fields ϕ and A with probability distribution

$$P(A, \phi) \propto \exp(-S_G - \phi^* M^{-1} \phi). \qquad (14)$$

To proceed I will assume that M is a positive matrix so this distribution is well defined.

Ref. 3 notes that while M^{-1} is the inverse of an enormous matrix, one really only needs $\phi^* M^{-1} \phi$, which is just one matrix element of this inverse. Furthermore, with a local fermionic action the matrix M is extremely sparse, the nonvanishing matrix elements only connecting nearby sites. In this case there exist quite efficient iterative schemes for finding the inverse of a large sparse matrix applied to a single vector. Thus it was proposed to directly simulate the partition function in Eq. (5.1) using a Gauss-Seidel algorithm to calculate $M^{-1}\phi$. Most recent work has turned to the conjugate gradient algorithm for this inversion.

The conjugate gradient algorithm has also been quite useful for fermionic studies in the "valence" or "quenched" approximation, where the Dirac equation is solved on gauge field configurations obtained by simulations ignoring the feedback of the dynamical quarks on the gauge fields themselves [4]. This approximation, discussed many places elsewhere in this book, circumvents the problem of simulating dynamical quarks, but still imposes substantial computational demands when the lattices are large.

The conjugate gradient method to find $\xi = M^{-1}\phi$ works by finding the minimum over ξ of the function $|M\xi - \phi|^2$. The solution is iterative; starting with some ξ_0, a sequence of vectors is obtained by moving to the minimum of this function along successive directions d_i. The clever trick of the algorithm is to choose the d_i to be orthogonal in a sense defined by the matrix M itself; in particular $(Md_i, Md_j) = 0$ whenever $i \neq j$. This last condition serves to eliminate useless oscillations in undesirable directions, and guarantees convergence to the minimum in a number of steps equal to the dimension of the matrix. There are close connections between the conjugate gradient inversion procedure and the Lanczos algorithm for tridiagonalizing sparse matrices. [5]

The procedure is a simple recursion. Select some arbitrary initial pair of nonvanishing vectors $g_0 = d_0$. For the inversion problem, convergence will be improved if these are a good guess to $M^{-1}\phi$. Then generate a sequence of further vectors by iterating

$$g_{i+1} = (Mg_i, Md_i)g_i - (g_i, g_i)M^\dagger Md_i$$
$$d_{i+1} = (Md_i, Md_i)g_{i+1} - (Md, Mg_{i+1})d_i$$

(15)

This construction assures that g_i is orthogonal to g_{i+1} and $(Md_i, Md_{i+1}) = 0$. It should also be clear that the three sets of vectors $\{d_0, ...d_k\}$, $\{g_0, ...g_k\}$, and $\{d_0, ...(M^\dagger M)^k d_0\}$ all span the same space.

The remarkable core of the algorithm, easily proved by induction, is that the set of g_i are all mutually orthogonal, as are Md_i. For an N dimensional matrix, there can be no more than N independent orthogonal vectors. Thus, ignoring roundoff errors, the recursion in Eq. (15) must terminate in N or less steps with the vectors g and d vanishing from then on. Furthermore, as the above sets of vectors all span the same space, in a basis defined by the g_i the matrix $M^\dagger M$ is in fact tri-diagonal, with (Mg_i, Mg_j) vanishing unless $i = j \pm 1$.

To solve $\phi = M\xi$ for ξ, simply expand in the d_i

$$\xi = \sum_i \alpha_i d_i.$$

The coefficients are immediately found from the orthogonality conditions

$$\alpha_i = (Md_i, \phi)/(Md_i, Md_i).$$

Note that if I start with the solution $d_0 = M^{-1}\phi$, then I have $\alpha_i = \delta_{i0}$.

This discussion applies for a general matrix M. If M is Hermitian, then one can work with better conditioned matrices by replacing the orthogonality condition for the d_i with (d_i, Md_j) vanishing for $i \neq j$.

In practice, at least when the correlation length is not too large, this procedure adequately converges in a number of iterations which does not grow severely with the lattice size. As each step involves vector sums with length proportional to the lattice volume, each conjugate gradient step takes a time which grows with the volume of the system. Thus the algorithm of Ref. 3 is expected to require computer time which grows as the square of the volume of the lattice. Such a severe growth has precluded use of this algorithm on any but the smallest lattices. Nevertheless, it does show the existence of an exact algorithm with considerably less computational complexity than would be required for a repeated direct evaluation of the determinant of the fermionic matrix.

Here and below when I discuss volume dependences, I ignore additional factors from critical slowing down when the correlation length is also allowed to grow with the lattice size. The assumption is that such factors are common for the local algorithms treated here. In addition, such slowing occurs in bosonic simulations, and I am primarily concerned here with the extra problems presented by the fermions.

This issue of critical slowing is potentially quite important, and has been mentioned in Sokal's chapter of this book. Cluster and multigrid type algorithms will inevitably become important to speed the above matrix inversions. At present, however, in lattice gauge theory the correlation lengths are rather small, and these acceleration algorithms are not yet used in major production runs. Thus I will not mention them further in this chapter.

Two other exact, but also volume squared, methods for fermionic simulation were presented in Ref. [6]. To use a Metropolis *et. al.* [7] scheme to find a configuration of A fields with distribution

$$P_{eq}(A) \propto |M(A)|e^{-S_G(A)}, \tag{16}$$

requires knowledge of how the determinant $|M|$ changes when A is replaced by a trial value A'. Actually one only needs the ratio of the old and the new determinants, and this can be calculated as an expectation value in two ways. First, if I construct an ensemble of complex scalar fields ξ with distribution

$$P(\xi) \propto e^{-\xi^* M(A)\xi} \tag{17}$$

then I have

$$\frac{|M(A')|}{|M(A)|} = \frac{1}{\langle \exp(-\xi^*(M(A') - M(A))\xi) \rangle}. \tag{18}$$

Alternatively, if I construct the fields ξ using the trial A'

$$P(\xi) \propto e^{-\xi^* M(A')\xi}, \tag{19}$$

then I have

$$\frac{|M(A')|}{|M(A)|} = \langle \exp(-\xi^*(M(A) - M(A'))\xi) \rangle. \tag{20}$$

Both approaches involve a Monte Carlo to find the ensemble of ξ fields inside the Monte Carlo determination of the A fields. Thus they are also volume squared algorithms.

Grady [8] found an intriguing variation on the second of these two approaches. In particular, the ensemble average over the ξ fields is unnecessary in this "look ahead" scheme where the probability for ξ is determined from the trial field A'. Consider a trial change A' chosen with a probability distribution $P_{T,A}(A')$. Solely to simplify the following equations, assume that this trial probability is symmetric under interchange of A and A'. Then generate a single ξ field with probability distribution as given in Eq. (19). The prescription is to accept the change with probability

$$P_{acc} = \min[1, \exp(S_G - S_G' + \xi^*(M' - M)\xi)]. \tag{21}$$

Here I use the shorthand notation S_G, S_G', M, and M' for $S_G(A)$, $S_G(A')$, $M(A)$, and $M(A')$, respectively.

282

To justify this procedure, consider the overall probability for taking A to A'

$$P(A \rightarrow A') = P_{T,A}(A')\frac{1}{Z_\xi}\int (d\xi^* d\xi)e^{-\xi^* M'\xi}P_{acc}. \tag{22}$$

Here I have defined the normalization factor for the ξ integral

$$Z_\xi = \int (d\xi^* d\xi)e^{-\xi^* M'\xi} \propto |M'|^{-1}. \tag{23}$$

Multiplying Eq. (22) by the equilibrium distribution in Eq. (16) and combining things gives

$$P_{eq}(A)\ P(A \rightarrow A') \propto |M||M'|P_{T,A}(A')$$
$$\int (d\xi^* d\xi)\min[e^{-S_a-\xi^* M'\xi},\ e^{-S'_a-\xi^* M\xi}]. \tag{24}$$

Remembering the symmetry of P_T, we see that this expression is symmetric under interchange of primed and non-primed variables. This symmetry is precisely the statement of detailed balance for the equilibrium distribution from Eq. (16).

The fact that a large ensemble of ξ fields is not required is a definite advantage of this approach. Nevertheless, it still contains a Monte Carlo inside a Monte Carlo to obtain the equilibrated ξ field. Thus as an exact algorithm it still requires computer time growing as the system volume squared.

The above exact approaches all require volume squared times. To avoid this, many fermionic schemes used in practice involve additional approximations. This usually involves an expansion in a step size for proposed changes. To eliminate systematic errors in principle requires an extrapolation to the limit of vanishing step. I will later return to exact algorithms and discuss how using a guided random walk can give a volume dependence intermediate between the above volume squared and the linear behavior of bosonic simulations.

The advantage of making small steps lies in the fact that a time consuming step such as a conjugate gradient inversion or a Monte Carlo generation of auxiliary fields need only be done once per sweep of all the gauge variables. In essence, this is an attempt to eliminate a Monte Carlo inside a Monte Carlo. Once one is making small steps anyway, there is no particular loss in using algorithms formulated in terms of a differential evolution. This is the basis of both the Langevin [9-10] and the microcanonical [11] methods discussed below. I begin this treatment of small-step algorithms with one of the oldest.

4. Pseudofermions

Fucito, Marinari, Parisi, and Rebbi [12] proposed a simple approximate method for calculating changes in the determinant of the matrix M. They begin by rewriting Eq. (12) in the form

$$Z = \int (dA)\ e^{-S_{PF}}. \tag{25}$$

where

$$S_{PF} = S_G(A) - \text{Tr} \log M(A) \tag{26}$$

For a Metropolis *et al.* [7] updating scheme one needs to know the change in the action upon a trial change of A. As a first approximation, consider making only small changes in the gauge field and then linearizing the change in the action

$$\frac{dS_{PF}}{dA} = \frac{dS_G}{dA} - \text{Tr}(M^{-1}\frac{dM}{dA}). \tag{27}$$

The quantity $\frac{dM}{dA}$ is easily calculated for a local M. The inverse of the matrix M is estimated using

$$(M^{-1})_{ij} = \langle \xi_j^* \, \xi_i \rangle \tag{28}$$

where the expectation value is over fields ξ, called pseudofermions, and distributed with weighting

$$P(\xi) \propto \exp(-\xi^* M \xi). \tag{29}$$

Note that this is the same distribution required in Eq. (17). A standard Monte Carlo simulation is used to give a set of N_c configurations of the ξ fields to estimate this expectation value. This simulation is normally done only once per full sweep of the lattice variables. This is not a major new assumption because a small-step-size approximation is already being made in using only the first derivative to calculate the changes in the action.

This algorithm, as several of the fermionic approaches discussed later, is not exact because of the approximation of small changes in A. The step size is merely a parameter in the standard applications of the Metropolis *et al.* [7] algorithm to bosons, but here it acquires a more significant role in characterizing an approximation. Whether the step size is sufficiently small can in principle be determined by comparing results for several values and doing an extrapolation to zero. Unfortunately, the amount of computer time necessary is sufficiently large that this check is rarely made.

Another source of error appears if the pseudofermionic fields are not calculated with the appropriate distribution. This would happen with insufficient equilibration time during the Monte Carlo simulation from which they are obtained. This error can in principle be eliminated by a trick if M is the square of a simple operator, say $M = DD^\dagger$. In this case consider first generating a random vector χ with a Gaussian distribution

$$P(\chi) \propto e^{-\chi^\dagger \chi} \tag{30}$$

As all components of χ are uncorrelated, it can be rather quickly generated. Then a simple change of variables gives a properly distributed pseudofermionic field

$$\chi = D^\dagger \xi \tag{31}$$

This equation can in principle be solved by some iterative algorithm such as the conjugate gradient method. This trick replaces the convergence of a Monte Carlo updating of the pseudofermionic fields with a potentially tedious inversion. I mention it here because this use of Gaussian random numbers is potentially quite useful with other fermionic algorithms as well.

The finite number of configurations of pseudofermionic fields used to estimate the expectation value in Eq. (28) introduces a random error into the estimate of the inverse of M. These errors, however, average out in the final extrapolation of observables to zero step size. This is a point I return to later when I discuss interpolation between pseudofermions and the Langevin approach.

5. Langevin, microcanonical, and hybrid schemes

Both the Langevin and microcanonical algorithms for lattice gauge theory are formulated as differential equations for evolution in a fictitious "time" τ. While these approaches are also applicable for the pure gauge theory, their main interest appears with fermionic simulations, where a differential evolution permits time consuming conjugate gradient inversions to be done only once per sweep of the lattice variables.

Rather than in terms of field theory, I frame this discussion in the context of a single degree of freedom, the coordinate x of a particle of mass m moving in one dimension. I do this because the basic ideas of these algorithms are nothing more than generalizations of Newton's equation. I will also treat the Langevin and microcanonical approaches together as limits of a more general hybrid formalism. I will be using a second order version of the Langevin equation, similar to that advocated by Horowitz. [13]

I begin by considering a particle moving in a potential $V(x)$. Newton's equation for the particle motion is

$$m\frac{d^2x}{d\tau^2} = -\frac{\partial V}{\partial x}.$$

(32)

I now doctor this motion by adding two terms. First I add a drag slowing the particle down with a force proportional to its velocity. This will tend to damp out any motion until the particle lies at a minimum of the potential. To keep things moving, I then add a random noise to the system. Thus consider the equation

$$m\frac{d^2x}{d\tau^2} = -\frac{\partial V}{\partial x} - \alpha\frac{dx}{d\tau} + (\frac{2\alpha}{\beta})^{1/2}\,\eta(\tau).$$

(33)

where α and β are parameters. Here the noise $\eta(\tau)$ formally satisfies

$$\langle\eta(\tau)\eta(\tau')\rangle = \delta(\tau - \tau').$$

(34)

How the $\eta(\tau)$ is actually defined will become clearer momentarily when I make the evolution discrete. I have written the coefficient of the noise as $(2\alpha/\beta)^{1/2}$ with hindsight. It is convenient to introduce the momentum p of the particle and rewrite this second order equation as two first order equations

$$\frac{dp}{d\tau} = -\frac{\partial V}{\partial x} - \frac{\alpha p}{m} + (\frac{2\alpha}{\beta})^{1/2} \, \eta(\tau), \tag{35}$$

$$\frac{dx}{d\tau} = \frac{p}{m}. \tag{36}$$

For simulation purposes the fictitious time is discreetly made discrete. Thus consider taking steps of size ϵ in τ. In one such step, p and x at time τ will become p' and x' at time $\tau + \epsilon$. Eq. 6.4 becomes

$$p' = p + \epsilon \, (-\frac{\partial V}{\partial x} - \frac{\alpha p}{m} + (\frac{2\alpha}{\beta})^{1/2} \, \eta), \tag{37}$$

$$x' = x + \frac{\epsilon p'}{m}. \tag{38}$$

Note that I have written the updating of x such as to use the new value p' of the momentum. This amounts to alternately updating the coordinates and the momenta of the system. Such a "leap frog" procedure effectively treats these variables at interleaved times. In the deterministic limit this advantageous technique serves to eliminate $\mathcal{O}(\epsilon^2)$ errors in the evolution. This transformation also preserves phase space volumes and is inverted by changing the sign of ϵ. These properties will be particularly helpful later when I return to exact algorithms.

The quantity η is obtained from a random number generator with probability distribution $\rho(\eta)$. The properties required of $\rho(\eta)$ are simply specified in terms of its moments

$$\int d\eta \, \rho(\eta) \, \eta^j = \begin{cases} = 1, & j = 0 \\ = 0, & j = 1 \\ = 1/\epsilon, & j = 2 \\ \leq \mathcal{O}(\epsilon^{-j/2}), & j \geq 3. \end{cases} \tag{39}$$

In this equation the first part indicates that the probability distribution is normalized, the second balances positive and negative noise, the third normalizes the delta function in Eq. (34), and the fourth eliminates nasty tails from the distribution. In many discussions the noise is considered as Gaussian, but this is not generally necessary.

To proceed, consider ensembles of particle coordinates and momenta. A necessary condition for any simulation algorithm is that it leave the equilibrium ensemble unchanged. Indeed, this condition is also sufficient if the algorithm is ergodic. Thus I am interested in finding ensembles of (x, p) pairs invariant under the evolution of Eq. (35), or its discretization in Eqs. (37-38).

Consider an ensemble with a probability density $P(x, p)$ of finding a state with given coordinate x and momentum p. Updating the states gives a new ensemble with probability distribution

$$P'(x',p') = \int dx\ dp\ P(x,p)P(x,p\to x',p')$$

$$= \int dx\ dp\ d\eta\ \rho(\eta)\ P(x,p)$$

$$\times \delta(p'-p-\epsilon\ (-\frac{\partial V}{\partial x}-\frac{\alpha p}{m}+(\frac{2\alpha}{\beta})^{1/2}\ \eta))$$

$$\times \delta(x'-x-\frac{\epsilon p'}{m}) \tag{40}$$

A little algebra gives the result

$$P'(x,p) = P(x,p)$$

$$+\epsilon\ [(\frac{\partial H}{\partial x}\frac{\partial P}{\partial p}-\frac{\partial H}{\partial p}\frac{\partial P}{\partial x})+\alpha(\frac{1}{\beta}\frac{\partial^2 P}{\partial p^2}+\frac{p}{m}\frac{\partial P}{\partial p}+\frac{1}{m}P)]+\mathcal{O}(\epsilon^2) \tag{41}$$

Here I have defined the Hamiltonian corresponding to the original Newton's equation of Eq. (32)

$$H = \frac{p^2}{2m}+V(x) \tag{42}$$

In deriving Eq. (41) it is necessary to keep terms of order η^2 because of the $1/\epsilon$ in the third part of Eq. (39).

Eq. (41) is a Fokker-Planck equation for the evolution of the probability density $P(x,p)$. It is now easily verified to order ϵ that a stationary distribution for this evolution is the simple Boltzmann weight

$$P(x,p) = \exp\{-\beta H(p,x)\}. \tag{43}$$

When α is non-zero, the algorithm is ergodic, and the solution is unique. Note that this distribution factors into a function of p times a function of x. Thus the equilibrium distributions of p and x are independent.

We see that repeated updating of an ensemble with the stochastic differential equation of Eq. (33) will eventually give thermal equilibrium at inverse temperature β. The source term can be thought of as a thermal bath coupled to the system. To apply these ideas to the gauge theory problem and obtain a distribution of fields as in Eq. (14), I merely generalize, replacing the variable x with the fields A and ϕ and replacing the potential $\beta V(x)$ with the action $S_G + \phi^* M^{-1}\phi$. Note that calculating the term involving $\frac{\partial V}{\partial x}$ will require, among other things, the evaluation of

$$\frac{\partial}{\partial A}\phi^* M^{-1}\phi = -\phi^* M^{-1}\frac{\partial M}{\partial A}M^{-1}\phi. \tag{44}$$

This involves $M^{-1}\phi$, which requires a conjugate gradient or equivalent inversion every time step.

It is interesting to consider various limits of this stochastic evolution. First, suppose that I had not included the drag term in Eq. (33). This situation follows from taking α to zero with β varying proportionally to keep the stochastic term. Thus noise without drag gives infinite temperature; that is, the random force will, on the average, increase the system energy without bound. Alternatively, if I include the drag but not the noise, the system will drop into a minimum energy state at effectively zero temperature. A finite temperature simulation requires the presence of both the drag and noise terms. The relation between the dissipative term proportional to α and the fluctuations of strength $(2\alpha/\beta)^{1/2}$ is the essence of the fluctuation dissipation theorem.

Note that the distribution in Eq. (43) is independent of the parameter α. Thus there is a class of algorithms. One of these corresponds to taking parameter m to zero. This can be effected by simultaneously adjusting α and rescaling the units of time. In this case Eq. (33) becomes first order and is the usual Langevin equation as used in Refs. [6] and [7].

$$\frac{dx}{d\tau} = -\frac{\partial V}{\partial x} + (\frac{2}{\beta})^{1/2} \eta(\tau). \tag{45}$$

For later reference, I write this in the discrete form for evolving x to x' in one time step of length ϵ

$$x' = x + \epsilon \times (-\frac{\partial V}{\partial x} + (\frac{2}{\beta})^{1/2} \eta). \tag{46}$$

Another interesting limit corresponds to taking α to zero while holding β constant. This case removes both the drag and noise terms, and returns simply to Newton's equation. This is the microcanonical approach, first advocated for pure gauge theories by Callaway and Rahman [14] and proposed for fermionic simulations in Ref. [11]. In this case the algorithm has no explicit dependence on β. Indeed, microcanonical algorithms require the temperature to be determined after the fact by some sort of thermometer. A convenient monitor is the average kinetic energy $\frac{1}{2}kT = \langle\frac{p^2}{2m}\rangle$. To change the temperature, one should start with a different total initial energy, which remains constant during the evolution.

Intermediate values of α give hybrid algorithms interpolating between the Langevin and microcanonical approaches. An alternative hybrid approach was proposed by Duane and Kogut, [15], who advocate updating with a microcanonical scheme for some number of iterations and then doing a step where all the momenta touch a heat bath. Eq. (43) shows that the momenta in equilibrium are Gaussianly distributed; thus, the latter step consists of replacing them all with new Gaussian random numbers.

This mixing of a molecular dynamics simulation with a randomizing noise on the momenta is is philosophically similar to keeping a small α in the above discussion. This would represent a small continuous refreshing of the momenta rather than a large refreshing of all momenta at the end of a "trajectory." Such an approach has been advocated in Ref. 13.

When the microcanonical trajectory length is short, the hybrid algorithm in fact becomes the first order Langevin equation. To see this, consider first replacing p with a Gaussianly distributed random number and then update the system microcanonically for a short time δ. This will take the coordinate x into

$$x' = x + \frac{p\delta}{m} - \frac{\delta^2}{2m}\frac{\partial V}{\partial x} + \mathcal{O}(\delta^3). \tag{47}$$

If the microcanonical updating time δ is small enough that the $\mathcal{O}(\delta^3)$ effects are negligible, then this evolution is identical to that in Eq. (46) where $\frac{\delta^2}{2m}$ plays the role of ϵ and $(\frac{\beta}{m\epsilon})^{\frac{1}{2}}p$ represents the noise η. Because the Langevin method is a special case of the hybrid approach, it will not in general represent the optimum choice of hybrid parameters.

In these hybrid approaches, one should adjust together both the step size ϵ and either the parameter α or the refreshing frequency in such a manner as to hold the finite step errors in observables at an acceptable size. The optimal algorithm minimizes the number of steps required to decorrelate lattices at a given error.

6. A Langevin-pseudofermion hybrid

In this section I discuss a fermion algorithm presented by Gavai and myself. [16] The approach has similarities with the Langevin evolution but is based on a small step-size limit of the Metropolis *et al.* [7] scheme. The interest in this approach is that a simple modification permits an interpolation to the pseudofermionic algorithm, thus clarifying the connection between them.

The goal is to generate an ensemble of configurations of fields A and ϕ distributed as in Eq. (14). To explicitly insure the positivity of the fermionic matrix M, assume that it is a square

$$M = DD^\dagger \tag{48}$$

This effectively doubles the number of fermionic species, one interacting with A via $D(A)$ and the other via $D^\dagger(A)$. I will later mention a possible way to remove this doubling. With this form for M, the desired probability distribution for A and ϕ is

$$P(A, \phi) \propto \exp(-S_G - \phi^*(D^\dagger)^{-1}D^{-1}\phi). \tag{49}$$

The algorithm consists of alternate sweeps through the ϕ and A fields. The ϕ updating is particularly simple, and represents a variation on eqs. (30) and (31). First generate a random vector χ with Gaussian weight

$$P(\chi) \propto e^{-\chi^\dagger \chi} \tag{50}$$

I now change variables and construct

$$\phi = D\chi. \tag{51}$$

This will be distributed with the desired probability

$$P_\phi \propto \exp(-\phi^*(D^\dagger)^{-1}D^{-1}\phi). \tag{52}$$

The Jacobian factor associated with the change of variables in Eq. (51) is irrelevant as the fields A are being held fixed during this step.

This construction is computationally fast because the individual components of χ are independent and because the matrix D is assumed to be local. Thus I can rapidly obtain a new ϕ field independent of its old value. This trick for updating ϕ is also used in the implementation of the Langevin algorithm in Ref. 9. Actually, the remainder of the algorithm does not explicitly need ϕ. Although I could eliminate this field and consider only χ, the discussion is simpler in terms of the coupled probability in Eq. (49).

In addition to the field χ, the updating of the gauge fields will require another quantity

$$\xi = (D^\dagger)^{-1}\chi = M^{-1}\phi. \tag{53}$$

This, unfortunately, is not so trivial to obtain, requiring a conjugate gradient inversion. Such a step is in common with the Langevin and microcanonical approaches.

I now come to the updating of the gauge field. What would be most desirable would be something like a Metropolis *et al.* [7] procedure where the acceptance of trial changes is governed by changes in the action

$$S(A,\phi) = S_G + \phi^*(D^\dagger)^{-1}D^{-1}\phi. \tag{54}$$

However, this is impractical because every time A is changed, D changes and its inverse on ϕ would have to be recalculated. To avoid this slow procedure, consider making only small changes in A. The changes in the action are then related to the first derivative with respect to A

$$\begin{aligned}
\frac{\partial S}{\partial A}\Big|_\phi &= \frac{\partial S_G}{\partial A} - 2\,\mathrm{Re}(\phi^*(D^\dagger)^{-1}D^{-1}\frac{\partial D}{\partial A}D^{-1}\phi) \\
&= \frac{\partial S_G}{\partial A} - \xi^*\frac{\partial D}{\partial A}\chi - \chi^*\frac{\partial D^\dagger}{\partial A}\xi.
\end{aligned} \tag{55}$$

where ξ is defined in Eq. (53).

Now consider the quantity

$$S_T(A,\chi,\xi) = S_G - \xi^*D\chi - \chi^*D^\dagger\xi. \tag{56}$$

Eq. (55) implies

$$\frac{\partial S}{\partial A}\Big|_\phi = \frac{\partial S_T}{\partial A}\Big|_{\chi,\xi}. \tag{57}$$

If I consider small changes in A, first order changes of the action S at constant ϕ equal the changes in S_T calculated at constant χ and ξ. As only changes in the action enter into the Metropolis *et al.* [7] algorithm, updating the A fields using S_T is equivalent in lowest order to using the exact action S. This is the proposal of Ref. 16, and is easily implemented because S_T is local.

As with the pseudofermion, Langevin, and microcanonical methods, this algorithm makes a small-step-size approximation. To have confidence that the errors induced by a finite step are small, one should study a desired measurable for a few values of this step size and extrapolate to the infinitesimal limit. Ref. 9 argued that with for the Langevin algorithm a finite step represents a simulation with an effective action which differs from the initial one by terms vanishing with the step size. If this new action has the same continuum limit, then these finite step simulations should give the same numerical results for physical observables. Nevertheless, an extrapolation to vanishing step is still necessary to compare results of different algorithms with a given set of parameters at a finite lattice spacing.

The solid points in figure 1, taken from Ref. 13, show the average plaquette $P = \langle \frac{1}{3} \text{Re Tr } U_p \rangle$ measured with this algorithm for the $SU(3)$ theory at $\beta = 4.5$. This is plotted versus the acceptance per hit, a simple measure of the step size. This simulation was done using the action from Ref. 14 with eight flavors and a fermion mass of 0.1 in lattice units. The zero step limit, which follows from extrapolation to unit acceptance, represents the correct plaquette value with the inclusion of the dynamical fermions. The crosses in this figure were obtained in a standard pseudofermionic run.

The Metropolis *et al.* [7] algorithm in the limit of small step size is quite close to the Langevin approach. Both cases involve small random changes in the field variables. A standard Metropolis *et al.* [7] program first tries unbiased changes about the old field, and then, to maintain the desired peaking of the distribution towards lower action, rejects a fraction of those changes which go towards larger action. In contrast, the Langevin approach always accepts the changes, but makes them in a direction biased towards lower action. This bias is determined by the same first derivative of the action with respect to A used above to construct S_T. The similarity of the approaches suggests that the finite step errors should be comparable. To directly make such a comparison, one should use a common definition of step size. One such measure would be to use the number of iterations needed to decorrelate lattices. I conjecture that the behavior of the solid points in figure (1) will mimic that of a Langevin simulation when plotted versus the decorrelation time.

We can now show the close connections of this algorithm with the pseudofermion method. To see this note that the field ξ has a probability distribution precisely the same as the pseudofermionic one in Eq. (29). Indeed, the present algorithm is equivalent to using but a single pseudofermionic field for the expectation value used in Eq. (28) to estimate M^{-1}. As mentioned earlier, the systematic errors

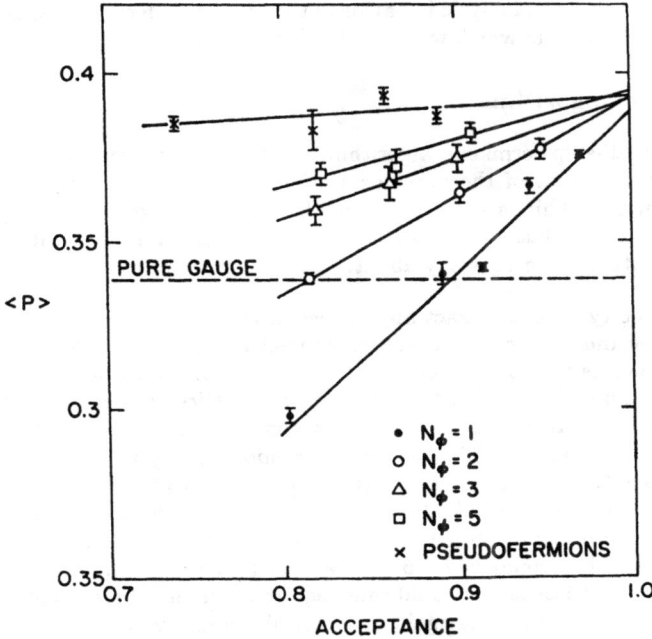

Fig. 1: The average plaquette as a function of the acceptance probability per Metropolis *et al.* [7] hit. The parameter N_ϕ is discussed in the text. Note the interpolation between the simple algorithm of this section as shown by the solid points and the pseudofermionic simulation shown by the crosses.

from using a finite number of pseudofermionic configurations average out after the extrapolation to zero step size.

Clearly the present algorithm represents an extreme case. One could interpolate between this and the pseudofermionic algorithm by averaging over some fixed number N_ϕ of ξ fields. This may also be thought of as considering N_ϕ species of fermions, each with its own ξ field, but then letting each species contribute only $1/N_\phi$ in the updating of the A field. As N_ϕ increases, I approach the pseudofermion algorithm. The remaining points in figure 1 exhibit this interpolation.

The allowing of each species to contribute only fractionally to the updating of the A field may provide a scheme to reduce the effective number of fermion species overall. Naively, this can remove the extra doubling introduced in Eq. (48) as well as any inherent doubling in the basic formulation of the fermions. Such a possibility has been frequently mentioned in the context all the approximate algorithms. There may, however, be some danger in this procedure because chiral symmetry breaking and anomalies suggest nonanalytic behavior as the number of fermionic species varies. This is an important issue which warrants further analytic study.

7. Exact global algorithms

These small-step fermionic algorithms, including pseudofermions, Langevin, microcanonical, and that of the previous section, all involve an extrapolation in a step size parameter. This is unfortunate in that lattice gauge calculations already involve tenuous extrapolations to zero lattice spacing and infinite volume, and this just gives us another thing to worry about.

The difficulty with the exact approaches discussed earlier is that a time consuming inversion must be done to test every trial change in the gauge field. The approximate schemes all work to reduce the frequency of such inversions to one per sweep. One could imagine making trial changes of all lattice variables simultaneously, and then accepting or rejecting the entire new configuration using the exact action. The problem with this approach is that a global random change in the gauge fields will generally increase the action by an amount proportional to the lattice volume, and thus the final acceptance rate will fall exponentially with the volume. The acceptance rate could in principle be increased by decreasing the step size of the trial changes, but then the step size would have to decrease with the volume. Exploration of a reasonable region of phase space would thus require a number of steps growing as the lattice volume. The net result is again an exact algorithm which requires computer time growing as volume squared.

So far this discussion has assumed that the trial changes are made in a random manner. If, however, one can properly bias these variations, it might be possible to reduce the volume squared behavior. An algorithm of this type was proposed in [17] For example, one could do either a Langevin or Metropolis *et al.* [7] sweep using the action S_T of the last section. By keeping track of all the probabilities for accepting changes along the way, one could in principle calculate the inverse probability for taking the new lattice back to the original in a similar sweep. Then one can construct a generalized acceptance for the entire lattice which will exactly restore detailed balance. If the changes in S_T are a good approximation to the changes in the true action, the factors in the acceptance criterion should tend to cancel, giving a reasonably large final acceptance rate. Attempts to use this approach in [18] were moderately successful, although those authors felt that standard hybrid techniques were superior. A variation on this idea where one does a global accept/reject step on the entire lattice after a microcanonical trajectory was presented in Ref. [19]. These algorithms have an interesting theoretical volume dependence that I now discuss. [20-22]

In some sense, the difficulties with fermions stem from the time consuming evaluation of $M^{-1}\phi$ appearing in the action of Eq. (13). For simplicity, let me not write ϕ explicitly and assume that I have some action which is particularly difficult to calculate; so, I want to evaluate it as rarely as possible. To further simplify the notation, I write the following equations in term of a single variable A.

To begin, consider a possible trial change of this variable A to

$$A' = A + p\delta + F(A)\delta^2. \tag{58}$$

Here δ is an adjustable step size parameter introduced for bookkeeping purposes. The "momentum" variable p represents a random noise, which for convenience I take to be Gaussianly distributed

$$P(p) \propto e^{-p^2/2}. \tag{59}$$

The function $F(A)$ represents a driving force or bias in the trial selection procedure and is for the moment arbitrary. It will soon correspond to the force term in the hybrid approach.

The Metropolis et al. [7] scheme accepts trial changes with a conditional probability chosen to maintain detailed balance when applied to an equilibrium ensemble. With an unbiased trial change this acceptance is determined entirely by the exponentiated change in the action. Here, however, the force term in the selection procedure must be corrected for in the acceptance condition. I can fully restore detailed balance by accepting the new value A' with probability

$$P_{acc} = \min[1, \exp(H(p, A) - H(p', A'))]. \tag{60}$$

Here H is a classical "Hamiltonian" analagous to that in Eq. (42)

$$H(p, A) = p^2/2 + S(A). \tag{61}$$

In Eq. 60 I introduce p' as the reverse noise, i.e. the noise which would be required for the selection of A as the trial had A' been the initial value

$$p' = -p - (F(A) + F(A'))\delta. \tag{62}$$

After the accept/reject step, the momenta should be refreshed; thus, they should be modified in a Monte Carlo or other fashion that the preserves the distribution in Eq. 59. Note that H is precisely the Hamiltonian used in the microcanonical algorithm [10] to describe evolution in "simulation time." Because of this analogy, I refer to H as the classical energy.

Note that with the second order terms in δ, the mapping defined by Eq. 58 and Eq. 62 exactly preserves areas in phase space

$$dA \, dp = dA' \, dp'. \tag{63}$$

Were this not so, the acceptance criterion would also need to depend on a ratio of measures.

Because of this preservation of areas, it is easy to show that the average change of energy is positive. In particular, I have

$$Z = \int dAdp \, \exp(-H(A,p))$$
$$= \int dA'dp' \, \exp(H(A',p') - H(A,p)) \, exp(H(A,p)) \tag{64}$$
$$= Z\langle\exp(H(A',p') - H(A,p))\rangle$$

where the expectation is with the weighting $e^{-H(p,a)}$. Using Jensen's inequality $\langle e^f \rangle \geq e^{\langle f \rangle}$, we find

$$\langle H(A',p') - H(A,p)\rangle \geq 0. \tag{65}$$

Thus regardless of the biasing force, on average the trial change will tend to increase the energy. A useful relation comes from expanding Eq. (64) in powers of the energy change. This gives immediately

$$\langle H(A',p') - H(A,p)\rangle = \frac{1}{2}\langle (H(A',p') - H(A,p))^2\rangle + \mathcal{O}((H' - H)^3) \tag{66}$$

To proceed I expand the energy change in the parameter δ. It is readily verified that

$$H' - H = (p\delta + \frac{1}{2}(p^2\frac{\partial}{\partial A} + 2F(A))\delta^2)(\frac{\partial S(A)}{\partial A} + 2F(A)) + \mathcal{O}(\delta^3). \tag{67}$$

This implies that the choice

$$F_L(A) = -\frac{1}{2}\frac{\partial S}{\partial A} \tag{68}$$

leads to an energy change

$$H' - H = \mathcal{O}(\delta^3) \tag{69}$$

Indeed, making this choice and ignoring the possibility of rejecting the trial change gives the usual Langevin algorithm [9,10], where the parameter δ is the square root of the step size used for discretization. In particular, compare Eq. (58) with (46), where ϵ corresponds to δ^2.

Let me first consider not making the Langevin choice for the driving force. In this case I use Eq. (66) to obtain

$$\langle H' - H\rangle_{A,p} = \frac{\delta^2}{2}\langle (\frac{\partial S(A)}{\partial A} + 2F(A))^2\rangle + \mathcal{O}(\delta^4). \tag{70}$$

Note that terms with odd powers of δ in the energy change expansion all involve odd powers of p and thus vanish on averaging. If I now consider updating some large number V of variables together, the positive $\mathcal{O}(\delta^2)$ quantities will coherently add and I expect to find a total energy change increasing linearly with V. By the central limit theorem, the fluctuations about this growth will become gaussian. Thus for large volumes I expect to find

$$H' - H \simeq C\delta^2 V + B\rho\delta V^{1/2} \tag{71}$$

where C and B are constants and ρ is a gaussian random variable which I normalize such that its probability distribution is

$$P(\rho) \sim e^{-\rho^2/2}. \tag{72}$$

If Eq. 71 were exact, then Eq. (64) would relate C and B

$$C = B^2/2. \tag{73}$$

With this explicit form for the energy change, I can obtain the expected acceptance in the large V limit

$$\langle P_{\text{acc}} \rangle = \langle \min[1, e^{H-H'}] \rangle = \frac{2}{\sqrt{\pi C V \delta^2}} e^{-CV\delta^2/4} \times (1 + \mathcal{O}(\frac{1}{CV\delta^2})). \tag{74}$$

The calculations required to derive Eq. (73) and Eq. (74), however, depend strongly on the tails of the distribution of the energy change and thus cannot be regarded as completely rigorous. For a further discussion on this point see [23]. For the following I will only assume that the expected acceptance is exponentially suppressed when $V\delta^2$ is large.

To avoid this exponential suppression and have a reasonable acceptance requires $\delta \sim V^{-1/2}$. However a small value for the step size raises the issue that the lattice will evolve only slowly from its original configuration. More precisely, consider taking N sweeps over the lattice. As the motion of A has both random and driven terms, the overall change in any given variable should go as

$$\Delta A = \mathcal{O}(\delta\sqrt{N}) + \mathcal{O}(\delta^2 N) = \mathcal{O}(\sqrt{N/V}) + \mathcal{O}(N/V). \tag{75}$$

The final result is that the number of sweeps required to obtain a substantially new configuration should grow as V. If V is proportional to the system volume, then the overall algorithm requires time growing as volume squared, one factor of volume from the number of sweeps, and the other from the fact that each sweep takes time proportional to the volume.

For a bosonic simulations this growth would be a disaster. The standard algorithms only grow as the system volume, and thus should be preferred over updating many variables simultaneously. On the other hand, for fermions this is the same overall growth observed above for the exact Monte Carlo inside a Monte Carlo approaches. That this is the same, however, indicates that there is no obvious additional penalty in going to global updates. Indeed, it might be possible to gain something by a judicious choice of F which will reduce the coefficient of this growth. This is the basis of the algorithms discussed below.

I now return to the Langevin choice of Eq. (68) for the driving force. Consider again updating only a single variable. At first glance one might think that since the exact action is so difficult to calculate, the requisite derivative for this force would be yet more intractable. Note, however, that this derivative was also needed to linearize the action for the approximate algorithms.

To proceed I slightly generalize this force and take

$$F(A) = -\frac{1}{2}\frac{\partial S}{\partial A} + g(A)\delta^2. \tag{76}$$

The $g\delta^2$ piece is included for the purpose of discussing possible higher order improvements. Using this, I calculate the next term in the expansion for the energy change

$$H' - H = -\frac{\delta^3}{12}(S_3\,p^3 - 3S_1\,S_2\,p - 24g\,p) + \mathcal{O}(\delta^4). \tag{77}$$

Here I use the notation

$$S_n = \frac{\partial^n S}{\partial A^n}. \tag{78}$$

Note that if S_3 is non-vanishing, i.e. if the theory is not harmonic, then no choice of $g(A)$ can make the $\mathcal{O}(\delta^3)$ term in this equation vanish for all p.

I now consider applying this procedure to a group of V variables simultaneously, as in the earlier discussion of unbiased changes. I am interested in any coherent addition of changes which could give an exponential suppression of the final acceptance. Because the $\mathcal{O}(\delta^3)$ term in Eq. 77 contains only odd powers of p, it will vanish on the average. The expectation for the energy change can again be most easily found using Eq. (66). With a little algebra and explicitly doing the average over p, I find

$$\langle H' - H \rangle = \frac{\delta^6}{96}\langle 2S_3^2 + 3(8g - S_3 + S_1\,S_2)^2 \rangle + \mathcal{O}(\delta^8). \tag{79}$$

This can be written in many forms; this expression as the sum of two squares emphasizes positivity.

I now return to updating a large number V of independent variables simultaneously. The positive contributions indicated in Eq. (79) will add coherently. Similar arguments to those leading to Eq. (74) now give an expected acceptance falling as

$$P_{\text{acc}} \sim e^{-CV\delta^6} \tag{80}$$

To have a reasonable acceptance requires only $\delta \sim V^{-1/6}$. This changes Eq. (75) to

$$\Delta A = \mathcal{O}(\delta\sqrt{N}) + \mathcal{O}(\delta^2 N) = \mathcal{O}(\sqrt{N/V^{1/3}}) + \mathcal{O}(N/V^{1/3}). \tag{81}$$

Thus, the number of sweeps for an independent lattice grows as $V^{1/3}$ and the overall computer time for decorrelation increases as

$$T \sim V^{4/3}, \tag{82}$$

This behavior is only slightly worse than the linear growth of the pure bosonic theory.

This algorithm was proposed in Ref. 17 and tested further with somewhat discouraging results in Ref. 18. Ref. [19] presents a quite promising variation, generally referred to as the "hybrid Monte Carlo" algorithm, which I now discuss. Recapitulating on the above treatment of biased updatings, I constructed both the trial new A and the noise needed to return

$$A' = A + p\delta + F(A)\delta^2 \tag{83a}$$

$$p' = -p - (F(A) + F(A'))\delta. \tag{83b}$$

This is an area preserving map of the (A, p) plane onto itself. The scheme proposed in Ref. 19 is to iterate the combination of this mapping with an inversion $p' \rightarrow -p'$ several times before making the accept/reject decision. This iterated map remains reversible and area preserving. The second order terms in this equation make it equivalent to the leap frog procedure with an initial half step as used in Ref. 19. The procedure thus generates a microcanonical trajectory.

The important point is that after each step the momentum remains exactly the negative of that which would be required to reverse the entire trajectory and return to the initial variables. If at some point on the trajectory I were to reverse all the momenta, the system would exactly reverse itself and return throughtr the same set of states from whence it came. Thus a final acceptance with the probability of Eq. (60) still makes the overall procedure exact. This slight modification of the hybrid algorithm of Ref. 15 makes it exact, just as the procedure with a single step removes the systematic errors of Langevin evolution. After each accept/reject step, the momenta p are refreshed, their values being replaced by new Gaussian random numbers. The fields ϕ could also be refreshed at this time, or less often, as turns out to be appropriate. The goal of the procedure is to use the microcanonical evolution as a way to restrict changes in the action so that the final acceptance will remain high for reasonable step sizes.

This procedure contains several parameters which can be adjusted for optimization. First is N_{mic}, the number of microcanonical iterations taken before the global accept/reject step and refreshing of the momenta p. Then there is the step size δ, which presumably should be set to give a reasonable acceptance. Finally, one can also vary the frequency with which the auxiliary scalar fields ϕ are updated.

The arguments for following a microcanonical trajectory for some distance before refreshing the momenta have been stressed in Ref. [15]. Refs. [21] and [22] show that this approach gives an algorithm where the computer time grows as $V^{5/4}$. I now review that argument.

The goal of the approach is to speed flow through phase space by replacing a random walk of the A field with a coherent motion in the dynamaical direction determined by the conjugate momenta. As long as the total microcanonical time for a trajectory is smaller than some characteristic time for the system, the net change in A will grow linearly with both N_{mic} and δ; thus Eq. 3.24 is replaced by

$$\Delta A \sim N_{mic}\delta. \tag{84}$$

which should be valid as long as

$$N_{mic}\delta < \mathcal{O}(1). \tag{85}$$

With large N_{mic}, the change in the classical energy will also grow. In any given microcanonical step the energy changes by an amount of order δ^3. For N_{mic} of order δ^{-1}, the total energy change will then be of order δ^2. Because the evolution preserves areas in phase space, Eq. (66) still applies to the overall evolution and I have for the expected energy change

$$\langle H' - H \rangle = \frac{1}{2}\langle (H' - H)^2 \rangle + \mathcal{O}((H' - H)^3) = \mathcal{O}(\delta^4). \tag{86}$$

Now if I update V independent variables together, these positive contributions can coherently add and earlier arguments give an overall acceptance falling as

$$P_{acc} \sim \exp(-CV\delta^4). \tag{87}$$

This means that δ should be taken to decrease with volume as $V^{-1/4}$. Correspondingly, N_{mic} should grow as $V^{1/4}$, the maximum allowed by Eq. (85). The final result is that the total time required to obtain a substantially changed lattice grows as

$$T \sim V^{5/4} \tag{88}$$

This may be only an asymptotic statement, valid for systems much larger than the correlation length. The main uncertainty lies in the unknown characteristic time scales that determine the $\mathcal{O}(1)$ right hand side of Eq. (85).

A variation on the hybrid Monte Carlo scheme in Ref. 13 essentially corresponds to keeping the parameter α of the earlier second order Lanbgevin discussion, and doing an acceptance after each step. Presumably this gives a similar time dependence to the above.

It has been argued [24] that the hybrid Monte Carlo algorithm will perform better if one does not always use a constant trajectory length. This will certainly be the case if the trajectory is some multiple of a fundamental frequency of the dynamical system being considered. Having variable trajectory lengths, in either a random of systematic manner, is a straightforward and worth while addition to the approach.

8. Higher order schemes

While the above theoretical volume dependence is quite promising, it is interesting to ask whether one can do better. Campostrini and Rossi [25] presented such a higher order scheme, while Ref. [26], generalized the scheme to reduce the errors to an arbitrarily small power of delta. This gives a higher order hybrid Monte Carlo scheme whose volume growth is arbitrarily close to that for bosonic simulations. On the other hand, the coefficient of this growth seems to increase with order, and in practice the above scheme appears to work adequately. I now review this higher order approach.

The change of variables in Eq. (83) is an area preserving map on (A, p) that conserves energy to order δ^3. It is often written in a "leapfrog" manner as a combination of the transformation

$$T_A(\delta) : (A, p) \longrightarrow (A + p\delta, \ p) \tag{89}$$

and

$$T_p(\delta) : (A, p) \longrightarrow (a, \ p - S'(A)\delta) \tag{90}$$

Eq. (83) is equivalent to the application of $T_p(\delta/2)T_A(\delta)T_p(\delta/2)$. I refer to this combintation as $T_2(\delta)$ because it preserves the energy through order δ^2).

I now wish to generalize this transformation giving a $T_n(\delta)$ which preserves the energy through order δ^n while maintaining the property of preserving areas in phase space. To apply the Metropolis *et al.* [7] algorithm one also needs the property of reversability

$$T^{-1}(\delta) = T(-\delta) \tag{91}$$

Then one can use this transformation to replace T_2 in the generation trial changes for the hybrid Monte Carlo algorithm discussed above, and presumably improve on the asymptotic volume dependence..

A simple construction of a suitable T_n is recursive; I consider combinations of lower order T_n which cancel out the errors at order δ^n. Since I have above an explicit T_2, I can recursively generate any higher order transformation required.

To proceed, consider the Hamiltonian as the generator of translations in time. In particular, for any function $F(t)$ which depends on the phase-space variables at time t, I write

$$e^{H\delta} : F(t) \longrightarrow F(t + \delta) \tag{92}$$

where $F(T + \delta)$ is the value of F after evolving along the exact classical trajectory for time δ. Suppose I now have an area preserving transformation T_n which gives the same evolution accuate through order δ^n. Thus I assume

$$e^{H\delta} = T_n(\delta) + \Delta\delta^{n+1} + \text{(higher order terms)} \tag{93}$$

Now consider two such transformations of different distances in time

$$T_n(\delta_2)T_n(\delta_1) = e^{(\delta_1+\delta_2)H} + \Delta e^{H\delta_1}\delta_2^{n+1} + e^{H\delta_2}\Delta\delta_1^{n+1} + \ldots \tag{94}$$

To order $n+1$ in the step sizes, this reduces to

$$T_n(\delta_2)T_n(\delta_1) = e^{(\delta_1+\delta_2)H} + (\delta_2^{n+1} + \delta_1^{n+1})\Delta + \ldots \tag{95}$$

Note that with the reversibility property of Eq. (91), $T(-\delta)T(\delta) = 1$ is exact. This implies that the operator Δ in the above equation must vanish when n is odd. Indeed, this is a simple way to understand why the simple leapfrog algorithm has errors that do not start until the third order in the step size. Indeed, only the elimination of odd powers of δ is a problem in going to higher orders. Thus I consider n even in the following.

For the inductive step, I now pick an arbitrary integer i. Consider taking i steps of size δ forward with our transformation T_n, and then one step backward with a different size $s\delta$, and finally i more forward steps of size δ. This combination results in a net motion of distance $(2i-s)\delta$, with the errors adding non-linearly

$$T_n(\delta)^i \, T_n(-s\delta) \, T_n(\delta)^i = e^{H(2i-s)\delta} + (2i - s^{n+1})\delta^{n+1}\Delta + \mathcal{O}(\delta^{n+3}).$$

If I pick

$$s = (2i)^{1/(n+1)},$$

the order δ^{n+1} terms cancel and I can write

$$T_{n+2}((2i-s)\delta) = T_n(\delta)^i \, T_n(-s\delta) \, T_n(\delta)^i \tag{96}$$

This algorithm is not unique, in particular there is the parameter i giving the number of forward steps before a backward one. In [26] it was argued that one should adjust i to minimize the largest step in a microcanonical trajectory of fixed length. Tests on simple models, however, showed that the simple leapfrog algorithm was quite adequate in most practical cases. One should bear in mind, however, in going to large systems that higher order schemes may eventually become useful.

9. Concluding remarks

So with all these algorithms, which is best? It appears that the local algorithms are all too time consuming for practical use. As the simple Langevin approach is a special case of the hybrid algorithm, one should certainly include the more general possibilities included in the latter. Adding the accept reject step to make the algorithm exact has the additional advantage that one need not worry about systematic biases beyond the lattice cutoff. These arguments are responsible for the current popularity of the hybrid Monte Carlo approach.

This method does, however, require the fermion matrix to be a square, requiring at least two species for Wilson fermions and eight for the Kogut Susskind case. Users of the hybrid algorithm without the global accept-reject step have argued for adjusting the number of fermion species by inserting a factor proportional to the number of flavors in front of the pseudofermionic term when the gauge fields are updated. Such a possibility was mentioned above in the discussion of the interpolation between Langevin and pseudofermionic methods. This modification is simple to make, but is not completely theoretically understood. Indeed, it may introduce spurious behavior if the physics is not smooth in the number of flavors. Nevertheless, this flexability of the hybrid algorithm has made it quite popular, although when the global accept condition can be applied, it is probably worthwhile for the extra confidence it supplies.

Despite the successes of these fermion algorithms, the overall procedure remains somewhat awkward, particularly when compared with the ease of setting up a pure bosonic simulation. This appears to be due to the non-local actions resulting from integrating out the fermions. Indeed, had one integrated out a set of bosons coupled quadratically to the gauge field, one would again have a non-local effective action, indicating that this analytic integration was not a good idea. Perhaps we should step back and explore algorithms before integrating out the fermions. One such attempt is the world line Monte Carlo method of [27]. Here one explicitly follows the paths of the fermions through the lattice and the Monte Carlo procedure involves random changes in these paths. The method has been shown to work well in two space-time dimensions, although sign problems have severely limited work in higher dimensionality. A recent attempt to treat these signs exactly using a recursive enumeration of paths appears to work well, although memory issues restrict the approach to extremely small systems. [28]

As Monte Carlo methods are particularly difficult with fermions, perhaps one should look for other numerical methods not based on stochastic processes. Here I include such possibilities as direct diagonalization of the Hamiltonian. So far these methods have been very constrained on possible system size, suggesting that one should look for new approximations to discard irrelevant information as the systems grow in volume.

An extremely difficult unsolved question is the simulation of fermionic systems when the corresponding determinant is not always positive. This situation is of considerable interest because it arises in the study of quark-gluon thermodynamics when a chemical potential is present. This issue is extensively discussed and referenced in Gavai's chapter of this book. All known approaches to this problem are extremely demanding on computer resources. One can move the phase of the determinant into the observables, but then one must divide out the average value of this sign. [29] This is a number which is expected to go to zero exponentially with the lattice volume;, thus, such an algorithm will require computer time growing exponentially with the system size. Another approach is to do an expansion about zero baryon density, but again to get to large chemical potential will require rapidly growing resources. New techniques are badly needed to avoid this growth; hopefully this will be a particularly fertile area for future algorithm development.

10. References

1. F.A. Berezin, The method of second quantization (Academic Press, NY, 1966).

2. P.T. Matthews and A. Salam, Nuovo Cimento 12 (1954) 563.

3. D. Weingarten and D. Petcher, Phys. Lett. 99B (1981) 333.

4. D. Weingarten and D. Petcher, Phys. Lett. 99B (1981) 333; H. Hamber and G. Parisi, Phys. Rev. Lett. 47 (1981) 1792.

5. A.N. Burkitt and A.C. Irving, Comp. Phys. Comm. 59 (1990) 447.

6. G. Bhanot, U.M. Heller, and I.O. Stamatescu, Phys. Lett. 129B (1983) 440.

7. N. Metropolis, A.W. Rosenbluth, M.N. Rosenbluth, A.H. Teller, and E. Teller, J. Chem. Phys. 21 (1953) 1087.

8. M. Grady, Phys. Rev. D32 (1985) 1496.

9. G.G. Batrouni, G.R. Katz, A.S. Kronfeld, G.P. Lepage, B. Svetitsky and K.G. Wilson, Phys. Rev. D32 (1985) 2736.

10. A. Ukawa and M. Fukugita, Phys. Rev. Lett. 55 (1985) 1854.

11. J. Polonyi and H.W. Wyld, Phys. Rev. Lett. 51 (1983) 2257; J. Kogut, J. Polonyi, H.W. Wyld, and D.K. Sinclair, Phys. Rev. Lett. 54 (1983) 1475.

12. F. Fucito, E. Marinari, G. Parisi, and C. Rebbi, Nucl. Phys. B180[FS2] (1981) 369.

13. A.M. Horowitz, Physics Letters 156B (1985) 89; Nucl. Phys. B280[FS18] (1987) 510; preprint Cern-Th-6172/91 (1991).

14. D. Callaway and A. Rahman, Phys. Rev. D28 (1983) 1506.

15. S. Duane and J. Kogut, Phys. Rev. Lett. 55 (1985) 2774; Nucl. Phys. B275 (1986) 398.

16. M. Creutz and R. Gavai, Nucl. Phys. B280 (1987) 181.

17. R. T. Scalettar, D.J. Scalapino and R.L. Sugar, Phys. Rev. B34 (1986) 7911.

18. S. Gottlieb, W. Liu, D. Toussaint and R.L. Sugar, Phys. Rev. D35 (1986) 2611.

19. S. Duane, A. D. Kennedy, B.J. Pendleton and D. Roweth, Phys. Lett., 195 (1987) 2.

20. H. Gausterer and S. Sanielevici, Phys. Rev.D38 (1988) 1220.

21. R. Gupta, G. Kilcup, and S. Sharpe, Phys. Rev. D38 (1988) 1278 (1988).

22. M. Creutz, Phys. Rev. D38 (1988) 1228.

23. A.D. Kennedy and B. Pendleton, Nucl. Phys. B (Proc. Suppl.) 20 (1991) 118.

24. P.B. Mackenzie, Phys. Lett. B226 (1989) 369; S. Gupta, preprint CERN-TH.6178/91.

25. M. Campostrini and P. Rossi, Nucl. Phys. B329 (1990) 753.

26. M. Creutz and A. Gocksch, Phys. Rev. Lett. 63 (1989) 9.

27. J.E. Hirsch, R.L. Sugar, and D.J. Scalapino, Phys. Rev. B26 (1982) 5033.

28. M. Creutz, preprint BNL-46782 (1991).

29. A. Gocksch, Phys. Rev. Letters 61 (1988) 2054.

SCALING, THE RENORMALIZATION GROUP
AND IMPROVED LATTICE ACTIONS

Rajan Gupta

T-8, Theoretical Division, MS-B285
Los Alamos National Laboratory
Los Alamos, N. M. 87545

1. Introduction

Many natural phenomena and mathematical models exhibit scaling which physicists exploit to make predictions. It has usually been the case that the origin of the observed scaling is not known. Fortunately, scaling in critical phenomena and in field theories is understood at a deep level; the successful underlying mathematical scaffold is the Renormalization Group (RG). In this chapter I attempt to show why scaling arises in lattice QCD, its consequences for numerical results, and how to improve scaling by changing the lattice action.

Scaling is formulated in a number of ways in different branches of physics even though the underlying concept is the same. First consider classical scaling based on dimensional analysis [1]. Let $\mathcal{F}(x_1, x_2, \ldots, x_n)$ be a dimensionful function of dimensional variables x_i. Let M, L and T be the fundamental units in terms of which $\mathcal{F}$ and x_i transform as

$$
\begin{aligned}
x_i &= M^{\alpha_i} L^{\beta_i} T^{\gamma_i} \\
\mathcal{F} &= M^{\alpha} L^{\beta} T^{\gamma} .
\end{aligned}
\tag{1.1}
$$

Under a change of units

$$
\begin{aligned}
M &\to M' = \lambda_M M \\
L &\to L' = \lambda_L L \\
T &\to T' = \lambda_T T
\end{aligned}
\tag{1.2}
$$

the variables x_i and the function $\mathcal{F}$ transform as

$$
x_i \to x_i' = \lambda_M^{\alpha_i} \lambda_L^{\beta_i} \lambda_T^{\gamma_i} x_i \equiv Z_i(\lambda) x_i
\tag{1.3}
$$

and

$$
\begin{aligned}
\mathcal{F} \to \mathcal{F}' &= \lambda_M^{\alpha} \lambda_L^{\beta} \lambda_T^{\gamma} \mathcal{F} \equiv Z(\lambda) \mathcal{F} \\
&= \mathcal{F}(Z_1(\lambda) x_1, Z_2(\lambda) x_2, \ldots, Z(\lambda) x_n)
\end{aligned}
\tag{1.4}
$$

The fact that physics cannot depend on the units used leads to three coupled partial differential equations obtained by considering variations with respect to λ_M, λ_L, and λ_T. The general form is ($I = M, L, T$)

$$\sum_{i=1}^{n} x_i \frac{\partial Z_i}{\partial \lambda_I} \frac{\partial \mathcal{F}}{\partial x_i'} = \frac{\partial Z}{\partial \lambda_I} \mathcal{F} . \tag{1.5}$$

Now using Eqs. (1.3) and (1.4) and taking the limit $\lambda_I = 1$, for example, gives

$$\sum_{i=1}^{n} x_i \alpha_i \frac{\partial \mathcal{F}}{\partial x_i} = \alpha \mathcal{F} . \tag{1.6}$$

The solutions to these equations are equivalent to constraints imposed by dimensional analysis and are just a formal mathematical expression of units invariance of physical laws.

Next consider the example of a ferromagnet in statistical mechanics. The magnetization M is in general a function of the temperature $t = T - T_c$ and the applied field h

$$M = M(t, h) \tag{1.7}$$

This equation cannot be arbitrary because it has to relate three dimensionful quantities. One expects a simplification on reformulating the equation as one between dimensionless variables. Close to the critical temperature the spontaneous magnetization for $h = 0$ is seen to behave as $M \sim t^\beta$, thus one can postulate the relation

$$\frac{M}{At^\beta} = \mathcal{Y}(\frac{h}{Bt^\Delta}) \tag{1.8}$$

where A and B are constants chosen to make the two scaled variables, M/At^β and h/Bt^Δ, dimensionless and Δ is the gap exponent. The resulting simplification now becomes clear; $\mathcal{Y}$ is a function of a single variable and more importantly it is independent of the particular ferromagnetic material chosen to define it. This statement of universality is borne out by experimental data.

In quantum field theories renormalization introduces a new hidden scale, the renormalization point μ. The invariance of physical quantities under the choice of μ is the basic content of the renormalization group. Unlike the example of dimensional analysis the invariance under change of μ has profound consequences. There is one additional complication compared to dimensional analysis, the transformation of field variables and operators under a change of scale μ is not given simply by the canonical dimensions. To make this explicit consider the constraint imposed by invariance under a scale change $\mu^2 \to \lambda\mu^2$ on a dimensionless correlation function $\Gamma(p^2/\mu^2, g(\mu^2))$, where p is a single external momentum,

$$\Gamma(\frac{p^2}{\lambda\mu^2}, g(\lambda\mu^2)) = Z(\lambda)\Gamma(\frac{p^2}{\mu^2}, g(\mu^2)) . \tag{1.9}$$

The generalization to many different momenta is straightforward. The additional factor $Z(\lambda)$ on the right is the aforementioned complication, introduced by the non-canonical scaling of the fields. It is a dimensionless finite number that can only depend on μ through the ratio Λ_{cutoff}/μ. We can once again derive the differential form of Eq. (1.9) by taking the derivative $\partial/\partial\lambda$ and setting $\lambda = 1$ at the end:

$$\mu^2 \frac{\partial\Gamma}{\partial\mu^2} + \mu^2 \frac{\partial g}{\partial\mu^2}\frac{\partial\Gamma}{\partial g} = \left.\frac{\partial Z}{\partial\lambda}\right|_{\lambda=1}\Gamma . \tag{1.10}$$

This typical renormalization group equation is usually written as

$$\left[\mu^2\frac{\partial}{\partial\mu^2} + \beta(g)\frac{\partial}{\partial g} - \gamma(g)\right]\Gamma(\frac{p^2}{\mu^2},g(\mu^2)) = 0 \tag{1.11}$$

where

$$\beta(g) \equiv \mu^2\frac{\partial g}{\partial\mu^2}$$
$$\gamma(g) \equiv \left.\frac{\partial Z}{\partial\lambda}\right|_{\lambda=1} = \mu^2\frac{\partial Z}{\partial\mu^2} \tag{1.12}$$

are respectively the β-function and anomalous dimension of the field. The general solution of this equation is

$$\Gamma(\frac{p^2}{\mu^2},g(\mu^2)) = \exp^{\int^g dg' \frac{\gamma(g')}{\beta(g')}} \mathcal{F}\left(\frac{p^2}{\mu^2} \exp^{\int^g \frac{dg'}{\beta(g')}}\right) \tag{1.13}$$

as can be checked by substitution in Eq. (1.12). Thus, invariance under renormalization scale change shows that Γ is not a function of two independent variables p^2/μ^2 and g but of just $\left(\frac{p^2}{\mu^2}\exp^{\int^g \frac{dg'}{\beta(g')}}\right)$. Later in Eq. (3.4) I show that to leading order in g the implicit dependence of $\mathcal{F}$ on μ through g cancels the explicit dependence, and in that case $\mathcal{F}$ scales trivially as it is independent of the hidden scale μ.

Now turning to lattice QCD, let me give a final example of scaling. In the pure gauge theory the glueball mass is given by the exponential fall off of the 2-point correlation function of Wilson loops. Lattice measurements give a dimensionless number Ma, the mass measured in lattice units, which in general is a function of the gauge coupling g

$$Ma = Cf(g) \tag{1.14}$$

In QCD a and g are not independent variables but are related by the renormalization group

$$\frac{f(g)}{a} = \Lambda_{QCD} \tag{1.15}$$

where Λ_{QCD} is a fundamental constant of the theory and characterizes the strength of strong interactions. The statement of scaling is that for sufficiently small a the lattice mass M changes with g because the scale used to measure it changes. Furthermore, this change of scale is given by a universal function $f(g)$, independent of the particular state used to define it.

Scaling, as introduced in the above examples, has corrections away from critical points where the correlation length ξ diverges. In this limit if ξ is the only relevant length scale then the problem simplifies. In lattice QCD this limit is at $g = 0$ or equivalently at $a = 0$ where one expects to recover the continuum field theory. Quantities measured at finite a have errors due to the discretization, for example the mass M of a typical hadron will behave, on dimensional grounds, as

$$Ma = Cf(g) \left(1 + \sum_{n=0}^{\infty} \alpha_n(\Lambda a)(g^2 \ln(\Lambda a))^n + \sum_{n=0}^{\infty} \beta_n(\Lambda a)^2(g^2 \ln(\Lambda a))^n + \ldots \right.$$
$$\left. + \sum_{n=0}^{\infty} \gamma_n(g^2 \Lambda a)(g^2 \ln(\Lambda a))^n + \ldots \right)$$

(1.16)

where Λ is some physical mass paramter. In this case the constant C is the non-perturbative number we wish to determine from lattice calculations. The hope is that the coefficients $\{\alpha_n, \beta_n, \gamma_n, \ldots\}$ are small in the region where simulations have been done so that the corrections can be estimated and corrected for. This is certainly not true at strong coupling, i.e. for $\beta \equiv 6/g^2 < 5.5$. Present non-perturbative lattice calculations using Monte Carlo methods have been done with β in the range $6.0-6.4$ where almost all correlation lengths are small, i.e. $\xi < 10$ (ξ_π is an exception as it can be tuned to an arbitrarily large value by decreasing the quark mass at any given β). When correlation lengths are small it is not a priori obvious that the scaling violations are small, therefore it becomes essential to ascertain whether these calculations represent continuum physics. To this end I will review ways to test and improve scaling.

The outline of this chapter is as follows. First I review the basic formulation of the renormalization group method followed by a discussion of its consequences for QCD, in particular why asymptotic freedom simplifies QCD. I then describe how scaling of lattice results can be tested and describe the Monte Carlo Renormalization Group (MCRG) method and the blocking transformations (RGT) used for 4-dimensional lattice gauge theories. Using these checks I analyze existing data for the string tension, T_c and glueball masses obtained in the pure gauge theory. I then discuss scaling violations and how to reduce these by improving both the gauge and fermion action. As examples I discuss the different improvement programs that are being explored at present. I end with a short discussion of the status of the improved results in the calculation of matrix elements.

2. Wilson's formulation of the Renormalization Group

Close to a critical point the correlation length in statistical mechanics models diverges. Because the degrees of freedom are correlated over macroscopic distances, the system looks very similar under a transformation that divides the lattice into small cells and replaces the degrees of freedom inside each cell by a cell average.

One can intuitively guess that the effects of this averaging are to cut down the correlation length by the scale factor b, where b is the linear dimension of the cell, and that the interactions between the new variables will be somewhat different. The correlation length decreases by a factor of b only because it is measured in lattice units, which in the above transformation changed from being the distance between the original lattice points to that between cells. If the block cell is small then one expects that the interactions between block spins will be local if the interactions between the starting spins were local. Using this very simple picture, Kadanoff, Fisher and Widom were able to derive phenomenological scaling laws for the behavior of thermodynamical quantities. For a detailed background on scaling see Refs. [2].

Wilson's formulation of the renormalization group method provided a systematic way of implementing the integration over a finite fraction of degrees of freedom in a system near its critical point, and quantifying the effect on the remaining variables. In so doing the method provides the mathematical infra-structure that explains scaling and universality observed in the behavior of physical systems near their critical points [3].

In order to understand the concepts it is very useful to first consider the case of an exactly soluble model, the $1 - d$ Ising model with nearest neighbor (NN) interactions. The averaging process is defined to consist of exact integration over every other spin in the chain. Clearly this process preserves the partition function, and in this very special case it turns out that the interaction between the surviving degrees of freedom remains NN, and is an analytic function of the original NN interaction. The process of integration can therefore be iterated, each time the new NN coupling K'_α is given in terms of the original coupling K_α by the same function $K'_{nn} = \mathcal{R}(K_{nn})$. The sequence of theories, specified by the couplings K^i_{nn}, generated under successive application of the RGT is called the RG flow. The critical behavior of the model can be unraveled from these RG flows [2] [4].

Now consider the simplest model in 2-dimensions, the 2-d Ising model with NN interactions [2]. In this case it is possible to integrate over spins on every even (or odd) site, however, this thinning of degrees of freedom induces next NN, third neighbor and 4-spin interactions between the remaining spins in addition to renormalizing the NN coupling. The new couplings are analytical functions of the original NN coupling. However, unlike the 1-d case, the process of integration can not be iterated to yield closed form solutions as at each step longer and longer range interactions get generated.

Wilson's formulation of the renormalization group is a generalization of these ideas. Given the dimensionality of the system and the symmetries of the variables, he, from the outset, considered the general class of theories defined by the infinite set, $\{K_\alpha\}$, of couplings corresponding to interactions of all possible range and complexity in the Hamiltonian H. He postulated that under a scale transformation, $\mathcal{R}_b$, performed by integrating over a fraction of the original degrees of freedom, the theory is mapped into another described by the renormalized couplings

$$K'_\alpha = \mathcal{R}_b(K_\alpha) . \qquad (2.1)$$

For most models we cannot determine this analytical relationship and have to resort to approximate numerical methods. The desired properties of a useful transformation are

1. The K'_α are analytical functions of the K_α.

2. If the starting K_α are local, the renormalized couplings also stay local under iteration of the renormalization group transformation. By local one now means that the strength of the couplings falls off exponentially with distance between spins in a given interaction term. This is a fundamental assumption of the RG method, whose validity has only been shown in the few solvable models and a *posteriori* by the consistency of numerical results.

3. The long distance properties of the original model are preserved under the RGT provided the integration is done in a sensible way.

4. The correlation length ξ measured in lattice units decreases by the factor b under each transformation, *i.e.* $\xi \to \xi' = \xi/b$. The ξ is preserved under the RGT in two special cases; if the starting Hamiltonian H has $\xi = 0$ or $\xi = \infty$. Theories with $\xi = 0$ are trivial, and examples of these are the $T = 0$ and $T = \infty$ limit of most statistical mechanics models.

The critical points H^c, with $\xi = \infty$, are special. Starting from a critical Hamiltonian H^c a RGT produces another H^c since $\xi' = \xi/b = \infty$. Thus the set of critical points define a hypersurface in the infinite dimensional space $\{K_\alpha\}$. The RG flows on this surface can (a) meander randomly, (b) go to some limit cycle or a strange attractor, or (c) converge to a fixed point H^*. At the fixed points the renormalized couplings are exactly equal to the original couplings, $K'_\alpha = K_\alpha$, and the theory reproduces itself at all length scales. This is in distinction to critical points where only the long distance behavior is reproduced. The location of the fixed point depends on the RGT and in general a given RGT will have more than one fixed point. To investigate critical phenomena we are interested in RGT which possess a critical fixed point *i.e.* $H^* = \mathcal{R}_b(H^*)$ with $\xi = \infty$. Each such fixed point has a basin of attraction *i.e.* the set of H^c that converge to it under the RGT. This basin of attraction defines the university class since the long distance behavior of all theories corresponding to these H^c is governed by the same fixed point.

Let me make the statement of universality more explicit with an example.

The Ising model is an idealization of a ferromagnet, yet it describes the critical behavior of ferromagnets like nickel, iron *etc.*, each of which has a different Curie temperature T^c. The reason for universality in their magnetic behavior is that all the H^c lie in the basin of attraction of the Ising fixed point. Unfortunately, even for the soluble $d = 2$ Ising model it has not been possible to determine H^*; the proof of its existence is based on consistency and circumstantial evidence. Thus one has to accept on faith that the Ising model has a fixed point based on the observation that it does an excellent job of describing ferromagnetic critical behavior.

Why do the H^c which lie in the basin of attraction of H^* have the same long distance physics? If we look at the behavior of the corresponding models at the level of each individual spin, then their correlation functions will be different at small separations. On the other hand, if we construct correlation functions out of effective spins that encapsulate the average behavior of all spins in a cell of size b^d, then the distinction between theories in the same universality class blurs as the cell size b is increased, *i.e.* the differences at short distances do not matter. A calculational tool that allows us to extract the long distance behavior of a substance by making the scale successively coarse is the RGT. The statement "same long distance behavior" is therefore equivalent to saying that scaled correlation functions at large separation are the same for the different models, or to saying that correlation functions of spins averaged over a sufficiently large block are the same at all scales larger than that block size.

Linearized Transformation Matrix and Classification of Exponents

Consider a point $\{K_\beta\}$ close to the fixed point $\{K_\beta^*\}$. Under the assumption that there are no singularities in the space of coupling constants, the transformation $K_\alpha'(K_\beta)$ can be written as a Taylor expansion about the fixed point

$$K_\alpha'(K_\beta) = K_\alpha(K_\beta^*) + \frac{\partial K_\alpha'}{\partial K_\beta}\bigg|_{K^*} \Delta K_\beta + \cdots . \qquad (2.2)$$

We shall restrict the discussion to small deviations from the fixed point, in which case one can define a "linear region" in which

$$\Delta K_\alpha' = \frac{\partial K_\alpha'}{\partial K_\beta}\bigg|_{K^*} \Delta K_\beta \qquad (2.3)$$

where $\Delta K_\beta = K_\beta - K_\beta^*$. The linearized transformation matrix

$$T_{\alpha\beta} \equiv \frac{\partial K_\alpha'}{\partial K_\beta}\bigg|_{K^*} \qquad (2.4)$$

controls the RG flows locally and is the starting point of MCRG analysis. Because of the truncation in (2.2) all statements will henceforth have the implicit assumption that they are only valid in some as yet undefined linear region close to H^*.

The critical exponents can be obtained from the eigenvalue equation $T\Phi = \Lambda_\Phi \Phi$. Here let me briefly state the properties of the matrix $T_{\alpha\beta}$ and its associated eigenvalues Λ_i and eigenvectors Φ_i (for more details see Refs. [3] and [2]):

1. The eigenvalues depend on the scale factor b. Since the theory obtained after two successive transformations by b should be the same as the one obtained after a single scale change by b^2, the general form of the eigenvalues is $\Lambda_i = b^{\lambda_i}$ where the λ_i are independent of the RGT.

2. The eigenvectors Φ are by definition the linear scaling fields. Consider a deviation u from H^* along a given eigenvector Φ, then under a RGT, the flow satisfies the equation

$$u'\Phi = T u\Phi = b^{\lambda*} u\Phi \quad . \tag{2.5}$$

Note the crucial use of the assumption of linearity in deducing this simple scaling of u.

3. Eigenvalues $\Lambda > 1$ lead to flows away from the fixed point as a deviation u grows under a RGT. These eigenvalues and the corresponding eigenvectors are called relevant.

4. Operators corresponding to eigenvalues $\Lambda < 1$ die out geometrically with the number of blocking steps. Therefore these operators do not contribute to the long distance properties of the system and are consequently called irrelevant. The associated exponents λ control corrections to scaling.

5. An eigenvalue that is exactly one is called marginal. To ascertain that an eigenvalue is truly marginal, one has to go beyond the linear approximation. Marginal operators do not flow under RGT. Typically marginal operators develop logarithmic corrections at higher order. Later we shall see that asymptotically free field theories are of this kind.

6. There is an additional class of eigenvectors, called redundant operators, that are not physical. Their eigenvalues can be < 1, $= 1$, or > 1. These eigenvalues depend on the choice of the *RGT*, *i.e.* different RGT with the same scale factor will give different λ_i. In fact one way to isolate them is to repeat the calculation of $T_{\alpha\beta}$ with a different *RGT*; the exponents λ_i that change with the RGT correspond to redundant operators. Redundant operators are given by the change in the Hamiltonian under a redefinition of fields and do not effect the physics. Since different non-linear RGT correspond to different rescaling of the fields, redundant operators are specific to the transformation. Even though they carry no physics information it is important to isolate these operators, especially in numerical calculations, in order to extract the physical ones.

7. Under certain conditions $T_{\alpha\beta}$ is block diagonal, *i.e.* the couplings break up into sub-sets that are closed under the *RGT*. For example, the spin flip symmetry $s \to -s$ causes the interactions to break up into odd and even sectors in the Ising model at zero magnetic field, and each sector has an independent eigenvalue spectrum.

Derivation of Exponents from the Eigenvalues of $T_{\alpha\beta}$

In order to derive the relationship between the eigenvalues of $T_{\alpha\beta}$ and the critical exponents ν and η I will assume that the reader is familiar with scaling relations in statistical mechanics. I will use the simple Ising model to illustrate the point and restrict the discussion by assuming that only the 2 relevant scaling fields t and h, corresponding to the temperature and the magnetic field, contribute. Then the relation between the correlation length exponent ν and the largest even eigenvalue of $T_{\alpha\beta}$ can be derived as follows. Consider the correlation length ξ at two temperatures with $t_1 < t_2$

$$\xi(t_1) \sim t_1^{-\nu}$$
$$\xi(t_2) \sim t_2^{-\nu} . \tag{2.6}$$

Using the RG relations $\xi(t_1) = b\xi(t_2)$ with $t_2 = R_b(t_1)$ we get

$$b = \left(\frac{t_1}{t_2}\right)^{-\nu}$$
$$\frac{t_2}{t_1} = b^{\frac{1}{\nu}} \tag{2.7}$$
$$\Rightarrow \frac{\partial t_2}{\partial t_1} = b^{\frac{1}{\nu}} = b^{\lambda_t} .$$

The generalization to the infinite dimensional space $\{K_\alpha\}$ is

$$\nu = \frac{1}{\lambda_t} = \frac{\ln b}{\ln \Lambda_t} , \tag{2.8}$$

where Λ_t is the largest eigenvalue of $T_{\alpha\beta}$ in the even sector.

The second independent exponent that can be obtained from MCRG is the correlation function exponent η. It is calculated from the largest odd (magnetic) eigenvalue Λ_h of $T_{\alpha\beta}$. To establish the connection between Λ_h and η involves using scaling and hyperscaling relations. These relations are given in [2] and I shall assume you are familiar with them. Starting with the scaling behavior for h given in Eq. (1.8) one gets

$$h \sim t^{\Delta} \sim t^{\beta\delta}$$
$$\Rightarrow h \sim \xi^{-\frac{\beta\delta}{\nu}} \tag{2.9}$$
$$\Rightarrow h' \sim \xi'^{-\frac{\beta\delta}{\nu}} = b^{\frac{\beta\delta}{\nu}} \xi^{-\frac{\beta\delta}{\nu}} = b^{\frac{\beta\delta}{\nu}} h .$$

Thus

$$\frac{\partial h'}{\partial h} = b^{\frac{\beta\delta}{\nu}} = \Lambda_h$$

$$\Rightarrow \frac{\beta\delta}{\nu} = \frac{\ln \Lambda_h}{\ln b} = \lambda_h .$$

(2.10)

Now using the scaling relations one can derive

$$\lambda_h = \frac{d+2-\eta}{2} .$$

(2.11)

Thus, for models with no hyperscaling violations, we can extract the two independent exponents ν and η from the leading relevant eigenvalues of $T_{\alpha\beta}$. The corrections-to-scaling exponents are related to the irrelevant eigenvalues. For a comprehensive review of these ideas, as applied to the 3D Ising model, see Ref. [5].

In order to highlight the difference between relevant and irrelevant operators I show two possible types of flows near a fixed point T^* in an idealized case of a single coupling T. In Fig. 1a, the fixed point is attractive under the RGT for all starting T which lie in the range $0 < T < T^*$ or in $\infty > T > T^*$. On the other hand the fixed point in Fig. 1b is repulsive, *i.e.* the flows go away from T^*. A relevant coupling is one for which deviations from the fixed point get magnified under RGT. Similarly, a coupling is called irrelevant if the deviations $\to 0$. Thus T is an irrelevant coupling in the example of Fig. 1a, and relevant for the case in Fig. 1b. Since flows on the critical surface converge to H^*, scaling fields contained on this hypersurface are irrelevant, while flows out of the critical surface define the relevant directions. For example, in the case of a ferromagnet H^* is unstable in two directions, *i.e.* with respect to variations in T and h. In fact T and h are the only two independent couplings for all ordinary second order transitions; as a consequence there exist scaling relations between the exponents. In general the number of independent couplings are equal to the number of unstable directions of H^*, accounting for the rich and complex cross-over behavior near multi-critical points.

Renormalized Trajectory

The trajectory flowing out of H^* is special and is called the *renormalized trajectory* (RT). The fixed point is unstable in the direction of the renormalized trajectory *i.e.* any deviation along it from H^* gets magnified under successive RGT. Yet along the RT, all irrelevant couplings are zero as was the case at H^*; so there are no scaling violations. Thus the long distance physics of a theory along the RT is the same as at the fixed point. The only difference is that when ξ is finite, all correlation lengths in the model transform as $\xi \to \xi/b$ under a RGT by a scale change b.

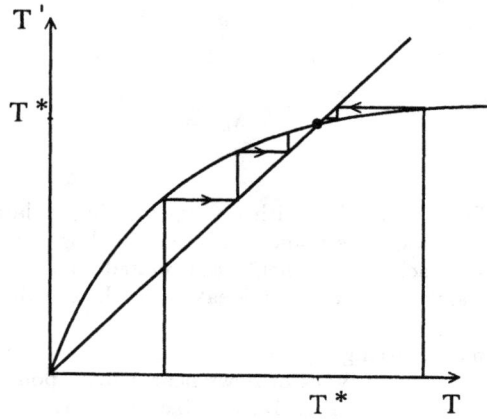

Fig. 1a: *Example of an attractive fixed point w.r.t. coupling T.*

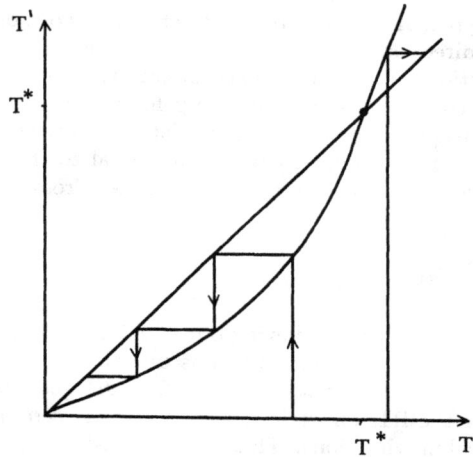

Fig. 1b: *Example of an repulsive fixed point w.r.t. coupling T.*

The flow along the RT can be parameterized by a single variable – the distance from the fixed point. Thus thermodynamic functions measured at different points along the RT, corresponding to different ξ, can all be expressed as a function of some reference H and the distance to it along the flow trajectory. For example, let t parameterize the distance along the RT from H^* in the even sector with $t' = b^{\lambda_t} t$ then the free energy transforms as

$$f(t,h) = \left(\frac{t}{t'}\right)^{2-\alpha} f\left(t', \left(\frac{t'}{t}\right)^{\Delta} h\right) \tag{2.12}$$

where $2 - \alpha = \nu d = d/\lambda_t$ and $\Delta = \lambda_h/\lambda_t$. From this scaling behavior of the free energy one deduces that the specific heat diverges as $c_v \sim t^{-\alpha}$.

Just as the H^* is the attractor for points on the critical surface, similarly the RT is the attractor for flows whose limit point is a H^c that lies in the domain of attraction of H^*. Another way to say this is that there is a one to one association between the H^c which lie in the basin of attraction of H^* and flows originating from them that are attracted by the RT. It is this association that we shall invoke later on to justify the extraction of long distance observables from lattice calculations done using the simple Wilson action.

Note that in principle one could work at any finite ξ along the RT and extract the critical behavior. Once $\xi \approx 1$ there is no distinction between short and long distance behavior and one expects calculations to become simple. On the other hand, for ξ small, the RT may be very weakly attractive and it may no longer be local. If this is the case, then a simple H (as is commonly used in simulations) will not lie in the domain of attraction of the fixed point and one cannot extract the desired critical behavior by simulating it. Towards the end of this chapter I will highlight the tug of war between simulating the simplest lattice action and being able to extract continuum physics at the shortest possible value of ξ.

3. Asymptotic Freedom Simplifies QCD

To analyze QCD using the renormalization group we want to first locate the relevant fixed point. Since pure gauge SU(3) in the continuum has only one coupling constant g, we should be able to find a critical point of the lattice theory by tuning g in Monte Carlo simulations. Historically, prior to the formulation of Lattice QCD, it proved very interesting to analyze the properties of QCD about $g = 0$, which in usual statistical mechanics systems is a trivial fixed point. This calculation can be done using perturbation theory, and the landmark result obtained by Politzer and by Gross and Wilczek is that the β-function (differential flow equation for g with respect to a change in momentum scale, i.e. $\frac{\partial g}{\partial \log \mu}$) is negative [6]. In terms of real space renormalization that we have discussed so far, this corresponds to following the flow along the RT backwards *i.e.* making the lattice spacing smaller. The importance of a negative β-function is that $g \to 0$ as $\mu \to \infty$, *i.e.* the perturbative fixed point at zero bare coupling is ultra-violet stable. This phenomena of the decrease in strength of $g(\mu)$ with increasing energy μ is called asymptotic freedom.

This result immediately makes a powerful case for QCD as the theory of strong interactions. The reason is that existing experimental data show that the strength of strong interactions decreases as the momentum exchanged in a process increases. This observation, along with other key features like the successful classification of hadrons using SU(3) symmetry, the need for 3 colors to resolve the $\pi^0 \to \gamma\gamma$ decay rate and the demonstration through scattering experiments of point like constituents inside hadrons, all follow naturally from QCD. However, detailed verification of QCD is hampered by the fact that only hadrons are seen as asymptotic states in nature. Thus, to demonstrate that QCD is the correct theory of strong interactions, one has to show that the observed hadrons arise as bound states of quarks and gluons and that quarks and gluons cannot exist as isolated states. One has also to show that the experimentally measured scattering rates can be predicted using QCD. In both cases the standard analytical tool – perturbation theory – runs into trouble because the coupling constant is of order $O(1)$ at typical hadronic scales. So, in order to verify QCD one has to develop non-perturbative methods to handle energy scales of 1 GeV or less where the coupling constant is of order unity.

Since the isolated observed states in nature are hadrons, even the highest energy scattering experiments involve non-perturbative parts, *i.e.* the initial distribution of quarks and gluons inside a hadron and the final formation of hadrons from the collision debris. In the intermediate region, which one hopes to describe in terms of hard scattering of quarks and gluons, it is not a priori true that the momentum exchanged in all sub-processes is large enough for perturbation theory to be valid. To make matters worse, the division into initial, final and "hard" regions has grey boundaries. So once again we are forced to consider non-perturbative techniques in order to verify QCD.

The important property of QCD that quarks and gluons are not seen as asymptotic states in nature is termed confinement. This aspect of QCD was addressed by Wilson (and earlier by Wegner for the simpler model – Z_2 gauge theory) using the strong coupling (g large) expansion. He showed that all gauge theories in 4 dimensions have an area law *i.e.* the expectation value of large Wilson loops is dominated by the area A of the loop, $\langle Wilson\ loop \rangle \sim \exp(-\sigma\ A)$, where the coefficient of the area term is the string tension σ [7] [8] [9]. An area law for Wilson loops implies that the potential has a linear piece. Such a potential would prevent a quark and an antiquark from separating over arbitrary distances because the energy stored in the string between them would grow linearly with the separation. If one were to try to separate a $q\bar{q}$ pair, then at some separation it would become energetically favorable to pop a $q\bar{q}$ pair out of the vacuum to form a meson pair that interacts through a screened short-range potential. Consequently one could never create isolated quarks in the laboratory. In this way the riddle of confinement is explained by a very simple and elegant mechanism.

Once it was shown that QCD incorporates asymptotic freedom (using perturbation theory which is valid around $g = 0$) and confinement (using strong coupling expansions around $g \to \infty$), one still had to demonstrate that both features co-exist. A "proof" of co-existence would be to show that the string tension σ in the lattice regularized theory is non-zero for all values of g. A non-zero value of σ in the continuum limit (defined as taking the lattice spacing $a \to 0$ or equivalently $g \to 0$) implies that linear confinement and asymptotic freedom occur simultaneously in non-abelian gauge theories. To extend the strong coupling result to $g \to 0$ a sufficient condition is that Lattice SU(3) gauge theory should not have a phase transition separating the weak coupling phase from the strong coupling phase. We shall return in section 7 to the status of the "proof" that $\sigma \neq 0$ as $g \to 0$ in numerical calculations.

If we assume that the fixed-point of Lattice QCD at $g = 0$ is the relevant fixed point at which the continuum theory describes strong interactions, then the perturbation theory result of Politzer and Gross and Wilczek predicts the scaling behavior of all observables. The basic relation is the 2-loop β-function, which for the SU(N) gauge theory with n_f light flavors is

$$\beta(g) \equiv \frac{\partial g}{\partial(\log \mu)} = -\beta_0 g^3 - \beta_1 g^5 - \cdots \tag{3.1}$$

where μ is the momentum scale and

$$\beta_0 = (11N - \frac{2n_f}{3})/16\pi^2$$
$$\beta_1 = (\frac{34N^2}{3} - \frac{10Nn_f}{3} - \frac{n_f(N^2 - 1)}{N})/(16\pi^2)^2 \ . \tag{3.2}$$

These first two terms in the expansion of $\beta(g)$ are gauge and regularization scheme invariant. (Note that the definition of $\beta(g)$ here is larger by a factor of 2 compared to the definition in Eq. (1.12).) Using this result we first show how QCD dynamically generates a mass scale. Integrating Eq. (3.1) from momentum scale μ_1 to μ_2 with $\mu_2 > \mu_1$ and keeping only the β_0 term to simplify the standard calculation, gives

$$\frac{1}{2\beta_0 g^2(\mu_2)} - \frac{1}{2\beta_0 g^2(\mu_1)} = \log \frac{\mu_2}{\mu_1} \ . \tag{3.3}$$

This solution shows that the coupling constant of non-abelian gauge theories runs, *i.e.* g depends on the momentum scale of the process, and $g(\mu) \to 0$ as $\mu \to \infty$. It also defines Λ_{QCD}, an invariant of the theory with dimensions of mass:

$$\frac{1}{2\beta_0 g^2(\mu)} - \log \mu = constant$$

$$\implies \quad \exp\left\{\frac{1}{2\beta_0 g^2(\mu)}\right\} = \frac{\mu}{\Lambda_{QCD}} \tag{3.4}$$

$$\implies \quad \alpha_s(\mu) = \frac{g^2(\mu)}{4\pi} = \frac{1}{8\pi\beta_0 \log \frac{\mu}{\Lambda_{QCD}}}$$

So we find that QCD, which is a theory with a dimensionless coupling constant and no intrinsic mass scale in the absence of quark masses, dynamically generates a mass scale. This happens because in order to specify g one has to specify a momentum scale. Furthermore, one can describe the strength of strong interactions either by the running coupling $g(\mu)$ or equivalently by the mass scale Λ_{QCD}. This phenomena is called dimensional transmutation. Note that this 1-loop result, Eq. (3.4), shows why the argument of $\mathcal{F}$ is independent of μ in Eq. (1.13).

The logarithmic change of $\alpha_s(\mu)$ with energy scale μ (see Eq. (3.4)) is the reason why it has been hard to quantify scaling in experimental data. The same slow change afflicts lattice calculations in that we can only hope to change the input bare coupling g in the range $1.0 - 0.90$ even with dedicated next generation supercomputers. Nevertheless, I hope to convince you that this interval is sufficient to test QCD.

To determine the value of Λ_{QCD} from experiments one equates a measured number to its prediction calculated in terms of Λ_{QCD} using QCD perturbation theory. For example, an estimate of $\Lambda_{QCD} \approx 200 MeV$ in MOM scheme, obtained using the ratio R measured in e^+e^- annihilation of the total hadronic cross-section to $\mu^+\mu^-$ production. Note that we had to specify the renormalization scheme, MOM, for otherwise the definition of Λ_{QCD} is not unique; the value of Λ_{QCD} depends on the the precise relation one chooses to use to specify how g varies with μ. For example, if we extend the above analysis to include β_1 in Eq. (3.2) then Eq. (3.4) is modified to

$$\Lambda_{QCD} = \mu \left(\frac{1}{\beta_0 g^2(\mu)}\right)^{-\frac{\beta_1}{2\beta_0^2}} \exp[-\frac{1}{2\beta_0 g^2(\mu)}] = \mu \, f_p(g(\mu)) \ . \tag{3.5}$$

However, once the value of Λ is determined in one scheme it can be related by a 1-loop calculation to that in any other perturbative scheme. For example in the lattice regularized theory $\Lambda_{lattice}$ is also defined by Eq. (3.5) but with μ replaced by $1/a$. Then to 1-loop

$$\frac{\Lambda_{QCD}}{\Lambda_{lattice}} = \mu a \, \exp\left\{-\frac{1}{2\beta_0}\left[\frac{1}{g^2(\mu)} - \frac{1}{g^2(a)}\right]\right\}. \tag{3.6}$$

In perturbation theory the two coupling constants are related as

$$g^2(\mu) = g^2(a)\left\{1 - \beta_0 g^2(a)\left(\log(\mu a)^2 - \log C^2\right) + O(g^4)\right\} \tag{3.7}$$

By substituting Eq. (3.7) into Eq. (3.6) one finds that

$$\Lambda_{QCD} = C\,\Lambda_{lattice} \tag{3.8}$$

i.e. the two constants, Λ_{QCD} and $\Lambda_{lattice}$, are related by a multiplicative constant. To calculate C requires knowing the finite part of the coupling constant renormalization to 1-loop in both the lattice and continuum regularization schemes [10]. The result of such a calculation, choosing the continuum scheme to be MOM, are [11]

$$\Lambda_{MOM} = 83.5\,\Lambda_{lattice} \qquad \text{pure gauge SU(3)}$$
$$\Lambda_{MOM} = 105.7\,\Lambda_{lattice} \qquad 3-\text{flavor QCD} \tag{3.9}$$
$$\Lambda_{MOM} = 117.0\,\Lambda_{lattice} \qquad 4-\text{flavor QCD}$$

It is important to note that even though the above definition and estimate of Λ_{QCD} is made using perturbation theory, it is intrinsically a non-perturbative quantity.

A simple consequence of asymptotic freedom is that the scaling law for all physical quantities with the dimensions of mass is the same, up to an unknown proportionality constant, as Λ_{QCD}. This means that in a lattice simulation

$$M_i a = c_i f(g) \equiv c_i \Lambda_{lattice} a . \tag{3.10}$$

Thus, for example, the string tension measured in lattice units, σa^2, scales like $(\Lambda_{lattice} a)^2$. An important consequence of confinement is that the constants c_i, one for every massive state in the theory, are fundamentally non-perturbative quantities. The stature of the function $f_p(g)$ defined in Eq. (3.5) is now elevated to that of a non-perturbative universal scaling function $f(g)$ defined in (3.10), i.e. independent of the physical quantity used to define it and whose domain of validity extends beyond perturbation theory.

In principle one could imagine that $f(g)$ depends on the choice of which physical quantity to hold fixed while taking the limit $a \to 0$. For sufficiently small g the function $f(g(a))$, which tells us how g changes with the cutoff a, is independent of the particular physical quantity chosen to define it. This interval of g is called the scaling region with $f(g)$ a universal scaling function. The more restrictive region, called the asymptotic scaling region, is defined as one where $f(g)$ is universal and given by the perturbative relation Eq. (3.5). Note that the scaling region includes the asymptotic scaling region, and it is possible that the two may coincide.

Both this scaling function $f(g)$ and its 2-loop asymptotic form $f_p(g)$ are universal in some, as yet undetermined, neighborhood of $g = 0$. It is precisely the existence of this universal scaling function in some finite neighborhood of the fixed point that justifies Monte Carlo calculations. In the scaling region all mass-ratios are by definition constant and equal to their values in the continuum. At the edge of the scaling region lattice artifacts will manifest themselves; unfortunately present numerical data lie in this grey region. Our goal therefore is to improve the quality of the data by either reducing the scaling violations mentioned in (1.16) or approximately calculating them.

To conclude, let me once again reiterate that it is the existence of a scaling region that allows a reliable estimate of mass-ratios in QCD from calculations done at finite and rather coarse values of a! The important quantitative questions, which we shall address later, are: what is the extent of the scaling region for QCD and how large are the scaling violations in present calculations?

In the next section I will review methods to calculate the non-perturbative β-function using data from lattice simulations and the Monte Carlo Renormalization Group method. Let me now address the question – why are non-abelian gauge theories non-trivial if the fixed point is at zero bare coupling? The answer is simple: even though the bare charge goes to zero in the continuum limit, the corresponding renormalized charge converges to a finite value there. To justify the statement that the vanishing of $g(a)$ as $a \to 0$ does not imply $g(\mu = 1\ GeV)$ is zero consider the following qualitative picture.

To calculate $g(\mu = 1\ GeV)$ we have to back off from $a = 0$ and calculate a physical quantity defined at scale $\mu = 1\ GeV$. Let this quantity be the mass Ma of a massive hadron (say glueball) measured in lattice units. Then at a fixed non-zero value of the lattice spacing a the mass Ma is a dimensionless non-zero number. Now if the limit $a \to 0$ can be taken with M held fixed, then we get a well defined non-trivial spectrum in the continuum limit. The limiting process simply states that the mass in lattice units goes to zero because we have taken the unit of measurement, a, to zero while holding M fixed. Now instead of the hadronic mass consider Λ_{QCD}, or equivalently the renormalized coupling constant at scale $1\ GeV$ say. The corresponding statement of the limiting process for the coupling constant is that $g(a)$, the coupling at scale a, goes to zero as $a \to 0$ according to Eq. (3.4) but with $g_R(\mu = 1\ GeV)$ held fixed. Note that $g_R(\mu = 1\ GeV)$ is related to the constant $\Lambda_{QCD} = \mu f(g(\mu))$ by dimensional transmutation.

One implication of our discussion is that if all states in QCD have zero mass in the continuum limit then $g = 0$ would be a trivial fixed point as g_R would also be zero. On the other hand one can ask whether even one of the states of QCD can be massless while the others are massive? It is interesting to note that we can actually say something about this aspect of the spectrum of QCD once the scaling behavior is fixed by Eq. (3.10). In fact under the following two assumptions QCD predicts a mass gap:

(1) There are no zero mass states at any finite non-zero value of g, $i.e.$ the dimensionless quantity ma does not vanish at any non-zero value of the lattice spacing a in a region that lies in the same thermodynamic phase as the continuum limit at $g = 0$.

(2) There exists only one relevant coupling g in QCD, and corresponding to it a single universal scaling function defined by say Eq. (3.5). If so then the scaling behavior of all observables is fixed and all mass-ratios have to stay finite as the continuum limit is taken.

These two assumptions are borne out by present numerical data, $i.e.$ there is no evidence for a mass-less state in the pure gauge sector. Also there is no indication of a phase boundary separating the region where simulations have been done and $g = 0$. Therefore, as things stand lattice QCD predicts that the lightest glueball state is massive.

Now let us introduce n_f flavors of almost massless quarks into the theory. Spontaneously broken chiral symmetry will give rise to $n_f^2 - 1$ almost massless pions in the spectrum. The masses of these quarks are additional parameters which have to be tuned to their value observed in the physical world. Alternatively, this tuning can be done using the associated pion. For example, in lattice calculations with three flavors, we can tune either the mass of the u, d, s quarks or that of the π, η and η' mesons to describe the physical world. We can also self-consistently take the chiral limit (with $n_f^2 - 1$ massless pions in the limit of zero mass quarks) without invalidating our previous conclusion about the mass-gap. The reason is that the pion mass is an independent parameter and not a prediction of the theory. We have to tune the pion mass to define the physical world at a given lattice scale. If this value is zero, then it is tuned to zero for all a.

To conclude, once we have verified that the scaling law given in Eq. (3.10) is valid then the only unknowns needed to predict the spectrum of QCD are the constants c_i. Asymptotic freedom has simplified QCD by allowing us to calculate the scaling behavior of all observables. The physics, however, lies in the coefficients

c_i which are intrinsically non-perturbative. The only first principle way we know of at present for determining them is numerical calculations. The justification of these Monte Carlo calculations rests in the RG machinery developed to handle critical phenomena.

For a review of the status of spectrum calculations see [12] [13] [14], and the chapter by DeGrand in this book. As a general reference the proceedings of the Lattice QCD conferences contain a yearly update on the progress made in solving field theories non-perturbatively.

4. Non-Perturbative β-function and Scaling

In the last section we showed how asymptotic freedom controls the scaling behavior of all physical quantities near $g = 0$. We defined scaling and the more restricted possibility of asymptotic scaling which occurs when the universal scaling function $f(g)$ converges to the 2-loop perturbative result. At $\beta \approx 6.0$, there is no theoretical reason for scaling to hold so one has to check whether lattice calculations reproduce continuum physics by, for example, showing that mass-ratios are independent of β for some $\beta \geq \beta_{scaling}$. We can do somewhat better; we can actually determine the non-perturbative function $f(g)$ using different physical quantities. Therefore, I will follow the following three fold approach to test lattice results:

(1) Check asymptotic scaling by plotting M/Λ_{2-loop}.

(2) Check scaling by plotting mass-ratios.

(3) Determine the scaling function by calculating its discrete version $\Delta\beta$.

In pure gauge theory three physical quantities have been measured in many lattice simulations – the string tension σ, the glueball spectrum and the deconfinement transition temperature T_c. For each of these observables we will construct $\Delta\beta$, the change in β for a scale change b, using data obtained at different values of the coupling. By comparing these different $\Delta\beta$ we can estimate the minimum value $\beta_{scaling}$ beyond which scaling sets in to within some accuracy.

The method to construct $\Delta\beta$ is very simple; let ma_1 and ma_2 be the calculated values of an observable, like the mass of a hadron measured in lattice units, at two values of the coupling, $\beta_1 \equiv 6/g_1^2$ and $\beta_2 \equiv 6/g_2^2$, corresponding to lattice scale a_1 and a_2 respectively. Then $\Delta\beta = \beta_1 - \beta_2$ for a scale change $b = ma_2/ma_1 = a_2/a_1$. Assuming that the variation of $\Delta\beta$ is essentially linear in the interval $\beta_1 - \beta_2$, the measured $\Delta\beta$ is the value at the midpoint of the interval, i.e. at $(\frac{\beta_1+\beta_2}{2})$. Note that in this definition a particular physical mass m is kept constant as the lattice scale a is changed, so this estimate is the value of $\Delta\beta$ for that observable.

The advantage of this method is that one works directly with physical observables calculated at essentially zero temperature (a finite lattice size is equivalent to a system at finite temperature). The disadvantage is that these long distance quantities have large statistical errors and in many cases it is not clear whether the values quoted by different authors are asymptotic. To minimize these errors one can select both data points from the same calculation, hoping that the systematic errors are similar. Also, in a Monte Carlo simulation the value of couplings are not selected to give $\Delta\beta$ for a given fixed scale change. Therefore, in order to compare different calculations we have to rescale data using $\Delta\beta(b_1)/\ln b_1 = \Delta\beta(b_2)/\ln b_2$. This rescaling assumes that $\frac{\partial\beta}{\partial(\ln a)}$ is either constant or essentially linear in β over some small interval prescribed by the scale change b, for only then is the β-function given by $\Delta\beta(\beta)/\ln b$.

In the analysis given below I rescale all data to $b = \sqrt{3}$. To minimize the systematic error introduced by this rescaling of data, I only use pairs of data points with a scale factor in the range $1.4 \leq b \leq 2.0$. I have checked that the resultant systematic error is small by taking different pairs of points, by considering a different range of scale change, and also by rescaling all data to $b = 2$.

The discrete version of the non-perturbative β-function can also be calculated using Wilson's 2-lattice Monte Carlo Renormalization Group (MCRG) method [15]. In principal this estimate of $\Delta\beta$ should be made along the renormalized trajectory (RT), and later we will discuss how well we do in this regard. First let me contrast the expected behavior of the flow along the RT between a normal second order transition and an asymptotically free field theory (AFFT). Let the RT be parameterized by a single scaling field K, then under a RGT by a scale factor b the behavior close to a second order transition is

$$(K^{(2)} - K^*) = b^{\frac{1}{\nu}} (K^{(1)} - K^*) \tag{4.1}$$

where the flow is from $K^{(1)}$ to $K^{(2)}$, ν is the correlation length exponent and K^* is the fixed point. The corresponding discrete flow relation is,

$$\Delta K = \frac{\partial K}{\partial(\ln b)} \ln b \approx \frac{1}{\nu} K \ln b \quad , \tag{4.2}$$

i.e. ΔK is proportional to the deviation K from the fixed point and to $\ln b$.

For an AFFT the leading term is independent of $K \equiv \beta$, and only a small β dependence gets generated by the 2-loop term. The relation in the continuum for pure gauge SU(3) theory is

$$\Delta\beta = \frac{\partial\beta}{\partial(\ln a)} \ln b$$

$$= (\frac{33}{4\pi^2} + \frac{459}{16\pi^4 \beta} + \cdots) \ln b \qquad (4.3)$$

in the limit of a small scale change $b = a_2/a_1$. Note that the analogue of temperature in spin models is the coupling g^2.

In practice we do not know the RT or K^* for a given model, nor do we know how to measure the couplings along the RT. Most simulations are done varying a single coupling (the temperature T in case of spin models and the coupling g for gauge theories). Wilson's 2-lattice method provides the link between the goal, to measure flows along the RT, and actual simulations done by varying just g [8].

The outline of the method is as follows: first simulate a system of size $L = N(b^n)^d$ with couplings K^L_α and calculate the expectation values of Wilson loops on the original lattice and the n block lattices. The integer N is usually 1 or 2. Next simulate a second system of size $S = N(b^{n-1})^d$ with couplings K^S_α chosen judiciously, and again calculate the expectation values on the n levels. Adjust the couplings K^S_α (which requires a new simulation each time) until the expectation values from the two simulations match on the same size lattices, $i.e.$ compare the mth blocked level on the larger starting lattice L with the $(m-1)^{th}$ on the smaller lattice S. The test of convergence of the two theories L^m and S^{m-1} is that the expectation values should match simultaneously on the last few levels. This ideal situation is shown in a two coupling constant space in Fig. 2. At matching, the correlation length on L (starting couplings K^L_α) is larger than on S (K^S_α) by the scale factor b since L has been blocked one more time. Thus $\Delta\beta$ for a scale change b is $(K^L - K^S)$. If the starting trajectory is taken to be the Wilson axis (or any 1 parameter line specified by β) then $\Delta\beta = (\beta^L - \beta^S)$.

Comparing expectation values is equivalent to matching the action. There is a one to one correspondence between the value of the couplings and the expectation values of Wilson loops. Under the assumption that the fixed point action is local ($i.e.$ a few short range couplings are sufficient to characterize the action along the RT), matching the expectation values of a few small Wilson loops is sufficient to guarantee that the two actions are equal. For this assumption to hold, the matching lattices should be large enough to accommodate all the important couplings and all the corresponding loops should be measured and matched. An important point to note is that while loop expectation values will have finite size effects, these effects are the same for the two simulations when matching occurs since the comparison is made on the same physical size lattices.

In practice it is sufficient to do two simulations S_1 and S_2 which bracket L and then use interpolation. Since simulations are done along some axis in the coupling constant space (for example the Wilson axis specified by the single coupling g) which may lie far from the RT, there will be scaling violations (transients) on the first few blocking levels. So matching should only be expected after some number of blocking steps. However, once matching takes place on lattices which are large enough to accommodate the important couplings, thereafter, the check that the two flows move together can be made on 1^4 lattices for non-abelian theories! Let me reiterate: the reason that $MCRG$ has good control over finite size effects and is a powerful method for critical phenomena is because empirically we find that the range of interactions at the fixed point falls off exponentially for models with short range interactions. Finite size effects in $MCRG$ are controlled by the range of the couplings and not the correlation length. So to match the two theories we only need to show equality of a few short range couplings or equivalently expectation values.

From a sequence of such calculations that give matching couplings one can determine $\Delta\beta(g)$. The advantage of this method is that since block expectation values are compared, scaling violations present on the original simulated lattices are to a large extent removed. Also, using simpler variables like Wilson loop expectation values in the matching is an advantage as they can be calculated with small statistical errors. The disadvantage is that it is hard to a priori judge whether the flows have converged to the RT. In practice one uses as large a starting lattice as possible and then *a posteriori* shows that there is matching at a number of blocked levels.

To check asymptotic scaling one has to show that $\Delta\beta$ measured by non-perturbative methods agrees with Eq. (4.3). To show scaling a sufficient condition is that $\Delta\beta$ calculated using different observables is the same for $\beta \geq \beta_{scaling}$. In the next section I discuss the popular RGT used for studying the pure gauge theory, before testing scaling using existing Monte Carlo data in section 6.

5. Block Transformations for 4-d SU(N) LGT

Blocking transformations for gauge theories in 4-d are to some extent non-intuitive. The reasons are (a) the degrees of freedom are SU(N) matrices and (b) these matrices are associated with links on the lattice rather than spins on a site. Nevertheless the steps one takes to define the blocking transformation are similar to those in spin models. First we define the geometry of the block lattice; the simplest example is the $b = 2$ transformation for which the basic cell is a 2^4 hypercube. Second, the block link between a pair of adjacent block sites should represent the average value of the gauge field A_μ in the cell. Lastly, in the construction of the

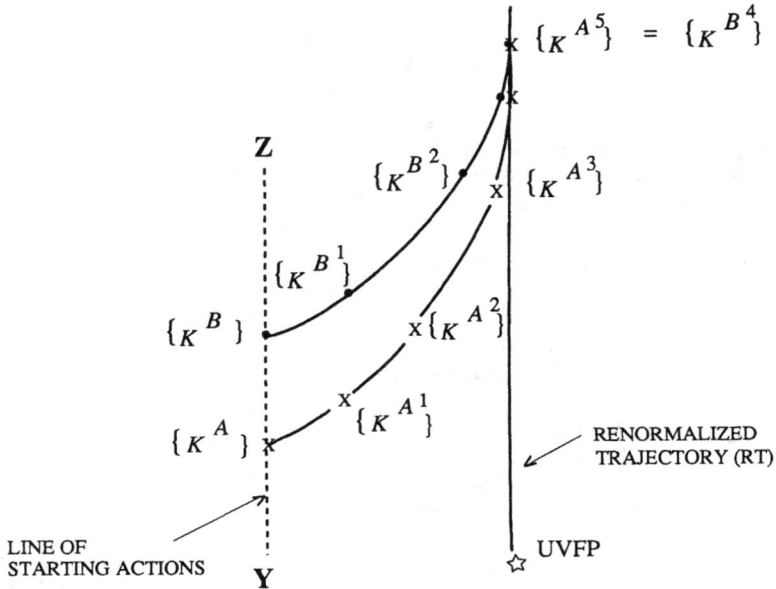

Fig. 2: *Schematic behavior of the two flows in Wilson's 2-lattice method at matching. As shown, the block actions match after 5/4 blockings.*

block link one should average over the maximum number of the original degrees of freedom in a cell, and if possible in a gauge covariant way. I now describe each of these steps in some detail.

The simplest gauge-covariant construction of the block link is a path ordered product of the original link variables between the block sites. This leaves intact the local gauge freedom at the block sites, precisely the freedom required on the block lattice. There are, however, many paths one can draw between any two adjacent block sites, so the question is which of these to use. It was first noted by Swendsen that a sum of paths also has the same gauge transformation property as any of the individual paths [16]. So, in the choice of paths we are guided by the following criteria: the transformation should be localized about the cell and at the same time use the maximum number of degrees of freedom in the cell. These features are expected to give faster convergence to the renormalized trajectory. Using too few paths to form the average is analogous to decimation transformation in spin-models which is known not to lead to convergent results.

When averaging a number of paths to construct the block link for an SU(N) gauge theory one faces the complication that the sum of SU(N) matrices is not proportional to a SU(N) matrix (SU(2) is an exception). The common practice is to project the sum of paths, a general $N \times N$ complex matrix Σ, back on to an SU(N) matrix. One prescription for doing this is to find the SU(N) matrix U that maximizes $Trace\ (\Sigma^\dagger U)$. Another is to generate the matrix U with probability $e^{-p\ Tr\Sigma^\dagger U}$, where p is a free parameter that can be tuned to optimize convergence. In either case we lose some dynamical information in the projection, and it still remains to be ascertained by numerical tests whether these constructions throw out some essential physics! While it is not necessary that the blocked links are group elements, projecting them on to SU(N) makes the transformation non-linear and this step has its origins in transformations used for non-linear sigma models. A practical simplification resulting from projecting the block link back to SU(N) is that it makes calculations on block lattices simple, and one does not have to worry about the relative normalizations of expectation values between the different block lattices.

A technical point for the interested reader who may wish to devise alternate blocking transformations is that the requirement of gauge covariance is not essential. One does have to make sure that the effects of the non-covariant operators generated by such a blocking transformation are small, *i.e.* these operators are irrelevant and have a small eigenvalue.

I now describe the geometry of the two popular methods used to construct the block link. Both methods are gauge covariant and differ primarily in the definition of the unit cell.

$b = 2$ **by Swendsen [16]:** This transformation, in its generalized form, is shown in Fig. 3. The coefficients α_i, one for each distinct topology of paths, are free parameters that need to be tuned to improve convergence. In this construction one can, in principle, include all possible paths that start and end at the two nearest neighbor block sites. As mentioned before, this preserves the local gauge invariance of the blocked theory. On the other hand, present calculations have used only up to 4 link paths in order to keep the construction local. In this case there is the two link path along the grid axis and two topologically distinct sets of 4-link paths. The weights α_i for these paths are not *a priori* specified and both numerical and analytical calculations show that the convergence of the RGT is improved considerably by tuning them. I shall return to this point later when discussing results for the non-perturbative β-function.

$b = \sqrt{3}$ **by Cordery, Gupta and Novotny [17]:** This transformation is specific to gauge theories in 4-dimensions. The basic cell consists of the block site and its eight neighbors. The geometry is based on the fact that on the original hypercubic

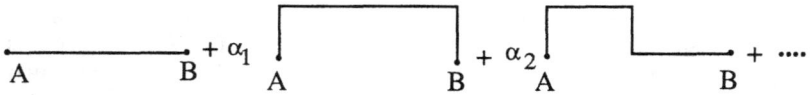

Fig. 3: *Swendsen's blocking transformation for gauge theories in 4-d with scale factor $b = 2$.*

lattice there are 4 positive 3-cubes associated with each site. The body diagonals of these cubes are orthogonal and of length $\sqrt{3}$ as shown in Fig. 4. On the original lattice the total number of degrees of freedom per block cell is 28 (the figure shows the seven associated with one of the 3-cubes). The block link is constructed as the average of the 6 paths $U_1 \cdots U_6$. Each of these six paths is a path-ordered product of three links. To explicitly show which links correspond to dynamical degrees of freedom, it is expedient to perform a local gauge transformation on each of the 8 nearest-neighbor sites such that the links coming out of the block site are set to the identity. For example note that the unshaded links U_a and U_c in the path U_1 in Fig. 4 can be gauge fixed to the identity, however, the value of modified link U_1 is identical to the original gauge-covariant path. Also, the surviving gauge freedom is now explicitly seen to be just at the two block sites on either end of the path. In this blocking one ignores the path U_7 since it connects block sites that become nearest-neighbor after two RGT. Thus, this transformation averages over 24/28 degrees of freedom per block site at each step.

The $\sqrt{3}$ transformation has several other attractive features in addition to having a smaller scale factor b. The first of these becomes obvious when one wishes to study SU(N) gauge theory coupled to scalar matter fields as in the case of the $SU(2) \times U(1)_Y$ theory . It is then easy and natural to define a simultaneous blocking transformation for both the link variables and the matter fields residing on the sites. The block gauge links are constructed as before, and to define the blocked scalar field one averages the nine degrees of freedom that form the block cell. Gauge invariance is preserved by using the parallel transport of the fields at the 8 nearest-neighbor sites in the construction of the block scalar field.

The extension to fermions is non-trivial since one integrates them out in order to do simulations. Thus, it is necessary to know the renormalized action produced under a given blocking transformation. One approximate method for calculating the renormalized Dirac operator is the block diagonalization process pioneered by Mütter and Schilling [18]. This has been done for both the $b = 2$ and for the $\sqrt{3}$ transformations. For the $\sqrt{3}$ blocking one finds that in the limit

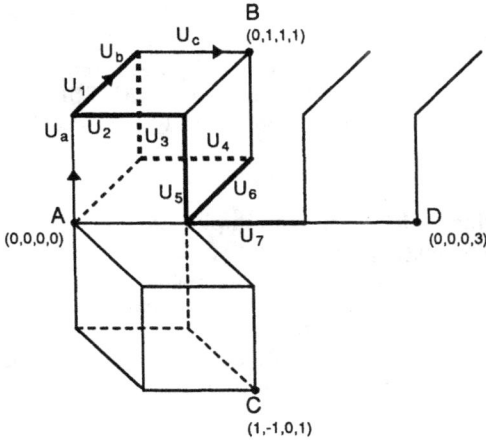

4-Dimensional Hypercubic Lattice

Fig. 4: $b = \sqrt{3}$ *blocking transformation for gauge theories in 4 dimensions*

of free field theory the interactions of the 9 original modes in the block cell can be approximated by a single light mode given by the arithmetic mean of the 9 original degrees of freedom (and taken to be the block field) and 8 degenerate heavy modes that can be treated as a correction. Blocking introduces longer range interactions even when starting with just a nearest-neighbor action. It is possible to iterate the blocking provided one keeps only tree level terms, in which case the iterations have a fixed point and the block action can be written in a closed form involving two terms. This approximate form can then be used in simulations [19]. These theoretical studies suggest that the $b = \sqrt{3}$ transformation leads to a better behaved renormalized action than the corresponding construction for the $b = 2$ case, but much more work is needed to show improved scaling using either improved action. The details of this improved action program are given in Refs. [18] [19].

There is one annoying feature of the $\sqrt{3}$ transformation. Under the first *RGT*, the new hypercubic lattice is rotated with respect to the old basis and the box has jagged boundaries. One can undo this on the second step by using a different set of basis vectors in the *RGT*. Thus the original box geometry is recovered after two blocking transformations, *i.e.* under a scale change by a factor of 3.

Other $b = 2$ transformations: The original RGT used by Wilson is described in

Ref. [8] and reviewed in [20]. The gauge invariant version of this transformation and a new $\sqrt{4}$ construction (block links formed by averaging the 24 paths connecting the opposite corners of 2^4 hypercubes) have recently been compared against Swendsen's method by Decker and Forcrand [21]. Their tests with SU(2) gauge theory show no advantage for these constructions. Consequently Swendsen's method is preferred by virtue of its simplicity.

Having defined a RGT, one can implement the calculation of critical exponents for non-abelian gauge theories using MCRG methods familiar from Statistical Mechanics. However, as discussed in the last section QCD has a single scaling function $f(g)$ whose form and the associated scaling exponent is known near the fixed point at $g = 0$. Thus the use of the MCRG method is restricted to the calculation of $f(g)$ at $g \approx 1$, i.e. in the non-perturbative region. The physical predictions involve the constants c_i which cannot be determined by MCRG, but the renormalization group is essential in order to justify their extraction from lattice calculations. In addition RG ideas have suggested techniques likes smearing that have been very successful in improving the signal to noise ratio in measurements. This application of the RG method has been discussed by DeGrand.

We are now ready to review results of the main application of MCRG to Lattice QCD, i.e. to calculate the β-function for both SU(2) and SU(3) pure-gauge theories and test scaling using the methods outlined here. We shall also compare the efficacy of different blocking transformations for obtaining reliable results.

6. Scaling of Monte Carlo Data and $\Delta\beta$

In this section I analyze the scaling behavior of observables calculated in SU(3) pure gauge theory. With this restricted set of data I circumvent the problems encountered in dealing with hadrons made up of quarks – the need to first extrapolate data to the physical quark masses and of adjusting the number of dynamical flavors used in the calculation. The other reason is that only very preliminary MCRG estimates for $\Delta\beta$ with dynamical fermions exist [22]. Since the methods are very similar, the extension to full QCD is straightforward and reliable results will become available once simulations with dynamical fermions can be carried out efficiently. The scaling analysis of the hadronic spectrum and finite temperature transition in the presence of quark loops has been presented by DeGrand and Gavai in this book.

The existing large lattice and high statistics data for the string tension σ, the glueball masses, and the deconfinement transition temperature T_c are given in

Table 1, 2 and 3 respectively. I summarize these results below and remark on the possible systematic errors in each. The conservative conclusion that I will draw from this data is that scaling is violated until $\beta \approx 6.1$ and that there are possibly large deviations (20%) from asymptotic scaling between $\beta = 6.2$ and 7.0.

The string tension data for SU(3) is obtained from the following references: [23] [24] [25] [26] [27] [28] [29] [30] [31]. The results suffer from two major systematic effects: 1) The results have been obtained on lattices of different size and finite size effects have not been accounted for, and 2) it has not been reliably demonstrated that the estimate of σ_W (string tension extracted from Wilson loops) is independent of loop size or whether σ_t is extracted from large enough separation in the 2-point correlation function of Polyakov lines. Empirically we find that the convergence of both σ_W as a function of loop size and of σ_t as a function of the separation is from above. Recent calculations have almost exclusively measured σ_t. I regard the difference between various estimates and the difference between σ_W and σ_t as an indicator of the magnitude of errors that may still exist.

$\frac{6}{g^2}$	σ_W	σ_W	σ_t	σ_t
5.5				0.31(1) [26]
5.6	0.279(9)*[24]		0.197(4) [23]	
5.7	0.195(10)*[27]		0.122(2) [23]	0.118(3) [26]
5.8	0.111(3) [24]	0.099(1)[25]		
5.9			0.061(3) [30]	0.054(3)[26][23]
6.0	0.061(2)* [24]	0.046 [28]	0.044(2) [30]	0.0346(22) [26]
6.0	0.0484(88)[27]			0.0343(27) [31]
6.1		0.046* [25]	0.0274(4)[23]	0.0267(24) [31]
6.2	0.036* [24]		0.0225(9)[30]	0.0215(18) [31]
6.3	0.0225(60)[27]	0.0173 [28]		

Table 1: *Estimates of the string tension extracted using Wilson loops (σ_W) and correlation of Polyakov Lines (σ_t). Lüscher's finite volume correction for data extracted from a lattice of size $L^3 \times \infty$, $\frac{\pi}{3L^2}$, is not included. The $*$ against values indicates that these estimates are definitely not asymptotic. The difference in σ extracted by different groups at the same coupling and the difference between σ_W and σ_t is still quite large.*

Finite volume corrections in σ_t extracted from a lattice of size $L^3 \times \infty$ have been analyzed by Lüscher in the context of a bosonic string model [32]; the leading correction gives $\sigma_t = \sigma_t(L) + \frac{\pi}{3L^2}$. Present data are not good enough to confirm this behavior or to ascertain whether L is large enough that we can extrapolate data to $L \to \infty$ using it. In Table 2 I show both $\sigma_t(L)$ and σ_t; again the $\approx 10\%$ difference between the two is an indicator of possible systematic errors due to finite volume. Note that $\sigma_t(L)$ approaches σ from below, so the two sources of systematic errors go in opposite directions. Nevertheless, the data suggest that the combined systematic error could still be as large as $10 - 20\%$ for $6/g^2 \geq 6.0$.

It has proven very hard to extract asymptotic estimates of glueball masses from a study of the decay of 2-point correlation functions. The data in Table 2 show sizable of variations between different measurements. (Note that this is true even after selecting what I consider the most reliable data.) Finite volume corrections in $M_G(L)$ are expected to be small once L is large enough such that the leading order result [33]

$$m(L) = m(\infty) \left[1 - \frac{3}{16\pi} \left(\frac{\lambda}{m(\infty)} \right)^2 \frac{1}{m(\infty)L} \exp\left(-\frac{\sqrt{3}}{2} m(\infty)L \right) \left(1 + \mathcal{O}(L^{-1}) \right) \right] \quad (6.1)$$

is valid. The coefficient $\frac{3}{16\pi} \left(\frac{\lambda}{m(\infty)} \right)^2$ is the three scalar glueball coupling constant α_{GGG} and could be as large as ~ 100. Even then the correction is at most a few percent for $mL \geq 8$.

An additional source of systematic error in the data is that the estimates are not asymptotic: wall source estimates approach the asymptotic value from below while the 2-point correlation function results converge from above [34]. The data from de Forcrand et al. and from the APE collaboration collected in Table 2 was obtained using a cold wall source, while that from Schierholz et al. and from Michael and Teper were obtained using the standard 2-point correlation function method. There is no estimate of the concomitant systematic error in either method, and this could be large. For example, if we require that at any given β at least two independent calculations give consistent results for estimates to be considered reliable, then very little data survive especially if each data point is taken at face value. One way to cancel some of the systematic errors in the calculation of $\Delta\beta$ using σ and glueball masses is to pick both data points from the same set. Alternatively, one can pick the best estimate at each β. I have tried both methods and find that the results are consistent, and in this review quote those from the latter method.

The deconfinement temperature in pure gauge SU(3), $T_c = 1/N_t a$, should also scale like a mass. For an extended review of the methods and analysis of QCD at finite temperature see the chapter by Gavai and Refs. [35] [36] [37]. The data for $N_t \leq 8$ are very precise; it has been obtained using a number of lattices with different spatial sizes and with very large statistics. Thus, even though SU(3) has a weak first order transition, the extraction of β_c is reliable. For larger N_t a finer scan

#	K_F	Lattice	$\sigma_t(L)$	$\sigma_t(\infty)$	$m_{0^{++}}(L)$	$m_{2^{++}}(L)$
1	5.5	$6^3 \times 12$	0.31(1)	0.339	1.07(3)	
2	5.7	$6^3 \times 16$	0.105(3)	0.134	0.66(4)	
3	5.7	$8^3 \times 16$	0.118(4)	0.134	0.86(4)	
4	5.9	$8^3 \times 20$	0.041(1)	0.0574	0.73(14)	
5	5.9	$10^3 \times 20$	0.052(1)	0.0625	0.68(8)	
6	5.9	$12^3 \times 20$	0.054(3)	0.0613	0.74(7)	
7	5.7	$9^3 \times 24$	0.122(2)	0.135	0.84(2)	
8	5.9	$10^3 \times 32$	0.047(3)	0.0545	0.65(3)	0.60(30)
9	5.9	$12^3 \times 32$	0.056(2)	0.0633	0.75(4)	0.80(20)
10	5.9	$16^3 \times 32$	0.054(1)	0.0581	0.76(3)	0.83(12)
11	6.0	$18^3 \times 32$	0.042(2)	0.0452	0.66(5)	0.98(7)
12	6.1	$16^3 \times 32$	0.027(1)	0.0311	0.54(4)	
13	5.85	$8^3 \times 16$	0.0613(15)	0.0777	0.54(4)	0.50(4)
14	6.0	16^4	0.0506(19)	0.0547	0.69(3)	1.10(7)
15	6.2	16^4	0.0282(6)	0.0323	0.48(4)	0.77(4)
16	5.9	12^4	0.061(2)	0.0683	0.82(5)	1.39(12)
17	6.0	$10^3 \times 20$	0.0315(16)	0.0420	0.67(8)	0.90(11)
18	6.0	16^4	0.0442(21)	0.0483	0.67(4)	1.07(6)
19	6.0	20^4	0.0445(30)	0.0471	0.70(9)	1.07(13)
20	6.2	20^4	0.0225(9)	0.0251	0.55(3)	0.86(4)

Table 2:   Monte Carlo data for the 0^{++} glueball mass and the string tension σ_t measured from the correlation of Polyakov lines. All data is for the the simple plaquette action. Entries 1 to 6 are from de Forcrand et al Ref. [26]; 7 to 12 are from the APE collaboration Ref. [23]; 13 to 15 are from Schierholz et al. Ref. [29]; and 16 to 20 are from Teper et al. Ref. [30].

in β and better statistics on lattices with larger spatial volumes are needed. I believe that the difference in β_c obtained using different confinement criteria (fraction of data in the deconfined phase in a given run) is an adequate measure of the possible size of systematic errors present in the data. At present T_c is the most accurately measured quantity of the three considered, and there exist data over the largest range, i.e. $5.0 < \beta < 6.5$.

To check asymptotic scaling I plot the ratio M/Λ_{2-loop} for the string tension, glueball mass and T_c data as a function of β in Fig. 5. For this analysis I

N_t	N_s	Kennedy	Gottlieb (3/4)	Others (1/2)	Columbia (1/2)
2	5	5.071(1)			
2	7	5.086(2)			
2	9	5.092(2)			
2	11	5.0945(10)			
2	∞	5.097(1)			
3	8			5.54(1)A	
4	7	5.669(5)			
4	9	5.680(5)			
4	11	5.690(3)			
4	∞	5.696(4)		5.6923(4)B	5.692(2)
5	10			5.79(1)B	
6	7	5.830(10)			
6	9	5.853(2)		5.865(2)C	
6	11	5.865(4)			
6	12			5.869(1)C	
6	16,24				5.889(8)
8	7		5.95 (3)		
8	11		5.985(15)		
8	13		6.05 (2)	6.00(1)D	
8	16				6.001(25)
8	19		6.02 (1)	6.01(1)D	
10	16				6.160(7)
10	17		6.18(2)	6.14(1)D	
12	16				6.268(12)
12	17		6.36(2)	6.31(2)D	
12	19		6.33(2)	6.28(1)D	
14	16				6.383(10)
14	19		6.47(4)	6.40(4)D	
14	21		6.45(2)	6.39(2)D	
16	24				6.45(5)

Table 3: *High statistics pure gauge T_c data. The data labelled "Kennedy" is from Ref. [38]; "Gottlieb" is compiled from Refs. [39]; "Columbia" is from Refs. [40]. The data marked (A) is from Ref. [41]; (B) is from Ref. [42]; (C) is from Ref. [43]; and (D) is from [39].*

have made a subjective choice of data taking the best/mean estimate at each β, and ignoring the systematic error that such a selection introduces. These estimates for σ and glueball masses are given in Table 4. Fig. 5 shows that both $\sqrt{\sigma}/\Lambda_{2-loop}$ and T_c/Λ_{2-loop} approach the asymptotic value from above and are essentially constant for $\beta > 6.1$. The present best estimates are

$$\sqrt{\sigma}/\Lambda_{2-loop} = 79(2) , \tag{6.2}$$

and

$$T_c/\Lambda_{2-loop} = 51 - 44 \tag{6.3}$$

where the two estimates are for 3/4 and 1/2 confinement criteria respectively (the data are shown separately for $\beta > 6.1$ in Figs. 5 and 6). One should consider the band formed by using the different confinement criteria as a measure of the systematic errors.

At first sight the $M_{0^{++}}$ data seem to show asymptotic scaling starting at $\beta = 5.5$, but this anomalous behavior is due to lattice artifacts. For example the mass ratio, $M_{0^{++}}/\sqrt{\sigma}$, shows a significant variation as a function of β; the ratio is ≈ 2 at $\beta = 5.5$ and flattens at ≈ 3.3 only for $\beta = 5.9$ (see Table 4). If one takes this onset of scaling at $\beta > 5.9$ seriously compared to evidence found above for asymptotic scaling at $\beta > 6.1$, then it can be argued that some of the systematic errors are cancelled in the mass ratios.

β	σa	$m_{0^{++}}a$	$m_{2^{++}}a$	$\dfrac{m_{0^{++}}}{\sqrt{\sigma}}$	$\dfrac{m_{2^{++}}}{m_{0^{++}}}$
5.5	0.31(1)	1.07(3)		1.9(1)	
5.6	0.197(4)	1.04(4)		2.3(1)	
5.7	0.120(4)	0.85(4)		2.5(1)	
5.8	0.099(1)				
5.85	0.061(2)				
5.9	0.054(1)	0.76(3)	1.39(12)	3.3(1)	1.8(3)
6.0	0.035(2)	0.69(4)	1.08(7)	3.7(3)	1.6(2)
6.1	0.027(1)	0.54(4)		3.3(3)	
6.2	0.0220(15)	0.48(4)	0.77(4)	3.2(3)	1.6(2)
6.3	0.0173(10)				

Table 4: My selection of best/mean Monte Carlo data for the string tension σ, the 0^{++} and 2^{++} glueball mass. No finite size corrections are included in these estimates. In order to translate these results into physical units one uses $\sqrt{\sigma} \approx 420$ MeV.

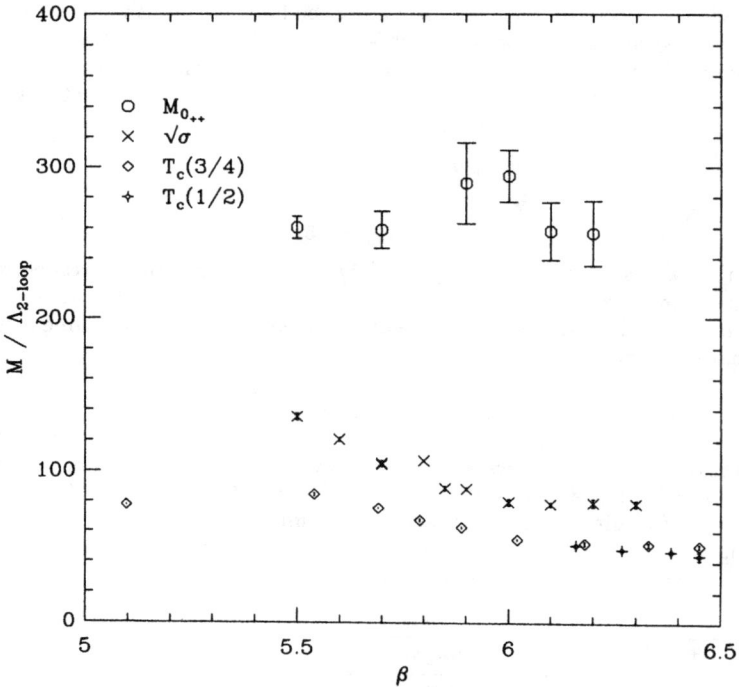

Fig. 5: *Plot of the ratio M/Λ_{2-loop} for the string tension, glueball mass and T_c data.*

It has been argued recently by Lepage and Mackenzie that perturbation theory is much better behaved provided an improved g^2 is used [44]. To explore the consequences of this improvement I have plotted in Fig. 6 the same data as in Fig. 5 but using the 1-loop Λ with the replacement $g^2 \rightarrow 1.75g^2$. (Note that Lepage and Mackenzie recommend determining the improved g^2 individually for each observable. The above estimate is roughly the average of the improved g^2 for the various quantities considered by them, and should suffice for the present qualitative discussion.) The plots in Fig. 6 are significantly different compared to those in Fig. 5 in that they do not show a plateau. This interlude should serve to underscore the point that evidence for asymptotic scaling is not conclusive.

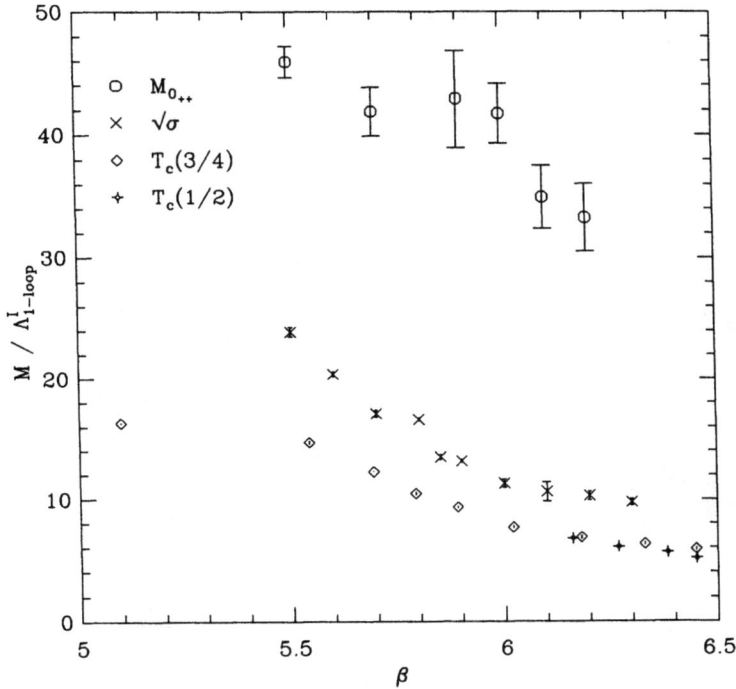

Fig. 6: *Plot of the ratio M/Λ^I_{1-loop} for the string tension, glueball mass and T_c data. Following the recommendation of Lepage and Mackenzie, I have used $1.75g^2$ as the improved estimate of g^2 in the 1-loop calculation of Λ^I_{1-loop}.*

So far I have limited the analysis to data obtained with the simple Wilson action. There also exist results for σ, M_{0++} and M_{2++} with an improved gauge action determined using MCRG [34]. In this calculation the estimate of mass-ratios are

$$\frac{M_{0++}}{\sqrt{\sigma}} = 3.5(3)$$
$$\frac{M_{2++}}{M_{0++}} = 1.6(2) ,$$

(6.4)

which are consistent within errors with estimates obtained using the simple Wilson

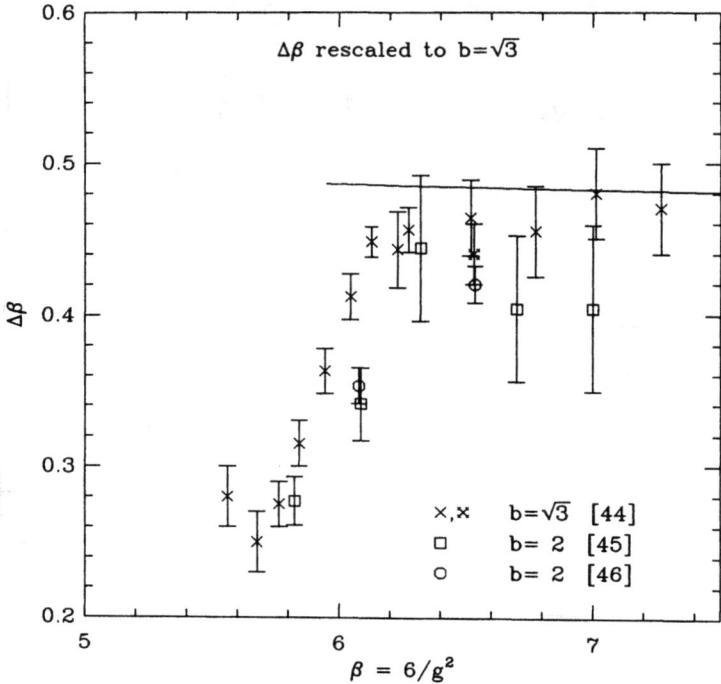

Fig. 7: *MCRG results for the non-perturbative* $\Delta\beta$. *The* $b = 2$ *results have been rescaled to* $b = \sqrt{3}$.

action, for example those given in Table 4. One can relate this extended action to the Wilson action by matching the string tension, $\sigma \approx 0.048$, against results in Table 4. A rough estimate of the equivalent Wilson action coupling then is $\beta \approx 5.94$.

Now turning to $\Delta\beta$ I show MCRG results for the $b = \sqrt{3}$ [45] and the $b = 2$ [46] [47] transformations in Fig. 7. Each value of $\Delta\beta$ is defined at the midpoint of the interval spanning a scale change b. The data can be roughly divided into three regions: $\beta < 6.2$, $6.2 < \beta < 7$, and $\beta > 7.0$. The main features in the data are (a) there is a large dip in the non-perturbative β-function below $\beta = 6.2$, (b) the minimum of the dip occurs at $\beta \approx 5.7$, and (c) in the range $6.2 < \beta < 7$ $\Delta\beta$ lies below the 2-loop perturbative value.

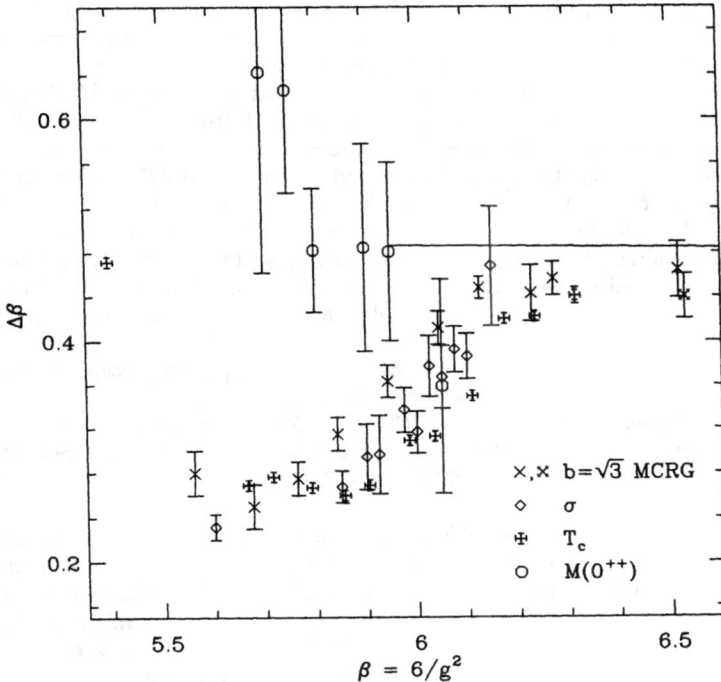

Fig. 8: *Comparison of $\Delta\beta$ obtained from observables σ, T_c and $M_{0^{++}}$ with the MCRG results.*

Before comparing this $\Delta\beta$ with scaling of observables let me first discuss possible systematic errors in MCRG results due to finite size effects. All the $b = \sqrt{3}$ data were obtained on $L = 9^4$ lattices (except for one run at $\beta = 6.75$ on $(9\sqrt{3})^4$ lattices). This means that for $\beta \geq 6.05$ the system is deconfined. The same holds for data on 16^4 lattices at $\beta \geq 6.4$. For example at $\beta = 7.5$, a 9^4 box is effectively at very high temperature, *i.e.* $T \approx 6T_c$. While it is true that the two systems L and S at matching are at the same physical temperature, it is important to determine if the measured $\Delta\beta$ is independent of temperature as we are interested in the zero temperature value. One way to check this is to compare results obtained on lattices of different size.

For the SU(2) theory, no significant difference was found in results obtained using 18^4 versus 9^4 lattices for the $b = \sqrt{3}$ transformation. These calculations were done over the range $2.5 \leq \beta \leq 3.5$ and the results were consistent with asymptotic scaling starting at $\beta \geq 3.0$ [48]. The deconfinement transition in SU(2) takes place at $\beta \approx 2.55$ for 9^4 lattices and at $\beta \approx 2.8$ for 18^4 lattices. The results for $\Delta\beta$ on 18^4 lattices are $\approx 1\sigma$ smaller over this range and this difference was not considered significant. Recently, Decker and de Forcrand have used 32^4 lattices to calculate $\Delta\beta$ using the $b = 2$ transformation at $\beta = 2.5$, 2.7 and 2.9 [49]. Their results lie below the $b = \sqrt{3}$ data, and even at starting $\beta = 2.9$ are $\approx 20\%$ below the 2-loop asymptotic value. A similar lack of asymptotic scaling in the string tension data has been reported by the UKQCD collaboration [50], although they find that the scalar glueball mass and the $q\bar{q}$ potential measured in the same run do show scaling.

In SU(3), the QCD_TARO collaboration [47] have presented new data at $\beta = 6.3$ on 16^4 lattices and at $\beta = 6.8$ on 32^4 lattices using the $b = 2$ transformation. For $\beta = 6.8$ their result, $\Delta\beta = 0.53(3)$ is $\approx 15\%$ below the 2-loop value. To first approximation this is consistent with the $b = \sqrt{3}$ data which lie about 10% below the 2-loop value over the range $6.35 \leq \beta \leq 7.0$.

Fortunately it is possible to make a detailed check for finite size effects because there exist three estimates at $\beta \approx 6.53$; $\Delta\beta = 0.464(25), 0.44(2)$ and $0.42(1)$ on lattices of size 9^4, $(9\sqrt{3})^4$ and 32^4 respectively. The difference can easily be due to statistical and other systematic errors. Taking the numbers at face value, however, the trend is clear: the finite box size leads to a systematic error that tends to increase $\Delta\beta$. The size of the effect in this case is $\approx 10\%$ assuming that the 32^4 data represents the zero temperature result. With this estimate of finite size effects I conclude that the deviation from asymptotic scaling is $10 - 20\%$ over the range $7.0 > \beta > 6.2$.

In section 5 I suggested that the $b = \sqrt{3}$ transformation is better behaved. The first criteria for judging the efficiency of the blocking procedure is the rate of convergence of $\Delta\beta$ as a function of the number of blockings. The results in [49] and [47] show that with their tuning of parameters in the generalized Swendsen blocking the convergence of the $b = 2$ transformation is from above and results have converged to within a few percent after 3 blockings (on 4^4 lattices starting from 32^4). The convergence of $\Delta\beta$ with the $b = \sqrt{3}$ transformation is not monotonic, the estimate oscillates about the final value. It converges to within 10% after 3 blocking steps. Thus the convergence of both blocking schemes as a function of the number of blocking steps is similar. The other possible advantage of the $b = \sqrt{3}$ transformation is the smaller scale factor; this is partially offset by the added complexity of the computer program. I conclude that the two schemes are of comparable utility and both should be used.

Now we are in a position to test whether the $\Delta\beta$ calculated from MC data for different physical observables agree with the $MCRG$ determination. I remind you that for purposes of comparison all data are rescaled to $b = \sqrt{3}$. At $\beta = 6.0$ the 0^{++} glueball mass, string tension $\sqrt{\sigma}$ and the deconfinement temperature T_c represent scales of $2, 5$ and 8 lattice units respectively. Over the range $6.0 < \beta < 7.0$ all scales are expected to increase by a factor of 4. Thus identical $\Delta\beta$ would be a reasonable test of scaling especially once data is available over this complete range. Such calculations are feasible with the emerging teraflop class computers.

The final results are shown in Fig. 8, and I stress that before drawing strong conclusions one should bear in mind all the known and unknown sources of systematic errors in the data. Close scrutiny of the present data lying between $\beta = 5.7$ and 6.2 shows a broad band with the $b = \sqrt{3}$ MCRG data showing the least amount of deviation from the 2-loop result. This is most likely a result of finite size effects as discussed above. The $\Delta\beta$ calculated from σ and T_c data agree. Thus, to first approximation these observables scale and are in agreement with MCRG results.

A big deviation from the above general behavior is shown by the $M_{0^{++}}$ data. The corresponding $\Delta\beta$ shows a rise rather than a dip for $\beta < 6.1$. The various $\Delta\beta$ seem to come together at $\beta \approx 6.1$, however, at this point we have also run out of data. Thus it is not possible to confirm scaling, *i.e.* whether the various $\Delta\beta$ stay together for $\beta > 6.1$.

The cause of the dip in the β-function and the lack of scaling for $\beta < 6.0$ can be explained on basis of the phase diagram of SU(3) theory defined in the two coupling plane -- fundamental and adjoint representation of the plaquette action. There is a line of first order transitions that ends in a spurious critical point above the Wilson axis [51]. If one were to extend the transition line below the end point, it would intersect the Wilson axis at $\beta \approx 5.7$, coincidental with the location of the minimum of the dip. At this spurious critical point numerical simulations show that the 0^{++} glueball mass vanishes while lines of constant σ and T_c get bunched up. This behavior would explain why $\Delta\beta$ calculated from σ and T_c show a dip while that from 0^{++} glueball shows a small rise. Thus, even though this spurious singularity lies away from the Wilson axis it modifies the scaling behavior in its vicinity.

I have offered conclusions at various places which a reader may find contradictory: in fig. 5 a plot of M/Λ_{2-loop} data suggests that asymptotic scaling sets in at $\beta > 6.1$ while the same data in fig. 6 does not (there is no plateau). The $\Delta\beta$ measurements imply there are $20 - 10\%$ deviations between $\beta = 6.2$ and 7.0, however, there is the possibility that scaling sets in for $\beta > 6.1$. These difference should be taken as a measure of the systematic errors and the uncertainty in the tests of scaling presented here. I believe that the situation will be clarified once better data is obtained over a larger range of β. The lesson I would like to draw from this analysis is that with the Wilson action one has to perform calculations at $\beta > 6.1$ in order to extract continuum physics to within 20%.

Overall, it seems miraculous that we can extract the infra-red behavior of QCD to better than 20% even though $g_{bare} \approx 1$. I would like to mention a few other calculations that provide corraborative evidence showing that it will be feasible to extract continuum physics from lattices of about 3 fermi across and with a lattice spacing of ≈ 0.05 fermi. These are (1) the restoration of staggered flavor symmetry to within a few percent by $\beta = 6.4$ [52]; (2) The value of B_K obtained with and without an improved Wilson action [53].

In the next section I switch from presenting a detailed scaling analysis of the spectrum to the calculation of matrix elements for which lattice artifacts are easier to demonstrate. These are precisely the quantities which we want to calculate in order to determine the unknown parameters of the Standard Model. It has become clear that for Wilson fermions the brute force approach of increasing the lattice size and β will proceed slowly and that one can gain a lot by improving the action [54]. For this reason there is considerable effort being devoted to exploring actions that are designed to have no leading order scaling violations $i.e.$ no $O(a)$ terms. This improvement program is discussed next.

7. Corrections to scaling in matrix elements

Mass ratios and matrix elements pertinent to continuum physics can be extracted if the conditions $L >> \xi >> a$ are satisfied in simulations. The ratio L/ξ should be large to control finite size effects, while ξ/a should be large to reduce scaling violations. The value of ξ depends on the quantity being measured. In practice one first chooses the lattice size L that can be simulated on a given computer and then g is picked to minimize finite size and finite lattice spacing errors. Thus, computational feasibility dictates that one should improve the action in order to extract continuum physics from simulations on rather coarse lattices, $i.e.$ from small values of the correlation length ξ/a. At the same time it is desirable to keep the lattice action as simple as possible to optimize update time and to allow perturbative calculations that are necessary to relate lattice results to their continuum counterparts. The result of the analysis presented in the last section suggests that the spectrum is not the most sensitive observable for testing scaling as one needs accurate data over a large range of β. I now show why calculations of matrix elements are a better tool for evaluating scaling and improvements under a change in the lattice action.

An instructive example of why, unlike mass-ratios, scaling violations in the calculation of matrix elements are large and can be reduced by improving the lattice action is the renormalization constant Z_V for the local vector current with Wilson fermions [55]. It is defined as

$$\hat{V}_\mu(x) = Z_V V_\mu^L(x) \tag{7.1}$$

where $V_\mu^L(x) = \overline{\psi}(x)\gamma_\mu\psi(x)$ is the local current and $\hat{V}_\mu(x)$ is the conserved current

$$\hat{V}_\mu(x) = \frac{1}{2}\left[\overline{\psi}(x)(\gamma_\mu - r)U_\mu(x)\psi(x+\hat{\mu}) + \overline{\psi}(x+\hat{\mu})(\gamma_\mu + r)U_\mu^\dagger(x)\psi(x)\right]. \tag{7.2}$$

which does not undergo any renormalization. There are two reasons that makes the vector current an excellent test case for evaluating improved actions. First, the fact that the vector current has no anomalous dimensions implies that Z_V is a finite renormalization and that $Z_V \to 1$ as $a \to 0$. The second reason is that the forward matrix elements of $\hat{V}_\mu$ do not have any $O(a)$ corrections.

By taking the ratio of the matrix elements of the two different lattice definitions of the vector current we can calculate Z_V for a variety of initial and final states. The results from the ELC collaboration are given in Table 5 and show a sizeable dependence on the external states for the Wilson action and do not agree with the perturbative estimate. The results for Z_V for the 1-link current are marginally better than those for the local current. This dependence on external states arises due to $O(a)$ corrections induced by mixing of the vector current with higher dimension operators. As an illustrative example of how interactions generate mixing I will analyze the 1-link vector current (the term independent of r in Eq. (7.2)) in section 10 and show that the coefficients of the higher dimension operators are powers of a and g^2. By comparison a significant improvement is found when an improved action is used. This is discussed in section 10.

Method	Z_V^{local}	Z_V^{1-link}
Pert. Theory	0.83	0.92
$\langle 0\|V_i\|\rho\rangle$	0.57(2)	0.79(4)
$\langle \pi\|V_\mu\|\pi\rangle$	0.70(1)	0.69(2)
$\langle N\|V_\mu\|N\rangle$	0.74(3)	0.84(1)
$\langle K\|V_\mu\|D\rangle$	0.98(10)	

Table 5: Estimates for the renormalization constant Z_V for the local and 1-link current using the Wilson action. The perturbation theory estimate is obtained using the bare lattice g^2.

In order to understand discretization errors due to finite lattice spacing, it is useful to start by considering the expectation value of a typical hadronic operator as given by the path integral. For example

$$\langle \overline{\psi}\Gamma_\alpha\psi \ldots \overline{\psi}\Gamma_\beta\psi \rangle = \langle M^{-1}\Gamma_\alpha \ldots M^{-1}\Gamma_\beta \rangle$$
$$= \int dU \left(M^{-1}\Gamma_\alpha \ldots M^{-1}\Gamma_\beta \right) \exp\left(-S_g + \text{Tr}\,\ln M \right) \qquad (7.3)$$

where M^{-1} is the quark propagator and Γ_α is one of the sixteen Dirac matrices. Scaling violations due to lattice artifacts will in general come from three possible sources:

(1) The vacuum, specified by the Boltzmann factor $\exp\left(-S_g + \text{Tr}\,\ln M \right)$, will have a modified weighting of background configurations. One can see why corrections arise by expanding the classical action in powers of a.

(2) The quark propagator M^{-1} calculated on the background gauge configurations will have corrections. These will be shown explicitly for the free field propagator when we derive the Feynman rules. In general the corrections begin at $O(a)$.

(3) Interactions give rise to a mixing between operators of different dimensions and even different symmetry as in the case of Wilson fermions which explicitly break chiral symmetry. Operators with different canonical dimensions will come with appropriate powers of a. Furthermore, there can be additional $\log a$ corrections coming from the relation between the lattice and continuum operators.

In order to improve the reliability of lattice calculations, it is necessary to evaluate these corrections and devise ways to reduce them. From our discussion of the renormalization group we know that there are no scaling violations along the renormalized trajectory. So, as originally proposed by Wilson, improvement can be achieved provided we carry out simulations using an action close to the RT. There are two ways to do this; (a) simulate with the simplest possible action on a very large lattice and block a couple of times before making measurements, (b) determine the renormalized action produced under blocking, i.e. use MCRG methods to find an action that lies closer to the renormalized trajectory, and then perform simulations using that. The drawback with the first approach is that one needs very large starting lattices to be able to block a couple of times and then measure observables. The much more serious problem with both methods is that one needs to know the renormalized fermion action in order to calculate the quark propagator M^{-1}. Renormalization Group methods have been used to determine such improved actions, but it has not yet been possible to demonstrate whether they lead to significant improvement because of the various systematic errors in the data. Their presence can overwhelm improvements in scaling. For more details on the progress made using this approach see Refs. [56] and [57]. It is clear that far more work needs to be done to first determine an action closer to the RT and then demonstrate that it improves scaling significantly.

The more popular and successful improvement program is the one initiated by Symanzik [58]. In this approach one determines all scaling violations for a given action as a power series in g^2 and a, and then systematically eliminates them. The improvement will require one to modify both the action and the operators. My goal in this review is to introduce the concepts and show how to remove these corrections at the lowest order. For this we shall consider

(1) the free quark propagator M^{-1},

(2) vertices and loop integrals,

(3) lattice operators, in particular non-local operators.

The next three sections outline the improvement program for (i) the gauge action (ii) the fermion action and (iii) lattice operators.

8. Improving the Gauge Action

There have been a number of methods proposed to calculate the blocked gauge action. These methods have been reviewed in [59]. A discussion comparing MCRG improved actions and Symanzik's program is given in [56]. In this chapter I only review Symanzik program in view of the later discussions of improved fermion actions.

The leading order term in the expansion of all Wilson loops is $\mathcal{O}^{(4)} = \sum_{\mu\nu} F_{\mu\nu}F_{\mu\nu}$ and corrections begin at $O(a^2)$ as there are no dimension 5 operators. Thus any lattice action written as a linear combination of Wilson loops will have the correct continuum limit with corrections at $O(a^2)$. There are three dimension

6 operators in the continuum:

$$\mathcal{O}_1^{(6)} = \sum_{\mu,\nu} \mathrm{Tr}\left(D_\mu F_{\mu\nu} D_\mu F_{\mu\nu} \right),$$

$$\mathcal{O}_2^{(6)} = \sum_{\mu,\nu,\rho} \mathrm{Tr}\left(D_\mu F_{\nu\rho} D_\mu F_{\nu\rho} \right), \tag{8.1}$$

$$\mathcal{O}_3^{(6)} = \sum_{\mu,\nu,\rho} \mathrm{Tr}\left(D_\mu F_{\mu\rho} D_\nu F_{\nu\rho} \right).$$

Also there are only 3 six-link loops that one can draw on the lattice. These are the planar $\mathcal{L}_1^{(6)}$, twisted $\mathcal{L}_2^{(6)}$ and the L shaped $\mathcal{L}_3^{(6)}$ respectively. Thus classical improvement of the lattice action to $O(a^2)$ can be achieved by taking a linear combination of the plaquette operator and these three six-link loops. Each of these loops has the expansion

$$\mathcal{L} = r^{(4)} \mathcal{O}^{(4)} + r_1^{(6)} \mathcal{O}_1^{(6)} + r_2^{(6)} \mathcal{O}_2^{(6)} + r_3^{(6)} \mathcal{O}_3^{(6)} + \ldots, \tag{8.2}$$

and Lüscher and Weisz have shown an elegent way of calculating the expansion coefficients $r_\alpha^{(d)}$ [60]. Their results are summarized in Table 6.

Loop	$r^{(4)}$	$r_1^{(6)}$	$r_2^{(6)}$	$r_3^{(6)}$
$\mathcal{L}^{(4)}$	$-\frac{1}{4}$	$\frac{1}{24}$	0	0
$\mathcal{L}_1^{(6)}$	-2	$\frac{5}{6}$	0	0
$\mathcal{L}_2^{(6)}$	-2	$-\frac{1}{6}$	$\frac{1}{6}$	$\frac{1}{6}$
$\mathcal{L}_3^{(6)}$	-4	$\frac{1}{6}$	0	$\frac{1}{2}$

Table 6: Coefficients of the small a expansion of the plaquette and the three six-link Wilson loops in terms of the continuum operators defined in Eq. (8.1).

The lattice action can be written as

$$A_g = \frac{6}{g^2}\left\{ c^{(4)}(g^2)\, \mathcal{L}^{(4)} + \sum_{i=1,3} c_i^{(6)}(g^2)\, \mathcal{L}_i^{(6)} \right\} \tag{8.3}$$

in terms of the plaquette and the three 6-link loops. One can impose the normalization condition

$$c^{(4)}(g^2) + 8c_1^{(6)}(g^2) + 8c_2^{(6)}(g^2) + 16c_3^{(6)}(g^2) = 1 \tag{8.4}$$

such that in the continuum limit the action reduces to $1/4 F_{\mu\nu} F_{\mu\nu} + O(a^2)$.

From Table 6 one can see that tree level improvement can be obtained by the choice

$$c^{(4)} + 20c_1^{(6)} = 0; \quad c_2^{(6)} = 0; \quad c_3^{(6)} = 0 . \tag{8.5}$$

This removes $\mathcal{O}_1^{(6)}$ by cancelling the contribution from the two planar loops. To go beyond tree-level one has to follow Symanzik's program systematically. In this approach, to achieve $O(a^2)$ improvement, one has to derive three relations between the c_i by requiring that on-shell quantities do not have $O(a^2)$ corrections. Lüscher and Weisz found that this produced only two constraints

$$c^{(4)} + 20c_1^{(6)} - 4c_2^{(6)} + 4c_3^{(6)} = 0$$
$$c_2^{(6)} = 0 . \tag{8.6}$$

Thus they were left with a 1-parameter family of improved actions. This freedom arises because there is a redundant operator present in this subspace of $\mathcal{O}^i$. In such cases one is at the liberty to choose the coefficient of an operator that is part of the redundant operator without affecting on-shell improvement. A convenient choice simplifying the action is $c_3^{(6)} = 0$, which leads us back to Eq. (8.5).

Alternatively, a third constraint can be obtained by requiring that some set of off-shell amplitudes be improved as well. For instance, the requirement that all Wilson loops be improved in perturbation theory (this corresponds to killing the $O(a^2)$ terms at all length scales instead of only in the long distance heavy quark potential) constrains the improved action to be [61]:

$$c^{(4)} + 20c_1^{(6)} = 0 \tag{8.7}$$

This agreement with the tree level result is fortuitous. Improving a different observable would in general give a different constraint that is consistent with and independent of Eq. (8.6). In conclusion, classical improvement is sufficient at $O(a^2)$.

It turns out that in the calculation of matrix elements the leading scaling violations are $O(a)$ and come from the fermion action, so for the remainder of this chapter it will suffice to use the simple plaquette gauge action as it improved to that order.

9. Improving the Dirac Action

We start with the Wilson fermion action (WF) defined as

$$
A_W = (m_q + \frac{4r}{a}) \sum_x \overline{\psi}(x)\psi(x)
$$

$$
+ \frac{1}{2a} \sum_x \overline{\psi}(x)[(\gamma_\mu - r)U_\mu(x)\psi(x + \hat{\mu}) - (\gamma_\mu + r)\, U_\mu^\dagger(x - \hat{\mu})\psi(x - \hat{\mu})]
$$

$$
\equiv \sum_{x,y} \overline{\psi}_L(x)M_{xy}^W \psi_L(y)
$$

(9.1)

where the rescaled lattice fields are $\psi_L = \psi/\sqrt{2\kappa}$ and the quark mass is given in terms of the lattice parameters κ and r as

$$
m_q = \frac{1 - 8\kappa r}{2\kappa a}
$$

$$
\equiv \frac{1}{2a}\left(\frac{1}{\kappa} - \frac{1}{\kappa_c}\right) .
$$

(9.2)

The interaction matrix M_W is

$$
aM_W[U]_{x,y} = \delta_{xy} + \kappa \sum_\mu \left[(\gamma_\mu - r)U_{x,\mu}\delta_{x,y-\mu} - (\gamma_\mu + r)U_{x-\mu,\mu}^\dagger \delta_{x,y+\mu}\right] \quad (9.3)
$$

from which one can derive the free propagator in momentum space;

$$
M_W^{-1}(p) = \frac{a}{1 - 2\kappa \sum_\mu (r \, \cos ap_\mu - i\gamma_\mu \, \sin ap_\mu)} .
$$

(9.4)

By adding the dimension 5 operator, i.e. the term proportional to r, Wilson eliminated the poles at $p_\mu = \pi$. This cured the doubling problem but at the expense of explicitly breaking the chiral symmetry. To expose the scaling violations expand (9.4) in powers of a:

$$
\frac{M_W}{2\kappa} = m_q + i\gamma_\mu p_\mu + \frac{ra}{2} \sum_\mu p_\mu^2 - \frac{ia^2}{3} \sum_\mu \gamma_\mu p_\mu^3 + \dots ,
$$

(9.5)

where $\frac{M_W}{2\kappa}$ is the inverse of the propagator normalized with respect to the fields ψ and has the dimensions of mass. Note that it is the r term that introduces corrections at $O(a)$ in the Wilson action. Thus, both the naive and the staggered action have only $O(a^2)$ violations, and for this reason they are already improved to the order I am working at and do not require any further discussion. Similarly, the one and two gluon vertices for the Wilson action are

$$
\begin{aligned}
V_\mu^{W(1)} &= 2g\kappa T^a \left[i\gamma_\mu \cos \frac{a(p+k)_\mu}{2} + r \sin \frac{a(p+k)_\mu}{2} \right] \\
V_\mu^{W(2)} &= ag^2 \kappa \frac{\{T^a, T^b\}}{2} \delta_{\mu\nu} \left[r \cos \frac{a(p+k)_\mu}{2} - i\gamma_\mu \sin \frac{a(p+k)_\mu}{2} \right]
\end{aligned}
\tag{9.6}
$$

with the following expansion in powers of a

$$
\begin{aligned}
\frac{V_\mu^{W(1)}}{2\kappa} &= gT^a \left[i\gamma_\mu + ra\frac{(p+k)_\mu}{2} - i\gamma_\mu a^2 \frac{(p+k)_\mu^2}{8} + \cdots \right] \\
\frac{V_\mu^{W(2)}}{2\kappa} &= \frac{a}{4}g^2 \{T^a, T^b\} \delta_{\mu\nu} \left[r - ia\gamma_\mu \frac{(p+k)_\mu}{2} - \frac{ra^2}{8}(p+k)_\mu^2 + \cdots \right].
\end{aligned}
\tag{9.7}
$$

Thus, the leading $O(a)$ corrections to the vertices are also proportional to ra. At this classical level there are two ways of improving the action to get rid of these leading terms. These are the Hamber and Wu (HW) and the Sheikholeslami and Wohlert (SW) improved actions, which I review next.

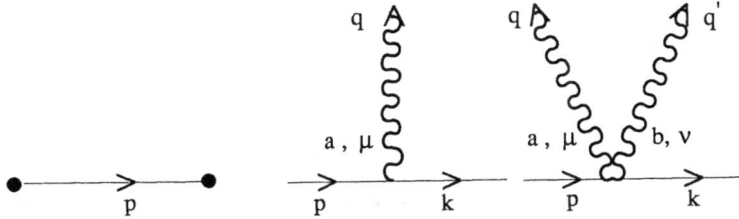

Fig. 9: *Feynman diagrams for the lattice propagator, 1-gluon and 2-gluon vertices.*

The simplest version of the Hamber and Wu improved action is [62]

$$
\begin{aligned}
A_{HW} &= A_W \\
&+ \frac{r\kappa}{4a} \sum_x \overline{\psi}^L(x) \left[U_\mu(x) U_\mu(x+\hat{\mu}) \psi^L(x+2\hat{\mu}) + \right. \\
&\qquad\qquad \left. U_\mu^\dagger(x-\hat{\mu}) U_\mu^\dagger(x-2\hat{\mu}) \psi^L(x-2\hat{\mu}) \right] \\
&\equiv \sum_{x,y} \overline{\psi}_L(x) M_{xy}^{HW} \psi_L(y)
\end{aligned}
\tag{9.8}
$$

for which the inverse of the free propagator is

$$aM_{HW}(p) = 1 - 2\kappa \sum_\mu \left(r \, \cos ap_\mu - i\gamma_\mu \, \sin ap_\mu \right) + \frac{kr}{2} \sum_\mu \cos 2ap_\mu \, . \qquad (9.9)$$

Expanding in powers of a gives

$$\frac{M_{HW}}{2\kappa} = m_q + i \sum_\mu \gamma_\mu p_\mu \left[1 - \frac{a^2 \, p_\mu^2}{3} \right] + \cdots \, . \qquad (9.10)$$

where

$$m_q = \frac{1 - 6\kappa r}{2\kappa a} \qquad (9.11)$$

The contribution of the 2-link term to the vertices is

$$
\begin{aligned}
V_\mu^{(1)} &= \frac{-g\kappa r T^a}{2} \left[\sin \frac{a(p + 3k)_\mu}{2} + \sin \frac{a(3p + k)_\mu}{2} \right] \\
V_\mu^{(2)} &= -a^2 g^2 \kappa r \frac{\{T^a, T^b\}}{2} \delta_{\mu\nu} \left[\cos \frac{aq_\mu}{2} \cos a(p + k)_\mu \cos \frac{aq'_\mu}{2} \right] \, .
\end{aligned}
\qquad (9.12)
$$

Consequently, the one and two gluon vertices of the HW action are

$$
\begin{aligned}
\frac{V_\mu^{HW(1)}}{2\kappa} &= g T^a i \gamma_\mu \left[1 - \frac{a^2 \, (p + k)_\mu^2}{8} \right] + \cdots \\
\frac{V_\mu^{HW(2)}}{2\kappa} &= \frac{a}{4} g^2 \{T^a, T^b\} \delta_{\mu\nu} \left[-ia\gamma_\mu \frac{(p + k)_\mu}{2} \right] + O(a^3) \, .
\end{aligned}
\qquad (9.13)
$$

Thus, by adding the 2-link term, Hamber and Wu were able to remove the leading term in both the free propagator and the gluon vertices. The remaining corrections start at $O(g^2 a)$, to remove these parameters like r has to be generalized to a power series in g^2.

The third ingredient in perturbative calculations is loop integrals. On the lattice there can be (i) power divergences proportional to $1/a$, (ii) $\log a$ divergences and (iii) finite terms that have a power series in a. The $\log a$ and $1/a$ divergences cannot occur simultaneously, i.e. there are no terms like $\log a/a$ from loop integrals. Thus, loop integrals can reduce the overall improvement only to $O(g^2 a)$ [55] in the case of mixing with higher dimension operators since the propagator and the vertices of the HW action are improved to $O(a^2)$. This argument holds to all orders; all terms of the form $a(g^2 \log a)^n$ in on-shell matrix elements are removed. This improvement can lead to a significant reduction in scaling violations even though $g^2 \approx 1$ because of the additional factors of 4π in loop integrals. This has been confirmed by the European Lattice Collaboration and I review their numerical evidence in Table 7 [63].

In general, to achieve improvement at $O(a)$, adding any gauge invariant term to the naive action that removes the doubling and introduces scaling corrections at $O(a^2)$ will do. A systematic procedure for improving the fermion lattice action was initiated by Sheikholeslami and Wohlert building on the work of Symanzik [58] and Lüscher and Weisz [60]. They incorporated the following ideas:

(1) Classical improvement of the action is equivalent to tree level improvement of on-shell quantities in the quantum theory. SW only concerned themselves with on-shell improvement since for non-abelian gauge theories it has not been demonstrated that Symanzik's program can be implemented for all off-shell quantities.

(2) In order to improve the action to $O(a^n)$ it is in general sufficient to consider all distinct operators of canonical dimension up to a^{4+n}. The coefficients of these operators in the action are, in general, a power series in g^2 and $m_q a$. From the point of optimizing simulation time it is useful to minimize the number of non-zero coefficients and keep only the shortest range interactions.

(3) Given an improved action to $O(a^n)$, one can generate a whole set of such improved actions by performing a transformation of the fields, A_μ, ψ and $\overline{\psi}$ in the functional integral, such that it preserves the spectrum to the given order. This is sufficient to ensure that all on-shell matrix elements of operators expressed in terms of the transformed fields, are improved to $O(a^n)$. To paraphrase, the set of iso-spectral transformations allow one to derive a whole class of improved actions; if the starting action is improved to $O(a^n)$, then all are improved to $O(a^n)$. It is possible that improvement of all on-shell quantities is not sufficient to fix the coefficients of all the terms in the action to that order. (An example of such a freedom was found in case of the gauge action where the presence of a redundant operator at $O(a^2)$ leads to a 1-parameter family of improved actions.) In that case one can try to derive additional conditions by improving some off-shell quantities or one can simply choose a value. In practice one wants follow the latter approach and set as many coefficients to zero in order to cut down the simulation time.

To improve the action to $O(a)$ SW considered an iso-spectral transformation starting with the naive action. The resulting change in the action is a redundant operator which lifts the doublers at $O(a)$, just like the Wilson action, and simultaneously reduces the leading correction to $O(g^2 a)$ as desired. The construction of the $O(a)$

improved SW action proceeds as follows.

Starting with the naive lattice action

$$
\begin{aligned}
A_N &= m_q \sum_x \overline{\psi}(x)\psi(x) + \sum_x \overline{\psi}(x)\gamma_\mu \overrightarrow{D}_\mu\psi(x) \\
&= m_q \sum_x \overline{\psi}(x)\psi(x) - \sum_x \overline{\psi}(x)\overleftarrow{D}_\mu\gamma_\mu\psi(x)
\end{aligned}
\tag{9.14}
$$

where the lattice derivatives are defined as

$$
\begin{aligned}
\overrightarrow{D}_\mu\psi(x) &= \frac{1}{2a}[U_\mu(x)\psi(x+\hat\mu) - U_\mu^\dagger(x-\hat\mu)\psi(x-\hat\mu)] \\
\overline{\psi}(x)\overleftarrow{D}_\mu &= \frac{1}{2a}[\overline{\psi}(x+\hat\mu)U_\mu^\dagger(x) - \overline{\psi}(x-\hat\mu)U_\mu(x-\hat\mu)] .
\end{aligned}
\tag{9.15}
$$

SW considered the field transformations

$$
\begin{aligned}
\psi(x) &\to \psi'(x) = \psi(x) + ia\epsilon\gamma_\mu \overrightarrow{D}_\mu\psi(x) \\
\overline{\psi}(x) &\to \overline{\psi}'(x) = \overline{\psi}(x) + ia\epsilon' \overline{\psi}(x)\overleftarrow{D}_\mu\gamma_\mu .
\end{aligned}
\tag{9.16}
$$

They showed that to $O(a)$ the measure is invariant and the change in the action is

$$
\begin{aligned}
\delta A &= am_q(\epsilon - \epsilon') \sum_x \overline{\psi}(x)\gamma_\mu \overrightarrow{D}_\mu\psi(x+\hat\mu) \\
&\quad + a(\epsilon - \epsilon') \sum_x \left\{ 2\Delta_W + \frac{ig}{2}\overline{\psi}(x)\sigma_{\mu\nu}F_{\mu\nu}\psi(x) \right\}
\end{aligned}
\tag{9.17}
$$

where

$$
\begin{aligned}
2(\epsilon - \epsilon')\Delta_W &= -\frac{r\kappa}{4a} \sum_x \overline{\psi}^L(x)\Big[U_\mu(x)U_\mu(x+\hat\mu)\psi^L(x+2\hat\mu) + \\
&\qquad U_\mu^\dagger(x-\hat\mu)U_\mu^\dagger(x-2\hat\mu)\psi^L(x-2\hat\mu) \Big]
\end{aligned}
\tag{9.18}
$$

is the same as the Hamber-Wu term provided we choose

$$
\epsilon = -\epsilon' = -\frac{r}{4} .
\tag{9.19}
$$

In other words the term Δ_W is such that it exactly cancels the HW term if we had performed the transformation (9.16) on the HW action and made the choice (9.19). Therefore one can "derive" the SW action starting from the HW improved action since the change in the 2 terms proportional to r in Eq. (9.8) is $O(a^2)$. Furthermore, the same result is obtained starting with the naive action since, to $O(a)$, Δ_W can be replaced by the 1-link Wilson term.

The final action, obtained by starting from either the naive or the HW action and making the transformation Eq. (9.16), is

$$
\begin{aligned}
A_{SW} &= m_q\overline{\psi}\psi + \left(1 - \frac{am_q r}{2}\right)\sum_x \overline{\psi}(x)\gamma_\mu \overrightarrow{D}_\mu\psi(x+\hat{\mu}) \\
&\quad - \frac{ar}{2}\sum_x\left\{\Delta_W + \frac{ig}{2}\overline{\psi}(x)\sigma_{\mu\nu}F_{\mu\nu}\psi(x)\right\} \\
&= A_W - \frac{iagr}{4}\overline{\psi}(x)\sigma_{\mu\nu}F_{\mu\nu}\psi(x) ,
\end{aligned}
\tag{9.20}
$$

where final SW form is obtained after a further redefinition of the fields to make the coefficient of the kinetic term unity. This is equivalent to starting out with the transformation

$$
\begin{aligned}
\psi(x) \to \psi'(x) &= \left[1 + \frac{arm_q}{4}\right]\left[1 - \frac{ar}{4}\gamma_\mu\overrightarrow{D}_\mu\right]\psi(x) \\
\overline{\psi}(x) \to \overline{\psi}'(x) &= \left[1 + \frac{arm_q}{4}\right]\overline{\psi}(x)\left[1 + \frac{ar}{4}\overleftarrow{D}_\mu\gamma_\mu\right] .
\end{aligned}
\tag{9.21}
$$

Note that this transformation breaks chiral symmetry, which is why the SW action removes the doublers. It is important to remember that this transformation has to be performed on all Dirac fields in the lattice operators when calculating matrix elements.

The change in the naive action under Eq. (9.21) can be summarized as

$$
A_N(\psi,\overline{\psi},m_q) \to A_{SW}(\psi,\overline{\psi},m_q + \frac{ar}{2}m_q^2)
\tag{9.22}
$$

To carry out perturbative calculations note that the free propagator for the SW action is the same as the Wilson action except for the definition of the quark mass. The vertices pick up an extra term, which for the 1-gluon case is

$$
\frac{V_\mu}{2\kappa} = \frac{-grT^a}{2}\sum_\nu \sigma_{\mu\nu}\sin a(q-q')_\mu \cos\frac{a(q-q')_\nu}{2}
\tag{9.23}
$$

Thus, at the classical level both the propagator and the vertices in the SW action have corrections at $O(a)$ and it is non-trivial to show that on-shell quantities are improved to $O(g^2 a)$ as with the HW action. Explicit examples of how this happens are given in [55]. A simple mnemonic is that since the SW action can be derived by an iso-spectral transformation (valid for states with E, $\vec{p} << \pi/a$) from the naive or the HW action, it is improved to the same order.

The ELC collaboration have chosen to use the SW action in their study of the improvement because of its computational advantage. The two link term in the HW action increases the CPU requirements by a factor of roughly six compared to the Wilson action, while the increase with SW action is a factor of about 2. The technical reasons for the large increase in CPU time with the HW action are (i) the extra multiply by the link matrices, (2) one cannot use the trick of spin projection which reduces the four component spinor to two independent components and (3) the interaction matrix cannot easily be factored into two parts corresponding to even and odd sites.

10. Improving the Operators

Assuming that the action has been improved to $O(a)$ we now consider how it impacts the extraction of matrix elements. A given lattice operator will mix with all possible operators that have the same symmetry properties and transform in the same way under the hypercubic group. Thus, the general relation of a matrix element $\langle f|\mathcal{O}|i\rangle$ calculated on the lattice to its desired value in the continuum is

$$
\begin{aligned}
a^d \langle f|\mathcal{O}|i\rangle_{cont} = {} & Z_{\mathcal{O}}\langle f|\mathcal{O}|i\rangle \\
& + \sum_{i=3}^{d-1} a^{-i} A^{(i-d)} \langle f|\mathcal{O}_l^{(i)}|i\rangle + \sum_{j=1} B^{(j)} \langle f|\mathcal{O}_s^{(j)}|i\rangle \\
& + \sum_{k=1}^{\infty} a^k C^{(k)} \langle f|\mathcal{O}_h^{(k)}|i\rangle \,.
\end{aligned}
\tag{10.1}
$$

where d is the canonical dimension of the operator $\mathcal{O}$ and a sum over all operators of a given dimension is implicit. The sum over i begins at 3 because the lowest canonical dimension for gauge invariant operators is 3 for fermions and 4 for gluons (excluding the Polyakov line operator). The various terms in the above expression are as follows:

1. The constant $Z_{\mathcal{O}}$ is the relative renormalization constant between the lattice and continuum operator. It and the constants A, B and C can be calculated as a power series in g^2 and in general have $\log a$ corrections. Logarithmic corrections in $Z_{\mathcal{O}}$ require a subtle matching between the lattice and the continuum.

2. The mixing with lower dimension operators $\mathcal{O}_l$ has to be removed in order to obtain physical results because otherwise the ME will diverge as powers of $1/a$. To remove these operators mandates non-perturbative methods. I do not address this problem here; a discussion can be found in [64] and in the reviews by Bernard [65] and Sharpe [66].

3. Naive dimensional analysis tells us that mixing with the same dimension operators ($\mathcal{O}_s$) is suppressed by powers of g^2 and that with higher dimensional operators ($\mathcal{O}_h$) is in addition suppressed by powers of a. For typical scales accessible in lattice simulations, $2 \ GeV < 1/a < 4 \ GeV$, the effects can be as large as 20%. As a result of this mixing, scaling violations including logarithms, will show up as a dependence of the matrix elements on the external states. The aim and scope of Symanzik's improvement program is to systematically remove the mixing with higher dimension operators.

This mixing with higher dimension operators can be demonstrated by the simple example of the symmetrized version of the 1-link vector current. It has the Taylor expansion

$$
\begin{aligned}
\overline{\psi}(x)\gamma_\mu U_\mu(x)\psi(x + \hat{\mu}) + \overline{\psi}(x + \hat{\mu})\gamma_\mu U_\mu^\dagger(x)\psi(x) & \\
= 2\overline{\psi}(x)\gamma_\mu\psi(x+) + a\partial_\mu\left[\overline{\psi}(x)\gamma_\mu\psi(x)\right] - \overline{\psi}(x)\gamma_\mu\left[a^2 g^2 A_\mu^2(x)\right]\psi(x) + \dots &
\end{aligned}
\tag{10.2}
$$

The leading correction is a total divergence which does not contribute to the on-shell ME, and the $\overline{\psi}gaA_\mu\psi$ term is killed by the symmetrization. (It is generally true that one can improve the scaling behavior of non-local operators by taking a spatially symmetrized version.) Thus, in this case the corrections at order a^2g^2 show that the leading operator it mixes with is $\overline{\psi}(x)\gamma_\mu A_\mu^2(x)\psi(x)$.

In the case of staggered fermions where the fields are spread over a 2^4 hypercube, the lattice transcription of operators that are local in the continuum (e.g. fermion bilinears and 4-fermi interactions) is non-local. This in general leads to $O(a)$ corrections that can be removed by symmetrization [66].

One can analyze the more serious corrections in both local and non-local operators, due to mixing induced by interactions, in perturbation theory. The highest superficial degree of divergence in loop integral arises at $O(a^{-3})$ from a closed fermion loop and at $O(a^{-2})$ from a closed gluon loop. In the case of staggered fermions symmetry of the momentum integrals imposes that only even powers of a occur [52], so the closed fermion loop also comes in at $O(a^{-2})$. The divergent terms lead to mixing with lower dimension operators, which as mentioned before have to be subtracted by non-perturbative methods. Since Symanzik's improvement program can only address mixing with higher dimension operators, therefore let us assume that it is possible to define finite lattice operators $\hat{O}$, such that

$$a^d \langle f|O|i \rangle_{cont} = \langle f|\hat{O}|i \rangle + \sum_{k=1}^{\infty} a^k C^{(k)} \langle f|\hat{O}_h^{(k)}|i \rangle \qquad (10.3)$$

where Z_O has been absorbed in the definition of $\hat{O}$. Improvement to $O(a)$ can then be achieved if the ME are calculated using $\hat{O} - aC^{(1)}\hat{O}^{(1)}$. A necessary condition for removing all $O(a)$ terms is that the action should also be improved to $O(a)$, otherwise the coefficients $C^{(1)}$ may contain terms like $(g^2 \log a)$ and have a dependence on external states [67].

To complete this discussion let us return to the numerical results for Z_V. The ELC have carried out an exploratory study of the ratios of matrix elements of four different vector currents with the SW action: (i) the local current (L), (ii) the conserved current C defined in Eq. (7.2), (iii) the improved current I obtained from the local current with the transformations (9.21), and (iv) the conserved improved current CI obtained from $\hat{V}_\mu$ under (9.21). The results are given in Table 7 [63].

Method	R_I^{CI}	R_I^C	R_L^{CI}	R_L^C	R_{CI}^C
Pert. Theory	0.92	0.92	0.87		
$\langle 0|V_i|\rho \rangle$	0.83	0.64	0.81	0.62	0.77
$\langle \pi|V_4|\pi \rangle$	0.85	0.85	0.79	0.79	1
$\langle \pi|V_i|\pi \rangle$	0.83	0.81(2)	0.78	0.75(2)	0.97(2)

Table 7: *Estimates for the renormalization constant Z_V from different ratios of the four currents defined in the text. The results are for the SW action and only forward matrix elements for the π are reproduced from [63]. The notation is, for example, $R_I^C = \langle f|V_\mu^C|i \rangle / \langle f|V_\mu^I|i \rangle$.*

The entries in the first and third column are expected to differ only at $O(g^2 a)$ and should be consistent within the accuracy of the calculation. Those in the second, fourth and fifth column have $O(a)$ corrections for non-forward matrix elements (see data for the ρ in the first row) as C is not improved. The forward matrix elements in the pion (columns 2 and 4) should also be improved. The results corroborate these expectations and show significant improvement over those given in Table 5. For further details the reader should consult Ref. [63].

In conclusion, the improvement program works and the quality of results improve significantly even with corrections decreasing from $O(a) \to O(g^2 a)$. Over the next couple of years a number of groups will try different forms of the improved action and it is very likely that the calculation of matrix elements will become reliable. As of this writing we are at an exciting juncture; lattice results are beginning to have an impact on standard model phenomenology.

ACKNOWLEDGEMENTS

I thank M. Creutz, D. Daniel, A. Patel and S. Sharpe for many discussions and for comments on the manuscript. The help of N. Christ, P. de Forcrand, A. Kronfeld, C. Michael, A. Nakamura and N. Stamatescu in compiling the data presented here is gratefully acknowledged. I also thank J. Kiskis and M. Lüscher for informative correspondance.

References

[1] G. B. West, in *Los Alamos Science* number 11, 1984

[2] M. Fisher, Lecture Notes in Physics, Vol 186, Springer-Verlag (1983) ;
P. Pfeuty and G. Toulouse, *Introduction to Renormalization Group and Critical Phenomena*, Wiley 1977

[3] K.G. Wilson and J.B. Kogut, *Phys. Rep.* **12C** (1984) 76

[4] R. Gupta, in *"From Actions to Answers"*, Proceedings of the 1989 TASI in Elementary Particle Physics, Boulder, Eds. T. DeGrand and D. Toussaint, World Scientific 1990173

[5] C. Baillie, R. Gupta, K. Hawick and S. Pawley, *Phys. Rev.* **B45** (1992) May 1.2 issue

[6] H. D. Politzer, *Phys. Rev. Lett.* **30** 1346 ;
D. Gross and F. Wilczek, *Phys. Rev. Lett.* **30** 1343

[7] K.G. Wilson, *Phys. Rev.* **D10** (1974) 2445

[8] K.G. Wilson, *Recent Developments in Gauge Theories*, Cargese (1979), eds. G. t'Hooft *et al.* (Plenum 1980)

[9] J. Kogut, *Reviews of Modern Physics* **55** (1982) 775

[10] W. Celmaster and R. J. Gonsalves, *Phys. Rev.* **D20** (1979) 1420

[11] A. Hasenfratz and P. Hasenfratz, *Phys. Lett.* **93B** (1980) 165 ;
H. Kawai, R. Nakayama and K. Seo, *Nucl. Phys.* **B189** (1981) 40

[12] R. Gupta, in *"LATTICE 89"*, Proceedings of the 1989 Symposium on Lattice Field Theory, Capri, Italy, 1989, Eds. N. Cabibbo *et al.*, *Nucl. Phys.* **B** (*Proc. Suppl.*) **17**, (1990)

[13] T. DeGrand, Int. Symp. *"LATTICE 90"*, Proceedings of the International Conference on Lattice Field Theory, Tallahassee, Florida, 1990, Eds. U. M. Heller *et al.*, *Nucl. Phys.* **B** (*Proc. Suppl.*) **20**, (1991)

[14] D. Toussaint, Int. Symp. *"LATTICE 91"*, To appear in the Proceedings of the International Symposium on Lattice Field Theory, Tsukuba, Japan, 1991

[15] K. G. Wilson, in *Recent Developments in Gauge Theories*, Cargese (1979),eds. G. t' Hooft, *et al.* (Plenum, New York, 1980)

[16] R. Swendsen, Phys. Rev. Lett. **47** (1981) 1775

[17] R. Cordery, R. Gupta and M. A. Novotny, *Phys. Lett.* **128B** (1983) 425

[18] K. H. Mütter and K. Schilling, *Nucl. Phys.* **B230** [FS10] (1984) 275

[19] R. Gupta, S. Güsken, K-H. Mütter, A. Patel, R. Sommer and K. Schilling, *Nucl. Phys.* **B314** (1989) 63 and *Phys. Lett.* **200B** (1988) 143.

[20] R. Gupta, in *"Lattice Gauge Theory Using Parallel Processors"*, CCAST Sym-

posium/Workshop Volume 1, Beijing, China, 1987, Eds. X. Li *et al.*, Gordon & Breach 1987, 567

[21] K. Decker and P. de Forcrand, in *"LATTICE 89"*, Proceedings of the 1989 Symposium on Lattice Field Theory, Capri, Italy, 1989, Eds. N. Cabibbo *et al.*, *Nucl. Phys.* **B** (*Proc. Suppl.*) **17**, (1990), 567

[22] R. Gavai and F. Karsch, *Phys. Rev. Lett.* **57** (1986) 40 ; K.M. Bitar, A.D. Kennedy and P. Rossi, *Phys. Rev. Lett.* **63** (1989) 2713

[23] APE collaboration, *Phys. Lett.* **192B** (1987) 163; *Phys. Lett.* **197B** (1987) 400; PLB205 (1988) 535

[24] D. Barkai, K. J. M. Moriarty and C. Rebbi, *Phys. Rev.* **D30** (1984) 1292

[25] S. W. Otto and J. D. Stack, *Phys. Rev. Lett.* **52** (1984) 2328

[26] P. de Forcrand *et al.*, *Phys. Lett.* **160B** (1985) 137 ; Ph. de Forcrand, unpublished

[27] A. Hasenfratz *et al.* , *Phys. Lett.* **143B** (1984) 193 ; K.C.Bowler *et al.*, *Phys. Lett.* **163B** (1985) 367

[28] P. De Forcrand, *Proceedings of the Metropolis Conference*, Los Alamos (1985), J. Stat. Phys. **43** (1986) 1077

[29] G. Schierholz, in *"Field theory on the Lattice"*, Proceedings of the International Symposium, Seillac, France, 1987, edited by A. Billoire *et al.*, *Nucl. Phys.* **B** (*Proc. Suppl.*) **4**, (1988)11 ; G. Schierholz, in *"LATTICE 88"*, Proceedings of the 1988 Symposium on Lattice Field Theory, Fermilab, Batavia, Eds. A.S. Kronfeld and P. B. Mackenzie, *Nucl. Phys.* **B** (*Proc. Suppl.*) **9** (1989)244.

[30] C. Michael and M. Teper, *Nucl. Phys.* **B314** (1989) 347

[31] Hong Ding, *Phys. Rev.* **D44** (1991) 2200

[32] M. Lüscher, Nucl. Phys. B180 [FS2] (1981) 317 ; R. Pisarski and O. Alvarez, Phys. Rev. D26 (1982) 3735

[33] M. Lüscher, in *Progress in Gauge Field Theory*, (Cargèse, 1983), p. 451, ed. G. 't Hooft *et al.*, Plenum Press (1984)

[34] R. Gupta, *et al.*, *Phys. Rev.* **D43** (1991) 2301

[35] A. Ukawa, in *"LATTICE 89"*, Proceedings of the 1989 Symposium on Lattice Field Theory, Capri, Italy, 1989, Eds. N. Cabibbo *et al.*, *Nucl. Phys.* **B** (*Proc. Suppl.*) **17**, (1990)118 ; A. Ukawa, Talk at the 25th International Conference on High Energy Physics, Singapore, 1990

[36] D. Toussaint, Int. Symp. *"LATTICE 91"*, To appear in the Proceedings of the

International Symposium on Lattice Field Theory, Tsukuba, Japan, 1991

[37] B. Svetitsky, *Phys. Reps.* **132** (1986) 1

[38] A. D. Kennedy, *Phys. Rev. Lett.* **54** (1985) 87.

[39] S. Gottlieb *et al.* , *Phys. Rev. Lett.* **55** (1985) 1958, ;
D. Toussaint, in *"Lattice Gauge Theory '86"*, Brookhaven, New York, 1986, NATO ASI Series B: Physics Vol. 159399, and private communications

[40] H. Ding and N. Christ, *Phys. Rev. Lett.* **60** (1988) 1367, ;
F. Brown, *et al.* , *Phys. Rev. Lett.* **61** (1988) 2058, ;
N. Christ, private communications

[41] B. Svetitsky and F. Fucito, *Phys. Lett.* **131B** (1983) 165.

[42] M. Fukugita *et al.* , *Phys. Lett.* **154B** (1985) 185, *Phys. Rev. Lett.* **63** (1989) 1768, and *Nucl. Phys.* **B337** (1990) 181.

[43] R. Gupta, in *Quark Confinement and Liberation*, Ed. F. Klinkhamer World Scientific, 1985.

[44] P. Lepage and P. Mackenzie, Int. Symp. *"LATTICE 90"*, Proceedings of the International Conference on Lattice Field Theory, Tallahassee, Florida, 1990, Eds. U. M. Heller *et al.*, *Nucl. Phys.* B (*Proc. Suppl.*) **20**, (1991) 173

[45] R. Gupta, G. Kilcup, A. Patel and S. Sharpe, *Phys. Lett.* **211B** (1988) 132

[46] K. Bowler, *et al.*, *Nucl. Phys.* **B257** [FS 14] (1985) 155; *Phys. Lett.* **179B** (1987) 375

[47] QCD_TARO Collaboration, Int. Symp. *"LATTICE 91"*, To appear in the Proceedings of the International Symposium on Lattice Field Theory, Tsukuba, Japan, 1991

[48] A. Patel, S. Otto and R. Gupta, *Phys. Lett.* **159B** (1985) 143

[49] K. M. Decker and P. de Forcrand, in *"LATTICE 89"*, Proceedings of the 1989 Symposium on Lattice Field Theory, Capri, Italy, 1989, Eds. N. Cabibbo *et al.*, *Nucl. Phys.* B (*Proc. Suppl.*) **17**, (1990)567

[50] UKQCD collaboration, Liverpool Preprint LTH 271

[51] G. Bhanot and M. Creutz, *Phys. Rev.* **D24** (1981) 3212

[52] S. Sharpe, Int. Symp. *"LATTICE 91"*, To appear in the Proceedings of the International Symposium on Lattice Field Theory, Tsukuba, Japan, 1991

[53] A. Vladikas, Int. Symp. *"LATTICE 91"*, To appear in the Proceedings of the International Symposium on Lattice Field Theory, Tsukuba, Japan, 1991

[54] A. Vladikas, Int. Symp. *"LATTICE 90"*, Proceedings of the International Conference on Lattice Field Theory, Tallahassee, Florida, 1990, Eds. U. M. Heller *et al.*, *Nucl. Phys.* B (*Proc. Suppl.*) **20**, (1991) 448

[55] G. Heatlie, C. Sachrajda, G. Martinelli, C. Pittori and G.C. Rossi, *Nucl. Phys.* **B352** (1991) 266

[56] A. Patel and R. Gupta, *Phys. Lett.* **183B**(1987) 193

[57] S. Güsken, *et.al. Nucl. Phys.* **B314** (1989) 63

[58] K. Symanzik, *Nucl. Phys.* **B226** (1983) 187 and 205

[59] R. Gupta, in *Lattice Gauge Theories: A Challenge in Large Scale Computing*, Wuppertal 1985, Eds. K-H Mütter and K. Schilling, Plenum Press (1986)

[60] M. Lüscher and P. Weisz, Commun. Math. Phys. **97** (1985) 59

[61] G. Curci, P. Menotti and G. Paffuti, *Phys. Lett.* **130B** (1983) 205 ;
P. Weisz and R. Wohlert, *Nucl. Phys.* **B236** (1984) 397

[62] H. Hamber and C.M.Wu, *Phys. Lett.* **133B** (183) 351

[63] G. Martinelli, C.T. Sachrajda and A. Vladikas, *Nucl. Phys.* **B358** (1991) 212

[64] L. Maiani, G. Martinelli, G. Rossi and M. Testa, *Nucl. Phys.* **B289** (1987) 505

[65] C. Bernard, in *"From Actions to Answers"*, Proceedings of the 1989 TASI in Elementary Particle Physics, Boulder, Eds. T. DeGrand and D. Toussaint, World Scientific 1990 233

[66] S. Sharpe, in *Standard Model, Hadron Phenomenology and Weak Decays on the Lattice*, Ed. G. Martinelli, to be published by World Scientific

[67] G. C. Rossi, in *"LATTICE 89"*, Proceedings of the 1989 Symposium on Lattice Field Theory, Capri, Italy, 1989, Eds. N. Cabibbo *et al.*, *Nucl. Phys.* **B** (*Proc. Suppl.*) **17**, (1990)607

LATTICE APPROACH TO ELECTROWEAK MATRIX ELEMENTS

CLAUDE BERNARD

Department of Physics, Washington University
St. Louis, MO 63130

and

AMARJIT SONI

Physics Department, Brookhaven National Laboratory
Upton, NY 11973, USA

ABSTRACT

Efforts towards developing a comprehensive framework for calculating electroweak amplitudes using lattice gauge methods are reviewed. The general formalism is outlined and its applications to K, D and B physics is discussed. Difficulties in handling $K \rightarrow 2\pi$ decays are recapitulated. The status of B parameters for K and B mesons, the nonleptonic D decays, semi leptonic form factors, pseudoscalar decay constant for the heavy light mesons and preliminary attempts at calculating $B \rightarrow K^*\gamma$ are reported. By and large this review covers the situation as of LATTICE '91.

1   Introduction: Where Do We Stand?

Efforts to calculate weak matrix elements using lattice gauge theory techniques have been receiving considerable attention for almost a decade.[1-3] While there is unanimous agreement that this is one of the most important applications of the lattice method, which essentially guarantees that the method will continue to be prominent, so far there are only a few real and tangible successes. Much progress has certainly been attained in clarifying the theoretical and numerical underpinnings and in crystallizing our understanding of what can and cannot be done._ In some areas, conceptual difficulties still need to be surmounted. Herein we review the status of these efforts.

Broadly speaking the electroweak matrix elements of interest can be classified in the following three categories (see Table 1.1):

1. $\langle M|\Theta|MM\rangle$ i.e. 4-point functions,

2. $\langle M|\Theta|M\rangle$ i.e. 3-point functions and

3. $\langle M|\Theta|\text{vac}\rangle$ i.e. 2-point functions.

Here M stands generically for a hadron, although throughout our discussion in this work we will confine ourselves to mesons, and Θ stands for some quark operator. As is indicated in Table 1.1, there appear to be very serious stumbling blocks preventing genuine, first-principle calculations of 4-point functions. In addition to nontrivial numerical problems

Table 1.1: Status at a Glance

Matrix Element Type	Physics Applications	Status of Theoretical Infrastructure	Attainable[a,b] Precision Now or very soon	Precision[a] Possible with ~ 5 times more computing power
1. $\langle M\|\Theta\|MM\rangle$	$K \to 2\pi$, $\Delta I = 1/2$, $\epsilon'/\epsilon\ldots$ $D \to K\pi$, $KK\ldots$ $B \to \rho\pi$	Incomplete		
2. $\langle M\|\Theta\|M\rangle$	$K - K$, ϵ $D \to K(K^*)e\nu\ldots$ $B \to B$	OK	~ 15% ~ 30% ~ 30%	~ 10%
	$B \to \pi(\rho)e\nu\ldots$		~ 30%	~ 10%
3. $\langle M\|\Theta\|\text{Vac}\rangle$	f_{su}, f_{cu}, f_{cs} f_{bu}, f_{bs}	OK	~ 20% ~ 20% ~ 20%	~ 10%

a) all numbers are in the quenched approximation
b) We define very soon to be $\lesssim 2$ years.

(such as maintaining laboratory kinematics) there are fundamental difficulties involving final state interactions between the two mesons in the final state and the consequent phases that characterize such strong interactions.[4,5] This has to be regarded as a great disappointment as these four point functions have repercussions for some of the most interesting issues in the phenomenology of K, D and B decays and for our understanding of CP violation.

Of course, we certainly would advocate efforts in this class of problems dealing with 4-point functions to continue. Considerable useful information can still be uncovered by examining these decays on the lattice in conjunction with some specific continuum models. To some limited extent this is illustrated by our study of $D \to K\pi$.[6,7] Quantitatively the situation looks more promising with regard to 3- and 2-point functions. These are also relevant to a host of very important phenomenological issues and lattice calculations are already able to get some useful numbers, in the quenched approximation, with an accuracy of about 30% or better. Issues pertaining to B-physics are still a bit of a problem.[8,9] However, this is unlikely to be a major difficulty for too long. Numerous simulations with $\beta = 6.3$ and 6.4 are being done and studied already. These should enlighten us about the dynamics of propagating "heavy" quarks on the lattice with masses in the neighborhood of 3.5 GeV or so. Also progress in static and non-relativistic heavy quarks on the lattice is steadily being made so those methods should help here too.

Clearly, it is difficult in a review such as this to encompass all the interesting and important developments on such a vast subject that have taken place over the course of almost a decade. Pressures of time and space did not permit inclusion of many interesting related topics. By and large attempt is made here to include results reported up to LATTICE '91. One of the important areas that is being left out is the use of improved action and operators for weak matrix elements calculations.[1,10–12] Also the use of fixed gauge subtraction methods will not be covered.[13]

2 Motivations: Why the Hooplah?

Attempts to calculate electroweak matrix elements using lattice techniques are important because success of such efforts will have lasting impact on three interwoven goals:

1. In experimental determination of the fundamental or "irreducible" parameters of the Standard Model (SM) and thereby in improving our understanding of the SM

2. In providing reliable tests of the SM

3. In the search for the effects of new physics.

Despite the fact that the SM continually scores impressive successes, it is a model with many (of order 20) fundamental, or irreducible, parameters. We are referring here to quantities such as masses of quarks, leptons and gauge fields, mixing angles etc. In the SM these parameters are irreducible because they can only be determined from experiments. Indeed much of the experimental effort is directed towards such determinations. However, more often than not, experimental information does not directly yield the parameters of fundamental interest, unless one knows the relevant "hadronic parameters", i.e. hadronic matrix elements. $B^\circ - \bar{B}^\circ$ is an important case in point which we discuss to illustrate features shared by countless other examples.

The mixing parameter $x_{bd} \equiv \Delta M/\Gamma$ is given by

$$x_{bd} = (\text{known factors}) \, h\,(m_t) \, |V_{td}|^2 \, f_{bd}^2 B_{bd}, \tag{1}$$

where $h(m_t)$ is a rapidly increasing function of m_t which is calculable in perturbation theory, V_{td} is a KM mixing angle, B_{bd} is the so-called "Bag factor" for $B^\circ - \bar{B}^\circ$ and f_{bd} is the B-meson pseudoscalar decay constant.

In this example m_t and V_{td} are fundamental parameters of the SM, whereas B_{bd} and f_{bd} are "hadronic parameters" not on the same footing as m_t or V_{td}. These hadronic parameters are of course dependent on the masses of the other light quarks and the QCD scale parameter and should be calculable in terms of these known fundamental parameters. The reason that the values of these hadronic parameters are not yet known is a reflection of the theoretical community's inability to do reliable calculations in the full SM, i.e. including QCD. As a result although x_{bd} has been experimentally[14] determined triumphantly, one cannot use that valuable information to constrain m_t and V_{td} effectively, or, for that matter, to deduce V_{td} once m_t becomes known.

Another simple example is the semi-leptonic transition: $B \to \rho\ell\nu$. Experimentalists are already claiming to have seen a few of these events.[15] As more B's get analyzed with improved detectors and especially if B factory becomes a reality, then many such events will surely be seen. From such observations one would like to deduce another fundamental parameter of the SM, namely V_{ub}, as precisely as possible.[16] Recall that much of the estimates for CP violation phenomena crucially involve this small mixing angle. However,

$$\Gamma(B \to \rho\ell\nu) \sim |V_{ub}|^2 \left| \langle \rho | J^\mu_{\text{quark}} | B \rangle J^\mu_{\text{lepton}} \right|^2 \tag{2}$$

where $\langle\rho|J|B\rangle$ is the unknown hadron matrix element determined again by light quark masses and QCD dynamics. So, once again, these matrix elements are unknown not because we lack knowledge of any of the relevant fundamental parameters of the SM but rather because they are governed by bound state QCD dynamics which we still are not able to handle.

It is easy to understand how efforts at these calculations could provide important tests of the SM. Basically these tests come through redundant measurements and self consistency checks. For example, V_{td} enters not just in $B^\circ - \bar{B}^\circ$ mixing but also it monitors the rate for $K \to \pi\nu\bar{\nu}$.[17] Different experiments had better yield the same value for V_{td} for the SM to stay viable. Precision tests of V_{ub} through, for example, exclusive semi-leptonic B decays could allow us to check the unitarity of the KM matrix via the famous unitarity triangle.[16] Furthermore, in the SM there is only one CP violating KM phase, so if we could calculate the matrix elements for various CP violating phenomena such as ϵ', $K \to \pi ee$ or B non-leptonic decays, then in conjunction with experimental studies we could test if all of these are accountable by the same numerical value for that unique phase.

Hadronic matrix elements enter crucially into the confrontation between experiment and non-standard theoretical constructs as well. Two examples where lattice methods have already been used will serve to illustrate the point. There is the $\Delta S = 2$ left-right matrix element:[18]

$$M_{LR} = \langle K^\circ|\bar{s}\gamma_\mu(1-\gamma_5)d\bar{s}\gamma_\mu(1+\gamma_5)d|\bar{K}^\circ\rangle \qquad (3)$$

This is relevant to left-right symmetric gauge theory wherein it controls the contribution of a W_L and a W_R exchange to the $K - \bar{K}$ mixing amplitude. Another example is the matrix element of the type[19]

$$\langle\pi^\circ e^+|\epsilon^{abc}(u^a C d^b_L)(\bar{e}^+ u^c_R)|p\rangle \qquad (4)$$

which controls the baryon number violating BR($p \to e^+ + \pi^0$) in SU(5) and other Grand unified theories.

3 Theoretical Ingredients

3.1 Continuum Effective Field Theory[3]

A generic feature shared by weak matrix elements is that the corresponding physical processes occur through the exchange of a W-boson with its concomitant distance scale of m_W^{-1}. At the quark level, without QCD, the weak amplitude is straight forward to calculate. For experimental purposes the relevant quantities are amplitudes for incoming and outgoing hadrons involving QCD intimately. QCD effects can cause significant corrections — as we all know from the famous example of the $\Delta I = 1/2$ rule. Invariably these calculations involve integrating out the W-boson (and the relevant heavy quarks, if necessary) by the use of the effective field theory apparatus of the operator product expansion and the renormalization group (OPE/RG). This becomes inescapable for two different reasons. Firstly it is due to the conceptual difficulty that still needs to be surmounted in latticization of the complete SM with its plethora of chiral fermions.[20] In addition, a practical reason for getting rid of the W-boson is that its mass is much larger than the typical mass (momentum)

366

scale $\sim m_K$, m_π at which the QCD effects are expected to be pronounced. Note that the cost of computer simulation scale as N^4, where N is the size of each side of the Euclidean box. Currently available computer resources limit this ratio, in the quenched approximation to be $\lesssim 32$. Thus, the lattice spacing, which is also approximately the ultra-violet cut-off, in current simulations is about $(N \times m_\pi) \approx 4$ Gev i.e. a factor of 20 less than m_W. So to have lattice spacing of $O(m_W)$ for simulation would require an increment in computing power by about five orders of magnitude, just in the quenched approximation! The need for integrating out the W-boson is therefore clear.

For convenience, we use the OPE/RG apparatus in the continuum first.[21-26] The amplitudes of interest, lowest order in weak interactions, can then be written as:

$$A_{IF} = \sum_i \frac{C_i(M_W/\mu, g(\mu))}{M_W^2} \langle F|\Theta_i(0)|I\rangle \tag{5}$$

where C_i, the Wilson coefficients, depend only on the short distance part of the theory, and are therefore perturbatively calculable and O_i are multiplicatively renormalizable operators. We give some examples below.

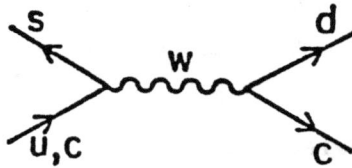

Fig. 3.1. Weak decay of the s quark in absence of QCD.

Perhaps the best known example is that of $K \to \pi\pi$ decays. Here, at the quark level, before QCD is switched on, the weak decay of the strange quark is described by graphs such as Fig. 3.1. In the $m_W \to \infty$ the figure generates the following 4-quark operator:

$$O_{FF} \equiv \bar{s}\gamma_\mu(1-\gamma_5)u\bar{u}\gamma_\mu(1-\gamma_5)d - (u \to c) \tag{6}$$

with the "free-field" coefficient:

$$C(M_W/\mu, g(\mu) = 0) = \frac{g_2^2}{8M_W^2} \sin\Theta_c \cos\Theta_c$$

$$= \frac{G_F}{\sqrt{2}} \sin\Theta_c \cos\Theta_c \qquad (7)$$

where g_2 is the coupling constant of the SU(2) gauge group of the SM and G_F the Fermi constant and Θ_c is the Cabibbo angle.

When gluonic corrections are turned on, the four-fermi operator O_{FF} is no longer multiplicatively renormalizable, rather one has the pair:

$$O_\pm = [\bar{s}\gamma_\mu(1-\gamma_5)d\,\bar{u}\gamma_\mu(1-\gamma_5)u \pm \bar{s}\gamma_\mu(1-\gamma_5)u\,\bar{u}\gamma_\mu(1-\gamma_5)d] - (u \to c) \qquad (8)$$

with the Wilson coefficient:

$$C_\pm(M_W/\mu, g(\mu)) = (\frac{g(\mu)}{g(M_W)})^{\gamma_\pm/b_0} \qquad (9)$$

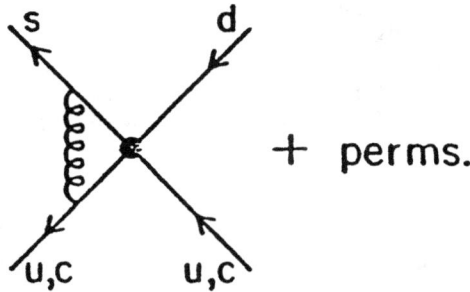

Fig. 3.2. Lowest order diagrams for the renormalization and mixing of the 4-quark operators.

Here $\gamma_\pm$ are the anomalous dimensions of the multiplicatively renormalizable operators arising from the infinite parts of the gluon exchange graphs shown in Fig. 3.2 and $b_0 = 11 - 2/3N_f$. We thus arrive at the $\Delta S = 1$ Hamiltonian for $K \to \pi\pi$ decays:

$$H_W = \frac{G_F}{2\sqrt{2}} \sin\Theta_c \cos\Theta_c \sum_\pm C_\pm(M_W/\mu, g(\mu))O_\pm(\mu) \qquad (10)$$

Note that O_- is a pure $\Delta I = 1/2$ operator whereas O_+ is a mixture of $1/2$ and $3/2$ pieces contributing to both types of $K \to \pi\pi$ transitions.

Note also that starting from the four-fermi operator for other non-leptonic transition (e.g., those relevant to D, B decays) one can similarly deduce the multiplicatively renormalizable 4-quark operators and write down the relevant effective Hamiltonian as in eqn. 10.

3.2   Operator Renormalization and Operator Mixing on the Lattice[3,4,27−29]

Since lattice operators are defined with the lattice spacing as the cut-off we cannot simply take the operator defined in the continuum, say in the $\overline{MS}$ scheme, as such to the lattice. Fortunately the relation between the lattice operators and their continuum counterpart is amenable to perturbation theory as it is governed by the regularization issues pertaining to the high energy regime. The procedure involves using lattice weak coupling perturbation theory (LWCPT) to calculate the renormalization of the operators and comparing with the corresponding continuum renormalization. The relation between the lattice and continuum operators can be generically written as:

$$O_i^{\text{cont}}(\mu) = Z_{ij}(\mu a, g(a)) O_j^{\text{latt}}(a) \tag{11}$$

Note that in general this relation involves "mixing" of operators. For example, for the case of the $(V - A) \times (V - A)$ four-fermi operator of the continuum theory, on the right hand side, one would, in general have in addition to the original LL operators, $S \times S$, $P \times P$, $T \times T$, $A \times A$ and $V \times V$ (dimension-6) contributions. Furthermore, for the four-quark LL operators of the continuum theory, one has in addition the contribution from lower-dimensional (i.e. two quark, dim-3) operators. Thus, for example, for the $O_\pm$ operator relevant to the $\Delta I = 1/2$ and $3/2$, the $K \to \pi\pi$ transitions one has:

$$
\begin{aligned}
O_\pm^{\text{cont}}(\mu) &= Z_\pm O_\pm^{\text{latt}} + Z'_\pm O'_\pm + Z_3 \bar{s} d + Z'_3 \bar{s} \gamma_5 d \\
&\quad + Z_5 \bar{s} \sigma_{\mu\nu} F^{\mu\nu} d
\end{aligned} \tag{12}
$$

As mentioned above, the off-diagonal operators $(O'_\pm)$ have the form:

$$Z'_\pm O'_\pm = Z_S S \times S + Z_P P \times P + Z_T T \times T + Z_V V \times V + Z_A A \times A \tag{13}$$

where S, P ... etc. are the usual Dirac quadratic forms. Thus these off-diagonal operators (eqn 13) do not have the same chiral structure as the LL form of the $O_\pm$. Their presence is a manifestation of the explicit violation of chiral symmetry by Wilson fermions. This has its origin in the "r-term" in the fermion action which is there to evade the fermion doubling problem. Thus even for massless quarks the lattice quark propagator of momentum p has the form:

$$Q(p) \sim \frac{1}{\not{p} - r\, ap^2} \tag{14}$$

As a result the off-diagonal operators $(O'_\pm)$ have the "wrong" chiral behavior i.e. $\langle O'\pm \rangle \to$ constant as the meson mass $\to 0$ in contrast to $\langle O_\pm^{\text{cont}} \rangle \to m^2$ for small m. Therefore, in the chiral limit, there has to be delicate cancellation among the constant pieces between $O'_\pm$

and $O_{\pm}^{\text{latt}}$ in order for the smaller m^2 piece to make its presence felt. This can clearly pose a serious numerical problem unless the lattice spacing is sufficiently small (i.e. unless the coupling is sufficiently weak) for the LWCPT to give coefficients with sufficient accuracy.

Of course the matching between the continuum and the lattice operators itself uses LWCPT which requires that the coupling be weak enough to be in the scaling regime and that it be perturbative. Numerical studies of the $K - \bar{K}$ mixing operator strongly suggests that $\beta > 6$ is required.[30,31] (See Section 6.1)

Let us now briefly discuss the renormalization coefficients

$$Z_{\pm} = 1 + \frac{g^2}{16\pi^2} [\gamma_{\pm} \ell_n a\mu + d_{\pm}] + \cdots \tag{15}$$

where, for our purpose, $\gamma_{\pm}$ are the one-loop anomalous dimensions and $d_{\pm}$ are constants typically $O(-50)$. The presence of the $\gamma_{\pm}$ guarantees that to the relevant order in the gauge coupling (here to order (g^2)) the μ dependence of the continuum coefficient $C_{\pm}(M_W/\mu, g(\mu))$ cancels so that the combination $C(\mu)Z(\mu a)C^{\text{latt}}(a)$ stays independent of μ, order by order in g^2.

The coefficients $Z'_{\pm}$, of course, start out at g^2 and have the form:

$$Z'_{\pm} = r^2 Z^*(r, g) \tag{16}$$

As $r \to 0$ chiral symmetry gets restored to the theory and $Z'_{\pm}$ vanish. For $r = 1$ these coefficients are much smaller than $Z_{\pm}$. This is easy to understand since $g^2/16\pi^2$ is $O(1/160)$. However, the smallness of these coefficients cannot be taken to mean that the contribution of $Z'_{\pm}O'_{\pm}$ for a given process would be necessarily small compared to $Z_{\pm}O_{\pm}$. In particular, for $K - \bar{K}$, $K \to \pi\pi$ transitions, for small meson masses $\langle O'_{\pm} \rangle$ can be considerably bigger than $\langle O_{\pm} \rangle$ as is implied by the constant versus m^2 behavior mentioned above.

3.3 Mixing with Lower Dimensional Operators[1,3,5,32,33]

Diagrammatically the mixing of the 4-quark operators with the lower dimensional operator arises (e.g.) from graphs of the type shown in Fig. 3.3. Clearly, these mixings are only relevant for those 4-quark operators that have a $q\bar{q}$ of the same flavor in them as is the case for $\bar{s}d\bar{u}u$, $\bar{b}d\bar{c}c$ etc. . .In the case of $\Delta S = 1$ transitions these lower dimensional operators are necessarily $\Delta I = 1/2$. In b quark decays, $\bar{b}d\bar{c}c$, $\bar{b}d\bar{u}u$ mix with $\Delta I = 1/2$ lower dimensional operators, whereas $\bar{b}s\bar{c}c$ or $\bar{b}s\bar{u}u$ mix with $\Delta I = 0$ ones.

Dimensional reasoning suggests that the coefficients of these lower dimensional operators can be proportional to inverse powers of a. Therefore, They gain importance as $a \to 0$, requiring delicate cancellations with O_i^{latt} to render O_i^{cont} independent of a. As mentioned above, for the case of $Z_{\pm}$ and $Z'_{\pm}$, perturbative estimates of Z_3, Z'_3 are even less likely to be reliable, especially at the current range of available gauge couplings, i.e. $\beta \sim 6.0$. For Z_3, Z'_3 even weaker gauge couplings may not be enough. Indeed, Martinelli has argued that perturbation theory may not be reliable for the lower dimensional operators at any gauge coupling. The point is that these coefficients may have non-perturbative

370

contributions of the form $\exp(-1/g^2(a))$, which, by the renormalization group, go like a as $g \to 0$. Presence of overall powers of a^{-1} thus mean that such contributions would survive even as $g \to 0$. It is therefore clear that mixing from lower dimensional operators must be eliminated either by non-perturbative subtraction methods or by the use of symmetries. These will be elaborated in later sections.

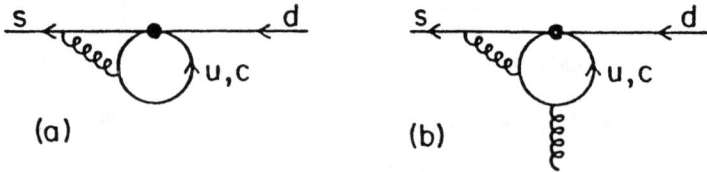

Fig. 3.3. Examples of diagrams that cause mixing of $0_\pm$ with lower-dimensional operators.

4 Implications of Discrete Symmetries[5,7,34]

In the context of hadronic matrix element calculations, discrete symmetries have proven very useful in three applications:

1. Reduction of noise

2. Elimination of certain graphs

3. Restrictions on lower dimensional operators that mix with 4-quark operators

We will illustrate some of these briefly here. For this purpose let us recall what the symmetry operations do to the quark propagators.

$$
\begin{array}{lll}
P : Q(x,y,[U]) &= \gamma_0 Q(x^P, y^P, [U]^P)\gamma_0 & (a) \\
C : Q(x,y,[U]) &= \gamma_0\gamma_2 Q^T(y, x, [U]^c)\gamma_2\gamma_0 & (b) \\
T : Q(x,y,[U]) &= \gamma_0\gamma_5 Q(x^T, y^T, [U]^T)\gamma_5\gamma_0 & (c)
\end{array}
\qquad (17)
$$

where $Q(x, y, [U])$ is the quark propagator from x to y in the configuration $[U]$. The key point is that these symmetries immediately tell us what the quark propagators, $Q[U]^R$ in the transformed configuration, U^R, is in terms of the propagators $Q[U]$ that has been obtained by explicit and costly numerical simulations, in the original configuration $[U]$. Thus the quark propagator in the transformed configuration comes in for free. We can, therefore, include it in our calculations of the Green's functions. Consequently, there is gain in statistics from averaging of the Green's functions over the combined sets $[U]$ and $[U]^R$ rather than only over the set $[U]$.

4.1 Applications to Meson 2-point Functions[7]

The simplest applications is to the meson 2-point functions. Consider, for example, the pion propagator in the fixed background gauge configuration U:

$$G_\pi^{(2)}(t, 0; \vec{p}, U) = \sum_{\vec{x}} e^{\frac{2\pi i}{L} \vec{p} \cdot \vec{x}} T_r[\gamma_5 D(0, x; U) \gamma_5 U(x, 0; U)] \tag{18}$$

where $D(x, y; U)$ is the propagator of the down quark etc.

Substituting equation (17a) we get the corresponding pion propagator, $G_\pi^{(2)}(t, 0; \vec{p})$ for the parity reversed gauge configuration U^P:

$$G_\pi^{(2)}(t, 0; \vec{p}, U^P) = \sum_{\vec{x}} e^{\frac{2\pi i}{L} \vec{p} \cdot \vec{x}} T_r[\gamma_5 D(0, x^P; U) \gamma_5 U(x^P, 0; U)] \tag{19}$$

Using $x^P \equiv (t, -\vec{x})$, $\vec{p} \cdot \vec{x} = \vec{p}^P \cdot \vec{x}^P$ and noting that the sum overall x in (19) can be replaced by a sum over $\vec{x}^P$, we get

$$G_\pi^{(2)}(t, 0; \vec{p}, U^P) = G_\pi^{(2)}(t, 0; -\vec{p}, U) \tag{20}$$

Thus the effect of averaging over the parity reversed configurations can be included by making the replacement:

$$G_\pi^{(2)}(t, 0; \vec{p}, U) \rightarrow \frac{1}{2} \left[G_\pi^{(2)}(t, 0; \vec{p}, U) + G_\pi^{(2)}(t, 0; -\vec{p}, U) \right] \tag{21}$$

Similarly, for the case of charge conjugation we can show that:

$$G_\pi^{(2)}(t, 0; \vec{p}, U^c) = \sum_{\vec{x}} e^{\frac{2\pi i}{L} \vec{p} \cdot \vec{x}} T_r [\gamma_5 D(0, x; U) \gamma_5 U(x, 0; U)]^* \tag{22}$$

Thus the combined effect of P and C on the meson propagator can be included by making the following replacement:

$$G_\pi^{(2)}(t, 0; \vec{p}, U) \rightarrow \frac{1}{2} \text{Re} \left[G_\pi^{(2)}(t, 0; \vec{p}, U) + G_\pi^{(2)}(t, 0; -\vec{p}, U) \right] \tag{23}$$

The effective set of configurations can be further extended by including the time reflected configuration, U^T in the configuration averages. As for P, C using eqn(17c) for T with $x^T \equiv (-t, \vec{x})$ we find:

$$G_\pi^{(2)}(t, 0; \vec{p}, U^T) = G_\pi^{(2)}(-t, 0; \vec{p}, U) \tag{24}$$

The effect of the time reflected configuration U^T can be included in the average by symmetrizing over time, i.e., by the replacement:

$$G_\pi(t, 0; \vec{p}, U) \to \frac{1}{2}[G_\pi(t, 0; \vec{p}, U) + G_\pi(-t, 0; \vec{p}, U)] \qquad (25)$$

Thus the combined effect of C, P, T on the meson propagator calls for the following replacement:

$$\begin{aligned} G_\pi(t, 0; \vec{p}, U) \to \ & \frac{1}{4} \mathrm{Re}\,[G_\pi(t, 0; \vec{p}, U) + G(t, 0; -\vec{p}, U) + \\ & G_\pi(-t, 0; \vec{p}, U) + G(-t, 0; -\vec{p}, U)] \end{aligned} \qquad (26)$$

4.2 Application to Three Point Functions: Semi-Leptonic Form Factors[34]

The discrete symmetries can be used in much the same way for all 3-point functions such as $K - \bar{K}$ mixing, $D \to K\pi$, $K \to \pi\pi$ etc. to improve statistics.[5] Here we will briefly discuss the semi-leptonic form-factors, say for the 0^- to 0^- transitions such as $D \to Ke\nu$, to serve as an explicit illustration.

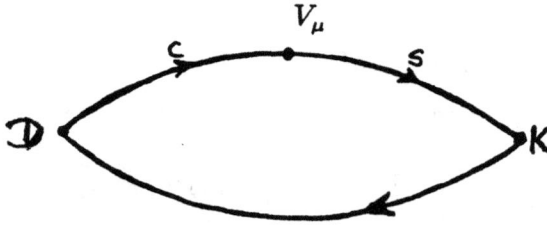

Fig. 4.1. Quark flow diagram for the semi-leptonic transition $D \to Ke\nu$, V_μ is the vector current.

The relevant 3-point function is given by (see Fig. 4.1):

$$\begin{aligned} G_\mu^{(3)}(t_D, 0; t_K; \vec{p}, U) = \ & \sum_{\vec{x}} \sum_{\vec{y}} e^{i p \cdot x} T_r \{\mathrm{U}(x, y; U)\gamma_5 S(y, 0; U) \\ & \gamma_\mu C(0, x; U)\gamma_5\} \end{aligned} \qquad (27)$$

As in the case of 2-point functions, we can use eqn. 17 to relate the expression for the 3-point functions $G^{(3)U^R}$ in the transformed configuration, U^R, to $G^{(3)U}$. Thus, we find, for $R = P$ (i.e. parity):

$$G_\mu^{(3)}(\vec{p}, U^P) = -(-1)^{\delta_{\mu,0}} G_\mu^{(3)}(-\vec{p}, U) \tag{28}$$

Similarly for $R = C$:

$$G_\mu^{(3)}(\vec{p}, U^C) = \left[G_\mu^{(3)}(-\vec{p}, U) \right]^* \tag{29}$$

and for $R = T$:

$$G^{(3)}(t_D, 0, t_K; \vec{p}, U^T) = (-1)^{\delta_{\mu,0}} G_\mu^{(3)}(-t_D, 0, -t_K; \vec{p}, U) \tag{30}$$

So, combining the effects of C, P and T we find:

$$G_0^{(3)}(t_D, 0, t_K; \vec{p}, U) \rightarrow \frac{1}{4} \text{Re} \left[G_0^{(3)}(t_D, 0, t_K; \vec{p}, U) - G_0^{(3)}(-t_D, 0, -t_K; \vec{p}, U) \right.$$
$$\left. + G_0^{(3)}(t_D, 0, t_K; -\vec{p}, U) - G_0^{(3)}(-t_D, 0, -t_K; -\vec{p}, U) \right] \tag{31}$$

$$G_i^{(3)}(t_D, 0, t_K; \vec{p}, U) \rightarrow \frac{1}{4} \text{Im} \left[G_i^{(3)}(t_D, 0; t_K; \vec{p}, U) \right.$$
$$+ G_i^{(3)}(-t_D, 0, -t_K; \vec{p}, U) - G_i^{(3)}(t_D, 0, t_K; -\vec{p}, U)$$
$$\left. - G_i^{(3)}(-t_D, 0, -t_K; -\vec{p}, U) \right] \tag{32}$$

where $i = 1, 2, 3$.

4.3 CPS symmetry[35,36,5,3]

A discrete symmetry that is of considerable importance is CPS where C and P are just the symmetries discussed above and S is the switching symmetry between s and d quarks, valid in the limit $m_s = m_d$. In that limit CPS is an exact symmetry in the continuum as well as on the lattice and as such is extremely useful for classification of operators. Note that when $m_s \neq m_d$ CPS gets softly broken so that terms that violate CPS must be proportional to a factor of $(m_s - m_d)$.

In the context of $K \rightarrow \pi\pi$ CPS has two important implications. Notice that altogether there are four lattice graphs (Fig. 4.2a–d) that contribute. The first implication of CPS is to make the contributions of Fig. b and d to $K_0 \rightarrow \pi\pi$ vanish. In particular, note that Fig. d is not calculable with current techniques so that the use of CPS in this regard is rather crucial. To prove the claim that these graphs vanish it is useful to adopt a more general version of S so that it allows switching of any d line in a graph to an s without renaming all of the lines. This way we see that switch of the s and d lines that occur on the left side of the 4-fermi operator in Fig. b and d changes the initial state from K^0 to $\bar{K}^0$. Applying CP first multiplies the whole graph by -1.

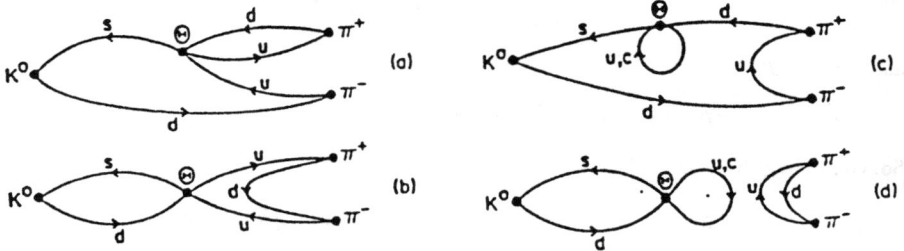

Fig. 4.2. Quark flow diagrams for $K_S^0 \to \pi^+ \pi^-$. Graphs (a) and (b) are "figure-eight"; (c) and (d) are "eyes". Graphs (b) and (d) vanish by CPS when $m_s = m_d$.

Another more important application of CPS is in eliminating the mixing between the 4-quark operators and the 2-quark operators relevant to the $\Delta I = 1/2$ amplitude. The point is that the weak operators $O_\pm$ relevant to $K \to \pi\pi$ decay are even under CPS so that they can only mix with the lower dimensional 2-quark operators that are also CPS even. Therefore, only the P-odd CPS even operator $(m_d - m_s)\bar{s}\gamma_5 d$ or $(m_d - m_s)\bar{d}\gamma_5 s$ can contribute to the $\Delta I = 1/2$ parity violating $K \to \pi\pi$ transition. Thus by choosing to work strictly in the SU(3) degenerate limit we can eliminate the mixing due to this lower dimensional operator.

Another elegant consequence of CPS symmetry is to eliminate the mixing between

$O^\pm$ as before and all $\Delta S = 1$, 4-quark, parity-odd operators. To order g^2 this is explicitly seen in the WCPT calculations that have been done. The original 4-quark operator to that order mixes only with 4-quark operators of the form $S \times S$, $P \times P$... which are all even under P. The fact that this is a general result that holds to all orders in strong interactions can be demonstrated by observing that the 4-quark operators have additional switching symmetries. By enumeration one can show that all 4-quark parity odd operators violate one or more of these symmetries. Consequently, the original 4-quark operators $O^\pm$ do not mix with any other parity violating 4-quark operator and therefore they require only a "diagonal" renormalization.

5   Brief Remarks on Staggered Fermions[37-39]

The theoretical tools for weak matrix elements with staggered fermions have been developed over the years by Kilcup and Sharpe. For completeness, we repeat a summary that was given in Ref. 3.

As is well known, 4 Dirac fields, each of a different "internal flavor", may be constructed out of the 16 staggered fermion fields on a 2^4 hypercube. One may write $\psi \rightarrow \psi_{\alpha i}$ where $\alpha = 1,\ldots,4$ is a Dirac index, and $i = 1,\ldots,4$ is an internal flavor index. On the lattice, there is one exact vector symmetry, which is a singlet in the internal flavor space, and (in the massless limit) one exact axial symmetry which is an internal flavor non-singlet. (Note that the singlet axial current must have an anomaly and therefore could not be exact on the lattice.) The currents corresponding to these symmetries may be written:

$$ j_\mu^V = \bar{\psi}\gamma_\mu\psi \quad , \quad j_\mu^A = \bar{\psi}T_5\gamma_\mu\gamma_5\psi , \tag{33} $$

where T_5 is the analogue of γ_5 and acts on the internal flavor indices.

Kilcup and Sharpe show that the good axial symmetry is enough to derive useful Ward identities. These imply that matrix elements of lattice operators in pseudoscalar meson states have all the desired chiral properties of the continuum matrix elements. This is only true, however, if the mesons are chosen to be the "good" mesons (true pseudo-Goldstone bosons) which couple to j_μ^A. The interpolating field of these mesons is

$$ \bar{\psi}T_5\gamma_5\psi. \tag{34} $$

(Note that $T_5\gamma_5$ is local in terms of the original fields on the fine lattice: $T_5\gamma_5 \rightarrow (-1)^x$.)

This result has important consequences for the organization of weak matrix element calculations with staggered fermions. First of all, since the internal flavor structure of the mesons is fixed, a new staggered fermion set $(\psi_{\alpha i})$ is needed for each physical flavor $(u, d, s, \ldots)$ in order to represent π's K's, etc. For example, the π^+ interpolating field is $\bar{d}T_5\gamma_5 u$. Secondly, each Wick contraction for an operator matrix element must be treated separately, to make sure both that the internal flavor structure of the operator matches with the internal flavor structure of the mesons and that the normalization due to internal flavor counting is adjusted correctly. As an illustration, consider the Wick contractions for the matrix element $\langle \bar{K}^0|(\bar{s}d\bar{s}d)_{LL}|K^0\rangle$. The meson interpolating fields are fixed to be $\bar{d}T_5\gamma_5 s$.

One may first try to transcribe the operator by

$$(sd\bar{s}d)_{LL} \to \frac{1}{N}(\bar{s}T_A\gamma_\mu(1-\gamma_5)d\bar{s}T_B\gamma_\mu(1-\gamma_5)d), \qquad (35)$$

where T_A, T_B are internal flavor matrices yet to be fixed, and N is a normalization factor. Now there are two types of contractions: "$M_{LL}^{(1)}$" in which the first $\bar{s}$ and last d in (35) go into one K and the middle two fields go into the other K, and "$M_{LL}^{(2)}$" in which the first $\bar{s}$ and first d (i.e., the first current) go into one K, and the other current goes into the other K. Then we have

$$\begin{aligned} M_{LL}^{(1)} &\propto tr(T_5T_AT_5T_B) \\ M_{LL}^{(2)} &\propto tr(T_5T_A)tr(T_5T_B). \end{aligned} \qquad (36)$$

Clearly one must choose $T_A = T_B = T_5$ so that $M_{LL}^{(2)} \neq 0$ in the continuum limit (where the internal flavor symmetry is exact). However, to take into account the internal flavor copies correctly, this requires $N = 16$ for $M_{LL}^{(2)}$ but $N = 4$ for $M_{LL}^{(1)}$. The eye contraction of the operator $\bar{s}T_A\gamma_\mu(1-\gamma_5)d\bar{u}T_B\gamma_\mu(1-\gamma_5)u$ is another good example. If one makes the same choice $T_B = T_5$ there, the contraction will vanish in the continuum limit. As a further restriction, some transcriptions of operators to staggered fermions may be incompatible with the Ward identities.

The conclusion is, therefore, that one cannot transcribe *operators* but only *contractions*. For a given contraction, it may be convenient to Fierz the operator in the continuum before transription. This is in general not equivalent to a Fierz transformation after transcription, since the lattice Fierz will involve internal flavor as well as spin indices. Because the choice of the internal flavor of the operator is often fixed for a given contraction, the operators are in general forced to be "non-local" within the 2^4 hypercube. The fields in the operators may be separated by as many as four links, and are then joined together by the gauge matrices along the links. This results in a practical difficulty: the presence of the gauge links can result in large fluctuations in a Monte-Carlo evaluation.

In practice, the K^0-$\bar{K}^0$ mixing matrix element is handled not by (35) with a contraction-dependent N, but by the introduction of two varieties of s and d quarks (s, s' and d, d') so that a given lattice operator can only have one possible contraction. One writes

$$\begin{aligned} \langle \bar{K}^0|\bar{s}d\bar{s}d)_{LL}|K^0\rangle \quad &\to \quad \frac{1}{8}\langle \bar{K}^{0\prime}|(\bar{s}_a'T_5\gamma_\mu(1-\gamma_5)d_b'\bar{s}_bT_5\gamma_\mu(1-\gamma_5)d_a \\ &\qquad + \bar{s}_a'T_5\gamma_\mu(1-\gamma_5)d_a'\bar{s}_bT_5\gamma_\mu(1-\gamma_5)d_b)|K^0\rangle \qquad (37) \\ &\equiv \quad M_{LL}^{(1)} + M_{LL}^{(2)} \\ &\equiv \quad M_{VV}^{(1)} + M_{AA}^{(1)} + M_{VV}^{(2)} + M_{AA}^{(2)}, \end{aligned}$$

where the color indices a, b are shown explicitly. $\bar{K}^{0\prime}$ is made out of the primed quarks (interpolating field $\bar{d}'T_5\gamma_5s'$). The first operator in the lattice matrix element is obtained by Fierzing in the continuum and then transcribing; it corresponds to contraction $M_{LL}^{(1)}$. The overall factor of $\frac{1}{8}$ is really $2/[tr(T_5^2)]^2$, where the denominator is just the internal-flavor counting factor, and the 2 is the usual symmetry factor which is absent from the lattice matrix elements of (37) because of the distinction between primed and unprimed quarks. The separation of $M_{LL}^{(1,2)}$ into their parity-even parts $M_{VV}^{(1,2)}$ and $M_{AA}^{(1,2)}$ in (37) is purely for convenience.

The fact that different contractions are transcribed differently, and, more importantly, the fact that the internal flavor symmetry is violated at finite a make the perturbative mixing problem for staggered fermions enormously complicated. Many different flavor structures mix together. However, the mixing problem is not especially "dangerous"—unlike the Wilson case—because all operators which mix should have the same chiral structure. Thus, at least among mixing operators of the same dimension, no subset is expected to have particularly large matrix elements in the chiral limit, and there should be no delicate cancellations required to restore the correct chiral behavior. The mixings among 4-quark operators have been calculated[39-41] at one loop and seem to conform to this expectation. However, just as in the Wilson case, there is mixing with operators of lower dimension ($\bar{s}d, \bar{s}\gamma_5 d \ldots$) through diagrams like those in Fig. 3.3. Such operators will of course have large coefficients proportional to inverse powers of a, and should be subtracted non-perturbatively. For staggered fermions, the existence of an internal non-singlet axial current which is conserved in the chiral limit is enough to guarantee that the procedure based on continuum CPTh goes through unscathed.[35] In principle, this allows one to calculate the $K \to 2\pi$ amplitudes (at least to lowest order in CPTh) by computing the $K \to \pi$ and (in the $\Delta I = 1/2$ case) $K \to$ vacuum amplitudes.

6 Kaon Physics

6.1 The B Parameter for $K\bar{K}$ Mixing

Over the years our attempts (using Wilson fermions) to calculate the $K - \bar{K}$ mixing matrix element M_{LL} defined by:

$$M_{LL} = \langle \bar{K}^0 | (\bar{s}d\bar{s}d)_{LL} | K^0 \rangle \tag{38}$$

where

$$(\bar{s}d\bar{s}d)_{LL} = [\bar{s}\gamma_\mu(1 - \gamma_5)d][\bar{s}\gamma_\mu(1 - \gamma_5)d] \tag{39}$$

have been hampered due to a nagging difficulty originating from the issue of operator mixing.[42,36.5] Let us briefly recapitulate that difficulty and then discuss its resolution.[43,3] For small 4-momenta of K and $\bar{K}$ chiral perturbation theory (CPTh) predicts the following behavior for the matrix element of the continuum operator $(\bar{s}d\bar{s}d)_{LL}^{\text{cont}}$:

$$\langle \bar{K}^0 | (\bar{s}d\bar{s}d)_{LL}^{\text{cont}} | K^0 \rangle = \gamma p_K \cdot p_{\bar{K}} + \cdots \tag{40}$$

which, for $K(\bar{K})$ at rest reduces to γm^2, with m the meson mass. However, as $m \to 0$ the matrix element computed on the lattice approaches a non-zero constant, in contradiction to CPTh.

We believe that this problem is caused by the failure of weak coupling perturbation theory, at couplings ($\beta \lesssim 6.0$) that have been used in the past several years, to predict with the necessary accuracy the coefficients of the "off-diagonal" 4-quark operators that mix with

$(\bar{s}d\bar{s}d)_{LL}^{\text{latt}}$. There are four independent, dimension 6 operators that are responsible for this mixing:

$$\begin{cases} (\bar{s}d\bar{s}d)_1' & = \bar{s}d\bar{s}d \\ (\bar{s}d\bar{s}d)_2' & = \bar{s}\gamma_5 d\bar{s}\gamma_5 d \\ (\bar{s}d\bar{s}d)_3' & = \bar{s}\sigma_{\mu\nu}d\bar{s}\sigma_{\mu\nu}d \\ (\bar{s}d\bar{s}d)_4' & = \bar{s}\gamma_\mu d\bar{s}\gamma_\mu d - \bar{s}\gamma_\mu\gamma_5 d\bar{s}\gamma_\mu\gamma_5 d \end{cases} \tag{41}$$

We assume that all violations of chiral symmetry (including those caused by order a corrections — a strong assumption!) are proportional to those caused by linear combinations of these operators.

The matrix elements of these operators have pieces that go as constants as $m \to 0$. Thus for small enough m these constant contributions have to cancel to a very good approximation for the $(\bar{s}d\bar{s}d)_{LL}^{\text{cont}}$ to exhibit the correct m^2 dependence expected from lowest order CPTh.

We will briefly describe here one (of the two) non-perturbative solutions to this problem that have been proposed.

As mentioned above the $(\bar{s}d\bar{s}d)_i'$, given in eq. 41, have the behavior:

$$\langle \bar{K}^0|(\bar{s}d\bar{s}d)_i'|K^0\rangle = \sigma_0^i + \sigma_1^i m^2 + \sigma_2^i p_K \cdot p_{\bar{K}} + \cdots \tag{42}$$

where σ's are constants and $\cdots$ represents higher order terms in m or p. The fact that $(\bar{s}d\bar{s}d)_i'$ has not been subtracted correctly makes $(\bar{s}d\bar{s}d)_{LL}^{\text{pert}}$ behave similarly:

$$\langle \bar{K}^0|(\bar{s}d\bar{s}d)_{LL}^{\text{pert}}|K^0\rangle = \gamma_0 + \gamma_1 m^2 + \gamma_2 p_K \cdot p_{\bar{K}} \tag{43}$$

where, of course,

$$(\bar{s}d\bar{s}d)_{LL}^{\text{pert}} = Z_+ \left[(\bar{s}d\bar{s}d)_{LL}^{\text{latt}} + \sum_{i=1}^{4} Z_i'(\bar{s}d\bar{s}d)_i' \right] \tag{44}$$

The procedure for "improving" $(\bar{s}d\bar{s}d)_{LL}^{\text{pert}}$ to give it the correct chiral behavior (eqn. 40 above) can now be summarized as follows:

1. Compute the matrix elements of $(\bar{s}d\bar{s}d)_{LL}^{\text{pert}}$ and $(\bar{s}d\bar{s}d)_i'$ for various values of the meson mass and then fit to the forms of eqns. 43 and 42. The simplest way to separate the m^2 and $p_K \cdot p_{\bar{K}}$ is to compute $\langle K|\Theta|\bar{K}\rangle$ as well as $\langle KK|\Theta|0\rangle$ for each operator Θ. Note that $p_K \cdot p_{\bar{K}}$ changes sign for the crossed matrix element, while m^2 of course does not.

2. Choose any two of $(\bar{s}d\bar{s}d)_i'$ and adjust their coefficients in eqn. 44 to cancel γ_0 and γ_1 of eqn. 43, and then compute the corresponding physical value of γ_2.

3. Use γ_2 from step 2) to compute the B_{LL}^{sd} parameter corresponding to the particular choices of $(\bar{s}d\bar{s}d)_i'$ under consideration.

4. Repeat 2) and 3) for all of the 6 possible choices

Table 6.1: B_{LL}^{sd} at scale $\mu = 1$ GeV and the renormalization group invariant $\hat{B}_{LL}^{sd}$ from various quenched lattice groups and from two continuum calculations. For definiteness, $N_F = 3$ and $\Lambda_{QCD} = 250$ MeV at two loops are used to get the renormalization group invariant quantity $\hat{B}_{LL}^{sd}$; a $\sim 10\%$ error is associated with these choices.

Group	Method	Parameters	$B_{LL}^{sd}(\mu = 1\text{GeV})$	$\hat{B}_{LL}^{sd}$
Kilcup et al.[44,30,31]	Quenched, staggered	$\beta = 5.7, 16^3$	$.91 \pm .01$	$1.15 \pm .03$
"	" "	$\beta = 6.0, 16^3$	$.73 \pm .02$	$.93 \pm .02$
"	" "	$\beta = 6.0, 24^3$	$.73 \pm .01$	$.92 \pm .01$
"	" "	$\beta = 6.2, 32^3$	$.65 \pm .01$	$.82 \pm .01$
"	" "	$\beta = 6.4, 32^3$	$.64 \pm .03$	$.80 \pm .04$
Gupta et al.[45]	Quenched, Wilson	$\beta = 6.0, 16^3$	$.76 \pm .26$	$.96 \pm .33$
Bernard, Soni[43,3]	Quenched, Wilson	$\beta = 5.7, 16^3$	$1.00 \pm .10 \pm .12$	$1.26 \pm .12 \pm .15$
"	" "	$\beta = 6.0, 16^3$	$.86 \pm .12 \pm .12$	$1.08 \pm .15 \pm .15$
"	" "	$\beta = 6.0, 24^3$	$.69 \pm .08 \pm .04$	$.86 \pm .11 \pm .05$
"	" "	$\beta = 6.1, 12^3$	$.79 \pm .21 \pm .22$	$1.00 \pm .26 \pm .27$
ELC[46,1]	Quenched, Wilson	$\beta = 6.0, 20^2 \times 10$ and $\beta = 6.2, 16^3$	$.68 \pm .16$	$.85 \pm .20$
Bardeen et al.[47]	Large N	————	$.56 \pm .06$	$.70 \pm .07$
Donoghue et al.[48,49]	Lowest order CPTh	————	$.26$	$.33$

5. Average the six results weighted appropriately by each of their statistical errors to compute a final B_{LL}^{sd}. The standard deviation of the six results may be taken as a measure of the systematic error associated with the procedure.

Our method differs from that of ELC in that they do not attempt to take into account the $p_K \cdot p_{\bar{K}}$ dependence of $(\bar{s}d\bar{s}d)'_i$ (i.e. σ_2 in eqn. 42). Our procedure, on the other hand, uses the unproven assumption that we can control the σ_2 term in the "statistical" manner described above. Fortunately, our results would only change by 5 to 10% with their method if we also use $f^2 m^2$ fits as they do (rather than our usual m^2 fits). However, their method does seem to be quite sensitive to the choice of fitting procedure: with m^2 fits, the results would change by $\sim 35\%$.

Recently, Gupta et al.[45] have performed another calculation of B_K with Wilson fermions at $\beta = 6.0$. Their procedure is similar to that of ELC in that they ignore the σ_2 term. They focus however on comparing processes with different 3-momenta rather than on comparing the on-shell ($\langle \bar{K}|\Theta|K \rangle$) and off-shell ($\langle 0|\Theta|KK \rangle$) processes. (ELC tried both approaches.) The momentum approach has the advantage that it does not introduce uncontrolled final state interactions. However, it is only in the recent calcualtion that the statistical errors have been reduced enough to make the method really viable—quantities with non-zero 3-momenta tend to have large fluctuations on the lattice.

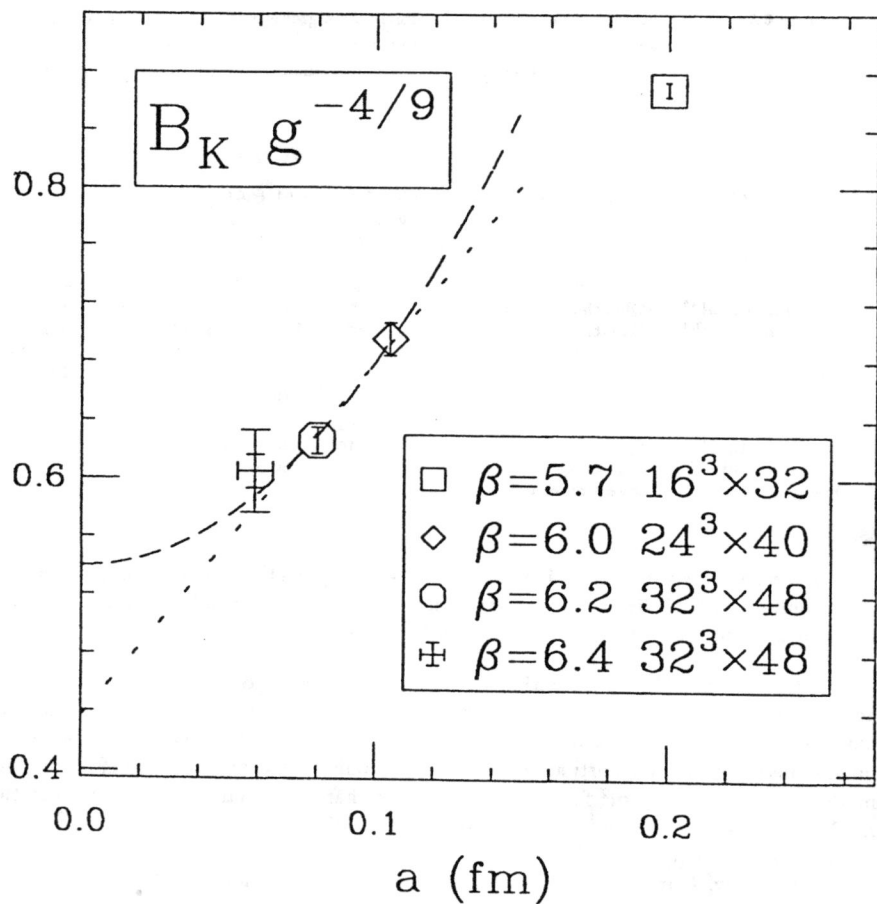

Fig. 6.1. Dependence of the Kaon B parameter on lattice spacing as reported by the staggered group in Ref. 31.

Table 6.1 compares our results for the continuum-normalized B_{LL}^{sd} at $\mu = 1$ GeV with those of ELC[46] and the recent result of Gupta et al[45] The results obtained by using staggered fermions are also shown.[44,30,31] There is remarkable agreement between our results and that of the staggered case for $\beta = 6.0$. Furthermore, the results of the two groups agree for $\beta = 5.7$. Both sets of results also show the scale breaking effects for $\beta = 5.7$ of about ~ 25–30% range. If one examines all the lattice results shown in this Table for $\beta \leq 6.0$ one would be tempted to conclude that lattice has finally given a value for this important matrix element—this seemed to be the case about two years ago. More recently, the staggered group has made considerable progress in calculating B_{LL}^{sd} at weaker couplings, i.e. $\beta = 6.2$ and $\beta = 6.4$.[30,31] Results from these simulations are compared with their earlier results from $\beta = 5.7$ and $\beta = 6.0$ in Fig. 6.1. The $\beta = 6.0$ results differ significantly from $\beta = 6.4$ and also from $\beta = 6.2$. It therefore seems that an exptrapolation to the continuum limit will be required. It is not clear as of this writing whether a linear or quadratic fit in a is more appropriate for the staggered calculation. (Wilson fermions—without "improvement"— almost certainly have a large order a term.) The difference between extrapolation methods gives a difference of $\sim 15\%$ in the continuum value of B_K.[31] Results with Wilson fermions may help to clarify these issues. The ELC group is currently running simulations at $\beta = 6.4$. We have accumulated considerable data in the past two years at $\beta = 6.3$. Both groups are using lattices of size 24^3. In the meantime, from the staggered numbers[31] so far available it would appear safe to quote the following tentative result:

$$B_{LL}^{sd}(\mu = 1\text{GeV}) = .58 \pm .05 \pm .07 \quad \text{i.e.} \quad \hat{B}_{LL}^{sd} = .72 \pm .06 \pm .08$$

The central value is the average of their linear and quadratic fits to a dependence and the systematic errors reflect the uncertainty in the a dependence. These numbers are also not inconsistent with the results from Wilson fermions (given in Table 6.1) given relatively large errors in the latter.

6.2 The $K \to \pi\pi$ $\Delta I = 1/2$ Amplitude

This problem, which was the first to receive attention among the weak matrix elements attempted with the lattice method, still remains unresolved. It is not terribly useful to describe here the many techniques that have been tried and the reasons for their failures. We will content ourselves with briefly describing one still-promising procedure which has up to now has not been adequately tested because of limited numerical We are referring here to the so called "direct" method. This method calls for a calculation of the full $K \to \pi\pi$ amplitude with all the mesons at rest, i.e. a 4-point function calculation on the lattice without recourse to CPTh.

The advantages of this method are the following:

1. Since this is a parity violating transition, CPS symmetry implies that the operators $O_\pm$ will not mix with any off-diagonal 4-quark operator to all orders in strong interactions. These operators then only require a diagonal renormalization, something that is far better under control by LWCPT.

2. There are two lower dimensional parity violating operators that can mix with $O_\pm$ namely $\bar{s}\gamma_5 d$ and $\bar{s}\gamma_5\sigma_{\mu\nu}G^{\mu\nu}d$. Again CPS symmetry requires that the coefficients of these operators be proportional to $(m_s - m_d)$ and consequently their contribution to the amplitude vanishes when one works in the SU(3) degenerate limit.

One disadvantage of the method is that it requires calculation of 4-point functions with two mesons in the final state. In addition to the usual drawback of requiring two inversions with the source method[50] (required in particular for the "eye-graph" of Fig. 4.2c) it is also essential to have a sufficiently large lattice to accommodate the two mesons in the final state with minimal distortions. Furthermore, the only kinematical point which is accessible is the point where all the mesons are at rest. If the outgoin pions are given momenta, the Euclidean amplitude ceases to correspond directly to any physical amplitude.[4]

We have studied this method at $\beta = 5.7$ and $\beta = 6.0$. The key difficulty that needs to be surmounted before this method can be used to extract physically meaningful results is to be able to separate out the contribution of the 0^{++} resonance in $I = 0$ final state of the two pions.[51] To illustrate this problem we will discuss the situation with regard to our $\beta = 6.0$, $24^3 \times 39$ lattice with 8 gauge configurations. Indeed a clean signal is seen in the $\Delta I = 1/2$ channel. However, a study of the time dependence of the amplitude reveals a serious problem, as can be seen in Fig. 6.2.[52] The K^0 and the π^+ are held at time slices -11 and $+11$ respectively. The 4-quark operator sits, as usual, at $t = 0$. Fig. 6.2 shows the amplitude as a function of the location of the π^-. ¿From similar studies of the time dependence of the pion correlator and also that of the $\Delta I = 3/2$ $K \to \pi\pi$ amplitude, we know that a reasonable plateau is reached for $t > 7$. In contrast, for the $\Delta I = 1/2$ case, Fig. 6.2 shows that as $t_{\pi-}$ approaches $t_{\pi+}$ ($c = 11$) from below the amplitude shows a rapid increase. Recall that the amplitude corresponds to the Green's function for $\langle \pi^+\pi^-|\Theta|K^0\rangle$ divided by the product of a kaon and two pion single-particle Green's functions. The fact that this amplitude changes with $t_{\pi-}$ therefore indicates that some other state(s) with energy different from twice the π energy must be contributing. Since the 3/2 amplitude (Fig. 6.3) shows no such anomalous time-dependence the most likely cause is a 0^{++} (SU(3)-Octet) σ particle that couples to the two pion state. The σ can affect only the (purely $\Delta I = 1/2$) eye-graph (via Fig. 4.2c) where it can appear as a two-quark intermediate state after the 4-quark operator acts.

It is clear that this resonance must make some contribution to the $\Delta I = 1/2$ part of $K^0_S \to \pi\pi$ via $\langle K^0_S|H_{wk}|\sigma\rangle$ followed by $\sigma \to 2\pi$. The critical issue is to quantify this contribution. The problem is compounded by the fact that the 0^{++} spectrum (for light quarks) is extremely difficult to study on the lattice since these $\ell = 1$ stattes couple very little to local interpolating operators. The 2-point functions are therefore difficult to handle, and the three point functions $\langle K^0_S|H_{wk}|\sigma\rangle$ and $\langle\sigma|\pi\pi\rangle$ required to assess the resonance contribution to $K_S \to \pi\pi$ are much harder still.

It is quite possible that in the real world the contribution of σ to the $\Delta I = 1/2$ amplitude for $K \to 2\pi$ is small. This is because we expect the σ to be quite heavy—perhaps about 1 GeV. In the laboratory it has never been clearly identified; this is consistent with the general expectation that its width is very broad. Since it is a $\ell = 1$ state it is natural to expect it be heavier than the ρ ($m_\rho \sim 770$ MeV). If the mass is really as high as 1 GeV

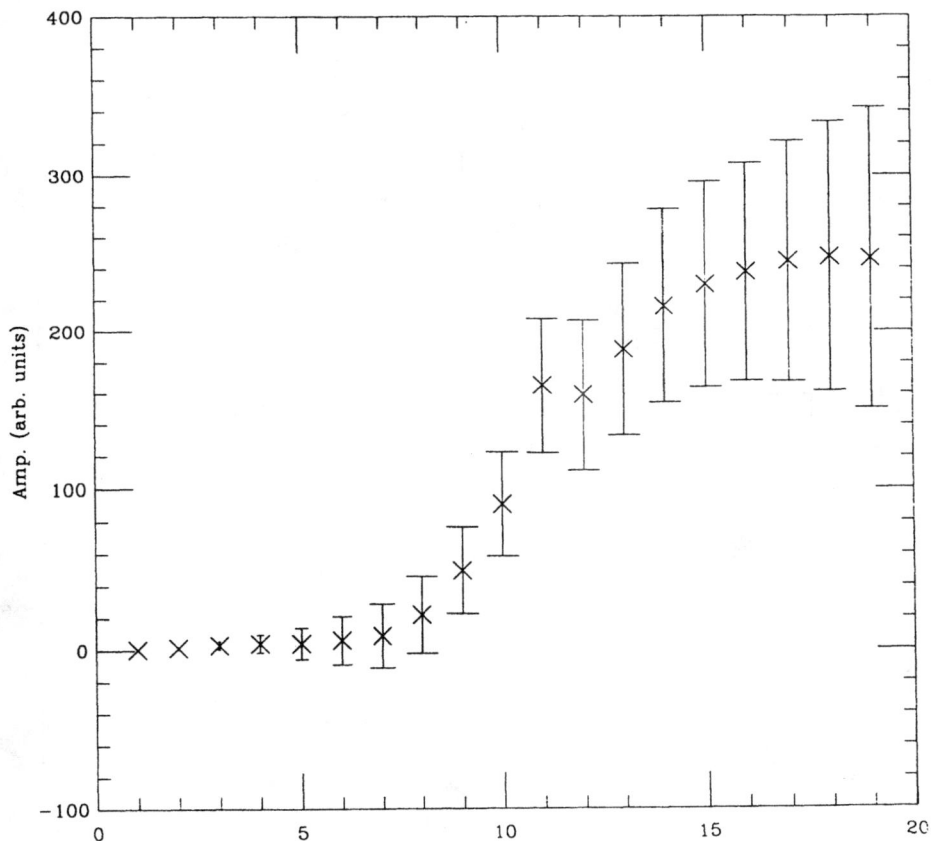

Fig. 6.2. Time dependence of the $\Delta I = 1/2$ $K \to \pi\pi$ amplitude ($24^3 \times 39$, $\beta = 6.0$, 8 gauge configurations, $K = .154$)

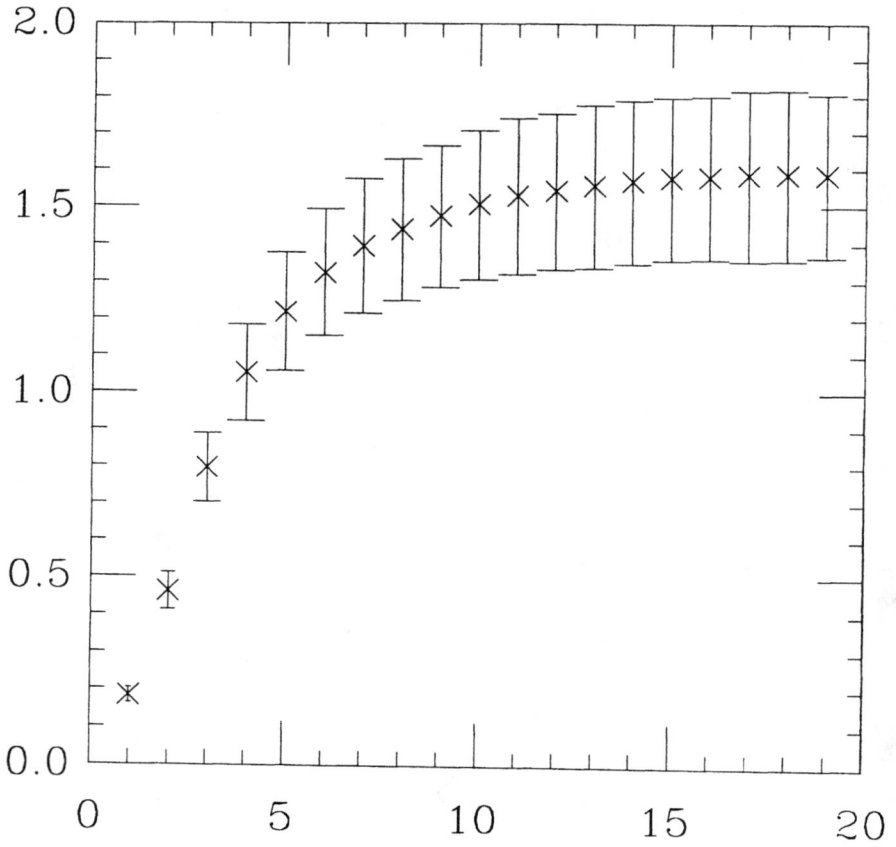

Fig. 6.3. Time dependence of the $\Delta I = 3/2\ K \to \pi\pi$ amplitude (same parameters as Fig. 6.2)

then it is significantly larger than the 2π threshold (for physical pions), and one ought, in principle, to be able to separate out its contribution. In lattice simulations to date this has not been possible because the light quarks available on the lattice are much heavier than physical quarks. For example, on the $\beta = 6.0$, $24^3 \times 40$ lattice that we have been using jointly with the staggered group, the lightest pion mass is ~ 400 MeV. At such relatively large quark masses, one expects $m_\sigma \sim 2m_\pi$; further, the σ is stable (from decays to 2π) in the quenched approximation in use here.

Thus in these lattice computations the σ can have appreciable effects on the $\Delta I = 1/2$ amplitude even if in the real world it may not. The σ has two effects. First, it weakens the relative coupling of the interpolating operator to the two pion state since the operator creates a mixture of that state with a σ. There is also the effect of the "resonance" i.e. the two pion state is created but it turns into a σ any time before the operator acts. The resonance corresponds to a pole enhancement in momentum space. In the configuration space lattice calculation it gives rise to an enhancement by a factor of t, where t is the time over which the two pions can convert to a σ and is equal to t_{π^-} for $t_{\pi^-} \leq t_{\pi^+}$. Since in the quenched approximation, the σ cannot change back to 2π, there are no $t^2, t^3 \ldots$ terms. This simple picture does not explain the rapid rise with t_{π^-} in Fig. 6.2. Perhaps one or more excited states of σ are also contributing. In passing we should mention that if one ignores the rapid rise and simply uses the "plateau region," say for $14 \leq t_{\pi^-} \leq 16$, then it gives an amplitude 5 to 10 times larger than experiment! Needless to say, we do not see any reason to ignore the linear rise and indeed regard it as a strong indication that the use of the plateau (and the numbers that emerge) as physically not meaningful.

It would appear that the direct $K \to 2\pi$ method, although very clean in principle, cannot overcome the difficulty of the σ resonance at least until appreciably more computing power allows us to simulate pseudoscalar masses that are considerably lighter than the current range of ~ 400 MeV.

6.3 The $K \to \pi\pi \; \Delta I = 3/2$ Amplitude

The $\Delta I = 3/2$ amplitude is simpler to deal with than the corresponding $\Delta I = 1/2$ amplitude. The major simplification is that the 4-quark operator(s) responsible for the transition cannot mix with lower dimensional 2-quark operator(s). This is so because the latter can only be of the type $\bar{s}\Gamma d$, which is purely $\Delta I = 1/2$. Both the direct $K \to \pi\pi$ and the reduced $K \to \pi$ approaches have been used. Each has its advantages and limitations. We briefly discuss these as well as the status of the numerical results.

The direct approach has the advantage that since the transition is parity violating the relevant 4-quark operator undergoes only diagonal renormalization. That is (as discussed before) the "off-diagonal" mixing with operators of the type $S \times S$, $P \times P \ldots$ etc. are all forbidden by CPS to all orders in strong interactions. From this point of view the method is very clean theoretically. However, as discussed for the $\Delta I = 1/2$ case, all $K \to 2\pi$ calculations suffer from some drawbacks. Firstly the presence of the two pions in the final state introduces non-trivial final state interaction (FSI) phases which are extremely difficult to handle in a lattice calculation.[51] Another somewhat minor theoretical problem is that CPTh has to be used to relate the ("at rest") amplitude, calculated on the lattice, to the experimental amplitude. The method, of course, also requires 4-point function calculations and therefore is numerically quite costly as it necessitates three inversions for the quark propagators.

Our results from such an approach are summarized in Figs. 6.4 and 6.5. Fig. 6.4 shows the $K^+ \to \pi^+\pi^0$ amplitude for degenerate mesons at rest. Fig. 6.5 has the physical amplitude obtained from 6.4 using lowest order CPTh. The at rest amplitude of Fig. 6.4 is consistent with the lowest order chiral expectation that it go as m^2. Also the physical amplitude in Fig. 6.5 appears roughly a constant with respect to m^2 although there are indications of departure from that behavior towards the higher mass points.

If one takes the constant behavior (Fig. 6.5) seriously then lowest order CPTh immediately tells us that $B_{LL}^{sd}(\mu = 1 GeV) = 0.69 \pm 0.06$ which is, again, roughly consistent with the direct determination of B_{LL}^{sd}. While this is encouraging it also implies that $K^+ \to \pi^+\pi^0$ amplitude seen on the lattice is too large compared to experiment by about a factor of two. This may well be due to higher order corrections in CPTh. Indeed FSI in this channel should manifest themselves as loop corrections to the lowest order chiral predictions. With this in mind we can fit the data (Fig. 6.4) to $a + bm^2$ leading to an appreciable value for the slope $b = 3.9 \pm 1.0$. Continuum studies also suggest large higher order corrections to CPTh expecially in the form of FSI[53]. These corrections are roughly in the right range to explain the discrepancy between Fig. 6.5 and experiment. Indications of scaling violations at $\beta = 6.0$ have also been reported by the staggered group.[30,31] In fact (as discussed for B_K) these would tend to reduce the B_K and $K \to 2\pi$ by about 25%.

6.4   The $\Delta I = 1/2$ Amplitude from the Reduced $K \to \pi$ Amplitude

Having described why the $K \to 2\pi$ method is not working we now turn to see if the simpler $K \to \pi$ amplitude could be used to get a handle on the $\Delta I = 1/2$ amplitude. In particular we will now briefly outline a procedure that the ELC group has advocated for calculating the $\Delta I = 1/2$ $K_S \to \pi\pi$ amplitude in terms of the matrix elements for $K \to \pi^{54}$. For simplicity, SU(3) limit is assumed with m as the degenerate meson mass. We recall that we are dealing with a parity conserving transition. In here all the relevant Z's given in eqn. 12 except for Z_3 are calculable by LWCPT. A non-perturbative method is required for evaluation of $Z_3^{\pm}$. Now CPTh requires[3]

$$\langle \pi^+|0_{\pm}^{cont}|K^+\rangle = \frac{4\alpha^{\pm}}{f^2}p_{\pi} \cdot p_K \qquad (45)$$

For the $\bar{s}d$ operator we have:

$$\langle \pi^+|\bar{s}d|K^+\rangle = \rho_1 + \rho_2 p_{\pi} \cdot p_K \qquad (46)$$

where $\rho_1 \to$ constant as $m^2 \to 0$. Thus the unsubtracted operators are expected to have the behavior

$$\langle \pi^+|0_{\pm}^{o}|K^+\rangle = \alpha_1^{\pm} + \alpha_2^{\pm}(p_{\pi} \cdot p_K) \qquad (47)$$

where $\alpha_1^{\pm}$ has both m^2 as well as constant terms. Thus the coefficient Z_3 in (12) can be obtained by requiring the cancellation of the constant pieces in (46) and (47) to restore the correct chiral behavior to $O_{\pm}^{cont}$. Thus

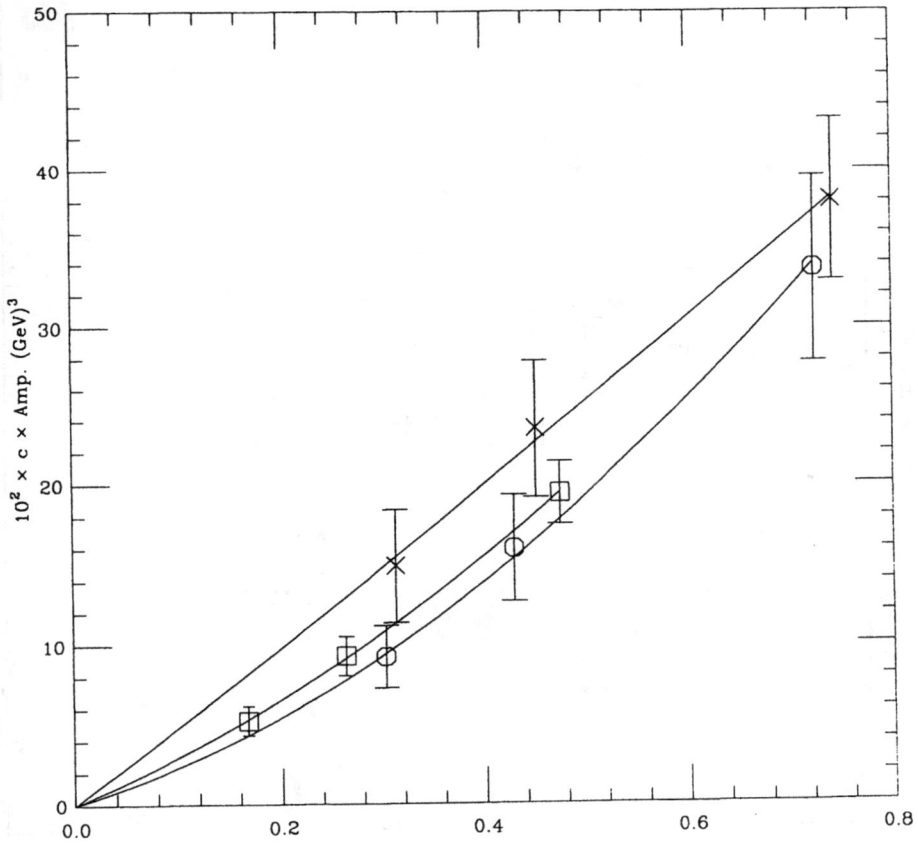

Fig. 6.4. The $K^+ \to \pi^+\pi^0$ amplitude for all mesons at rest vs. m^2. Key: octagon $\to 24^3 \times 39$, 8 configs ($\beta = 6.0$); cross $\to 16^3 \times 39$, 19 configs ($\beta = 6.0$); squares $\to 16^3 \times 25$ or $16^3 \times 33$ ($\beta = 5.7$), 32 configs.

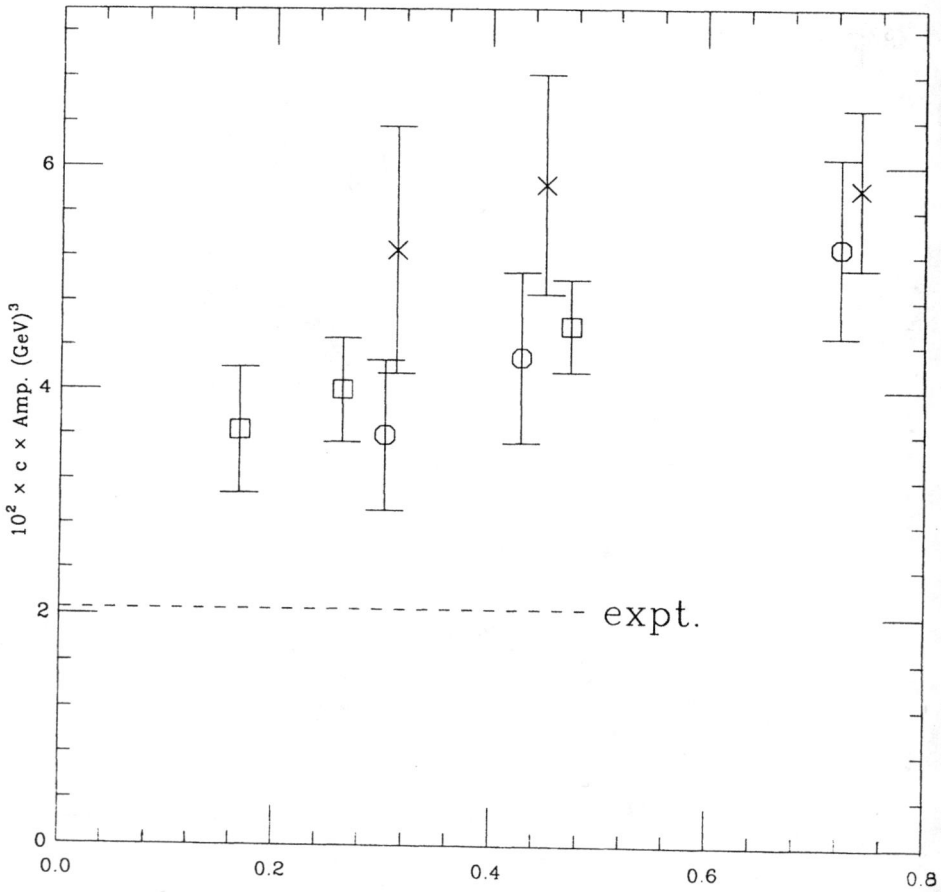

Fig. 6.5. The physical $K^+ \rightarrow \pi^+ \pi^0$ amplitude compared to experiment (see Fig. 6.4 for notations).

$$Z_3^{\pm} = -\alpha_1^{\pm}/\rho_1 \qquad (48)$$

The desired coefficient $\alpha^{\pm}$ in (45) is thus given by:

$$\frac{4}{f^2}\alpha_2^{\pm} = \alpha_2^{\pm} - \alpha_1^{\pm}\rho_2/\rho_1 \qquad (49)$$

To separate the $p_\pi \cdot p_K$ term in (46) and (47) necessary for deducing $\alpha_{1,2}^{\pm}$ or $\rho_{1,2}$ some additional input is needed. Two choices that have been proposed are[54,55]

(1) Calculate $\langle\pi^+|0_\pm^q|K^+(\vec{p})\rangle$ and $\langle\pi^+|\bar{s}d|K^+(\vec{p})\rangle$ where $\vec{p}$ is a non-zero 3-momentum.

(2) Calculate the rest amplitudes $\langle 0|0_\pm^q|\pi^-K^+\rangle$ and $\langle 0|\bar{s}d|\pi^-K^+\rangle$. The factor $p_\pi \cdot p_K$ has the opposite sign here from the original matrix elements, while m^2 is, of course, unchanged.

These methods are interesting but also have some practical difficulties. In the first method the problem is that momentum injection, especially to light mesons tends to introduce large statistical fluctuations. This, however, is probably not an insurmountable problem. The second method seems to have more substantiative difficulties due to the presence of the two mesons in the final state. In this sense the difficulties are similar to the direct $K \to \pi\pi$ method: large finite size effects and resonance (i.e. σ) and/or final state interaction effects.[51,52,5] In particular, the σ problem may well be the cause for the current failure of this approach. The point is that the amplitude $\langle 0|\Theta|\pi K\rangle$ can have an intermediate σ; whereas $\langle\pi^+|\Theta|K^+\rangle$ cannot. Thus a subtraction of lower dimensional operators which depends crucially on the CPTh relation between these two amplitudes is unlikely to succeed. As discussed earlier in the case of $K \to 2\pi$, this $K \to \pi$ method is also not likely to work unless $2m_\pi << m_\sigma$ that is, unless we have enough computing resources to enable us to treat light enough m_π.

6.5 The CP Violating Parameter ϵ' [56,3]

A very important class of matrix elements for $K \to 2\pi$ are those that are relevant for the CP violating parameter ϵ'. Recall that ϵ' is related to the difference in the CP violating amplitudes for $K_L \to \pi^+\pi^-$ and $K_L \to \pi^0\pi^0$. Thus

$$\eta_{+-} \equiv \frac{\text{Amp}(K_L \to \pi^+\pi^-)}{\text{Amp}(K_S \to \pi^+\pi^-)} \simeq \epsilon + \epsilon' \qquad (50)$$

$$\eta_{00} \equiv \frac{\text{Amp}(K_L \to \pi^0\pi^0)}{\text{Amp}(K_S \to \pi^0\pi^0)} \simeq \epsilon - \epsilon' \qquad (51)$$

where

$$\epsilon' \equiv \frac{i}{\sqrt{2}} \exp[i(\delta_2 - \delta_0)]\text{Im}(A_2/A_0) \qquad (52)$$

with A_I and δ_I being the $K \to 2\pi$ amplitudes and phase shifts in the $I = 0$ or 2 channels:

$$\text{Amp}(K^0 \to \pi\pi(I)) = A_I \exp(i\delta_I) \qquad (53)$$

It is useful to understand the relation of ϵ' to the $\Delta I = 1/2$ rule. As equation (52) indicates $\epsilon' = 0$ unless $A_2 \neq 0$. The statement of the $\Delta I = 1/2$ rule is that $A_2 \ll A_0$. Thus ϵ' is driven by the (small) violation in the $\Delta I = 1/2$ rule. Indeed, recent experiments indicate that ϵ'/ϵ is extremely small.

CERN experiment NA31 reports[57]: $(3.3 \pm 1.1) \times 10^{-3}$

whereas FNAL experiment E731 reports[58]: $(-0.4 \pm 1.4 \pm 0.6) \times 10^{-3}$

In the SM the major contribution to ϵ' arises from the penguin graph. Since it is CP violating the t, c quarks in the loop make the nonvanishing contribution.

The effects of the t quark go as $V_{ts}^* V_{td} \ell n(\frac{m_t}{\mu})$ and that of the charm quark as $V_{cs}^* V_{cd} \ell n(\frac{m_c}{\mu})$. The top contribution is likely to dominate. Indeed the major contribution is driven by a single operator (Q_6):

$$\left| \frac{\epsilon'}{\epsilon} \right| \sim \text{Im}(V_{ts}^* V_{td}) C_6 \langle \pi\pi(I = 0)|Q_6|K^0 \rangle \tag{54}$$

where C_6 is the perturbative coefficient and Q_6 is one of the two "SVZ" LR operators[23], i.e.:

$$Q_6 \equiv (\bar{s}_a \gamma_\mu(1 - \gamma_5)d_b) \left[\bar{u}_b \gamma_\mu(1 + \gamma_5)u_a + \bar{d}_b \gamma_\mu(1 + \gamma_5)d_a + \bar{s}_b \gamma_\mu(1 + \gamma_5)s_a \right] \tag{55}$$

Note also:

$$Q_5 \equiv (\bar{s}_a \gamma_\mu(1 + \gamma_5)d_a) \left[\bar{u}_b \gamma_\mu(1 + \gamma_5)u_b + \bar{d}_b \gamma_\mu(1 + \gamma_5)d_b + \bar{s}_b \gamma_\mu(1 + \gamma_5)s_b \right] \tag{56}$$

where a, b are color indices. The general expectation is that these LR operators have larger matrix elements compared to the LL operators as under Fierz transformations they go over to products of pseudoscalar currents. Q_6 then should dominate over Q_5 since C_6 is larger than C_5 and also since Q_6 will have unmixed color indices after the Fierz transformation.

For the case of Wilson fermions CPS again protects this operator from mixing with lower dimensional operators enabling, in principle, the use of the direct $K \rightarrow 2\pi$ approach to calculating matrix elements of Q_6 (or Q_5 for that matter). The problem, so far, is exactly the same as for the $\Delta I = 1/2$ rule namely the Euclidean time dependence the relevant Green's function does not show a plateau. Indeed the indications are that this amplitude is also suffering from the presence of a σ resonance and use of lighter quarks will be required to make progress.

The staggered group[59] has had more success with the calculation of the matrix elements for Q_5 and Q_6. Again for their case the matrix elements for $K \rightarrow 2\pi$ can be obtained by calculating simpler quantities namely $K \rightarrow \pi$ and $K \rightarrow 0$. They express the matrix elements for Q_5 and Q_6 in terms of "B parameters" i.e., B_5 and B_6. Thus deviation of these from unity is a measure of the departure from vacuum saturation.

Although they have considerable problems with the eye diagrams for the $\Delta I = 1/2$ amplitude, they see a good signal for the B_5 and B_6. This is attributed to a cancellation between the contribution of eyes, eights and a subtraction term. Working on the $24^3 \times 40$ lattice, with $\beta = 6.0$, 15 gauge configuration, (generated under the DOE Grand Challenge and used extensively by the staggered and the Wilson groups) they find $B_5 \simeq B_6 \approx 1.15 \pm .15$. We note that most phenomenological calculations for ϵ'/ϵ tend to set B_5 and $B_6 = 1$.[60,61]

7 Form Factors

7.1 Semi-leptonic Form Factors

Calculation of semi-leptonic form factors is a very interesting application of the lattice method.[62-69] Conceptually these calculations are considerably easier than the non-leptonic decays, and they have important applications towards allowing a determination of the CKM mixing angles from the experimental data and of course for testing continuum (phenomenological) models of semi-leptonic form factors.[70-74] Since no "deep conceptual" issues are involved numerical accuracy becomes all the more important for phenomenological impact.

The calculations involve matrix elements of the type $\langle i|J|f\rangle$, where J is the hadronic current (i.e. a 2-quark operator), i is the initial pseudoscalar meson (such as D for the charm sector) and f may be a pseudoscalar or a vector (e.g. K or K^* resulting from D decays). (See Fig. 4.1) Such a basic treatment can, in principle, handle a host of interesting reactions:

$$
\begin{aligned}
K &\rightarrow \pi e\nu \\
D &\rightarrow K(K^*)e\nu, \pi(\rho)e\nu \cdots \\
D_S &\rightarrow \eta'(\phi)e\nu, K(K^*)e\nu \cdots \\
B &\rightarrow D(D^*)e\nu, \pi(\rho)e\nu \cdots \\
B_S &\rightarrow D_S(D_S^*)e\nu, K(K^*)e\nu \cdots
\end{aligned}
\tag{57}
$$

From a phenomenological point of view the most important, perhaps, are those in the $b \rightarrow u$ sector, namely $B \rightarrow \pi(\rho)e\nu$. However, these would require sufficiently weak coupling, for an accurate treatment of the propagating b-quark on the lattice or some technique like static or non-relativistic heavy quarks. Clearly all of these approaches are worth trying.

Perhaps the easiest transitions for the lattices that are currently available ($\beta \simeq 6.0$) are the decays of the charm mesons, i.e. D and D_s. Even in this limited sector very interesting results—which we will now describe—have already been obtained.

For 0^- to 0^- transitions Table 7.1 shows the resulting form factors obtained by our group[67] and by the ELC group both with $\beta = 6.0$.[65,68] The systematic errors at this point are quite substantial, our estimate is that a large fraction of the total errors are due to scale breaking errors. We have estimated these by comparing of the $\beta = 6.0$ results with those at $\beta = 5.7$. For $D \rightarrow K$ experimental results are also shown. Our results for $f_1(0)$ tend to be considerably bigger than the experiment although they are within about $1\,\sigma$ of the rather large systematic errors. The ELC number seems in better agreement with experiment. Note though that although ELC does not state its systematic errors they are unlikely to be significantly smaller than ours. So to a certain extent the agreement of their number with experiment may be fortuitous.

Many effects due to SU(3) breaking have also been studied. At the moment numerical accuracy limits their use. However, it is still useful to know the capabilities of lattice calculations in this regard. Undoubtedly, efforts for improvements will continue. As one example of a concrete SU(3) breaking effect we note here the direct calculation of[67]

$$
\frac{f_1^\pi(0)}{f_1^K(0)} = 0.93 \pm 0.10 \pm 0.13
\tag{58}
$$

process	model	$f_+(0)$	$f_+(q_m^2)$	$f_0(0)$	$f_0(q_m^2)$
$D \to K$	BSW, KS [70,71]	0.76	1.32	0.76	1.05
	GISW, AW [72,74]	0.77	1.15	0.77	0.80
	AEK, DP [73]	0.6 − 0.75		0.6 − 0.75	
	Experiment	0.69 (04)			
	ELC [65,68]	.63 ± .08			
	BKS [67]	0.90 (08)	1.64 (36)	0.70 (08)	0.95 (11)
	sys. error	±0.21		±0.24	
$D \to \pi$	BSW, KS	0.69	2.62	0.69	1.35
	GISW, AW	0.51	1.28	0.51	0.53
	AEK, DP	0.6 − 0.75		0.6 − 0.75	
	ELC	.38 ± .04			
	BKS	0.84 (12)	2.44 (80)	0.62 (06)	0.95 (10)
	sys. error	±0.35		±0.34	
$D_s \to \eta$	BSW	0.72	1.32	0.72	1.03
	ELC	.58 ± .09			
	BKS	0.89 (09)	1.59 (36)	0.70 (08)	0.95 (12)
	sys. error	±0.30		±0.25	

Table 7.1: Various D decays in comparison

Although the errors are large it is useful to note that even at this early stage we are able to learn that SU(3) breaking at the mass scale of charm works to better than about 20%.

Table 7.2 shows the results for the form factors for $D \to K^*$.[69] Recall that there are now three form factors (in the limit of zero lepton mass), two are for the axial current and one for the vector current. Results from the ELC group,[66,68] other theoretical calculations[70–74] as well as from experiment are shown also. There is no strong disagreement with regard to $A_1(0)$ and $V(0)$ amongst various calculations and experimental results.[76,77] However, the form factor $A_2(0)$ is somewhat controversial amongst the two lattice groups as well as among the experiments. ELC finds it vanishingly small in agreement with the FNAL experiment E691 but in direct contrast with the continuum models, as well as our calculation and the recent results of experiment E653. The reasons for the discrepancy between the two lattice results for A_2, especially for A_2/A_1, are not clear but may be due to finite size effects that are affecting the ELC extrapolation to the chiral limit.[64] Our results for A_2/A_1 are in very good agreement with the recent results reported by E653; less so with E691.

Table 7.3 shows more results for 0^- to 1^- transitions of D and D_s mesons.[69] Unfortunately there is no experimental data for these to compare with. The two lattice groups again are in rough agreement for A_1 and V and tend not to agree for A_2. In light of the discussion above for $D \to K^*$ this is clearly understandable as the same data set is really being used to perform slightly different extrapolations and interpolations.

group	$A_1(0)$	$A_2(0)$	$A_2/A_1(0)$	$V(0)$	$V/A_1(0)$
E691 [76]	0.46 ± 0.05	0.0 ± 0.2	0.0 ± 0.5	0.9 ± 0.3	2.0 ± 0.6
sys. error	± 0.05	± 0.1	± 0.2	± 0.1	± 0.3
E653 [77]			$0.82 \,{}^{+\,0.22}_{-\,0.23}$		$2.0 \,{}^{+\,0.34}_{-\,0.32}$
sys. error			± 0.11		± 0.16
BSW [70]	0.88	1.15		1.27	
KS [71]	0.82	0.82	1.0	0.82	1.0
AW/GS [74,73]	0.8	0.6		1.5	
BBD [74]	0.50 ± 0.15	0.60 ± 0.15	1.2 ± 0.2	1.10 ± 0.25	2.2 ± 0.2
ELC [68]	0.53 ± 0.03	0.19 ± 0.21		0.86 ± 0.10	
BKS [69]	0.83 ± 0.14	0.59 ± 0.14	0.70 ± 0.16	1.43 ± 0.45	1.99 ± 0.22
sys. error	± 0.28	${}^{+\,0.24}_{-\,0.23}$	${}^{+\,0.20}_{-\,0.15}$	${}^{+\,0.48}_{-\,0.49}$	${}^{+\,0.31}_{-\,0.35}$

Table 7.2: The form factors for $D \to K^*$ from various experiments and model calculations

Table 7.3: The form factors for $D \to \rho$ and $D_s \to \phi$ on the $24^3 \times 40$ lattice at $\beta = 6.0$

process	$A_1(0)$	$A_2(0)$	$A_2/A_1(0)$	$V(0)$	$V/A_1(0)$	$A_0(0)$	$A_0/A_1(0)$
$D \to \phi$	0.73	0.55	0.78	1.30	2.00	0.71	1.14
stat. error	0.12	0.10	0.08	0.32	0.19	0.13	0.04
sys. error	± 0.24	${}^{+\,0.24}_{-\,0.20}$	${}^{+\,0.17}_{-\,0.13}$	± 0.43	${}^{+\,0.20}_{-\,0.25}$	± 0.23	${}^{+\,0.13}_{-\,0.19}$
$D \to \rho$	0.65	0.59	0.89	1.07	2.01	0.64	1.21
stat. error	0.15	0.31	0.37	0.49	0.40	0.17	0.16
sys. error	${}^{+\,0.24}_{-\,0.23}$	${}^{+\,0.28}_{-\,0.25}$	${}^{+\,0.22}_{-\,0.19}$	± 0.35	${}^{+\,0.30}_{-\,0.33}$	± 0.21	${}^{+\,0.25}_{-\,0.26}$

7.2 Form factors at $q^2 = 0$ and at $q^2 = q^2_{max}$ [64]

One drawback of the lattice calculations to date is that they have to assume pole dominance to extract the value of the form factors at 0 momentum transfer that are quoted in, for example, Table 7.1. In principle lattice calculations at various values of q^2 should be possible, enabling us to deduce the shape and the normalization of the form factor. Numerical limitations have, so far, not allowed us to do that very meaningfully. Having said that it is also important to stress that for the 0^- to 0^- case (in the approximation of ignoring the lepton mass) the experimental differential rate depends on only one form factor.

$$\frac{d\Gamma}{dq^2}(A \to B\ell\nu_\ell) = \frac{G_F^2}{192\pi^3 m_A^3}|V|^2\lambda^{3/2}|F_+(q^2)|^2 \tag{59}$$

where

$$\lambda \equiv \lambda(m_A, m_B, q^2) = (m_A^2 - m_B^2 - q^2)^2 - 4m_B^2 q^2$$

Thus knowledge of $f_1(q^2)$ at one value of q^2, in conjunction with an experimental measurement of the differential rate at that q^2 is sufficient to allow one to deduce the relevant CKM angles (i.e. V in eqn. (59). There is nothing terribly magical about $q^2 = 0$, except that it is a convenient place to normalize form factors in the context of pole models. Indeed, $q^2 = 0$ is not accessible to experiment (leptons are hard to detect there); it also has not been feasible in our lattice calculation at $\beta = 6.0$. In our calculation, we have been able to get a few values of $q^2/m_D^2 \leq 0.2$. ELC[65] has been able to get near $q^2 \approx 0$ by injecting a minimum unit of momentum in their application of the source method to the D. Each calculation has some advantages.

A more interesting value of q^2 is the end-point region, i.e. $q^2 = q^2_{max}$. For the lattice calculations this should be the best determined data value as both the initial and the final mesons are at rest. It is also the place where some continuum models based on quark model ideas probably give their best determination of the form factors.[72] It is also not a bad place for experiment to deduce the differential rate. Indeed just as in the case of pseudoscalar final states (see eq. 59) for vector final states the rate at the end-point is controlled by a single form factor, namely A_1 (as $q^2 \to q^2_{max}$):

$$\frac{d\Gamma}{dq^2}(A \to B^*\ell\nu_\ell) \to \frac{G_F^2}{64\pi^3 m_A^3}|V|^2\lambda^{1/2}(m_A + m_{B^*})^2 q^2 |A_1(q^2)|^2 \tag{60}$$

Thus knowledge of $A_1(q^2 = \text{max})$ can lead to a model independent determination of the relevant CKM angle. Motivated by these considerations we show a subset of our data for the A_1 at the end-point.[64] (See Fig. 7.1). From this figure as well as from a large body of data that we have[69] it appears that $A_1(q^2 = \text{max})$ shows little dependence on κ's (i.e. light quark mass) and is always close to unity, within about 20%. Thus for $D \to K^*$ we deduce:

$$A_1(q^2 \text{ max}) = 1.26 \pm .17 \pm .43$$

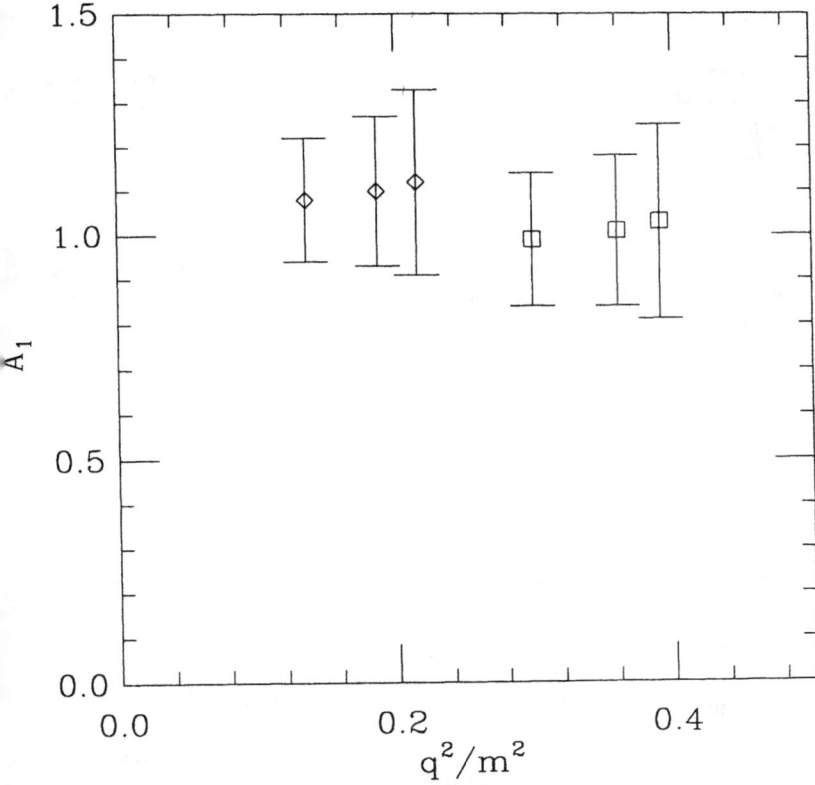

Fig. 7.1. A_1 at the "end-point" i.e., $q^2 = q^2_{max}$. Points on the left are $\kappa_{charm} = .135$, those on the right for $\kappa_{charm} = .118$. Data shown is for SU(3) degenerate light quarks. The non-degenerate data shows a very similar behavior.

At this point the systematic error being quoted is very preliminary, crude and very likely an overestimate. It is based on studies of other form factors at $\beta = 5.7$. Indeed, note also that using the ELC data[79] we have deduced:

$$A_1(q^2 \, max) = 1.10 \pm .26$$

in good agreement with our result.

It is very interesting that $A_1(q^2 \, max)$ is coming out to be so close to unity. For a heavy quark Q transitions to another heavy quark Q', $A_1(q^2 \, max)$ should be approximately one.[80] It is rather curious that this aspect of the heavy quark symmetry is becoming operative so "precociously" since the initial and final quark masses are not particularly heavy.

At this point we should also mention that our value for $A_1(q^2\,\mathrm{max})$ shows a strong disagreement with E691 which had reported $0.54 \pm 0.06 \pm 0.06$.[76] Hopefully this important parameter will be measured in other experiments soon to settle the current controversy.

Since the CKM angles for the charm sector (V_{cs} and V_{cd}) are already known to such a high degree of accuracy,[81] lattice study of the form factors for charm are no longer very enlightening for that purpose. Their role is more mundane, namely to test continuum models.[70-74] In that context test of pole dominance and other approximate concepts is useful. Of course, if we could actually determine the corresponding form factor for $B \to \pi$ decays then lattice calculations would start to have immediate impact on deduction of V_{ub} from the experimental data.

8 $D \to K + \pi$

In the class of 4-point function calculations another process that has been attempted is the study of exclusive charm meson decays: $D \to K + \pi$.[82-85] The motivations for a lattice study of this process should be quite clear. There are many issues in the dynamics of charm decays that need to be clarified. A better understanding is especially needed as it is likely to have repercussions for B decays and for CP violation observability.

An important issue in charm decay is the origin of the lifetime difference between charged and neutral D mesons, i.e. the experimental ratio:[86]

$$\frac{\tau(D^+)}{\tau(D^0)} = 2.5 \pm 0.1. \tag{61}$$

Some of the explanations that have been offered are:[87-92]

1. in the inclusive decays of the D^+ (via $c \to su\bar{d}$) presence of the two $\bar{d}$ quarks in the final state means that the amplitudes that result from their interchange will have a relative negative sign leading to destructive interference. Since the final state in D^0 does not contain identical quarks it does not suffer such a cancellation. Consequently this effect tends to increase the lifetime of D^+ relative to D^0.[87]

2. A second source for the life time difference is that the D^0 has the "annihilation" channel available to it. This results from the exchange of the W-boson across the quark and the antiquark in the D^0. This channel is not available to the D^+. As a result it can shorten the lifetime of the D^0 relative to the D^+.[88]

Clearly both of these mechanisms are simultaneously operational. The important issue is a quantitative one as to what their relative contribution is.

Inclusive decays are hard to deal with on the lattice. A simple set of exclusive decay modes that can yield information on these issues are generically of the type $D \to K + \pi$:

$$D^0 \;\to\; K^-\pi^+$$
$$D^0 \;\to\; \bar{K}^0\pi^0$$
$$D^+ \;\to\; \bar{K}^0\pi^+$$

The effective weak Hamiltonian responsible for these transitions is:

$$H_{weak}^{eff} = \frac{G_F V_{cs}^* V_{ud}}{2\sqrt{2}} \{c_+(\mu)0_+(\mu) + c_-(\mu)0_-(\mu)\}$$

where

$$0_\pm = \frac{1}{2}[0_1 \pm 0_2]$$

with

$$0_1 = \bar{s}_a \gamma_\mu (1 - \gamma_5) c_a \bar{u}_b \gamma_\mu (1 - \gamma_5) d_b$$
$$0_2 = \bar{s}_a \gamma_\mu (1 - \gamma_5) d_a \bar{u}_b \gamma_\mu (1 - \gamma_5) c_b$$

Note that these operators do not contain a $q\bar{q}$ of the same flavor i.e. all the four quark (antiquarks) have distinct flavor. This has the important consequence that these operators cannot mix with lower dimensional (2-quark operators). For the same reason these operators do not lead to eye-graphs. Another point to emphasize is that since these processes are parity violating mixing of the $(V - A) \times (V - A)$ operators with the "off-diagonal" operators such as $S \times S$, $P \times P$...does not take place, at least to one loop order.

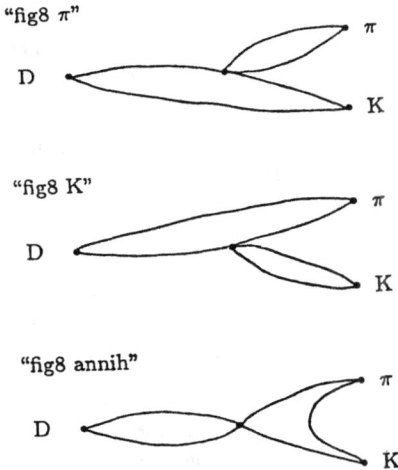

Fig. 8.1. Quark flow diagrams for $D \to K + \pi$

There are three lattice graphs that contribute to these processes. (See Fig. 8.1): Fig. 8An, Fig. 8K and Fig. 8π. Thus:

$$A(D^0 \to K^- \pi^+) = \text{Fig}8An + \text{Fig}8\pi$$
$$A(D^0 \to \bar{K}^0 \pi^0) = -\text{Fig}8An + \text{Fig}8K$$
$$A(D^+ \to \bar{K}^0 \pi^+) = \text{Fig}8\pi + \text{Fig}8K$$

Over the years a large body of experimental information has been gathered on these modes. Based on that one finds, e.g.[86]

$$\frac{|A(D^0 \to K^- \pi^+)|}{|A(D^+ \to \bar{K}^0 \pi^+)|} = 1.83^{+.30}_{-.25}$$

which is indicative of the inclusive picture mentioned above.

One nice feature of these amplitudes is that although they are 4-point functions they require only one and not two applications of the source method. However, there are still numerous other difficulties that these calculations represent:

1. Since the available momenta are discretized, laboratory kinematics cannot be implemented on the lattice.

2. Presence of the two mesons in the final state signifies that there may be nontrivial phases due to rescattering effects. Indeed, in $D \to K + \pi$ the relative phase between the $I = 3/2$ and $I = 1/2$ amplitudes has been determined from experiment to be 77 ± 11 degrees.[93] These final state interactions can introduce unphysical contributions to Euclidean amplitudes when the three-momenta of the final state particles are non-zero;[4] thus only "at-rest" amplitudes were considered in the work. This in turn means that the direct comparison of the lattice calculation with experiment is very hard. A useful strategy is to work with some continuum models. Since in general it is relatively easier to use a given continuum model and make predictions for the at rest amplitudes, that allows the lattice calculations to provide a useful check of the continuum models.

3. The lattice amplitudes relevant to these reactions vary with lattice spacing as a^{-3}. Thus even a 20% uncertainty in the lattice spacing translates into a rather hefty one in the amplitude of about a factor of two.

As already mentioned above it is often difficult to make direct comparisons between the lattice calculation and experiment. However, qualitative and conceptual information can still be obtained. Following is a brief account of some of the results that have been obtained.[82,83,85]

An approximation method that is used very extensively in the continuum is the vacuum saturation method. Lattice study of $D \to K + \pi$ gives considerable insight to the limitation and applicability of the vacuum saturation approximation (VSA). As an example, consider the case where all the three mesons are at rest, specifically say for the case of Fig. $8\pi_2$ (i.e. Fig. 8π for the operator O_2). Inserting the vacuum in all possible ways and using a Fierz rearrangement:

$$A(\text{Fig. } 8\pi_2) = c_2 \langle K | \bar{s}^b \gamma_\mu c^a | D \rangle \langle \pi | - \bar{u}^a \gamma_\mu \gamma_5 d^b | 0 \rangle \tag{62}$$

$$= \frac{c_2}{N} f_{DK}^0 ((m_D - m_K)^2) f_\pi m_\pi (m_D + m_K) \tag{63}$$

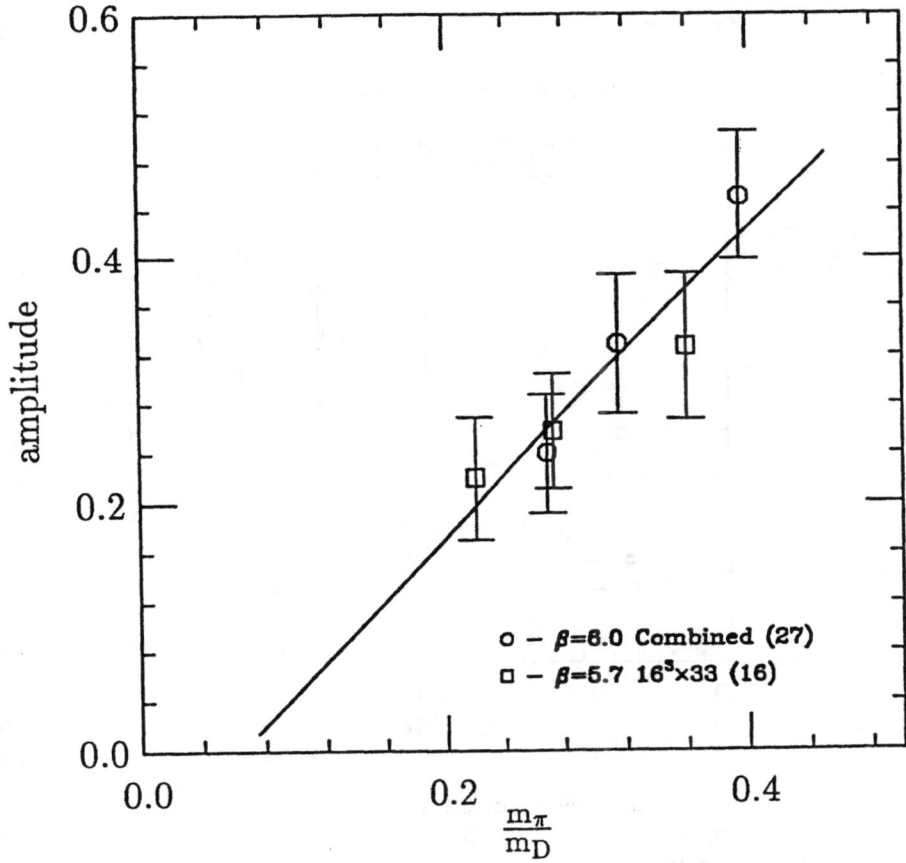

Fig. 8.2a. Testing vacuum saturation for the "spectator" amplitude i.e., via Fig. 8π and Fig. $8K$. (a) A(Fig. $8\pi_2$) vs. m_π/m_D.

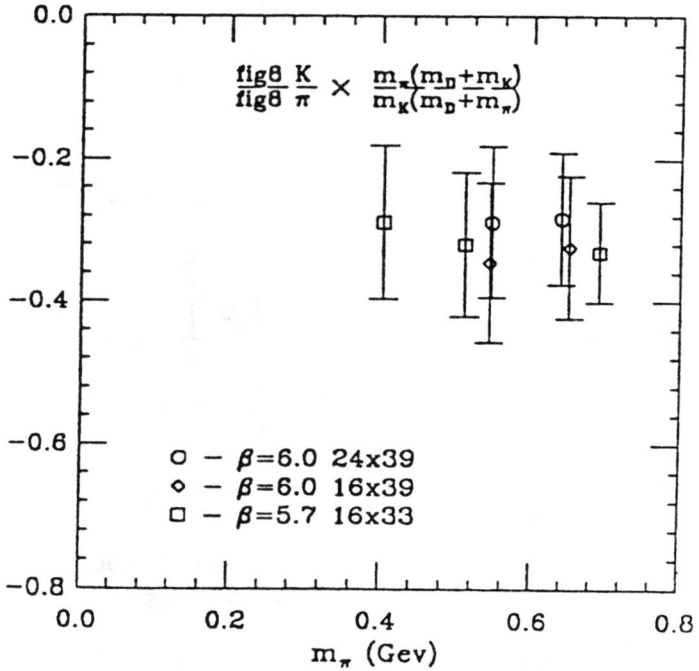

Fig. 8.2b. Testing vacuum saturation for the "spectator" amplitude i.e., via Fig. 8π and Fig. $8K$. (b) Ratio of Fig. $8K$/Fig. 8π amplitudes with mass factors to cancel the mass dependence predicted by vacuum saturation resulting in an approximately constant behavior. For further details to Fig. 8.1–8.5 see Refs. 6,7,85.

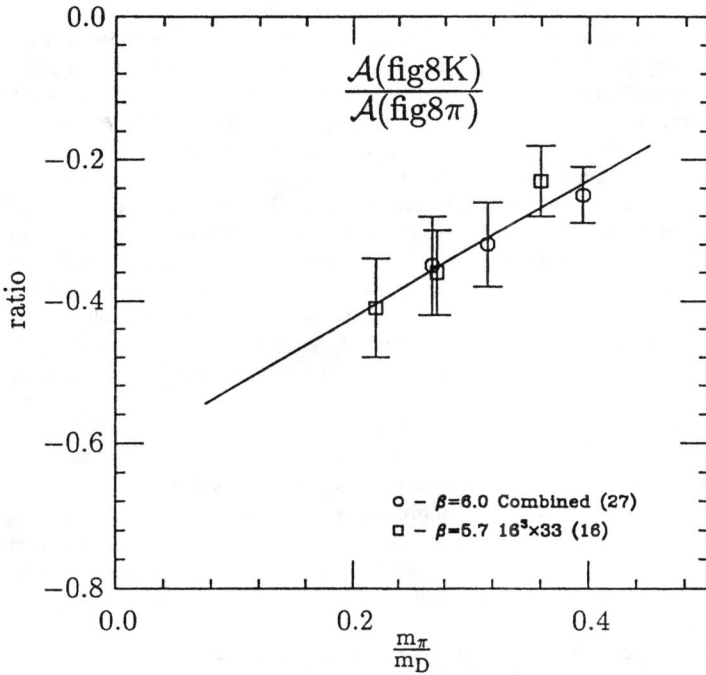

Fig. 8.3. The Fig. illustrates destructive interference between Fig. $8K$ and Fig. 8π that contribute to $D^+ \to \bar{K}^0 \pi^+$.

Thus vacuum saturation implies that the at rest amplitude through a typical specta-tor graph depends linearly on m_π.[94] The data (see Fig. 8.2a and b) supports this expectation of the VSA.

In Table 8.1 we present another test of VS. Herein we show that the ratio of the actual 4-point function to its value obtained by using VSA. We must emphasize, in this connection, that our use of VS means VS performed on the lattice amplitudes. This means that (e.g. for Fig. $8\pi_2$) the values of the 3-point and 2-point functions that are needed to test VS (through an equation like 62) are also directly computed on the lattice. This is a much better test of the concept of VSA rather than testing it by substituting the numerical values of the two and three-point functions given by other continuum methods and/or experiment.

Now it is instructive to consider the behavior of various amplitudes under large N. For that purpose we recall that Fig. $8\pi_1$ (Fig. $8K_2$) are of $O(1)$ whereas Fig. $8\pi_2$ (Fig. $8K_1$) are of $O(1/N)$. Thus Table 8.1 shows that vacuum saturation works very well for the spectator amplitudes that are of $O(1)$ in the large N approximation whereas vacuum saturation does not do well for the non-leading terms of $O(1/N)$. Recall that terms of $O(1)$ represent factorizable contributions whereas those of order $O(1/N)$ contain both factorizable as well as non-factorizable contributions. The fact that VS works poorly for Fig. $8\pi_2$ may be due to the fact there are other $1/N$ non-factorizable contributions to the four-point functions that nearly cancel the factorizable contributions. Such a cancellation amongst $1/N$ terms would mean that the effective value of $1/N$ would be smaller than expected (i.e. 1/3). Indeed Table 8.2 shows the effective value of $1/N$ calculated on the lattice and it appears to be consistent with being much less than 1/3.[95,96]

The effect of Pauli interference anticipated by continuum arguments[87] is substanti-ated by the lattice data. This is illustrated for the ratio [Fig. $8K$/Fig. 8π] shown in Fig. 8.3. Notice, in particular, that this ratio is negative indicating destructive interference. Also the dependence on m_π can be qualitatively understood using arguments similar to those given above. Thus at the physical m_π this ratio approaches -0.6, indicating that destructive in-terference between the two graphs for D^+ reduces the amplitude by as much as 40%. This, of course, has the effect of decreasing the D^+ partial width into the exclusive channels relative to the D^0.

One very important issue in the dynamics of weak decays is the role of the annihi-lation graph.[88,97] As mentioned previously, the annihilation channel is only available to the neutral D's and not to the charged ones. Lattice study of $D \to K + \pi$ indicates that the annihilation amplitude for this mode could be substantial. In this regard see Fig. 8.4 which compares the annihilation amplitude to the dominant spectator amplitude. Unfortunately, the size of the error bars in the current data make it difficult to make very precise quan-titative statements. Indeed, unlike the spectator amplitudes, the annihilation amplitude does not show a clear plateau region for large Euclidean times. In fact the latter amplitude seems to indicate a linear time-dependence instead. The linear rise with time becomes quite clear when one examines the ratio of the $I = 1/2$ amplitude to the $I = 3/2$ amplitude as shown in Fig. 8.5. (Note that the annihilation amplitude contributes exclusively to the 1/2 channel).

Such a linear time dependence is very likely due to the presence of a resonance in the $I = 1/2$ channel.[4,5] In fact a particle that could enter this channel and have the

Table 8.1: Ratio of Fig. 8π lattice functions to the corresponding function computed in lattice vacuum saturation. A ratio near one indicates that vacuum saturation is a good approximation.

Lattice	$\dfrac{\mathcal{A}^{Lat}(fig8\pi_1)}{\mathcal{A}^{Lat}_{vs}(fig8\pi_1)}$	$\dfrac{\mathcal{A}^{Lat}(fig8\pi_2)}{\mathcal{A}^{Lat}_{vs}(fig8\pi_2)}$
$\beta = 6.0\ 24^3 \times 39$ $\kappa_{u,d} = .155\ \kappa_s = .154$	$1.2 \pm .1$	$.1 \pm .4$
$\beta = 6.0\ 16^3 \times 39$ $\kappa_{u,d} = .155\ \kappa_s = .154$	$1.5 \pm .2$	$-.2 \pm .7$

Table 8.2: Effective $1/N$ values. These values are to be compared to $1/3$.

Lattice	$\dfrac{\mathcal{A}^{Lat}(fig8\pi_2)}{\mathcal{A}^{Lat}(fig8\pi_1)}$	$\dfrac{\mathcal{A}^{Lat}(fig8K_1)}{\mathcal{A}^{Lat}(fig8K_2)}$
$\beta = 6.0\ 24^3 \times 39$ $\kappa_{u,d} = .155\ \kappa_s = .154$	$.03 \pm .13$	$-.02 \pm .11$
$\beta = 6.0\ 16^3 \times 39$ $\kappa_{u,d} = .155\ \kappa_s = .154$	$-.04 \pm .15$	$-.09 \pm .14$

right quantum numbers and approximately the right mass is listed in the Particle Data Tables (PDT).[98] This is the κ^* particle with $I = 1/2$, $J^P = 0^+$ and mass ≈ 1.43 GeV and decay width ≈ 287 MeV. Furthermore, from the PDT one finds that the resonance decays predominantly to $K + \pi$.

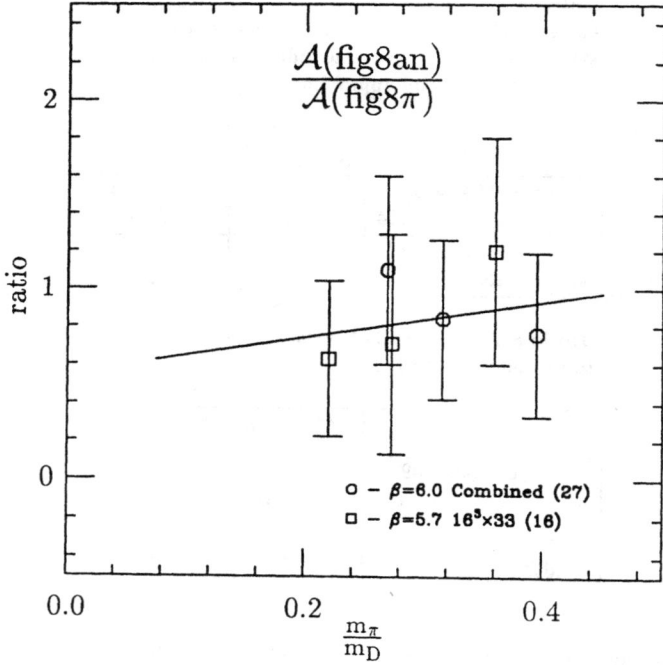

Fig. 8.4. Ratios of the annihilation amplitude to the dominant specta-tor amplitude vs. m_π/m_D.

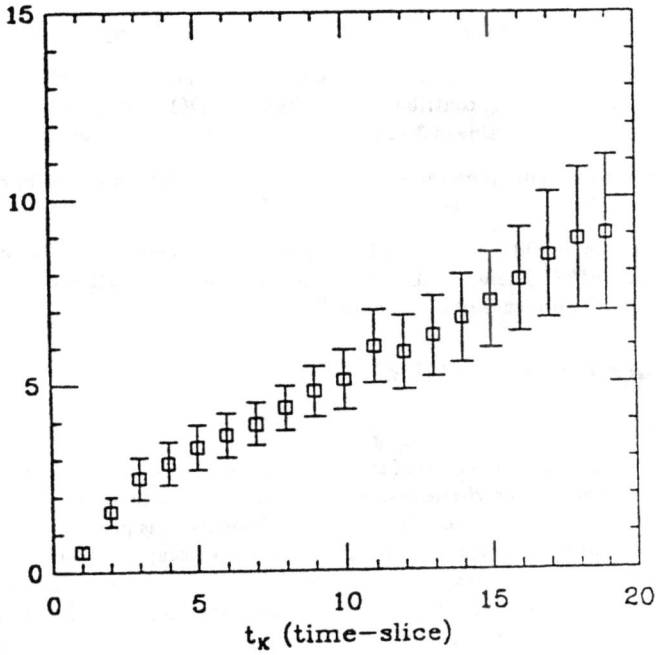

Fig. 8.5. Time dependence of the ratio of the $I = 1/2$ to the $3/2$ amplitudes vs. t_K (Kaon time slice).

This resonance hypothesis can be studied further on the lattice. The point is if the resonance really dominates the channel of $D \to K\pi$ then that decay can be understood as a two stage process: $D \to \kappa$ followed by $\kappa \to K + \pi$. Each of these two steps is in principle calculable on the lattice. Therefore, one should be able to verify the extent to which the resonance dominates the channel.[99]

Here is a very brief summary of the points made above emerging from our study of $D \to K + \pi$:

1. For the spectator graphs vacuum saturation appear to work very well.

2. In application of vacuum saturation to D decays, it seems that the effective value of N, in so far as non-leading contributions (i.e. terms of $O(1/N)$) is concerned, is much larger than the nominal value of 3—suggesting that such terms should be ignored.[95,96]

3. Clear evidence of Pauli interference effects at the level of 40% are seen in conformity with estimates from simple vacuum saturation.[87]

4. The annihilation amplitude seems to be large i.e. an appreciable fraction of the spectator amplitude.[88,97] There are also indications that the annihilation channel may be dominated by the presence of a resonance.[99]

9 B-Physics on the Lattice[8,9]

9.1 Introduction

In this section we will review the status of the various attempts for calculations of the weak matrix elements relevant to the B-mesons. Although in the introduction we discussed the general motivation for the calculation of weak matrix elements. it is perhaps of some use to very briefly re-emphasize the unique position B-physics now occupies in our thinking.

In the past decade several experimental and theoretical findings regarding the properties of the b-quark have accentuated its role in our understanding of the SM. Chief among these were: (1) A long b life-time. This discovery provided a crucial clue in the specific hierarchy amongst CKM angles.[100] (2) Discovery of a significant amount of mixing[14] between $B - \bar{B}$ which provided an early hint for the heaviness of the top quark and made us revise our expectations for the prospects for observability of CP violation in b-decays. (3) The experimental finding that the top quark is very heavy.[101] Along with the long b lifetime. this implies large branching ratios (BRs) for rare decays via flavor changing neutral currents (FCNC). This again has repercussions for observability of CP violation and for stringent tests of the SM.

In so far as lattice computation of B amplitudes is concerned, it must be noted that the excitement with regard to B-physics is due to the general expectation that it will allow us to have a precise handle on some of the important parameters of the SM[16,102]. Therefore any program for lattice calculation of matrix elements of B mesons that is not geared towards precision (at least in the long-run) is likely to be only transitory at best and is unlikely to serve much of a useful, long-lasting role.

9.2 Methods

In the general program for lattice calculation of electroweak matrix elements, treatment of the b quark on the lattice presents a new obstacle: the mass of the b-quark ($\sim$ 4.5 GeV) is uncomfortably large for the β values that have been used in simulations so far. In the roughly five years leading up to Lattice '90 most simulations that had been done for weak matrix elements used β = 5.7 to 6.0 with reasonable size lattices. The lattice spacing for these ranged roughly between 1 to 2 GeV. In the last two years, activity has shifted to the β = 6.2 to 6.4 range with $a^{-1} \sim$ 2.5 to 3.5 GeV. The suitability of these β's for b-quarks is unclear as one does not know what the true ultra-violet cut-off in simulations is. One in general expects errors of order $m_Q a$ (m_Q is the heavy quark mass), but the size of the coefficient of such terms is not known a priori. One may take a pragmatic point of view and try to ascertain the systematic effects due to simulating a heavy quark on the lattice by studying the quantity of physical interest at several different β values.

Since a b-quark of mass 4.5 GeV or so may not be suitable for simulations, one is naturally led to ask whether one really needs to simulate a quark with such a mass to learn about its properties. Is it possible that one could study how the quantities of physical interest depend on the quark mass by using quark masses that are "relatively heavy", say $am \lesssim 1$, but still not as heavy as m_b? If one could do that then one could hope to extract the value of that quantity by extrapolating to m_b. Needless to say, for this method to be practically useful one will need to study the amplitudes at several β values and verify that quantitative precision will not be jeopardized to an unacceptable level. We give below the reason why one may expect this procedure to be at all useful. After all the mass scale for nonperturbative effects in QCD is not more than a few hundred MeV. Lattice simulations are primarily designed to address the nonperturbative effects. The asymptotically free nature of the theory coupled with the fact that the QCD scale parameter is of a few hundred MeV suggest that large QCD corrections are unlikely to extend beyond a quark mass of $\approx 2 - 3$ GeV. Once kinematic dependence on the heavy quark mass of a given physical amplitude is factored out, the dynamics of QCD are likely to have, at most, a mild dependence on the heavy quark mass. This reasoning formed the basis for the first attempt to do B physics on the lattice.[103] This method is now often called the extrapolation method or the conventional method or simply the propagating quark method. It should be noted that the rationale for that work was one of the numerous examples in which the particle physics community was already using the existence of a heavy quark limit of QCD even before the associated symmetry was put on a formal and systematic basis by Isgur and Wise.[104,105]

Two other methods have been proposed to overcome the obstacle of the heavy b-mass. Lepage and Thacker (LT) proposed the use of the non-relativistic approximation[106] and Eichten[107] proposed the use of the static approximation for treating the heavy quark mass. These two methods also rely on the existence of the heavy quark limit of QCD. LT use the non-relativistic action rather than the full QCD action for treating the b-quark mass. In the non-relativistic (NR) approach of course the b mass becomes an irrelevant parameter and the kinetic energy of the b quark plays the important, dynamical role. Eichten instead proposed freezing out the motion of the b quark altogether by using the static action.

408

In the LT approach the relevant Lagrangian of QCD, in the continuum, is the NR one, called non-relativistic QCD and given by:[8]

$$
\mathcal{L}_{NRQCD} = \psi^+ \left\{ iD_t + \frac{\vec{D}^2}{2m_Q} + \frac{c_1}{2m_Q}\sigma \cdot g\vec{B} + \frac{c_2}{8m_Q^2} i\vec{\sigma} \cdot (\vec{D} \times g\vec{E} \right.
$$
$$
\left. -g\vec{E} \times \vec{D}) + \frac{c_3}{8m_Q^3}\vec{D}^4 + \frac{c_4}{8m_Q^2}[\vec{D}\cdot, \vec{E}] \right\} \psi
$$
$$
+ \cdots . \tag{64}
$$

The quark field ψ is now a two-component Pauli spinor. The first two terms are reminiscent of the Schrodinger equation, next comes the Pauli spin interaction term followed by the other relativistic correction terms. LT argue that these correction terms would have a smaller effect on the physics of heavy-light systems. This obviously needs to be checked quantitatively. The couplings g, M, c_1, c_2... are unknown, but it is suggested that these can be chosen so that NRQCD gives results that are equivalent to those given by QCD in NR systems.

One of the advantages of using the NRQCD for b-quarks on the lattice is that the b-quark propagator can be obtained without the need for matrix inversion. (It should be noted, though, that this is not a huge advantage. In the program for doing weak matrix element calculations, most of the computer time is spent on inversions for the light flavors. Heavy, propagating quarks add a modest amount ($\sim 20\%$) to the total cost.) This is a consequence of the decoupling of the antiparticles from particles in NRQCD; negative energy components are removed from the quark propagator in this approximation. The NR quark propagator can therefore be obtained by using the evolution equation:[8]

$$
S(x+t) = U_4^\dagger(x)\left[S(x) + \frac{1}{2m_Q}\sum_{i=1}^3 (S(x+\hat{i}) - 2S(x) \right.
$$
$$
\left. +S(x-\hat{i})) + \cdots \right] \tag{65}
$$

with a single pass through the gauge configuration.

Although the NR approach has been used to obtain many interesting results[108,109] in the physics of quarkonia (i.e. heavy-heavy system) so far it has not been used in the heavy-light system which is the main concern of this review.

The static-quark limit of Eichten is the third method that has been proposed for doing B physics on the lattice.[107,110,111] It is closely related to the non-relativistic one and it is obtainable from $\mathcal{L}_{NRQCD}$ by retaining only the $\psi^+ D_t \psi$ term. In leading order (in $1/m_Q$) the heavy quark propagator on the lattice is simply given by:

$$
S_Q^{latt}(x,y) = \delta^3(\vec{x}-\vec{y})\exp[-am_Q|x_0-y_0|] *
$$
$$
\left\{ \theta(x_0-y_0)\left(\frac{1+\gamma_0}{2}\right)\prod_{t=x_0-1}^{y_0} U_{\hat{0}}^\dagger(\vec{y},t) \right.
$$
$$
\left. +\theta(y_0-x_0)\left(\frac{1-\gamma_0}{2}\right)\prod_{t=x_0}^{y_0-1} U_{\hat{0}}(\vec{y},t) \right\} . \tag{66}
$$

9.3 Applications to the Calculation of the Pseudoscalar Decay Constant

The most extensive application of the static and the conventional approach has been, so far, to the calculation of the pseudoscalar decay constant of the B mesons, i.e. f_B.[112-116,124,8,9]

In the conventional treatment the basic methodology is very similar to other matrix elements and we do not need to spell out anything further here.

For the static case the situation is considerably more involved.

In principle, the calculation of f_B follows simply from the computation of the axial-axial correlator with the use of eqn. (66) for the static propagator for the heavy-quark propagator. It was realized quite early on that this simple approach fails to produce a clear signal for the ground state.[117] The contribution from the excited states contaminates the signal for long Euclidean times. As a result the binding energy fails to exhibit a clear plateau.

Considerable progress has been made in the last two years in overcoming this initial difficulty. Following the successful use of "smeared operators" by the APE group[118] in the context of light hadron spectroscopy, several groups have adopted similar techniques to enhance the overlap with the ground state in the context of the static heavy quark calculation of the decay constant.

Smeared operators are used on the source side ($t = 0$) or the sink side ($t_{sink} = T$) or both. The "source-smeared" correlation then reduces to:[9,117]

$$\langle A_0 A_0^S \rangle^{latt} = \left\langle \frac{T_r}{n^3} \sum_i^{(n)} \left[\frac{1 \pm \gamma_0}{2} P^{\pm} S_q(\vec{i}, t_x; \vec{0}, t_0) \right] \right\rangle \tag{67}$$
$$\rightarrow \xi_S \exp[-a(M_P - m_Q)|t_x - t_0|]$$

where $P^{\pm}$ stands for the appropriate product of links in eqn. (66). Similarly we can define the correlation with smearing on both the source and the sink side as $\langle A_0^S A_0^S \rangle^{latt}$. The decay constant can then be extracted as:

$$f_P \sqrt{M_P} = Z_A^{stat} \sqrt{2K_q} \sqrt{\frac{2\xi_S^2}{\xi_{SS}}} a^{-3/2} \tag{68}$$

where Z_A^{stat} is the renormalization constant for the axial-vector current that is calculated by use of lattice weak coupling perturbation theory.

At the time of Lattice'90 several groups had used such an approach with results that were qualitatively very similar: $f_B^{stat} \approx 300\text{-}350$ MeV, i.e. roughly a factor of three larger than the results obtained by the conventional approach.[112-114,9]

During the past year, considerable progress was made in understanding the origin of the disagreement. One major culprit is found to be simply the stability and the quality of the signal on which the earlier static results were based. Two independent studies showed a significant dependence of the results on the size of the volumes used for smearing of the operators.[115,116] Fig. 9.1 is from the paper by Hashimoto and Saeki illustrating this dependence.[116] As they change the smearing size from 5^3 to 13^3 they find the decay constant change from 548 ± 28 MeV to 270 ± 24 MeV. Results of a similar nature from our collaboration (with Jim Labrenz) are shown in Fig. 9.2[115] Again a systematic decrease

410

in f_B is seen as the smearing volume is decreased. This seems to be especially noticeable when results are extracted at earlier times. When later time interval is used there may well be an indication in the data that the dependence on smearing size has tapered off at volumes $\geq 7^3$. However, the fluctuations have also increased so that no reliable conclusion can be drawn at present. Of course, the physical result can in principle be extracted with any smearing volume as long as large enough Euclidean times are used. The point however is that fluctuations at large times can wipe out the the signal or, worse, make it appear that a plateau is present when it is in fact not.

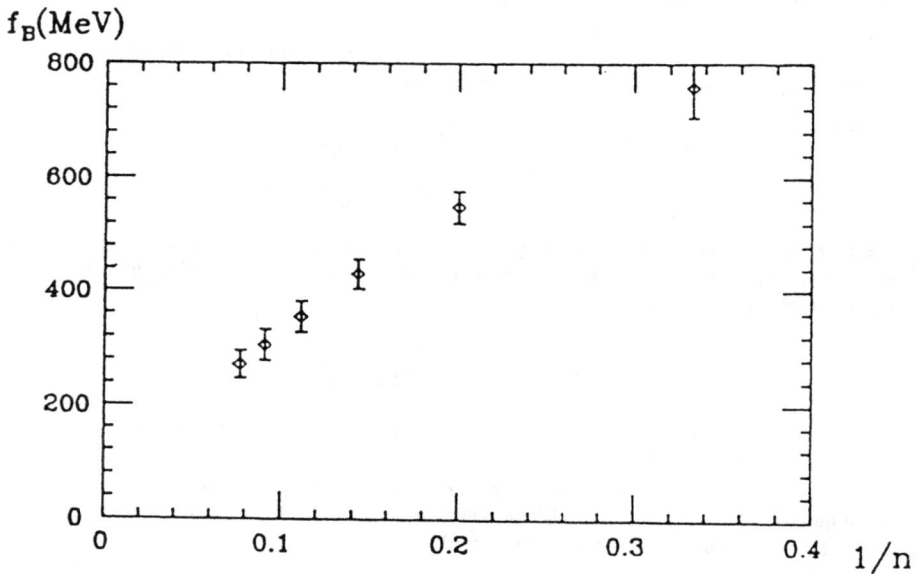

Fig. 9.1. Dependence of f_B^{static} on the size (n) of the cube used for smearing reported by Hashimoto and Saeki.[116] Data points correspond to $n = 3, 5 \ldots 13$ on a $\beta = 6.2$, $20^3 \times 40$ lattice. For further details see Ref. 116.

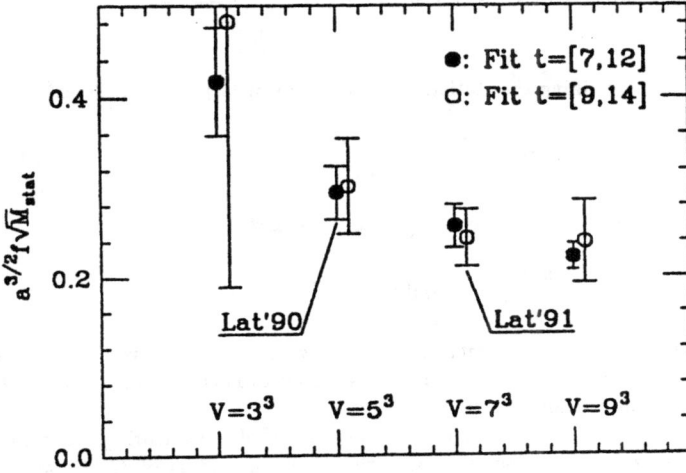

Fig. 9.2. Dependence of static results on smearing volumes and fitting
intervals.[115] ($\beta = 6.0$, $24^3 \times 39$, 8 configurations).

The problem with the quality of the static signal, especially at large times, can be
qualitatively understood by comparing signal to noise as a function of time.[8,115,119] At large
times the heavy-light meson propagator falls off exponentially as $\exp(-E_{Qq}t)$, E_{Qq} being
the binding energy, a large fraction of which is just the static energy:[8]

$$E_{St} \approx 0.167g^2/a \tag{69}$$

Now the statistical fluctuation $\sigma^2_{Qq}(t)$ in the meson propagator can be estimated as:

$$\sigma^2_{Q\bar{q}}(t) \sim \left\langle 0|T_r\hat{S}_q(t)\hat{S}_{St}^\dagger T_r\hat{S}_{St}(t)\hat{S}_q^\dagger|0\right\rangle \tag{70}$$

Since the two static propagators in this expression tend to cancel each other out the

Table 9.1: Results for the decay constants at $\beta = 6.3$.

meson	$\hat{f}(\text{GeV}^{3/2})$	$f(\text{GeV})$	$\sigma_f[\text{fits}]$	$\delta f[N]$
D	.310(40)	.225(25)	$\pm.01$	$-.035$
D_s	.335(30)	.240(20)	$\pm.01$	$-.040$
B	.455(20)	.195(10)	$\pm.03$	$-.060$
B_s	.480(20)	.205(10)	$\pm.03$	$-.060$

pion mass ends up determining the statistical fluctuations at large times. Thus:[8]

$$\sigma^2_{Q\bar{q}}(t) \sim \left\langle 0|T_r \hat{S}_q(t) \hat{S}_q^\dagger(t)|0\right\rangle$$
$$\sim \exp[-m_\pi t] \tag{71}$$

and, at large times, the signal to noise is expected to go as:[8,115,119]

$$\left.\frac{S_{Q\bar{q}}(t)}{\sigma_{Q\bar{q}}(t)}\right|_{St} \sim \exp[(-E_{Q\bar{q}} + (m_\pi/2))t]$$

Since $E_{Q\bar{q}}$ contains the linearly divergent static energy, E_{St} given above, it is very likely greater than half the pseudoscalar mass. Thus the signal to noise ratio is expected to fall rapidly with increase in Euclidean time.

While this explanation is highly qualitative, studies of the smearing volume dependence of f_B have at least led to the conclusion that "largish" values of f_B^{static} (≈ 300–350 MeV) that were being reported at the time of Lat90 must be regarded as completely unreliable.

Further indication that these very high values of f_B^{static} are most likely incorrect comes from the use of wall source propagators and "wall-smeared" interpolating operators at $\beta = 6.3$ on a $24^3 \times 55$ lattice. Decay constants were calculated using both static and conventional methods. Fig. 9.3 and Table 9.1 show preliminary results reported by our group at LAT91. Interestingly with these wall sources the static method gives now $f_B^{static} \sim 200$ MeV and the discrepancy between the static and the conventional method is drastically reduced.[115]

There has been an interesting development with regard to the conventional method also. Traditionally, in normalizing the lattice Green functions to the continuum, each quark field ψ is naively rescaled by a factor of $\sqrt{2\kappa}$. Although this normalization works fine for light quarks, it may not be suitable for heavier quarks, esp. for quarks with $ma \sim 1$. The point is that the quark propagator for zero momentum on the lattice satisfies the normalization condition:[120]

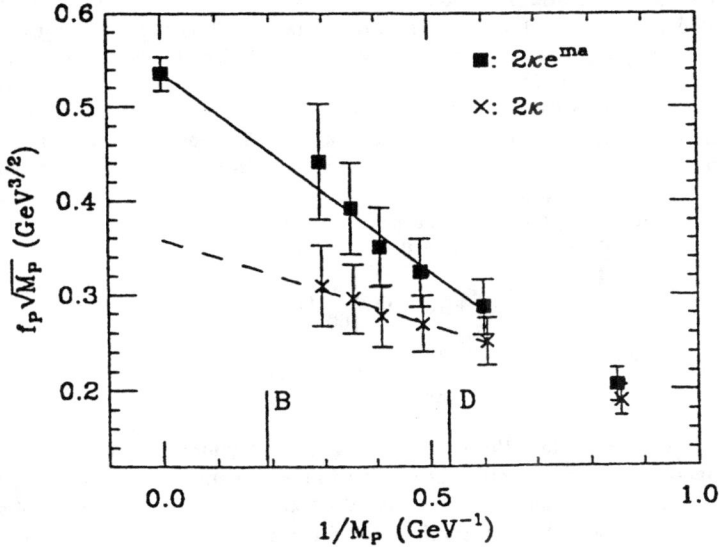

Fig. 9.3. Data[115] after extrapolation of the light quark ($K_L = 1.49$, .150, .1507) to K_c; $\beta = 6.3$, $24^3 \times 55$, 20 configurations. Results quoted in Table 9.1 are from the solid points and associated fit.[115]

$$2\kappa \exp(am) \sum_x \langle \psi_x \bar{\psi}_0 \rangle^{latt} = \int d^3 x \, \langle \psi_x \bar{\psi}_0 \rangle^{cont} \tag{72}$$

where

$$ma = \ell n \left[1 + \frac{1}{2\kappa} - \frac{1}{2\kappa_c} \right]. \tag{73}$$

Equations (72) and (73) with κ_c replaced by its renormalized value may provide a more suitable (although rather ad-hoc) normalization for dealing with heavy quarks.

Note that this procedure gives the correct results in the limit of small quark mass as well as in the limit of very heavy (static) quark mass. Furthermore, following Lepage and Mackenzie's observation that the dominant perturbative corrections on the lattice arise from tadpole diagrams,[121] which are momentum independent, these perturbative corrections can automatically be included by a shift of parameters in the free-quark relations.

Fig. 9.3 shows the static data point along with the results of the conventional method, presented here with and without the exp ma factor. Although this normalization factor is admittedly quite *ad hoc*, the conventional data with this factor seems to fit smoothly on to the static result: fits have $\chi^2/d.o.f \approx 1$. Such is not the case if the exp ma is left out: such fits have $\chi^2/d.o.f \sim 4 - 5$. Fig. 9.3 and Table 9.1 also indicate what happens if the exp ma and the static data point are both left out and the linear fit through the data point of the conventional method is retained. The resultant drop in the decay constant is shown as $\delta f[N]$, i.e. shift due to normalization, in the last column of this Table. Pending further study, this shift was reported as a source of systematic error.[115]

Thus, our (incomplete) study at $\beta = 6.3$ data leads to a preliminary result for f_B of:[115]

$$f_B = 195 \pm 10 \pm 30 \,^{+\ 0}_{-\ 60} \text{ MeV} \tag{74}$$

It is useful to recall our old result:[103]

$$f_B^{conv,old} = 105 \pm 17 \pm 35 \text{ MeV} \tag{75}$$

Although the central value (195 MeV) now is appreciably higher than the old result we note that a large portion of the difference is due to the $\exp(ma)$ normalization factor. The old analysis yielding the result (75) was with the traditional normalization of simple factor of $\sqrt{2\kappa}$ which in hindsight appears ill-suited for quarks with $am_Q \simeq 1$ that were being used. Note that, if one disregards the shift of 60 MeV due to this factor then the remaining value for the B decay constant of 135 MeV from the current study of $\beta = 6.3$ is close to our older result of 105 MeV above which was based on $\beta = 6.1$. The presence of the $\exp(ma)$ factor gives a much larger $1/m_Q$ correction than was previously apparent and is important in the current agreement between conventional and static approaches. Alternatively, merely going to $\beta = 6.3$ from $\beta = 6.0$ or 6.1 without including the $\exp(ma)$ factor already considerably improves the situation since the values of ma needed are smaller.

We stress again that the numbers reproduced here from our contribution at LAT'91[115] are from an ongoing and incomplete study.[122,123] The analysis is incomplete in many respects. In particular, we hope to study the static method at 6.3 with smearing over cubes of varying volumes to pursue the issue of the dependence of f_B on cube-size. We feel it is important to reconcile the results we are getting with the wall-smearing method with cube-smearing. This is particularly important in view of the appreciable dependence that we[115] (and HS[116]) have seen on the smearing cube volumes (in our case at $\beta = 6.0$). The results that we have described above for $\beta = 6.3$ with wall sources should only be taken seriously after we are able to show that in the cube smearing method, as smearing is done over cubes of increasing volumes, there comes a stage that f_B becomes essentially independent of that volume with quantitative consistency with the numbers that we are getting with the wall-smearing method. In addition to this reconciliation, in our ongoing analysis we would also like to incorporate all of the data that we have, i.e. $\beta = 5.7, 6.0$ and 6.3 in a consistent fashion before the results can be considered reliable and final.[122,123] For an alternative discussion of f_B, see Ref. 124.

9.4 Difficulties in $1/M_Q$ Corrections to the Static Method due to Mixing of the Kinetic Operator with Lower Dimensional Operators[125,1]

In addition to the difficulty in extracting a reliable signal that we have discussed above, the corrections to the static limit may also be afflicted with the problem of non-perturbative subtractions. Maiani, Martinelli and Sachrajda have made the very important observation that such formidable subtractions become necessary to enable one to calculate $1/m_Q$ corrections to matrix elements of hadronic operators that contain a heavy quark field.[125,1]

The basic point is very simple. The effective action for heavy quarks, when expanded to order $1/m_Q$, contains a kinetic energy term that goes as

$$\frac{1}{2m_Q}\bar{Q}(x)\vec{D}^2Q(x)$$

where $Q(x)$ represents the two-component field of the heavy quark. Under renormalization, this dimension-5 kinetic energy operator mixes with two lower dimensional operators. One is a dim-3 operator $(\bar{Q}(x)Q(x))$ with a coefficient which goes as $1/a^2$ and the other is a dim-4 operator $\bar{Q}(x)\gamma_4 D_4 Q(x)$ with a coefficient which diverges linearly as $1/a$. Although the quadratic divergence is the most severe one of the two, it gets absorbed in the renormalization of the heavy quark mass leaving our ability to compute matrix element intact. The linear divergence, on the other hand, is more serious requiring a non-perturbative subtraction method. The point is that this dim-3 operator will give rise to a power-divergent contribution to the wave-function renormalization constant, and thereby to the renormalization constant of any operator containing the heavy quark field. The standard method of relating matrix elements of interest in the continuum to those computed with lattice simulations through the use of lattice weak coupling perturbation theory cannot be trusted when dealing with power divergences.

The situation here is basically similar to that when dealing with the $\Delta I = 1/2$ amplitude in K decays where the corresponding culprit are the lower dimensional operators of the form $\bar{s}\gamma_5 d$ and $\bar{s}d$ that mix with the usual 4-quark operators. As explained in that context, factors of the form[125,1]

$$\exp\left[-\int^{g_0(a)} dg'/\beta(g')\right] = a\Lambda_{QCD}$$

originating from non-perturbative effects, in conjunction with coefficients that diverge as inverse powers of lattice spacing, can (and therefore are expected to) give non-vanishing contributions. While for some "fortuitous reasons" it is possible that such a contribution could end up being tolerably small this cannot be taken for granted and will need to be demonstrated by giving up the ability to compute a physical quantity.

9.5 Other Applications of the Conventional Method

Needless to say, in addition to the pseudoscalar decay constant, there are numerous other matrix elements involving B mesons that can and should be computed on the lattice. Perhaps the most interesting ones are semi-leptonic, especially $B \to \rho(\pi)e\nu$, B parameter for $B - \bar{B}$ mixing, and radiative transitions such as $B \to K^*\gamma$. We will briefly review here some of the work that has been done, or is in progress, involving two of these topics.

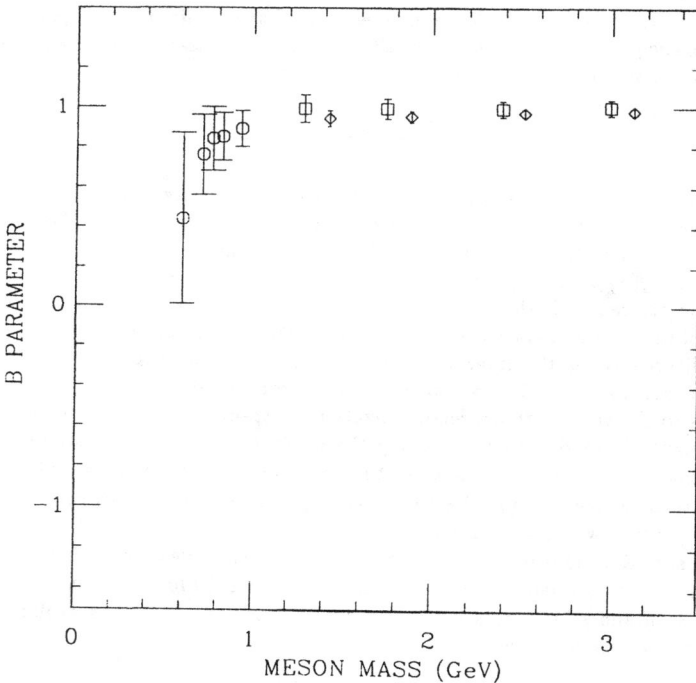

Fig. 9.4. B parameter for heavy light mesons. Key: octagons $\to$ light-light, squares $\to$ heavy-light$_1$, diamonds $\to$ heavy-light$_2$. ($\beta = 6.1$, $12^3 \times 33$, 30 gauge configuration[103])

9.5.1   B Parameter for $B - \bar{B}$ Mixing[103,126]

The B parameters for B_d and B_s mesons involves calculations of matrix elements such as

$$\langle B|[\bar{b}\gamma_\mu(1 - \gamma_5)d]^2|\bar{B}\rangle$$

This was the first three point function for the heavy-light system that was tried using the extrapolation method.[103] A set of 30 quenched gauge configurations at $\beta = 6.1$ on a $12^3 \times 33$ lattice formed the core of the study. Data at $\beta = 5.7$ on a $16^3 \times 25$ lattice was used to study systematic errors, especially those due to scale breaking effects and to order a effects. The key finding was that, as one varies the heavy quark mass with the light quark fixed, the B parameter (at a fixed scale proportional to a^{-1}) stays essentially constant. Indeed, it is 1 within about 15%, for all heavy quark masses greater than about 1.5 GeV (corresponding roughly to the heavy-light pseudoscalar mass around $m_D \approx 1.8$ GeV). See Fig. 9.4. The fact that the B-parameter for heavy light mesons of mass $\geq m_D$ is unity means of course that vacuum saturation works for these amplitudes. This finding was quite nontrivial: Since the reduced mass of the heavy-light system is just about the light quark mass, the validity of vacuum saturation is nontrivial and quite interesting. The fact that the B-parameter stays constant as we vary the heavy quark mass is extremely interesting and was in fact qualitatively understood as due to the existence of a heavy quark limit of QCD. The results of this study were

$$\hat{B}_{cu} \simeq \hat{B}_{bd} \simeq \hat{B}_{bs} \approx 1.3$$

where $\hat{B}_{Q_q}$ is the RG invariant B parameter for a heavy (Q) and a light (q) pseudoscalar ($\hat{B} \equiv B(\mu)(\alpha_s(\mu))^{\frac{-6}{25}}$), the exponent is the relevant one for 4 propagating quarks, $\mu \sim 5$GeV, and we use $\Lambda^{(4)} = 200$MeV.

Similar results have been found by other groups.[126]

9.5.2   Flavor Changing, Radiative B-Decays[127]

Flavor Changing radiative B-decays can provide a crucial test of the electroweak SM to one-loop as these decays proceed through the electroweak penguin graph.[102] At the inclusive (i.e. quark) level the calculations are reliable since the loop is short-distance dominated (most of the contribution arises due to the virtual top or charm quarks in the loop). The virtual up quark in the loop makes a completely negligible contribution due to the fact that $V_{ub}^* V_{us}$ is much smaller than $V_{tb}^* V_{ts} \sim V_{cb}^* V_{cs}$. There is also an additional suppression as the amplitudes tend to grow with virtual quark mass. This decoupling of the up quark from the flavor changing transitions makes rare (i.e. those that go via a loop) b-decays extremely potent test of the SM as it renders the QCD corrections (at the inclusive level) much more amenable to traditional, perturbative methods. This is in sharp contrast to (most) flavor changing K decays ($K \to \pi\nu\bar{\nu}$ is a well known important exception[17]). Unfortunately, it is much more difficult to search for these inclusive decays experimentally than it is to search some of the corresponding exclusive modes. For example, $B \to K^* + \gamma$ is considerably easier to search for experimentally rather than $B \to \gamma + X$ (where X has to be a "charmless" final state). Unfortunately, what is difficult for experimentalists is easy for the theorists (namely

$b \to s + \gamma$ or $B \to \gamma + X$), and what is easy for the experimentalists is extremely difficult for the theorists (namely $B \to K^* + \gamma$). Indeed this is particularly so for the ratio:[102,127-129]

$$R \equiv \frac{\Gamma(B \to K^* + \gamma)}{\Gamma(b \to s + \gamma)} \qquad (76)$$

which monitors the fraction of the flavor changing radiative (inclusive) transitions that materialize in the exclusive channel $K^* + \gamma$. In the past several years there have been many attempts to calculate this ratio by using a variety of models.[102] The results of these calculations span the range of 4% to 98% for R—indicating how challenging this calculation is for traditional, continuum based models.[128] Bearing in mind all that as well as the progress that lattice methods have already made in calculating semi-leptonic form factors—which are essentially similar—it is clearly desirable to initiate a computation of the matrix elements for the radiative B decays.

The hadronic matrix elements relevant to decays such as $B \to K^* \gamma$ can be generically parametrized using three form factors:[128]

$$\langle V(k) | J^\mu | B(p) \rangle = \sum_{i=1}^{3} c_i^\mu(k,p) T_i(q^2) \qquad (77)$$

where J^μ is the current:

$$J^\mu = \bar{\ell} \sigma^{\mu\nu} q^\nu b_R$$

where ℓ is the final light quark (i.e. s or d in b decays) and q is the momentum transfer, $p - k$. The c_i's are given by:

$$c_1^\mu = \epsilon^{\mu\nu\lambda\sigma} \epsilon^\nu(b) p^\lambda k^\sigma \qquad (78)$$

$$c_2^\mu = \epsilon^\mu(k)(m_B^2 - m_V^2) - (\epsilon \cdot q)(p+k)^\mu \qquad (79)$$

$$c_3^\mu = \epsilon \cdot q \left[(p-k)^\mu - \frac{q^2}{m_B^2 - m_V^2}(p+k)^\mu \right] \qquad (80)$$

where $\epsilon(k)$ is the polarization of the final state vector meson V.

In this notation the ratio R defined in eq (76) is a simple function of $|T_1(q^2 = 0)|^2$ [130]. Thus for $B \to K^* \gamma$ only one form factor needs to be evaluated at $q^2 = 0$. In general, this would require momentum injection to the initial and/or final meson, extraction of T_1 for several values of q^2 and then extrapolation to $q^2 = 0$. For us the most convenient way to do this is by injecting momentum to the final vector meson. We apply the source method to the initial "B" meson at rest. We have done a preliminary study of $T_1(q^2)$ on our $\beta = 6.0$, $24^3 \times 39$ lattice with 8 gauge configurations. This lattice, of course, has been extensively used in various other studies. For the heavy quark we used $\kappa_{light} = .152, .154$ and $.155$. (Recall $\kappa_c = .157$). It turns out that for these heavy-light combinations $q^2 \simeq 0$ is readily accessible and no serious extrapolation in q^2 is called for to deduce $T_1(q^2 = 0)$. However, the heavy-light pseudoscalar masses that are available are $\lesssim 2.5$ GeV so an extrapolation to m_B will be necessary. So far the data suggests $T_1(q^2 = 0) \leq 0.5$. We are in the process of analyzing more data involving 19 gauge configurations at $\beta = 6.0$, $16^3 \times 39$ lattice and 20 gauge configurations at $\beta = 6.3$ on a $24^3 \times 61$ lattice.

10 The Effects of Quenching

Essentially all weak matrix element calculations performed to date have used the quenched approximation. In those few cases where "full QCD" configurations have been examined,[36,30] the dynamical quark masses have been so large ($m_\pi \gtrsim 400$MeV) that the lack of an observable difference between the quenched and full QCD lattice results cannot be taken as a serious bound on the physical effects of quenching. It is thus very important to look for other methods to estimate the error introduced by the quenched approximation.

One such method is chiral perturbation theory. The utility of the effective, low-energy, chiral theory of full QCD is well known. Recently, it has become possible to perform chiral perturbation theory calculations in the quenched approximation. Morel[131] and Sharpe[132] did the first such calculations in the strong coupling and $1/d$ expansions; Sharpe and collaborators[133,134] then worked in the "quark-flow" approach in which one removes by hand the diagrams of ordinary chiral perturbation theory which correspond to quark flow diagrams with virtual quark loops. Bernard and Golterman[135] have advocated a Lagrangian approach which systematizes quenched chiral perturbation theory, making such calculations straightforward.

Comparing the chiral logarithms in the full and quenched theories gives some indication of the effects of quenching. Such logarithms are in general completely different in the two cases, and indeed typically have completely different physical origins. The quenched logarithms often come exclusively from "two-hairpin" diagrams[135], associated with the quenched remnant of the η'; whereas the logarithms in the full theory can be caused by arbitrary meson loops. Thus the quenched approximation can in general only be expected to be good in cases where the chiral logarithms give small corrections. An important exception is the case of B_K. The 1-loop chiral logarithms there are identical in the full and quenched theories;[133,134] the unwanted two-hairpin diagrams cancel in the ratio which defines B_K. Thus the effect of quenching in B_K is expected to be quite small, perhaps 5% (*i.e.*, like a 2-loop chiral correction). It is very important to study other weak matrix elements and classify them according to the expected errors in the quenched approximation.

Arguments based solely on chiral logarithms of course are really just guesses, since one has not taken into account the "finite" contributions coming from the order p^4 terms in the chiral Lagrangian. However, it is possible to write down relations in which all the unknown coefficients cancel. An example is:[136]

$$\frac{\langle \bar{d}d \rangle}{\langle \bar{u}u \rangle} - \frac{(m_{K^0}^2 - m_{K^+}^2)}{(m_{K^0}^2 - m_{\pi^+}^2)} \frac{\langle \bar{s}s \rangle}{\langle \bar{u}u \rangle} \tag{81}$$

Calculating this quantity in both the quenched and full chiral perturbation theories. one finds an $\sim 10\%$ difference. There are, in principle, many such calculable quantities. A key task for the future is to search for them and perform the appropriate chiral perturbation theory calculations. This should give a very useful working guide to the effects of quenching.

Summary and Epilogue

The last several years have been exciting for those of us who have been involved in attempts to calculate weak matrix elements using lattice methods. For two and three

420

point functions many quantitative results have become available with an accuracy in the quenched approximation, at the level of about 30% or better. In the near future we can anticipate progress in improved calculations of these quantities. We can also look forward to an increase in activity in these calculations using dynamical fermions, as well as more analytic estimates of the effects of quenching. Progress in calculating four-point functions has been extremely slow; important conceptual hurdles still remain. The difficulties are, in part, due to the problem posed by operator renormalization and operator mixing on the lattice. Increased numerical resources should help here at least in dealing with some of the interesting calculations. One of the areas in which more efforts need to be directed is towards making contact of lattice calculations to phenomenological models. Such efforts can surely educate us in the dynamics of QCD.

Acknowledgements

We must thank our students and collaborators on this project: Terry Draper, Mike Heard, George Hockney, Dan Murphy, Aida El-Khadra, James Simone, James Labrenz, Paul Hsieh and Ken Yee. We have benefitted from inputs and discussions with Luciano Maiani, Guido Martinelli, Chris Sachrajda and with Rajan Gupta, Greg Kilcup, Apoorva Patel and Steven Sharpe. The computing was done at the NERSC (Livermore) and at SDSC (San Diego). In particular we thank the US DOE for extensive support for this project. Finally we thank Michael Creutz for inviting us to write this chapter and to Fern Simes for her invaluable help in TeXing. C.B. was partially supported by the DOE under grant number DE2FG02-91ER40628 and A.S. under DOE contract number DE-AC02-76CH00016.

References

1. For a very recent review of the status see G. Martinelli, INFN preprint no. 854, to appear in Lattice '91.

2. For another recent review see S. Sharpe, "Staggered Fermions on the Lattice...", University of Washington preprint DOE/ER/40614-5, to appear in "Standard Model, Hadron Phenomenology and Weak Decays on the Lattice", Ed. G. Martinelli, to be published by World Scientific.

3. For a pedagogical review see C. Bernard in Proceedings of the 1989 Theoretical Advanced Study Institute in Elementary Particle Physics in Boulder, Colorado, 1989, eds. T. Degrand and D. Toussaint (World Scientific, 1990). References to earlier literature may be found in this article.

4. L. Maiani, M. Testa, Phys. Lett. **B245**, 585 (1990).

5. C. Bernard and A. Soni, Nucl. Phys. (Proc. Suppl.) **9**, 155 (1989).

6. C. Bernard, J. Simone and A. Soni, Nucl. Phys. B (Proc. Suppl.) **20**, 434 (1991) and in preparation.

7. James N. Simone, UCLA PhD Thesis (1991), "Lattice Calculation of Non-Leptonic Charm Decays", BNL preprint #46890.

8. For a recent review see Cornell preprint CLNS-92-1131, "Simulating Heavy Quarks", by G.P. Lepage to be published in Lattice '91.

9. See also E. Eichten, Nucl. Phys. (Proc. Suppl.) **20**, 475 (1991).

10. B. Sheikholeslami and R. Wohlert, Nucl. Phys. **B259**, 572 (1985).

11. G. Heatlie et al., Nucl. Phys. **B352**, 266 (1991).

12. G. Martinelli, C.T. Sachrajda, and A. Vladikas, Nucl. Phys. **B358**, 211 (1991).

13. C. Bernard, A. Soni and K. Yee, Nucl. Phys. B (Proc. Suppl.) **20**, 410 (1991) and in preparation.

14. e.g., H. Schroder DESY91–139 and references therein.

15. For the ARGUS Collaboration, see: H. Albrecht et al., Phys. Lett. **B255**, 297 (1991); Phys. Lett. **234B**, 409 (1990). For the CLEO collaboration see R. Fulton et al., Phys. Rev. Lett. **64**, 16 (1990).

16. See, e.g., A.I. Sanda in the Proceedings of the BNL Summer Study on CP Violation (1990), p. 115, eds. S. Dawson and A. Soni, published by World Scientific.

17. See, e.g., L. Littenberg in the Proceedings of the 14th International Symposium on Lepton and Photon Interaction, Stanford (1989).

18. For the relevance of this operator to the underlying theory see G. Beall, M. Bander and A. Soni, Phys. Rev. Lett. **48**, 848 (1982). For calculations on the lattice see Ref. 5.

19. M.B. Gavela et al., Nucl. Phys. **B306**, 865 (1988).

20. For recent reviews see J. Smit, Nucl. Phys. B (Proc. Suppl.) **4**, (1988), p. 451; M.F.L. Goelterman, Nucl. Phys. B (Proc. Suppl.) **20**, 528 (1991).

21. The topic has a very long and rich history, The key references are 22–26.

22. M.K. Gaillard and B.W. Lee, Phys. Rev. Lett. **33**, 108 (1974); G. Altarelli and L. Maiani, Phys. Lett. **52B**, 351 (1974).

23. M.A. Shifman, A.I. Vainshtein and V.I. Zakharov, Pis'ma Zh. Eksp. Teor. Fiz. **22**, 123 (1975) [JE TP Lett. **22**, 55 (1975); Nucl. Phys. **B120**, 316 (1977).

24. F.J. Gilman and M.B. Wise, Phys. Rev. **D20**, 2392 (1979).

25. M.B. Wise, SLAC Report 227 (1980). (Available from National Technical Information Service).

26. G. Altarelli, G. Curci, G. Martinelli, S. Petrarca, Phys. Lett. **99B**, 141 (1981).

27. G. Martinelli, Phys. Lett. **141B**, 395 (1984).

28. C. Bernard, T. Draper and A. Soni, Phys. Rev. **D36**, 3224 (1987).

29. G. Curci, E. Franco, L. Maiani and G. Martinelli, Phys. Lett. **202B**, 363 (1988).

30. See, e.g., G. Kilcup, Nucl. Phys. B (Proc. Suppl.) **20**, 417 (1991).

31. S. Sharpe, CEBAF preprint TH-92-01 to be published in the Proceedings of LATTICE '91.

32. M. Bochicchio et al., Nucl. Phys. **B262**, 331 (1985).

33. L. Maiani, G. Martinelli, G.C. Rossi and M. Testa, Phys. Lett. **179B**, 445 (1986); Nucl. Phys. **B289**, 505 (1987).

34. A. El-Khadra, "Lattice Calculation of Meson Form Factors for Semi-Leptonic Decays", Ph.D. Thesis, UCLA (1989).

35. C. Bernard et al., Phys. Rev. **D32**, 2343 (1985).

36. C. Bernard, T. Draper, G. Hockney and A. Soni, Nucl. Phys. (Proc. Suppl.) **4,**483 (1988).

37. G. Kilcup and S. Sharpe, Nucl. Phys. **B283**, 493 (1987).

38. S.R. Sharpe et al., Nucl. Phys. **B286**, 253 (1987).

39. S.R. Sharpe in Proceedings of the Ringberg Workshop on Hadronic Matrix Elements and Weak Decays, p. 255, eds. A.J. Buras, J.-M. Gérard and W. Huber, (1989).

40. D. Daniel and S.N. Sheard, Nucl. Phys. **B302**, 471 (1988).

41. S.N. Sheard, Nucl. Phys. **B314**, 238 (1989).

42. C. Bernard, et al., Phys. Rev. Lett. **55**, 2770 (1985).

43. C. Bernard and A. Soni, Nucl. Phys. (Proc. Suppl.) **17**, 495 (1990).

44. G. Kilcup, S. Sharpe, R. Gupta and A. Patel, Phys. Rev. Lett. **64**, 25 (1990).

45. R. Gupta, et al., Los Alamos preprint LA-UR-91-3522 (1991) and talk to appear in Lattice '91.

46. M.B. Gavela et al., Nucl. Phys. **B306**, 677 (1988).

47. W. Bardeen, A. Buras and J.-M. Gerard, Phys. Lett. **211B**, 343 (1988); Phys. Lett. **192B**, 128 (1987).

48. J. Donoghue, E. Golowich and B. Holstein, Phys. Lett. **119B**, 412 (1982).

49. J. Bijnens, H. Sonoda and M.B. Wise, Phys. Lett. **137B**, 245 (1984).

50. For literature on this method see Ref. 3.

51. Difficulties in lattice studies of 4 point functions such as $K^0 \to 2\pi$, due to final state interactions are emphasized in L. Maiani and M. Testa, Ref. 4.

52. See Ref. 43.

53. W. Bardeen, A. Buras and J.M. Gerard, Phys. Lett. **211B**, 343 (1988) and Phys. Lett. **192B**, 128 (1987); N. Isgur, K. Maltman, J. Weinstein and T. Barnes, Phys. Rev. Lett. **64**, 161 (1990).

54. L. Maiani, G. Martinelli, G.C. Rossi and M. Testa, Phys. Lett. **176B**, 445 (1986); Nucl. Phys. **B289**, 505 (1987).

55. M.B. Gavela, Nucl. Phys. (Proc. Suppl.) **4**, 466 (1988).

56. M.B. Wise, "CP Violation", Lectures in Proceedings of the Branff Summer Institute, (1988).

57. H. Burkhardt *et al.*, Phys. Lett. **B206**, 169 (1988).

58. J.R. Patterson *et al.*, Phys. Rev. Lett. **64**, 1491, (1990).

59. See, e.g., S.R. Sharpe in Nucl. Phys. B (Proc. Suppl.) 20 (1991).

60. See, e.g., G. Buchalla, A. Buras and M. Harlander, Nucl. Phys. **B337**, 313 (1990) and references therein.

61. See also: E.A. Paschos, T. Schneider and Y.L. Wu, Fermilab-Conf-90/48-T.

62. C. Bernard, A. El-Khadra and A. Soni, Proceedings of Ringberg Workshop on Hadronic Matrix Elements and Weak Decays, Eds. A.J. Buras, J.M. Gerard and W. Huber, p. 277 (1989).

63. See Ref. 34.

64. For a recent review see, C. Bernard, A. El-Khadra, and A. Soni, BNL preprint 45101, to appear in the proceedings of LAT '91.

65. M. Crisafulli, G. Martinelli, V. Hill and C.T. Sachrajda, Phys. Lett. **B223**, 90 (1989).

66. V. Lubicz, G. Martinelli, and C.T. Sachrajda, Nucl. Phys. **B356**, 301 (1991).

67. C. Bernard, A. El-Khadra and A. Soni, Phys. Rev. **D43**, 2140 (1991).

68. V. Lubicz, G. Martinelli, M.S. McCarthy and C.T. Sachrajda, Phys. Lett. **B274**, 415 (1992).

69. C.W. Bernard, A.X. El-Khadra, and A. Soni. Phys. Rev. **D45**, 869 (1992).

70. M. Bauer, B. Stech and M. Wirbel, Z. Phys. **C29**, 637 (1985); M. Bauer and M. Wirbel *ibid* **42**, 671 (1989).

71. J. Körner and G. Schuler, Z. Phys. **C38**, 511 (1988); J. Körner and G. Schuler, Z. Phys. **46**, 93 (1990); J. Körner, K. Schilcher, M. Wirbel and Y.L. Wu, Z. Phys. **C48**, 663 (1990).

72. B. Grinstein, N. Isgur, D. Scora and M. Wise, Phys. Rev. **D39**, 799 (1989); N. Isgur and D. Scora *ibid* **40**, 1491 (1989).

73. F.J. Gilman and R.R. Singleton, Phys. Rev. **D41**, 142 (1990); T.M. Aliev, V.L. Eletskii, Ya.I. Kogan, Sov. J. Nucl. Phys. **40**, 527 (1984); C. Dominguez, N. Paver, Phys. Lett. **B207**, 499 (1988); Z. Phys. **C41**, 217 (1988).

74. T. Altomari and L. Wolfenstein *ibid* **37**, 681 (1988); P. Ball, V.M. Braun, H.G. Dosch, and M. Neubert, Phys. Lett. **B259**, 481 (1991); P. Ball, V.M. Braun, and H.G. Dosch, Phys. Rev. **D44**, 3567 (1991).

75. J. Adler *et al.* (MARK III Collaboration), Phys. Rev. Lett. **62**, 1821 (1989); See also R.H. Schindler *et al.*, Munich High Energy Physics 1988, p. 484.

76. J. Anjos *et al.* (Fermilab E691 Collaboration), Phys. Rev. Lett. **65**, 2630 (1990).

77. K. Kodama *et al.* (E653 Collaboration), Phys. Lett. **B274**, 246 (1992).

78. See, e.g., Ref. 72.

79. We took the data given in Ref. 68.

80. N. Isgur and M. Wise, Phys. Lett. **232B**, 113 (1989).

81. Particle Data Group, Review of Particle Properties, Phys. Lett. **239B** (1990), p. III.61.

82. See Ref. 7.

83. C. Bernard, J. Simone and A. Soni, Nucl. Phys. B (Proc. Suppl.) **17**, 504 (1990); Nucl. Phys. B (Proc. Suppl.) **20**, 434 (1991).

84. A. Abada *et al.*, Nucl. Phys. B (Proc. Suppl.) **17**, 518 (1990).

85. C. Bernard, J. Simone and A. Soni in preparation.

86. Ref. 81, p. II.10–11.

87. B. Guberina, S. Nussinov, R. Peccei and R. Rückl, Phys. Lett. **89B**, 111 (1979).

88. M. Bander, D. Silverman and A. Soni, Phys. Rev. Lett. **44**, 7 (1980); *errata* **44**, 962 (1980).

89. P. Rosen, Phys. Rev. Lett. **44**, 4 (1980).

90. H. Lipkin, Phys. Rev. Lett. **44**, 710 (1980).

91. B. Blok, M. Shifman, Yad. Fiz. **45**, 211 (1987).

92. For a review see: I.I. Bigi in the Proceedings of the Heavy Quark Symposium, P. 18, Cornell University (1989).

93. See also: D. Hitlin in Proceedings of Particles and Fields Meeting at Banff (1988) p. 607; S. Stone, Syracuse preprint HEPSY 1–92 to appear in the "Heavy Flavors" (World Scientific, 1992), eds. A.J. Buras and H. Lidner.

94. We thank Gustavo Burdman for suggesting the use of vacuum saturation to understand the pion mass dependence of the $D \rightarrow K\pi$ amplitudes.

95. For a continuum based discussion of the use of factorization to D decays see: M. Bauer, B. Stech and M. Wirbel, Z. Phys. **C34**, 103 (1987).

96. For the relevance of long N to D decays see: A.J. Biras, J.-M. Gerard and R. Rückl, Nucl. Phys. **B268**, 16 (1986).

97. For a recent discussion of the role of the annihilation amplitudes to charm decays see: I.I. Bigi and N.G. Uraltsev, U.N.D-HEP-91-BIG06.

98. Ref. 81, p. II.9.

99. The influence of resonance on D decays was emphasized long ago by H. Lipkin, see Ref. 90.

100. L. Wolfenstein, Phys. Rev. Lett. **51**, 1945 (1983).

101. F. Abe *et al.* (CDF Collaboration), Phys. Rev. Lett. **68**, 447 (1992).

102. For a recent general review of B-Physics see: A. Soni in Proceedings of the Workshop on High Energy Physics Phenomenology, India (1991), BNL preprint 46579.

103. C. Bernard, T. Draper, G. Hockney and A. Soni, Phys. Rev. **D38**, 3540 (1988).

104. For a recent review see: N. Isgur and M. Wise, Preprint, CEBAF-TH-92-10 (1991) and references therein.

105. See Refs. 106 and 107 for different approaches to the heavy quark limit on the lattice. For another example of an early use of heavy quark symmetry see: S. Nussinov and W. Wetzel, Phys. Rev. **D36**, 130 (1987).

106. G.P. Lepage and B.A. Thacker, Nucl. Phys. B (Proc. Suppl.) **4**, 199 (1988); Phys. Rev. **D43**, 196 (1991).

107. E. Eichten, Nucl. Phys. B (Proc. Suppl.) **4** (1988) 170.

108. B.A. Thacker and G.P. Lepage, Nucl. Phys. B (Proc. Suppl.) **20** (1991) p. 509.

109. C.T.H. Davies, B.A. Thacker, Phys. Rev. **D45**, 915 (1992); G.P. Lepage *et al.*, Cornell preprint CLNS-92-1136 (1992).

426

110. See also E. Eichten and B. Hill, Phys. Lett. **234B**, 511 (1990); Phys. Lett. **240B**, 193 (1990); E. Eichten, G. Hockney and H. Thacker, Nucl. Phys. B (Proc. Suppl.) **17** (1990) 529.

111. Ph. Boucard, C.L. Lin and O. Pene, Phys. Rev. **D40**, 1529 (1989); Erratum, *ibid* **41** (1990) 3541.

112. C. Alexandrou, F. Jegerlehner, S. Gusken, K. Schilling and R. Sommer, Phys. Lett. **B256**, 60 (1991); R. Sommer *et al.*, preprint WU-B-91-2, Nucl. Phys. (Proc. Suppl.); CERN-TH-6371-92, to appear in Lattice '91.

113. C.R. Alton, C.T. Sachrajda, V. Lubicz, L. Maiani and G. Martinelli, Nucl. Phys. **B349**, 598 (1991).

114. C. Bernard, J. Labrenz and A. Soni, Nucl. Phys. B (Proc. Suppl.) **20**, 488 (1991).

115. C. Bernard, J. Labrenz and A. Soni, BNL preprint to be published in LATTICE '91.

116. S. Hashimoto and Y. Saeki, Mod. Phys. Lett. **A7**, 387 (1992). See also the contribution by these authors in the Proceedings of LATTICE '91.

117. E. Eichten, G. Hockney and H.B. Thacker, Nucl. Phys. B (Proc. Suppl.) **17** (1988) 529.

118. A. Billoire, E. Marinari and G. Parisi, Phys. Lett. **162B**, 160 (1985).

119. E. Eichten, to appear in the proceedings of LATTICE '91.

120. Paul Mackenzie (private communication); See also Ref. 8.

121. G.P. Lepage and P.B. Mackenzie, Nucl. Phys. (Proc. Suppl.) **20** (1991) p. 173.

122. The study forms the basis of Ph.D. Thesis of Jim Labrenz to be completed by Sept. 1992.

123. See also C. Bernard, J. Labrenz and A. Soni in preparation.

124. Abada, *et al.*, Rome preprint 823, (1991).

125. L. Maiani, G. Martinelli and C.T. Sachrajda, Nucl. Phys. **B368**, 281 (1992).

126. The conventional method was also used to obtain B_D (but not B_B) in M.B. Gavela, *et al.*, Phys. Lett. **206B** (1988) 113. The B parameter has been calculated in the static limit in Refs. 107, 117, 113, and 124.

127. C. Bernard, P.F. Hsieh and A. Soni, preprint BNL-45033 to appear in Proceedings of LATTICE '91.

128. Radiative B decays have a long history. First suggestion of their importance was made in B.A. Campbell and P.J. O'Donnell, Phys. Rev. **D25**, 1989 (1982); see also P.J. O'Donnell and H.K. Tung, Phys. Rev. **D44**, 741 (1991) and references therein; N.G. Deshpande and J. Trampetic, Mod. Phys. Lett. **A4**, 2095 (1989) and references therein; A. Ali and T. Mannel, Phys. Lett. **B264**, 447 (1991); *Erratum-ibid.* **B274**, 526 (1992).

129. For QCD corrections see: B. Grijanis, H. Navelet, P.J. O'Donnell and M. Sutherland, Phys. Lett. **B237**, 252 (1990); B. Grinstein, R. Springer and M. Wise, Phys. Lett. **200B**, 138 (1988).

130. See e.g., N.G. Deshpande and J. Trampetic, Ref. 125.

131. A. Morel, J. Physique **48** (1987) 111.

132. S. Sharpe, Phys. Rev. **D41** (1990) 3233.

133. S. Sharpe, Nucl. Phys. B (Proc. Suppl.) **17** (1990) 146 and DOE/ER/40614-5, to be published in *Standard Model, Hadron Phenomenology and Weak Decays on the Lattice*, ed. G. Martinelli, World Scientific; G. Kilcup *et al.*, Phys. Rev. Lett. **64** (1990) 25.

134. S. Sharpe, preprint CEBAF-TH-92-12 (1992).

135. C. Bernard and M. Golterman, Wash. U. HEP/92-60 (submitted to Phys. Rev. D) and Wash. U. HEP/91-31, to appear in Proceedings of LATTICE '91.

136. C. Bernard and M. Golterman, in preparation.

www.ingramcontent.com/pod-product-compliance
Lightning Source LLC
Chambersburg PA
CBHW061613220326
41598CB00026BA/3748